Communications in Computer and Information Science 638

Commenced Publication in 2007
Founding and Former Series Editors:
Alfredo Cuzzocrea, Dominik Ślęzak, and Xiaokang Yang

More information about this series at http://www.springer.com/series/7899

Alexander Dudin · Alexander Gortsev
Anatoly Nazarov · Rafael Yakupov (Eds.)

Information Technologies and Mathematical Modelling

Queueing Theory and Applications

15th International Scientific Conference, ITMM 2016
named after A.F. Terpugov
Katun, Russia, September 12–16, 2016
Proceedings

 Springer

Editors
Alexander Dudin
Belarusian State University
Minsk
Belarus

Anatoly Nazarov
Tomsk State University
Tomsk
Russia

Alexander Gortsev
Tomsk State University
Tomsk
Russia

Rafael Yakupov
Kemerovo State University
Anzhero-Sudzhensk
Russia

ISSN 1865-0929   ISSN 1865-0937   (electronic)
Communications in Computer and Information Science
ISBN 978-3-319-44614-1   ISBN 978-3-319-44615-8   (eBook)
DOI 10.1007/978-3-319-44615-8

Library of Congress Control Number: 2016947942

Printed on acid-free paper

This Springer imprint is published by Springer Nature
The registered company is Springer International Publishing AG Switzerland

Preface

The series of scientific conferences on Information Technologies and Mathematical Modelling (ITMM) was started in 2002. In the beginning, its status was that of a national conference but in 2012 become international. The conference series is named after Alexander Terpugov, one of the first organizers of the conference, an outstanding scientist of the Tomsk State University, a leader of the famous Siberian school on applied probability, queueing theory, and applications.

Traditionally, the conferences has about 10 sections in various fields of mathematical modelling and information technologies. Throughout the years, the sections on probabilistic methods and models, queueing theory, telecommunication systems, and software engineering have been the most popular ones at the conference. International participation is presented by researchers from many countries: Austria, Azerbaijan, Belarus, Bulgaria, China, Germany, Hungary, India, Italy, Kazakhstan, Korea, The Netherlands, Poland, UK, USA, etc. Many of our foreign participants come to this Siberia conference every year because there is a warm acceptance and serious scientific discussions.

This volume presents selected papers devoted to new results in queueing theory and its applications. It is aimed at specialists in probabilistic theory, random processes, operations research, and mathematical modelling as well as engineers engaged in logical and technical design and operational management of telecommunication and computer networks, contact centers, manufacturing systems, etc.

September 2016

Alexander Dudin
Alexander Gortsev
Anatoly Nazarov
Rafael Yakupov

Organization

The ITMM conference was organized by the National Research Tomsk State University together with the Kemerovo State University, Trapeznikov Institute of Control Sciences of the Russian Academy of Sciences, People's Friendship University of Russia, and Anzhero-Sudzhensk Branch of Kemerovo State University.

International Program Committee

A. Dudin, Belarus (Chair)
M. Lugachev, Russia
A. Gortsev, Russia (Co-chair)
Y. Malinkovsky, Belarus
A. Nazarov, Russia (Co-chair)
G. Medvedev, Belarus
R. Yakupov, Russia (Co-chair)
Q.-L. Li, China
I. Atencia, Spain
A. Melikov, Azerbaijan
S. Chakravarthy, USA
R. Nobel, Netherlands
B.D. Choi, Korea
E. Orsingher, Italy
T. Czachórski, Poland
M. Pagano, Italy

D. Efrosinin, Austria
K. Samuylov, Russia
M. Farhadov, Russia
J. Sztrik, Hungary
E. Gelenbe, UK
H. Tijms, The Netherlands
K. Al-Begain, UK
O. Tikhonenko, Poland
C.S. Kim, Korea
I. Tsitovich, Russia
Y. Koucheryavy, Finland
V. Vishnevsky, Russia
A. Krishnamurthy, India
N. Yevtushenko, Russia
U. Krieger, Germany

Local Organizing Committee

A. Dudin (Chair)
M. Matalytsky
A. Gortsev (Co-chair)
A. Moiseev
A. Nazarov (Co-chair)
S. Moiseeva
R. Yakupov (Co-chair)
Y. Naumkina
E. Daniliuk
S. Paul
E. Fedorova
S. Rozhkova
I. Garayshina
V. Rykov
Y. Gaydamaka
D. Semionova

E. Glukhova
A. Shkurkin
A. Gudov
I. Shmyrin
R. Ivanovskiy
S. Suschenko
V. Ivnitsky
G. Tsitsiashvili
T. Kabanova
K. Voytikov
Y. Kostyuk
O. Zmeev
E. Lisovskaya
A. Zorin
K. Livshits
V. Zadorozhny

Contents

Maximum Likelihood Estimation of the Dead Time Period Duration in the Modulated Semi-synchronous Generalized Flow of Events

Maria Bakholdina[✉] and Alexander Gortsev

Department of Operations Research, Faculty of Applied Mathematics
and Cybernetics, National Research Tomsk State University,
36 Lenina Avenue, Tomsk 634050, Russia
maria.bakholdina@gmail.com

Abstract. This paper is focused on studying the modulated semi-synchronous generalized flow of events which is one of the mathematical models for incoming streams of events (claims) in computer communication networks and is related to the class of doubly stochastic Poisson processes (DSPPs). The flow is considered in conditions of its incomplete observability, when the dead time period of a constant duration T is generated after every registered event. This paper is devoted to the maximum likelihood estimation of the dead time period duration on monitoring the time moments of the flow events occurrence.

Keywords: Modulated semi-synchronous generalized flow of events ·
Doubly Stochastic Poisson Process (DSSP) · Markovian Arrival Process
(MAP) · Maximum likelihood estimation · Likelihood function · Dead
time period duration

1 Introduction

In the recent literature, the problem of studying doubly stochastic Poisson processes (DSPPs) [1–6] has been of great interest, since DSPPs have found applications in many fields such as network theory, peer-to-peer streaming networks and adaptive data streaming, optical communication systems, statistical modeling, quantitative finance, spatial epidemiology, etc. [7–11]. In real situations the input flow parameters can be unknown or partially known or, worse, may vary in time in a random way. That is why the central problems faced when modeling these processes are: (1) flow states estimation on monitoring the time moments of events occurrence (the filtering of the underlying and unobservable intensity process) [12–15]; (2) flow parameters estimation on monitoring the time moments of events occurrence [16–21].

It is worth noting that in most cases researchers consider the mathematical models of flows where time moments of the flow events occurrence are observable. In practice, however, any recording device (a server in this context) spends some finite time on event measurement and registration, during which the server

© Springer International Publishing Switzerland 2016
A. Dudin et al. (Eds.): ITMM 2016, CCIS 638, pp. 1–17, 2016.
DOI: 10.1007/978-3-319-44615-8_1

cannot handle the next event correctly. In other words, every event registered by a server causes a period which is called the period of dead time [19, 20, 22–26], during which no other events are observed (they are lost). We may suppose that this period has a fixed duration.

In this paper we continue to study the modulated semi-synchronous generalized flow of events [15, 27, 28], which belongs to the class of doubly stochastic Poisson processes (DSPPs) with a piecewise constant intensity process. There are a large number of references in the literature which may be found in [15, 28, 29] and contain studies of similar flows of events (synchronous, asynchronous and semi-synchronous flows), herewith we shall note that these flows of events can be presented as mathematical models of MAP-flows of events [29]. The present paper is devoted to the maximum likelihood estimation of the dead time period duration on monitoring the time moments of events occurrence. The rest of the paper is organized as follows. In Sect. 2 we present the modulated semi-synchronous generalized flow of events, which provides our modelling framework. In Sect. 3 we describe the optimization problem consisting of maximizing the likelihood function and, finally, find the solution of the optimization problem in Sect. 4.

2   Problem Statement

We consider the modulated semi-synchronous generalized flow of events (further flow or flow of events), whose intensity process is a piecewise constant stationary random process $\lambda(t)$ with two states 1, 2 (first, second correspondingly). In the first state $\lambda(t) = \lambda_1$ and in the second state $\lambda(t) = \lambda_2$ ($\lambda_1 > \lambda_2 \geqslant 0$). During the time interval of a random duration when the process $\lambda(t)$ is in state λ_i ($\lambda(t) = \lambda_i$), a Poisson flow of events with intensity λ_i, $i = 1, 2$, arrives. The transition of the process $\lambda(t)$ from the first state to the second state is possible at any moment of a Poisson event occurrence in state 1 of the process $\lambda(t)$, herewith the process $\lambda(t)$ can change its state to the second one with probability p ($0 \leqslant p \leqslant 1$) or continue to stay in state 1 with complementary probability $1-p$. The transition of the process $\lambda(t)$ from state 1 to state 2 is also possible at any moment that does not coincide with the moment of a Poisson event occurrence, herewith the duration of the process $\lambda(t)$ staying in the first state is distributed according to the exponential law with parameter β: $F(\tau) = 1 - e^{-\beta\tau}$, $\tau \geqslant 0$. Then the duration of the process $\lambda(t)$ staying in the first state is distributed according to the exponential law with parameter $(p\lambda_1 + \beta)$: $F_1(\tau) = 1 - e^{-(p\lambda_1+\beta)\tau}$, $\tau \geqslant 0$. The transition of the process $\lambda(t)$ from the second state to the first state at the moment of a Poisson event occurrence in state 2 is impossible and can be done only at a random time moment. In this case the duration of the process $\lambda(t)$ staying in state 2 is distributed according to the exponential law with parameter α: $F_2(\tau) = 1 - e^{-\alpha\tau}$, $\tau \geqslant 0$. At the moment when the state changes from the second to the first one, an additional event is assumed to be initiated with probability δ ($0 \leqslant \delta \leqslant 1$). Such flows with additional events initiation are called generalized flows. In accordance with these assumptions we can assert that $\lambda(t)$ is a Markovian process.

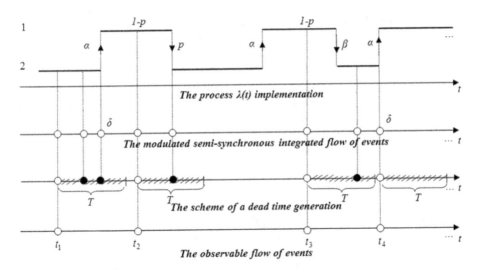

Fig. 1. The formation of an observable flow of events

The registration of flow events is considered in conditions of a constant (unextendable) dead time. The dead time period of a constant duration T begins after every registered at the moment t_k, $k \geqslant 1$, event. During this period no other events are observed. When the dead time period is over, the first coming event causes the next interval of dead time of duration T and so on. Figure 1 shows the possible variant of the flow operation and observation. Here 1, 2 are the states of the process $\lambda(t)$; additional events, which may occur at the moment of the process $\lambda(t)$ transition from state 2 to state 1, are marked with letter δ; dead time periods of duration T are marked with hatching; unobservable events are displayed as black circles, observable events t_1, t_2, ... are shown as white circles.

The process $\lambda(t)$ is considered in steady-state conditions, which is why we may neglect transient processes at the interval of observation $(t_0, t]$, where t_0 is the instant of beginning the observations, t is the instant of ending the observations (the moment of a decision making). In steady-state conditions we may take $t_0 = 0$. We should note that the process $\lambda(t)$ and possible events (events of Poisson flows with intensity λ_i, $i = 1, 2$, and additional events) are basically unobservable; we register only time moments t_1, t_2, ..., t_k of events occurrence in observable flow during the interval of observation $(t_0, t]$. We assume that the flow parameters $\lambda_1 > \lambda_2 \geqslant 0$, $0 \leqslant p \leqslant 1$, $\beta > 0$, $\alpha > 0$, $0 \leqslant \delta \leqslant 1$ are known and the duration of the dead time period T is not known. In that way, the main problem is to obtain the estimate $\hat{T}$ of the dead time period duration at the moment t of ending the observations on monitoring the time moments t_1, t_2, ..., t_k of events occurrence in an observable flow using the maximum-likelihood technique.

3 Likelihood Function Construction

Let us denote by $\tau_k = t_{k+1} - t_k$, $k = 1, 2, ...$, the value of the k interval length between two consecutive flow events ($\tau_k > 0$). In steady-state conditions we may accept that the probability density function of the k interval length is $p_T(\tau_k) = p_T(\tau)$, $\tau \geqslant 0$, for any k (the index T stresses that the probability density depends on the dead time period duration). Thereby we may also accept that the time moment t_k is equal to zero, i.e. the moment of the event occurrence is $\tau = 0$. Then one-dimensional probability density function $p_T(\tau)$ can be written as [28]

$$p_T(\tau) = 0, \ 0 \leqslant \tau < T,$$

$$p_T(\tau) = \frac{z_1}{z_2 - z_1} \left[z_2 - \frac{1}{\beta_1 + \beta_2} f(T) \right] e^{-z_1(\tau - T)}$$

$$- \frac{z_2}{z_2 - z_1} \left[z_1 - \frac{1}{\beta_1 + \beta_2} f(T) \right] e^{-z_2(\tau - T)}, \ \tau \geqslant T;$$

$$f(T) = \lambda_1 \alpha + (p\lambda_1 + \beta)(\lambda_2 + \alpha\delta) + \alpha(\lambda_1 - \lambda_2 - \alpha\delta) \tag{1}$$

$$\times \{(p\lambda_1 + \beta)[\lambda_1(1 - p + p\delta) - \lambda_2 + \delta\beta] - p\lambda_1\alpha\} \frac{e^{-(\beta_1 + \beta_2)T}}{F(T)},$$

$$F(T) = z_1 z_2 - q e^{-(\beta_1 + \beta_2)T}, \ \beta_1 = p\lambda_1 + \beta, \ \beta_2 = \alpha,$$

$$q = \lambda_1[\lambda_2 - p(\lambda_2 + \alpha\delta)],$$

$$z_{1,2} = \frac{1}{2} \left[\lambda_1 + \lambda_2 + \alpha + \beta \mp \sqrt{(\lambda_1 - \lambda_2 - \alpha + \beta)^2 + 4\alpha\beta(1 - \delta)} \right].$$

Now let $\tau_1 = t_2 - t_1$, $\tau_2 = t_3 - t_2$, ..., $\tau_k = t_{k+1} - t_k$, $\tau_1 \geqslant 0$, $\tau_2 \geqslant 0$, ..., $\tau_k \geqslant 0$, be the sequence of values of the interval lengths between consecutive flow events measured during the interval of observation $(0, t]$. Arrange the variables $\tau_1, \tau_2, ..., \tau_k$ in ascending order: $\tau_{min} = \tau^{(1)} < \tau^{(2)} < ... < \tau^{(k)}$. Under the made assumptions we can assert that the sequence of the time moments $t_1, t_2, ..., t_k, ...$ corresponds to an embedded Markov chain $\{\lambda(t_k)\}$, i.e. the flow has the Markov property if the evolution of the flow is considered from the time moment t_k, $k = 1, 2, ...$, of the event occurrence.

As the main problem is to estimate the dead time period duration on the assumption that the flow parameters $\lambda_1 > \lambda_2 \geqslant 0$, $0 \leqslant p \leqslant 1$, $\beta > 0$, $\alpha > 0$, $0 \leqslant \delta \leqslant 1$ are known, then according to the maximum-likelihood technique we should maximize the likelihood function $L\left(T | \tau^{(1)}, ..., \tau^{(k)}\right)$ and solve the following task of optimization:

$$L\left(T | \tau^{(1)}, ..., \tau^{(k)}\right) = \prod_{j=1}^{k} p_T\left(\tau^{(j)}\right) \Longrightarrow \max_T, \ 0 \leqslant T \leqslant \tau_{min}, \ \tau_{min} > 0, \tag{2}$$

where $p_T\left(\tau^{(j)}\right)$ is defined by (1) for $\tau = \tau^{(j)}$. The value of T, when the likelihood function (2) reaches its global maximum, will be the estimate $\hat{T}$ of the dead time period duration.

4   The Solution of the Optimization Problem (2)

Let us denote by $\tau_m = \tau_{min}$. Since the likelihood function (2) is different from zero when $0 \leqslant T \leqslant \tau_m$, assume that $p_T\left(\tau^{(j)}\right) = 0$, $j = \overline{2,k}$, $T > \tau_m$ $(\tau_m > 0)$. The situation continues when $\tau_m = 0$, refers to the extension of defining the functions being studied at the boundary points. Let us study the behaviour of the function $p_T(\tau_m)$ as a function of a variable T $(0 \leqslant T \leqslant \tau_m)$. We shall note that $p_T(\tau_m) \geqslant 0$, as $p_T(\tau_m)$ is the probability density function. Turn now to studying the first derivative $p'_T(\tau_m)$ with respect to the argument T. We obtain

$$\begin{aligned} p'_T(\tau_m) = &\frac{1}{(z_2 - z_1)(\beta_1 + \beta_2)} \\ &\times \{z_1 e^{-z_1 \tau_m}[z_1 z_2(\beta_1 + \beta_2) - z_1 f(T) - f'(T)]e^{-z_1 T} \\ &- z_2 e^{-z_2 \tau_m}[z_1 z_2(\beta_1 + \beta_2) - z_2 f(T) - f'(T)]e^{-z_2 T}\}, \\ f'(T) = &-\alpha(\lambda_1 - \lambda_2 - \alpha\delta)(\beta_1 + \beta_2)z_1 z_2 \\ &\times \{(p\lambda_1 + \beta)[\lambda_1(1 - p + p\delta) - \lambda_2 + \delta\beta] - p\lambda_1\alpha\} \\ &\times \frac{e^{-(\beta_1 + \beta_2)T}}{F^2(T)}, \; 0 \leqslant T \leqslant \tau_m, \end{aligned} \tag{3}$$

where $f(T)$, $F(T)$ are defined by (1), $f'(T)$ is a derivative of $f(T)$ with respect to the argument T.

Lemma 1. *The derivative $p'_T(\tau_m)$ is a positive function of a variable τ_m for $T = 0$, $0 \leqslant \tau_m < \infty$ $(p'_{T=0}(\tau_m) > 0)$.*

Proof. Since τ_m is an arbitrary nonnegative number $(\tau_m \geqslant 0)$, we can consider $p'_{T=0}(\tau_m)$ as a function of a variable τ_m. Substituting $T = 0$ into (3) and carrying out some transformations, we obtain

$$\begin{aligned} p'_{T=0}(\tau_m) = &\frac{C}{A^2(z_2 - z_1)} \\ &\times \left[z_2(C - z_1 A)e^{-z_2 \tau_m} - z_1(C - z_2 A)e^{-z_1 \tau_m}\right], \; \tau_m \geqslant 0, \\ C = &\lambda_1\alpha[\lambda_1(1 - p + p\delta) + \delta\beta] + (\lambda_2 + \alpha\delta)[\lambda_2(p\lambda_1 + \beta) + p\lambda_1\alpha] \\ = &-z_1 z_2(\beta_1 + \beta_2) + (z_1 + z_2)A > 0, \\ A = &\lambda_1\alpha + (p\lambda_1 + \beta)(\lambda_2 + \alpha\delta) = z_1 z_2 - q > 0. \end{aligned} \tag{4}$$

Let us consider the equation $p'_{T=0}(\tau_m) = 0$ for the purpose of finding roots, which can be transformed to the following form using (4):

$$B = e^{-(z_2 - z_1)\tau_m}, \; B = \frac{z_1(z_2 A - C)}{z_2(z_1 A - C)}. \tag{5}$$

Taking into account (4) we also find

$$p'_{T=0}(\tau_m = 0) = (C/A)^2 > 0, \; \lim_{\tau_m \to \infty} p'_{T=0}(\tau_m) = \pm 0. \tag{6}$$

Substituting the expressions for z_1, z_2, that are defined by (1), into (5), we obtain

$$B = \frac{z_1^2 \left[-2A + (\beta_1 + \beta_2)(z_1 + z_2) + (\beta_1 + \beta_2)\sqrt{D} \right]}{z_2^2 \left[-2A + (\beta_1 + \beta_2)(z_1 + z_2) - (\beta_1 + \beta_2)\sqrt{D} \right]} = \frac{B_1}{B_2}.$$

where $D = (\lambda_1 - \lambda_2 - \alpha + \beta)^2 + 4\alpha\beta(1 - \delta)$.

It can be shown that $B_1 > 0$ irrespective of the expression $(\lambda_1 - \lambda_2 - \alpha + \beta)$ sign. There are two possible variants for the expression (5):

(1) $B_1 = z_1 (z_2 A - C) > 0$; $B_2 = z_2 (z_1 A - C) > 0$. Then $B > 0$ and the difference $B_1 - B_2 = C(z_2 - z_1) > 0$. Consequently, $B_1 > B_2$ and $B > 1$. Then the Eq. (5) does not have a solution and therefore $p'_{T=0}(\tau_m) > 0$, $\tau_m \geqslant 0$, since $p'_{T=0}(\tau_m = 0) > 0$ and $\lim\limits_{\tau_m \to \infty} p'_{T=0}(\tau_m) = +0$ due to (6).

(2) $B_1 = z_1 (z_2 A - C) > 0$; $B_2 = z_2 (z_1 A - C) < 0$. Then $B < 0$ and the Eq. (5) does not have a solution. Therefore $p'_{T=0}(\tau_m) > 0$, $\tau_m \geqslant 0$, similarly to the previously described variant.

It remains to consider the case when $z_1 A - C = 0$. Substituting $z_1 A - C = 0$ into (4), we have

$$p'_{T=0}(\tau_m) = \frac{C}{A^2(z_2 - z_1)}(z_2 A - C) z_1 e^{-z_1 \tau_m} > 0, \ \tau_m \geqslant 0,$$

since $B_1 = z_1 (z_2 A - C) > 0$ and, consequently, $z_2 A - C > 0$.

Thus, if $z_2 (z_1 A - C) \geqslant 0$ or $z_2 (z_1 A - C) < 0$, then $p'_{T=0}(\tau_m) > 0$, $\tau_m \geqslant 0$ *Lemma 1 is proved.*

Remark 1. $q > q_1$.

Proof. As has already been pointed out in the course of the proof of Lemma 1, $z_2 A - C > 0$. It can be shown that $z_2 A - C = z_1 (q - q_1)$ and, consequently, $q > q_1$.

Lemma 2. *The derivative $p'_T(\tau_m)$ is strictly greater than zero for $T = \tau_m$ ($p'_{T=\tau_m}(\tau_m) > 0$), $0 \leqslant \tau_m < \infty$.*

Proof. Substituting $T = \tau_m$ into (3), we obtain

$$p'(\tau_m) = p'_{T=\tau_m}(\tau_m) = \frac{1}{\beta_1 + \beta_2} \left\{ C + \varphi(q)\Psi(\tau_m)\frac{e^{-(\beta_1 + \beta_2)\tau_m}}{F^2(\tau_m)} \right\},$$

$$\Psi(\tau_m) = C + (z_1 + z_2) q \left[1 - e^{-(\beta_1 + \beta_2)\tau_m} \right], \qquad (7)$$

$$\varphi(q) = (\beta_1 + \beta_2) C - A^2 = -q^2 + [2z_1 z_2 - (\beta_1 + \beta_2)(z_1 + z_2)] q$$
$$- z_1 z_2 (z_1 - \beta_1 - \beta_2)(z_2 - \beta_1 - \beta_2),$$

where q, $F(\tau_m)$ are defined by (1); A, C are defined by (4). It can be shown that $\varphi(q) = -(q - q_1)(q - q_2)$, where $q_1 = z_2(z_1 - \beta_1 - \beta_2)$, $q_2 = z_1(z_2 - \beta_1 - \beta_2)$ are the roots of the equation $\varphi(q) = 0$. Let us note that

$$p'(\tau_m = 0) = (C/A)^2 > 0, \quad \lim_{\tau_m \to \infty} p'(\tau_m) = \frac{C}{\beta_1 + \beta_2} > 0,$$

$$p'(\tau_m = 0) - \lim_{\tau_m \to \infty} p'(\tau_m) = \frac{C}{A^2(\beta_1 + \beta_2)}\varphi(q). \tag{8}$$

It follows from $C > 0$ that $\Psi(\tau_m) > 0$, $\tau_m \geqslant 0$. Then the sign of derivative $p'(\tau_m)$ depends on the sign of function $\varphi(q)$. Then, if $\varphi(q) \geqslant 0$, the derivative (7) is strictly greater than zero $(p'(\tau_m) > 0)$ for $\tau_m \geqslant 0$.

Let us assume that $\varphi(q) < 0$. Introduce into consideration the second derivative $p''(\tau_m)$ of a variable T at the point $T = \tau_m$. Using (3), we obtain

$$p''(\tau_m) = -z_1 z_2 \frac{\varphi(q)}{F^3(\tau_m)} e^{-(\beta_1 + \beta_2)\tau_m} \left[z_1 z_2 (z_1 + z_2 - \beta_1 - \beta_2) \right.$$

$$\left. - (z_1 + z_2 + \beta_1 + \beta_2) q e^{-(\beta_1 + \beta_2)\tau_m} \right], \quad \tau_m \geqslant 0,$$

where q, $F(\tau_m)$ are defined by (1); $\varphi(q)$ is defined by (7). Taking into account that $\varphi(q) < 0$, we have $\left[-z_1 z_2 \dfrac{\varphi(q)}{F^3(\tau_m)} e^{-(\beta_1 + \beta_2)\tau_m} \right] > 0$, therefore the sign of $p''(\tau_m)$ depends on the sign of function

$$y(\tau_m) = z_1 z_2 (z_1 + z_2 - \beta_1 - \beta_2) - (z_1 + z_2 + \beta_1 + \beta_2) q e^{-(\beta_1 + \beta_2)\tau_m}, \quad \tau_m \geqslant 0.$$

Let us turn now to studying the function $y(\tau_m)$ as a function of a variable τ_m $(\tau_m \geqslant 0)$. We have

$$y(0) = C - q(\beta_1 + \beta_2),$$

$$\lim_{\tau_m \to \infty} y(\tau_m) = y(\infty) = z_1 z_2 (z_1 + z_2 - \beta_1 - \beta_2) = z_1 z_2 [\lambda_1(1 - p) + \lambda_2] > 0,$$

where q is defined by (1); C is defined by (4). At this point there are several variants possible:

(1) $q < 0$ and, consequently, $y(0) = C - q(\beta_1 + \beta_2) > 0$. Then $y(\tau_m)$ is a decreasing function of a variable τ_m $(\tau_m \geqslant 0)$; it decreases from $y(0) = C - q(\beta_1 + \beta_2) > 0$ to $y(\infty) = z_1 z_2 (z_1 + z_2 - \beta_1 - \beta_2) > 0$. Therefore $y(\tau_m) > 0$, $\tau_m \geqslant 0$. It follows that $p''(\tau_m) > 0$, $\tau_m \geqslant 0$, and $p'(\tau_m)$ is an increasing function of a variable τ_m $(\tau_m \geqslant 0)$. So, taking into account (8), we have $p'(\tau_m) > 0$, $\tau_m \geqslant 0$.

(2) $q = 0$. Then $y(\tau_m) = z_1 z_2 (z_1 + z_2 - \beta_1 - \beta_2) > 0$, $\tau_m \geqslant 0$. The result is identical to the result for variant 1.

(3) $q > 0$. Then $y(\tau_m)$ is an increasing function of a variable τ_m $(\tau_m \geqslant 0)$; it increases from $y(0) = C - q(\beta_1 + \beta_2)$ to $y(\infty) = z_1 z_2 (z_1 + z_2 - \beta_1 - \beta_2) > 0$. At this variant there are three possible cases: (a) $y(0) > 0$; (b) $y(0) = 0$; (c) $y(0) < 0$. Let us look at these cases one by one.

Case (a): $y(0) = C - q(\beta_1 + \beta_2) > 0$. Then $y(\tau_m) > 0$, $\tau_m \geqslant 0$. It follows that $p''(\tau_m) > 0$, $\tau_m \geqslant 0$, and $p'(\tau_m)$ is an increasing function of a variable τ_m ($\tau_m \geqslant 0$). So, taking into account (8), we have $p'(\tau_m) > 0$, $\tau_m \geqslant 0$.

Case (b): $y(0) = C - q(\beta_1 + \beta_2) = 0$. Then $y(\tau_m) \geqslant 0$, $\tau_m \geqslant 0$, at that the equality of zero ($y(\tau_m) = 0$) holds only at the point $\tau_m = 0$. The result is identical to the result for case (a).

Case (c): $y(0) = C - q(\beta_1 + \beta_2) < 0$. Then: (1) $y(\tau_m) < 0$, $0 \leqslant \tau_m < \tau_m^*$; (2) $y(\tau_m) = 0$, $\tau_m = \tau_m^*$; (3) $y(\tau_m) > 0$, $\tau_m^* < \tau_m < \infty$. Hence, the derivative $p'(\tau_m)$ reaches its minimal value $p'(\tau_m^*)$ at the point $\tau_m = \tau_m^*$. The point of minimum $\tau_m = \tau_m^*$ is found from the equation $p'(\tau_m) = 0$:

$$\tau_m^* = -\frac{1}{\beta_1 + \beta_2} \ln \frac{z_1 z_2 (z_1 + z_2 - \beta_1 - \beta_2)}{(z_1 + z_2 + \beta_1 + \beta_2) q}, \quad \frac{z_1 z_2 (z_1 + z_2 - \beta_1 - \beta_2)}{(z_1 + z_2 + \beta_1 + \beta_2) q} > 0.$$

Next, we calculate the $p'(\tau_m)$ value at the point τ_m^*:

$$p'(\tau_m^*) = \frac{1}{\beta_1 + \beta_2} \left\{ C + \varphi(q) \frac{(z_1 + z_2 - \beta_1 - \beta_2)^2}{4(\beta_1 + \beta_2) q} \right\}.$$

It can be shown that $p'(\tau_m^*) > 0$. Finally we conclude $p'(\tau_m) > 0$, $\tau_m \geqslant 0$. Thus, if $\varphi(q) < 0$, then the derivative $p'(\tau_m)$ is strictly greater than zero ($p'(\tau_m) > 0$), $\tau_m \geqslant 0$. *Lemma 2 is proved.*

Let us turn now to studying the derivative $p'_T(\tau_m)$ as a function of a variable T, $0 \leqslant T \leqslant \tau_m$. Let us consider the equation $p'_T(\tau_m) = 0$ for the purpose of finding roots, which can be transformed to the following form using (3):

$$e^{-(z_2 - z_1)(\tau_m - T)} = f_1(T) / f_2(T), \quad 0 \leqslant T \leqslant \tau_m,$$
$$f_1(T) = z_1 [z_1 z_2 (\beta_1 + \beta_2) - z_1 f(T) - f'(T)], \tag{9}$$
$$f_2(T) = z_2 [z_1 z_2 (\beta_1 + \beta_2) - z_2 f(T) - f'(T)].$$

Then $\chi(T) = e^{-(z_2 - z_1)(\tau_m - T)}$ is an increasing function of a variable T, $0 \leqslant T \leqslant \tau_m$; it increases from $\chi(0) = e^{-(z_2 - z_1)\tau_m}$ to $\chi(T = \tau_m) = 1$; its maximum value is 1.

Since τ_m can be equal to infinity, let us now study functions $f_1(T)$, $f_2(T)$ as functions of a variable T, $T \geqslant 0$.

Proposition 1. *The function* $f_1(T) = z_1 [z_1 z_2 (\beta_1 + \beta_2) - z_1 f(T) - f'(T)]$, $T \geqslant 0$, *satisfies the following relations:*

(1) $f_1(0) = (\beta_1 + \beta_2) \frac{z_1 C}{A^2} (z_2 A - C) = (\beta_1 + \beta_2) \frac{z_1^2 C}{A^2} (q - q_1) > 0$,   *where*
$q_1 = z_2 (z_1 - \beta_1 - \beta_2)$, $q > q_1$;

(2) $f_1(\infty) = \lim\limits_{T \to \infty} f_1(T) = z_1 (z_2 A - C) = z_1^2 (q - q_1) > 0$;

(3) $f_1'(T) = \frac{z_1^2 z_2 (\beta_1 + \beta_2)}{F^3(T)} \varphi(q) e^{-(\beta_1 + \beta_2)T}$

$$\times \left[z_1 z_2 (z_1 - \beta_1 - \beta_2) - (z_1 + \beta_1 + \beta_2) q e^{-(\beta_1 + \beta_2)T} \right], \quad T \geqslant 0;$$

(4) $T_1(q)$ is a bending point (or an inflexion point) of function $f_1(T)$:

$$T_1(q) = -\frac{1}{\beta_1 + \beta_2} \ln \frac{z_1 z_2 (z_1 - \beta_1 - \beta_2)}{(z_1 + \beta_1 + \beta_2) q}, \quad \frac{z_1 - \beta_1 - \beta_2}{q} > 0; \quad (10)$$

(5) $f_1(T_1(q)) = \dfrac{z_1 (z_1 + \beta_1 + \beta_2)^2}{4 (\beta_1 + \beta_2) q} [q - z_2 (z_1 - \beta_1 - \beta_2)]$

$$\times \left[q - \frac{z_1 (z_1 - \beta_1 - \beta_2)^2 (z_2 - \beta_1 - \beta_2)}{(z_1 + \beta_1 + \beta_2)^2} \right], \quad \frac{z_1 - \beta_1 - \beta_2}{q} > 0,$$

where $q_{11} = q_1 = z_2 (z_1 - \beta_1 - \beta_2)$, $q_{12} = \dfrac{z_1 (z_1 - \beta_1 - \beta_2)^2 (z_2 - \beta_1 - \beta_2)}{(z_1 + \beta_1 + \beta_2)^2}$ are

the roots of the equation $f_1(T_1(q)) = 0$.

Here q, $F(T)$ are defined by (1); A, C are defined by (4); $\varphi(q)$ is defined by (7).

Lemma 3. $f_1(T)$ is a positive function ($f_1(T) > 0$) of a variable T, $T \geqslant 0$.

Proof. Let $q = q_1$, then

$$f_1(T, q = q_1) = z_1^2 [z_2 (\beta_1 + \beta_2) - z_1 z_2 + z_2 (z_1 - \beta_1 - \beta_2)] = 0;$$

let $q = q_2$, then

$$f_1(T, q = q_2) = z_1^2 (z_2 - z_1)(\beta_1 + \beta_2) > 0.$$

The points $q = q_1$, $q = q_2$ can be excluded from further consideration as $\varphi(q) = 0$ at these points. Let us discuss later all possible cases:

Case 1. $\varphi(q) > 0$, $0 < q_1 < q < q_2$. The sign of derivative $f_1'(T)$ depends on the sign of function $y_1(T) = z_1 z_2 (z_1 - \beta_1 - \beta_2) - (z_1 + \beta_1 + \beta_2) q e^{-(\beta_1 + \beta_2)T}$, $T \geqslant 0$. In this case $y_1(T)$ is an increasing function of a variable T, $T \geqslant 0$; it increases from $y_1(T = 0, q) < 0$ to $y_1(\infty) = z_1 z_2 (z_1 - \beta_1 - \beta_2) > 0$, i.e. it intersects only once the axis of abscisses at the point $T_1(q)$, that is defined by (10). It follows that the function $f_1(T)$ has only one point of minimum at $T = T_1(q)$. It can be shown that $f_1(T_1(q)) > 0$, then $f_1(T) > 0$, $T \geqslant 0$. Finally, if $\varphi(q) > 0$, $0 < q_1 < q < q_2$, then $f_1(T) > 0$, $T \geqslant 0$.

Case 2. $\varphi(q) > 0$, $q_1 = 0 < q < q_2$. In this case $y_1(T) = -(z_1 + \beta_1 + \beta_2) \times q e^{-(\beta_1 + \beta_2)T} < 0$, $T \geqslant 0$. Then $f_1'(T) < 0$, $T \geqslant 0$, i.e. $f_1(T)$ is a decreasing function of a variable T, $T \geqslant 0$; it decreases from $f_1(0) > 0$ to $f_1(\infty) > 0$. It follows that $f_1(T) > 0$, $T \geqslant 0$. Finally, if $\varphi(q) > 0$, $q_1 = 0 < q < q_2$, then $f_1(T) > 0$, $T \geqslant 0$.

Case 3. $\varphi(q) > 0$, $q_1 < 0 < q < q_2$. In this case $y_1(T) = z_1 z_2 (z_1 - \beta_1 - \beta_2) - (z_1 + \beta_1 + \beta_2) q e^{-(\beta_1 + \beta_2)T} < 0$, $T \geqslant 0$. Then $f_1'(T) < 0$, $T \geqslant 0$; further the idea is basically the same as in the proof of case 2. Finally, if $\varphi(q) > 0$, $q_1 < 0 < q < q_2$, then $f_1(T) > 0$, $T \geqslant 0$.

Case 4. $\varphi(q) > 0$, $q_1 < q = 0 < q_2$. In this case $y_1(T) = z_1 z_2 (z_1 - \beta_1 - \beta_2) = $ const < 0, $T \geqslant 0$. Then $f_1'(T) < 0$, $T \geqslant 0$; further the idea is basically the same as in the proof of case 2. Finally, if $\varphi(q) > 0$, $q_1 < q = 0 < q_2$, then $f_1(T) > 0$, $T \geqslant 0$.

Case 5. $\varphi(q) > 0$, $q_1 < q < 0 < q_2$. In this case $y_1(T) = z_1 z_2 (z_1 - \beta_1 - \beta_2) - (z_1 + \beta_1 + \beta_2) q e^{-(\beta_1 + \beta_2)T}$, $T \geqslant 0$, and $y_1(T)$ is a decreasing function of a variable T, $T \geqslant 0$; it decreases from $y_1(T = 0, q)$ to $y_1(\infty) = z_1 z_2 (z_1 - \beta_1 - \beta_2) < 0$. Let us look at $y_1(T = 0, q) = z_1 z_2 (z_1 - \beta_1 - \beta_2) - (z_1 + \beta_1 + \beta_2) q$, $q_1 \leqslant q \leqslant 0$. We have: (a) $y_1(T = 0, q = q_1) = -z_2 (\beta_1 + \beta_2)(z_1 - \beta_1 - \beta_2) > 0$; (b) $y_1(T = 0, q = 0) = z_1 z_2 (z_1 - \beta_1 - \beta_2) < 0$; (c) $y_1(T = 0, q = q_1^*) = 0$, where $q_1^* = z_1 z_2 (z_1 - \beta_1 - \beta_2) / (z_1 + \beta_1 + \beta_2)$, $q_1 < q_1^* < 0$. Then for $q_1 < q < q_1^* < 0$ we have: (1) $y_1(T) > 0$, $0 \leqslant T < T_1(q)$; (2) $y_1(T) = 0$, $T = T_1(q)$; (3) $y_1(T) < 0$, $T > T_1(q)$; $T_1(q)$ is defined by (10). Therefore, for $q_1 < q < q_1^* < 0$ we have: (1) $f_1'(T) > 0$, $0 \leqslant T < T_1(q)$; (2) $f_1'(T) = 0$, $T = T_1(q)$; (3) $f_1'(T) < 0$, $T > T_1(q)$. It follows that the function $f_1(T)$ has only one point of maximum at $T = T_1(q)$ and thus $f_1(T) > 0$, $T \geqslant 0$. For $q = q_1^*$ we have $y_1(T) = z_1 z_2 (z_1 - \beta_1 - \beta_2) \left[1 - e^{-(\beta_1 + \beta_2)T} \right] < 0$, $T \geqslant 0$. Then $f_1'(T) < 0$, $T \geqslant 0$, i.e. $f_1(T)$ is a decreasing function of a variable T, $T \geqslant 0$; it decreases from $f_1(0) > 0$ to $f_1(\infty) > 0$. It follows that $f_1(T) > 0$, $T \geqslant 0$. For $q_1^* < q < 0$ the extreme point $T = T_1(q)$ does not exist ($T \geqslant 0$), herewith $y_1(T = 0, q) < 0$ and, consequently, $y_1(T) < 0$, $T \geqslant 0$. Further the proof is entirely analogous to that of case 2. Finally, if $\varphi(q) > 0$, $q_1 < q < 0 < q_2$, then $f_1(T) > 0$, $T \geqslant 0$.

Case 6. $\varphi(q) > 0$, $q_1 < q < q_2 = 0$. In this case $y_1(T) = -z_1 z_2 (z_2 - z_1) - (z_1 + z_2) q e^{-z_2 T}$, $T \geqslant 0$, and $y_1(T)$ is a decreasing function of a variable T, $T \geqslant 0$; it decreases from $y_1(T = 0, q)$ to $y_1(\infty) = -z_1 z_2 (z_2 - z_1) < 0$. Let us look at $y_1(T = 0, q) = -[z_1 z_2 (z_2 - z_1) + (z_1 + z_2) q]$, $q_1 \leqslant q \leqslant q_2 = 0$. We have: (a) $y_1(T = 0, q = q_1) = -z_2^2 (z_1 - \beta_1 - \beta_2) > 0$; (b) $y_1(T = 0, q = q_2 = 0) = -z_1 z_2 (z_2 - z_1) < 0$; (c) $y_1(T = 0, q = q_1^*) = 0$, where $q_1^* = -z_1 z_2 (z_2 - z_1) \times (z_1 + z_2)^{-1}$, $q_1 < q_1^* < 0$. Further the idea is basically the same as in the proof of case 5. Finally, if $\varphi(q) > 0$, $q_1 < q < q_2 = 0$, then $f_1(T) > 0$, $T \geqslant 0$.

Case 7. $\varphi(q) > 0$, $q_1 < q < q_2 < 0$. In this case $y_1(T) = z_1 z_2 (z_1 - \beta_1 - \beta_2) - (z_1 + \beta_1 + \beta_2) q e^{-(\beta_1 + \beta_2)T}$, $T \geqslant 0$, and $y_1(T)$ is a decreasing function of a variable T, $T \geqslant 0$; it decreases from $y_1(T = 0, q)$ to $y_1(\infty) = z_1 z_2 (z_1 - \beta_1 - \beta_2) < 0$. Let us look at $y_1(T = 0, q) = z_1 z_2 (z_1 - \beta_1 - \beta_2) - (z_1 + \beta_1 + \beta_2) q$, $q_1 \leqslant q \leqslant q_2 < 0$. We have: (1) $y_1(T = 0, q = q_1) = -z_2 (\beta_1 + \beta_2)(z_1 - \beta_1 - \beta_2) > 0$; (2) $y_1(T = 0, q = q_2) = -z_1 (\beta_1 + \beta_2) [z_2 - z_1 + (z_2 - \beta_1 - \beta_2)]$, here $z_2 - \beta_1 - \beta_2 < 0$. At this point there are some variants possible. *Variant (a)*: $y_1(T = 0, q = q_2) > 0$. In this variant the function $y_1(T)$ intersects the axis of abscisses once at the point $T_1(q)$, that is defined by (10). Then we have: (1) $f_1'(T) > 0$, $0 \leqslant T < T_1(q)$; (2) $f_1'(T) = 0$, $T = T_1(q)$; (3) $f_1'(T) < 0$, $T > T_1(q)$. It follows that the function $f_1(T)$

has only one point of maximum at $T = T_1(q)$ and thus $f_1(T) > 0$, $T \geqslant 0$, as $f_1(0) > 0$, $f_1(\infty) > 0$, $f_1(0) > f_1(\infty)$. *Variant (b):* $y_1(T = 0, q = q_2) = 0$. The result is identical to the result obtained in variant (a). *Variant (c):* $y_1(T = 0, q = q_2) < 0$. The result is identical to the result obtained in case 5. Finally, if $\varphi(q) > 0$, $q_1 < q < q_2 < 0$, then $f_1(T) > 0$, $T \geqslant 0$.

Summarizing the results obtained in cases 1–7, we have: if $\varphi(q) > 0$, $q_1 < q < q_2$, then $f_1(T) > 0$, $T \geqslant 0$. We may now moce on to consider the cases when $\varphi(q) < 0$. Since $q > q_1$, the region where $q \leqslant q_1$ can be excluded from further consideration.

Case 8. $\varphi(q) < 0$, $0 < q_1 < q_2 < q$. In this case $y_1(T) = z_1 z_2 (z_1 - \beta_1 - \beta_2) - (z_1 + \beta_1 + \beta_2) q e^{-(\beta_1+\beta_2)T}$, $T \geqslant 0$, and $y_1(T)$ is an increasing function of a variable T, $T \geqslant 0$; it increases from $y_1(T = 0, q)$ to $y_1(\infty) = z_1 z_2(z_1 - \beta_1 - \beta_2) > 0$. Let us look at $y_1(T = 0, q) = z_1 z_2 (z_1 - \beta_1 - \beta_2) - (z_1 + \beta_1 + \beta_2) q$, $0 < q_1 < q_2 \leqslant q$. Then $y_1(T = 0, q = q_2) < 0$ and (1) $y_1(T) < 0$, $0 \leqslant T < T_1(q)$; (2) $y_1(T) = 0$, $T = T_1(q)$; (3) $y_1(T) > 0$, $T > T_1(q)$; $T_1(q)$ is defined by (10). Therefore, we have: (1) $f_1'(T) > 0$, $0 \leqslant T < T_1(q)$; (2) $f_1'(T) = 0$, $T = T_1(q)$; (3) $f_1'(T) < 0$, $T > T_1(q)$. It follows that the function $f_1(T)$ has only one point of maximum at $T = T_1(q)$, herewith $f_1(0) > 0$, $f_1(\infty) > 0$, $f_1(0) < f_1(\infty)$. Thus, $f_1(T) > 0$, $T \geqslant 0$. Finally, if $\varphi(q) < 0$, $0 < q_1 < q_2 < q$, then $f_1(T) > 0$, $T \geqslant 0$.

Case 9. $\varphi(q) < 0$, $q_1 = 0 < q_2 < q$. In this case $y_1(T) = -(z_1 + \beta_1 + \beta_2) \times q e^{-(\beta_1+\beta_2)T} < 0$, $T \geqslant 0$. Then $f_1'(T) > 0$, $T \geqslant 0$, i.e. $f_1(T)$ is an increasing function of a variable T, $T \geqslant 0$; it increases from $f_1(0) > 0$ to $f_1(\infty) > 0$. It follows that $f_1(T) > 0$, $T \geqslant 0$. Finally, if $\varphi(q) < 0$, $q_1 = 0 < q_2 < q$, then $f_1(T) > 0$, $T \geqslant 0$.

Case 10. $\varphi(q) < 0$, $q_1 < 0 < q_2 < q$. In this case $y_1(T) = z_1 z_2 (z_1 - \beta_1 - \beta_2) - (z_1 + \beta_1 + \beta_2) q e^{-(\beta_1+\beta_2)T} < 0$, $T \geqslant 0$. Then $f_1'(T) > 0$, $T \geqslant 0$; further the idea is basically the same as in the proof of case 9. Finally, if $\varphi(q) < 0$, $q_1 < 0 < q_2 < q$, then $f_1(T) > 0$, $T \geqslant 0$.

Case 11. $\varphi(q) < 0$, $q_1 < q_2 < 0 < q$. In this case $y_1(T) = z_1 z_2 (z_1 - \beta_1 - \beta_2) - (z_1 + \beta_1 + \beta_2) q e^{-(\beta_1+\beta_2)T} < 0$, $T \geqslant 0$. Then $f_1'(T) > 0$, $T \geqslant 0$; further the idea is basically the same as in the proof of case 9. Finally, if $\varphi(q) < 0$, $q_1 < q_2 < 0 < q$, then $f_1(T) > 0$, $T \geqslant 0$.

Case 12. $\varphi(q) < 0$, $q_1 < q_2 < q = 0$. In this case $y_1(T) = z_1 z_2 (z_1 - \beta_1 - \beta_2) = $ const < 0, $T \geqslant 0$. Then $f_1'(T) > 0$, $T \geqslant 0$; further the idea is basically the same as in the proof of case 9. Finally, if $\varphi(q) < 0$, $q_1 < q_2 < q = 0$, then $f_1(T) > 0$, $T \geqslant 0$.

Case 13. $\varphi(q) > 0$, $q_1 < q_2 < q < 0$. In this case $y_1(T) = z_1 z_2 (z_1 - \beta_1 - \beta_2) - (z_1 + \beta_1 + \beta_2) q e^{-(\beta_1+\beta_2)T}$, $T \geqslant 0$, and $y_1(T)$ is a decreasing function of a variable T, $T \geqslant 0$; it decreases from $y_1(T = 0, q)$ to $y_1(\infty) = z_1 z_2(z_1 - \beta_1 - \beta_2) < 0$. Let us look at $y_1(T = 0, q) = z_1 z_2 (z_1 - \beta_1 - \beta_2) - (z_1 + \beta_1 + \beta_2) q$, $q_2 \leqslant q \leqslant 0$. We have: (1) $y_1(T = 0, q = q_2) = -z_1(\beta_1 + \beta_2)[z_2 - z_1 + (z_2 - \beta_1 - \beta_2)]$; (2)

$y_1 (T = 0, q = 0) = z_1 z_2 (z_1 - \beta_1 - \beta_2) < 0$. Now three subcases are possible:

Subcase 13.1. $y_1 (T = 0, q = q_2) > 0$. At this point there are some variants possible. *Variant (a):* $y_1 (T = 0, q) > 0$, $q_2 < q < q_1^* < 0$. In this variant the function $y_1 (T)$ intersects the axis of abscisses once at the point $T_1(q)$, that is defined by (10). Then we have: (1) $f_1' (T) < 0$, $0 \leqslant T < T_1(q)$; (2) $f_1' (T) = 0$, $T = T_1(q)$; (3) $f_1' (T) > 0$, $T > T_1(q)$. It follows that the function $f_1 (T)$ has only one point of minimum at $T = T_1(q)$. It can be shown that $f_1 (T_1(q)) > 0$, $q_2 < q < q_1^* < 0$, then $f_1 (T) > 0$, $T \geqslant 0$. *Variant (b):* $y_1 (T = 0, q) = 0$, $q = q_1^* < 0$. In this variant $y_1 (T)$ is a decreasing function of a variable T, $T \geqslant 0$; it decreases from $y_1 (T = 0, q = q_1^*)$ to $y_1 (\infty) = z_1 z_2 (z_1 - \beta_1 - \beta_2) < 0$. Then: (1) $y_1 (T) < 0$, $T > 0$; (2) $y_1 (T) = 0$, $T = 0$. Therefore, we have: (1) $f_1' (T) > 0$, $T > 0$; (2) $f_1' (T) = 0$, $T = 0$. It follows that the function $f_1 (T)$ is an increasing function of a variable T, $T > 0$, and nondecreasing function at the point $T = 0$; $f_1 (T)$ increases from $f_1 (0) > 0$ to $f_1 (\infty) > 0$. Thus, $f_1 (T) > 0$, $T \geqslant 0$. *Variant (c):* $y_1 (T = 0, q) < 0$, $q_1^* < q < 0$. In this variant $y_1 (T)$ is a decreasing function of a variable T, $T \geqslant 0$; it decreases from $y_1 (T = 0, q) < 0$ to $y_1 (\infty) = z_1 z_2 (z_1 - \beta_1 - \beta_2) < 0$. Then $y_1 (T) < 0$, $T > 0$, and, consequently, $f_1' (T) > 0$, $T > 0$. It follows that the function $f_1 (T)$ is an increasing function of a variable T, $T \geqslant 0$; it increases from $f_1 (0) > 0$ to $f_1 (\infty) > 0$. Thus, $f_1 (T) > 0$, $T \geqslant 0$. Here $q_1^* = z_1 z_2 (z_1 - \beta_1 - \beta_2) / (z_1 + \beta_1 + \beta_2)$.

Subcase 13.2. $y_1 (T = 0, q = q_2) = 0$. In this subcase $y_1 (T)$ is a decreasing function of a variable T, $T \geqslant 0$; it decreases from $y_1 (T = 0, q) < 0$ to $y_1 (\infty) = z_1 z_2 (z_1 - \beta_1 - \beta_2) < 0$. Then $y_1 (T) < 0$, $T \geqslant 0$, and further the proof is entirely analogous to that of case 9.

Subcase 13.3. $y_1 (T = 0, q = q_2) < 0$. Further the proof is entirely analogous to that of subcase 13.2.

Summarizing the results obtained in subcases 13.1–13.3, we have: if $\varphi (q) < 0$, $q_1 < q_2 < q < 0$, then $f_1 (T) > 0$, $T \geqslant 0$. Summarizing the results obtained in cases 8–13, we have: if $\varphi (q) < 0$, $q_1 < q_2 < q$, then $f_1 (T) > 0$, $T \geqslant 0$.

Finally, we have: (1) if $\varphi (q) > 0$, $q_1 < q < q_2$, then $f_1 (T) > 0$, $T \geqslant 0$; (2) if $\varphi (q) = 0$, $q = q_2$, then $f_1 (T) = const > 0$, $T \geqslant 0$; (3) if $\varphi (q) < 0$, $q_1 < q_2 < q$, then $f_1 (T) > 0$, $T \geqslant 0$, where $q_1 = z_2 (z_1 - \beta_1 - \beta_2)$, $q_2 = z_1 (z_2 - \beta_1 - \beta_2)$ are arbitrary real numbers. *Lemma 3 is proved.*

Proposition 2. *The function* $f_2 (T) = z_2 [z_1 z_2 (\beta_1 + \beta_2) - z_2 f (T) - f' (T)]$, $T \geqslant 0$, *satisfies the following relations:*

(1) $f_2(0) = (\beta_1 + \beta_2) \dfrac{z_2 C}{A^2}(z_1 A - C) = (\beta_1 + \beta_2)\dfrac{z_2^2 C}{A^2}(q - q_2)$, where $q_2 = z_1(z_2 - \beta_1 - \beta_2)$; herewith $f_2(0)$ may be: (a) $f_2(0) > 0$, (b) $f_2(0) = 0$, (c) $f_2(0) < 0$; $q > q_1$, $q_1 < q_2$;

(2) $f_2(\infty) = \lim\limits_{T \to \infty} f_2(T) = z_2(z_1 A - C) = z_2^2(q - q_2)$; herewith $f_2(\infty)$ may be: (a) $f_2(\infty) > 0$, (b) $f_2(\infty) = 0$, (c) $f_2(\infty) < 0$;

(3) $f_2'(T) = \dfrac{z_1 z_2^2(\beta_1 + \beta_2)}{F^3(T)} \varphi(q) e^{-(\beta_1+\beta_2)T}\left[z_1 z_2(z_2 - \beta_1 - \beta_2) - (z_2 + \beta_1 + \beta_2)\right]$
$\times q e^{-(\beta_1+\beta_2)T}$, $T \geqslant 0$;

(4) $T_2(q)$ is a bending point (or an inflexion point) of function $f_2(T)$:
$$T_2(q) = -\frac{1}{\beta_1 + \beta_2}\ln\frac{z_1 z_2(z_2 - \beta_1 - \beta_2)}{(z_2 + \beta_1 + \beta_2)q}, \quad \frac{z_2 - \beta_1 - \beta_2}{q} > 0;$$

(5) $f_2(T_2(q)) = \dfrac{z_2(z_2 + \beta_1 + \beta_2)^2}{4(\beta_1 + \beta_2)q}[q - z_1(z_2 - \beta_1 - \beta_2)]$
$$\times\left[q - \frac{z_2(z_1 - \beta_1 - \beta_2)(z_2 - \beta_1 - \beta_2)^2}{(z_2 + \beta_1 + \beta_2)^2}\right], \quad \frac{z_2 - \beta_1 - \beta_2}{q} > 0, \text{ where } q_{22} =$$
$q_2 = z_1(z_2 - \beta_1 - \beta_2)$, $q_{21} = \dfrac{z_2(z_1 - \beta_1 - \beta_2)(z_2 - \beta_1 - \beta_2)^2}{(z_2 + \beta_1 + \beta_2)^2}$ are the roots
of the equation $f_2(T_2(q)) = 0$.

Here q, $F(T)$ are defined by (1); A, C are defined by (4); $\varphi(q)$ is defined by (7).

Lemma 4. *The function $f_2(T)$ is a negative function $(f_2(T) < 0)$ of a variable T, $T \geqslant 0$, if $\varphi(q) > 0$, $q_1 < q < q_2$, and $f_2(T)$ is a positive function $(f_2(T) > 0)$ of a variable T, $T \geqslant 0$, if $\varphi(q) < 0$, $q_1 < q_2 < q$, where $q_1 = z_2(z_1 - \beta_1 - \beta_2)$, $q_2 = z_1(z_2 - \beta_1 - \beta_2)$ are arbitrary real numbers.*

Proof. The proof is very similar to the proof of Lemma 3.

Let us turn now to studying the function $\Phi(T) = f_1(T) - f_2(T)$ as a function of a variable T, $T \geqslant 0$.

Proposition 3. *The function $\Phi(T) = f_1(T) - f_2(T)$, $T \geqslant 0$, satisfies the following relations:*

(1) $\Phi(T) = (z_2 - z_1)\left[-z_1 z_2(\beta_1 + \beta_2) + (z_1 + z_2)f(T) + f'(T)\right]$, $T \geqslant 0$;

(2) $\Phi(0) = (z_2 - z_1)(\beta_1 + \beta_2)(C/A)^2 > 0$, consequently, $f_1(0) > f_2(0)$;

(3) $\Phi(\infty) = \lim\limits_{T \to \infty}\Phi(T) = (z_2 - z_1)C > 0$, consequently, $f_1(\infty) > f_2(\infty)$;

(4) $\Phi(0) - \Phi(\infty) = (z_2 - z_1)\varphi(q)C/A^2$;

(5) $\Phi'(T) = -(z_2 - z_1)\varphi(q)\dfrac{z_1 z_2(\beta_1 + \beta_2)e^{-(\beta_1+\beta_2)T}}{F^3(T)}$
$\times\left[z_1 z_2(z_1 + z_2 - \beta_1 - \beta_2) - (z_1 + z_2 + \beta_1 + \beta_2)q e^{-(\beta_1+\beta_2)T}\right]$, $T \geqslant 0$;

(6) $T^{*}(q)$ is a bending point (or an inflexion point) of function $\Phi(T)$:

$$T^{*}(q) = -\frac{1}{\beta_1 + \beta_2} \ln \frac{z_1 z_2 (z_1 + z_2 - \beta_1 - \beta_2)}{(z_1 + z_2 + \beta_1 + \beta_2) q}, \quad \frac{z_1 + z_2 - \beta_1 - \beta_2}{q} > 0;$$

(7) $\Phi(T^{*}(q)) = (z_2 - z_1) \left\{ C + \varphi(q) \frac{(z_1 + z_2 - \beta_1 - \beta_2)^2}{4(\beta_1 + \beta_2) q} \right\},$

$\dfrac{z_1 + z_2 - \beta_1 - \beta_2}{q} > 0.$

Here q, $F(T)$ are defined by (1); A, C are defined by (4); $\varphi(q)$ is defined by (7).

Lemma 5. *The function $\Phi(T)$ is a positive function ($\Phi(T) > 0$) of a variable T, $T \geqslant 0$, i.e. $f_1(T) > f_2(T)$, $T \geqslant 0$.*

Proof. The proof is very similar to the proof of Lemma 3.

Theorem 1. *The derivative $p_T'(\tau_m)$ is a positive function ($p_T'(\tau_m) > 0$) of a variable T, $0 \leqslant T \leqslant \tau_m$, $0 \leqslant \tau_m < \infty$, for any set of parameters $\lambda_1 > \lambda_2 \geqslant 0$, $0 \leqslant p \leqslant 1$, $\beta > 0$, $\alpha > 0$, $0 \leqslant \delta \leqslant 1$.*

Proof. The proof is based on consistent application of Lemmas 1–5:

(1) If $\varphi(q) > 0$, $q_1 < q < q_2$, $q \neq q_2$, then $f_1(T) > 0$, $f_2(T) < 0$, $T \geqslant 0$. Thus, $f_1(T)/f_2(T) < 0$, $T \geqslant 0$, including $0 \leqslant T \leqslant \tau_m$. It follows that the Eq. (9) does not have a solution and therefore $p_T'(\tau_m)$ does not reach zero point for $0 \leqslant T \leqslant \tau_m$. Since $p_{T=0}'(\tau_m) > 0$, $p_{T=\tau_m}'(\tau_m) > 0$, then $p_T'(\tau_m) > 0$, $0 \leqslant T \leqslant \tau_m$.

(2) If $\varphi(q) = 0$, $q = q_1$, $q \neq q_2$, then $f_1(T) = 0$, $f_2(T) < 0$, $T \geqslant 0$. Thus, $f_1(T)/f_2(T) = 0$, $T \geqslant 0$, including $0 \leqslant T \leqslant \tau_m$. It follows that the Eq. (9) does not have a solution and, consequently, $p_T'(\tau_m) > 0$, $0 \leqslant T \leqslant \tau_m$.

(3) If $\varphi(q) = 0$, $q \neq q_1$, $q = q_2$, then $f_1(T) > 0$, $f_2(T) = 0$, $T \geqslant 0$. In this case the Eq. (9) can be converted into the following equation to avoid divide-by-zero error:

$$e^{(z_2 - z_1)(\tau_m - T)} = f_2(T)/f_1(T), \quad 0 \leqslant T \leqslant \tau_m. \tag{11}$$

Thus, $f_2(T)/f_1(T) = 0$, $T \geqslant 0$, including $0 \leqslant T \leqslant \tau_m$. It follows that the Eq. (11) does not have a solution and, consequently, the Eq. (9) also does not have a solution. Finally, $p_T'(\tau_m) > 0$, $0 \leqslant T \leqslant \tau_m$.

(4) If $\varphi(q) < 0$, $q > q_2$, then $f_1(T) > 0$, $f_2(T) > 0$, $f_1(T) > f_2(T)$, $T \geqslant 0$. Thus, $f_1(T)/f_2(T) > 1$, $T \geqslant 0$, including $0 \leqslant T \leqslant \tau_m$. It follows that the Eq. (9) does not have a solution and therefore $p_T'(\tau_m) > 0$, $0 \leqslant T \leqslant \tau_m$. The Theorem 1 is proved.

Theorem 2. *The probability density function $p_T(\tau_m)$ is an increasing function of a variable T, $0 \leqslant T \leqslant \tau_m$, $0 \leqslant \tau_m < \infty$, for any set of parameters $\lambda_1 > \lambda_2 \geqslant 0$, $0 \leqslant p \leqslant 1$, $\beta > 0$, $\alpha > 0$, $0 \leqslant \delta \leqslant 1$, and reaches its global maximum at the point $T = \tau_m$, $0 \leqslant \tau_m < \infty$.*

Proof. The proof follows from the proof of Theorem 1.

Corollary 1. *By Theorem 1, functions* $p_T\left(\tau^{(j)}\right)$, $j = \overline{2,k}$, *are increasing functions of a variable* T, $0 \leqslant T \leqslant \tau^{(j)}$, $0 \leqslant \tau^{(j)} < \infty$, $j = \overline{2,k}$, *for any set of parameters* $\lambda_1 > \lambda_2 \geqslant 0$, $0 \leqslant p \leqslant 1$, $\beta > 0$, $\alpha > 0$, $0 \leqslant \delta \leqslant 1$.

Corollary 2. *By Theorem 2, the likelihood function* $L\left(T \mid \tau^{(1)}, ..., \tau^{(k)}\right)$ *reaches its global maximum at the point* $\hat{T} = \tau_m$, *i.e. the solution of optimization task (2) is the estimate of the dead time period duration:* $\hat{T} = \tau_m$.

5   Conclusion

The obtained results provide the possibility of solving the problem of parameters estimation of the modulated semi-synchronous generalized flow of events in conditions of a constant dead time without using numerical methods: during the observation of the flow (during the time interval $(0, t]$) the values τ_k, $k = \overline{1, n}$, are calculated; $\tau_{min} = min\,\tau_k$, $k = \overline{1, n}$, is found and the estimate of the dead time period duration is assumed to equal $\hat{T} = \tau_{min}$. Let us note that the maximum likelihood estimate of the dead time period duration will always be shifted for a finite value of t ($\tau_{min} > T$); the unbiasedness of the estimate will take place only in an asymptotic case ($t \to \infty$).

Acknowledgments. The work is supported by Tomsk State University Competitiveness Improvement Program.

References

1. Cox, D.R.: Some statistical methods connected with series of events. J. Roy. Stat. Soc. B **17**, 129–164 (1955)
2. Kingman, J.F.C.: On doubly stochastic Poisson processes. Proc. Camb. Philos. Soc. **60**(4), 923–930 (1964)
3. Basharin, G.P., Kokotushkin, V.A., Naumov, V.A.: Method of equivalent substitutions for calculating fragments of communication networks for digital computer. Eng. Cyber. **17**(6), 66–73 (1979)
4. Neuts, M.F.: A versatile Markov point process. J. Appl. Probab. **16**, 764–779 (1979)
5. Last, G., Brandt, A.: Marked Point Process on the Real Line: The Dynamic Approach, 1st edn. Springer, New York (1995)
6. Gortsev, A.M., Nezhelskaya, L.A.: An asynchronous double stochastic flow with initiation of superfluous events. Discrete Math. Appl. **21**(3), 283–290 (2011)
7. Basharin, G.P., Gaidamaka, Y.V., Samouylov, K.E.: Mathematical theory of teletraffic and its application to the analysis of multiservice communication of next generation networks. Autom. Control Comput. Sci. **47**(2), 62–69 (2013)
8. Adamu, A., Gaidamaka, Y., Samuylov, A.: Discrete Markov chain model for analyzing probability measures of P2P streaming network. In: Balandin, S., Koucheryavy, Y., Hu, H. (eds.) NEW2AN 2011 and ruSMART 2011. LNCS, vol. 6869, pp. 428–439. Springer, Heidelberg (2011)

9. Bouzas, P.R., Valderrama, M.J., Aguilera, A.M., Ruiz-Fuentes, N.: Modelling the mean of a doubly stochastic Poisson process by functional data analysis. Comput. Stat. Data Anal. **50**(10), 2655–2667 (2006)
10. Centanni, S., Minozzo, M.: A Monte Carlo approach to filtering for a class of marked doubly stochastic Poisson processes. J. Am. Stat. Assoc. **101**, 1582–1597 (2006)
11. Hossain, M.M., Lawson, A.B.: Approximate methods in Bayesian point process spatial models. Comput. Stat. Data Anal. **53**(8), 2831–2842 (2009)
12. Gortsev, A.M., Shmyrin, I.S.: Optimal estimation of states of a double stochastic flow of events in the presence of measurement errors of time instants. Autom. Remote Control **60**(1), 41–51 (1999)
13. Bushlanov, I.V., Gortsev, A.M.: Optimal estimation of the states of a synchronous double stochastic flow of events. Autom. Remote Control **65**(9), 1389–1399 (2004)
14. Gortsev, A.M., Nezhelskaya, L.A., Solovev, A.A.: Optimal state estimation in MAP event flows with unextendable dead time. Autom. Remote Control **73**(8), 1316–1326 (2012)
15. Bakholdina, M.A., Gortsev, A.M.: Optimal estimation of the states of modulated semi-synchronous integrated flow of events in condition of its incomplete observability. Appl. Math. Sci. **9**(29), 1433–1451 (2015)
16. Gortsev, A.M., Klimov, I.S.: Estimation of the parameters of an alternating Poisson stream of events. Telecommun. Radio Eng. **48**(10), 40–45 (1993)
17. Gortsev, A.M., Nezhelskaya, L.A.: Estimation of the parameters of a synchro-alternating Poisson event flow by the method of moments. Radiotekhnika **40**(7–8), 6–10 (1995)
18. Vasileva, L.A., Gortsev, A.M.: Estimation of parameters of a double-stochastic flow of events under conditions of its incomplete observability. Autom. Remote Control **63**(3), 511–515 (2002)
19. Gortsev, A.M., Nezhelskaya, L.A.: Estimation of the dead-time period and parameters of a semi-synchronous double-stochastic stream of events. Measur. Tech. **46**(6), 536–545 (2003)
20. Gortsev, A.M., Nissenbaum, O.V.: Estimation of the dead time period and parameters of an asynchronous alternative flow of events with unextendable dead time period. Russ. Phys. J. **48**(10), 1039–1054 (2005)
21. Bushlanov, I.V., Gortsev, A.M., Nezhelskaya, L.A.: Estimation parameters of the synchronous twofold-stochastic flow of events. Autom. Remote Control **69**(9), 1517–1533 (2008)
22. Apanasovich, V.V., Koljada, A.A., Chernjavski, A.F.: The Statistical Analysis of Series of Random Events in Physical Experiment. University Press, Minsk (1988). (in Russian)
23. Normey-Rico, J.E.: Control of Dead-time Processes. Advanced Textbooks in Control and Signal Processing. Springer, London (2007)
24. Gortsev, A.M., Klimov, I.S.: Estimation of intensity of Poisson stream of events for conditions under which it is partially unobservable. Telecommun. Radio Eng. **47**(1), 33–38 (1992)
25. Vasileva, L.A., Gortsev, A.M.: Estimation of the dead time of an asynchronous double stochastic flow of events under incomplete observability. Autom. Remote Control **64**(12), 1890–1898 (2003)
26. Gortsev, A.M., Solovev, A.A.: Joint probability density of interarrival interval of a flow of physical events with unextendable dead time period. Russ. Phys. J. **57**(7), 973–983 (2014)

27. Bakholdina, M., Gortsev, A.: Joint probability density of the intervals length of the modulated semi-synchronous integrated flow of events and its recurrence conditions. In: Dudin, A., Nazarov, A., Yakupov, R., Gortsev, A. (eds.) Information Technologies and Mathematical Modelling. Communications in Computer and Information Science, vol. 487, pp. 18–25. Springer, Switzerland (2014)
28. Bakholdina, M., Gortsev, A.: Joint probability density of the intervals length of modulated semi-synchronous integrated flow of events in conditions of a constant dead time and the flow recurrence conditions. In: Dudin, A., Nazarov, A., Yakupov, R. (eds.) Information Technologies and Mathematical Modelling - Queueing Theory and Applications. Communications in Computer and Information Science, vol. 564, pp. 13–27. Springer, Switzerland (2015)
29. Gortsev, A.M., Nezhelskaya, L.A.: On connection of MC flows and MAP flows of events. Tomsk State Univ. J. Control Comput. Sci. 1(14), 13–21 (2011). (in Russian)

Sign CUSUM Algorithm for Change-Point Detection of the MMPP Controlling Chain State

Yulia Burkatovskaya[1,2,3], Tatiana Kabanova[2(✉)], and Olga Tokareva[1]

[1] Institute of Cybernetics, Tomsk Polytechnic University,
30 Lenin Prospekt, 634050 Tomsk, Russia
tracey@tpu.ru, ostokareva@gmail.com
[2] Department of Applied Mathematics and Cybernetics,
Tomsk State University, 36 Lenin Prospekt, 634050 Tomsk, Russia
tvk@bk.ru
[3] International Laboratory of Statistics of Stochastic Processes
and Quantitative Finance, Tomsk State University,
36 Lenin Prospekt, 634050 Tomsk, Russia
http://www.tpu.ru, http://www.tsu.ru

Abstract. The authors consider the Markov modulated Poisson process with two states of the Markovian controlling chain. The flow intensity of the observed process depends on the unobserved controlling chain state. All the process parameters are supposed to be unknown. The paper develops a new sequential change-point detection method based on the cumulative sum control chart approach to determine the switching points of the flow intensity. Usage of special sign statistics allows the obtaining of theoretical characteristics of the proposed algorithm.

Keywords: Markov modulated Poisson process · Jump intensity · Change-point detection · Sign statistics · CUSUM algorithm

1 Introduction

The Markov-modulated Poisson process (MMPP) has been extensively used for modeling Poisson processes whose arrival intensities vary randomly over time. It qualitatively models the time-varying arrival rate and captures some of the important correlations between the interarrival times while still remaining analytically tractable [1]. It can be described as a Poisson process whose intensity is determined by a controlling chain state. Transition between the states occurs in unknown random instants, and the time of being in a state is distributed exponentially.

MMPP arise in many applications of interest, such as Web-servers, multimedia traffic, call-centers, cell phone call activity, product demand, etc. Mingrui Zou and Jianqing Liu in [2] use MMPP to investigate an unsaturated IEEE 802.16 network with the contention-based access mechanism. The authors model packet arrivals at each subscriber station as a MMPP) and derive analytical expressions

© Springer International Publishing Switzerland 2016
A. Dudin et al. (Eds.): ITMM 2016, CCIS 638, pp. 18–33, 2016.
DOI: 10.1007/978-3-319-44615-8_2

for the network throughput and packet delay subject to the MMPP parameters, i.e., the steady-state probabilities and the average arrival rate.

Nogueira et al. in [3] use a superposition of discrete time MMPP model (dMMPP) for the modeling of network traffic on multiple time scales. Two Markovian models are proposed: the fitting procedure of the first model matches the complete distribution of the arrival process at each timescale of interest, while the second proposed model is constructed using a hierarchical procedure that decomposes each MMPP state into new MMPPs that incorporate a more detailed description of the distribution at finer time scales. The traffic process is then represented by a MMPP equivalent to the constructed hierarchical structure. Both approaches use estimators of the characterizing parameters of each MMPP, that is, the matrices corresponding to the transition probabilities and the Poisson arrival rates for each state.

Giacomazzi in [4] develops a method for using traffic sources modeled as a MMPP in the framework of the bounded-variance network calculus, a novel stochastic network calculus framework for the approximated analysis of end-to-end network delay. The mean and the variance of the cumulative traffic are analyzed for two traffic envelopes, the first, the two-moment envelope, is an approximation of traffic with the same first two moments of the actual source traffic. The second, the linear envelope, provides a less precise approximation but it permits the closed-form analysis of single-node and end-to-end delay with several types of important schedulers.

Choi et al. in [5] consider MMPP as model for traffic streams with bursty characteristic and time correlation between interarrival times. Traffics such as voice and video in ATM networks have these properties. By using the embedded Markov chain method, the authors derive the queue length distribution at departure epochs. They also obtain the queue length distribution at an arbitrary time by the supplementary variable method. The authors apply the results for preventive congestion control in telecommunication networks.

There exist a number of rather complicated flow models based on the MMPP. Vasil'eva and Gortsev in [6] study an asynchronous double stochastic flow of events where each event results in a dead time period when other events cannot be observed. The authors determine the Laplace transformation of the event-event interval probability density in the observed flow and derive the equations of moments for the estimation of dead time and initial event flow parameters. Gortsev and Nezhelskaya in [7] study the stationary mode of an asynchronous double stochastic flow with initiation of superfluous events. The authors determine important properties of the flow studied as interval probability density and joint probability density of neighboring intervals lengths. Also this work specifies conditions in which the flow either becomes recursive or degenerates to an elementary one.

The papers surveyed above describe some applications of the MMPP process but of course not all of them. To take decisions concerning process behavior and to develop dispatching rules one needs to fit a model and to evaluate the model parameters. There are two classical approaches to the MMPP parameter estimation problem: maximum likelihood estimation and its implementation via the EM (expectation-maximization) algorithm and matching moment method. A detailed survey of former methods is given in [8]. The survey [9] with a huge bibliography is focused on the latter methods. These approaches are connected with complicated numerical calculations. It implies difficulties in their theoretical investigations and, hence, the necessity of their study via simulation. That stresses the urgency of the MMPP parameter estimation problem and necessity of simple efficient estimation algorithms, whose non-asymptotic properties can be investigated theoretically.

In [10], we a the sequential analysis approach to parameter estimation for MMPP process. The time intervals between the observed flow events were considered as the values of a stochastic process. The mean of the process was supposed to change in some unknown instants, or change points.

The problem of sequential change-point detection can be formulated as follows. A stochastic process is observed. Several parameters of the process change in random points. The problem is to detect the change points when the process is observed online. Sequential methods include a special stopping rule that determines a stopping time. At this instant, a decision on the change point can be made. There are two types of errors typical for sequential change-point detection procedures: a false alarm, when one makes a decision that change has occurred before a change point (type 1 error), and delay, when the change is not detected (type 2 error).

At the first stage of the algorithm, these change points were detected by using CUSUM (cumulative sum control chart) algorithms. After that the intensity parameters were estimated under the assumption that the intensity was constant between detected change points. The quality of the proposed algorithms was studied via simulation.

In this paper, a modification of the CUSUM algorithm is proposed. Instead of the lengths of the interval between events, special sign statistics are used at the change-point detection procedure. It allows us to investigate theoretically the properties of the algorithm and give recommendations concerning the parameter choice.

2   Problem Statement

We consider a Markov-modulated poisson process, i.e. a flow of events controlled by a Markovian chain with a continuous time. The chain has two states and transition between the states happens at random instants. The time of sojourn of the chain in the i-th state is exponentially distributed with the parameter α_i, $i = 1, 2$.

The flow of events has exponential distribution with the intensity parameter λ_1 or λ_2 subject to the state of the Markovian chain. We suppose that the intensity of the switch between the controlling chain states is sufficiently smaller then the intensity of the arrival process, i.e. $\alpha_i << \lambda_i$. In this case, several events commonly occur before the change of the controlling chain state. The parameters of the system λ_1, λ_2 and the instants of switching between the states are supposed to be unknown. The sequence of instants of arriving events is observed. The problem is to estimate the parameters λ_1, λ_2, α_1, α_2.

3   Sign CUSUM Algorithm

Consider the process $\{\tau_i\}_{i\geq1}$, where $\tau_i = t_i - t_{i-1}$ is the length of the i-th interval between arriving events in the observed flow. Figure 1 demonstrates the construction of the sequence $\{\tau_i\}$.

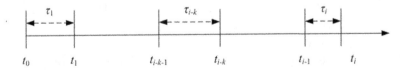

Fig. 1. Construction of the sequence $\{\tau_i\}$.

If the controlling chain is in the l-th state then the mean length between events is equal to $1/\lambda_l$. So at the first stage of our procedure we try to detect the instants of the chain transition from one state to another as the instants of change in the mean of the process $\{\tau_i\}_{i\geq1}$ using the Sign CUSUM procedure. At the first time such algorithm was proposed in [11] and developed in [12] for the case of the single change point. In this paper we consider the multiple change-point detection problem which has some special features.

Now we describe the procedure. Let the parameters λ_1, λ_2 satisfy the condition

$$0 < \lambda_2 < \lambda_1;$$
$$\frac{1}{\lambda_2} - \frac{1}{\lambda_1} > \Delta, \tag{1}$$

where Δ is a certain known positive parameter. Choose then an integer parameter k describing the memory depth. The idea is to compare the values τ_i and τ_{i-k}. If there are no changes in the controlling chain state within the interval $[t_{i-k-1}, t_i]$ then the variables τ_i and τ_{i-k} have the identical exponential distribution with the mean $1/\lambda_1$ or $1/\lambda_2$. If the chain state changes within the interval $[t_{i-k-1}, t_i]$ then the expectations of the values τ_i and τ_{i-k} are different.

For the case of the single change-point detection the parameter k is taken commonly as large as possible in order to guarantee that the decision is taken while the means of the variables are different. For our case of multiple change-point detection the parameter k should not be too large so as to contain more than one chain state change within the interval $[t_{i-k-1}, t_i]$. In paper [10] we recommended choosing $k \approx r/2$, where $\lambda_i \geq r\alpha_i$. Further we consider the choice of the algorithm parameters in more detail.

As the initial state of the chain is unknown, we shall consider two CUSUM procedures simultaneously. The first procedure is set up to detect an increase in the mean of the process and, hence, decrease of the intensity, and the second procedure is set up to detect an decrease in the mean and, hence, increase of the intensity. For the first procedure we introduce the sequence of the statistics

$$z_i^{(1)} = n \left(\text{sign}(\tau_i - \tau_{i-k}) - \delta\right), \quad i > k. \tag{2}$$

For the second procedure we introduce the sequence of the statistics

$$z_i^{(2)} = n \left(\text{sign}(\tau_{i-k} - \tau_i) - \delta\right), \quad i > k. \tag{3}$$

These statistics are calculated at the instant t_i. Here $\delta = m/n$, m and n are integers, $m < n$, and the fraction m/n is irreducible.

Consider then four hypothese concerning the state of the controlling chain:

- $H_1(t_{i-k-1}, t_i)$ – the intensity of the arrival process on the interval $[t_{i-k-1}, t_i]$ is constant and equal to λ_1;
- $H_2(t_{i-k-1}, t_i)$ – the intensity of the arrival process on the interval $[t_{i-k-1}, t_i]$ is constant and equal to λ_2;
- $H_{1,2}(t_{i-k}, t_{i-1})$ – the intensity of the arrival process on the interval $[t_{i-k}, t_{i-1}]$ changed once from λ_1 to λ_2, i.e., decreased;
- $H_{2,1}(t_{i-k}, t_{i-1})$ – the intensity of the arrival process on the interval $[t_{i-k}, t_{i-1}]$ changed once from λ_2 to λ_1, i.e., increased.

Note that in the conditions of the hypothesis $H_l(t_{i-k-1}, t_i)$ the random variables τ_i and τ_{i-k} have the same mean and the functions $\text{sign}(\tau_i - \tau_{i-k})$ and $\text{sign}(\tau_{i-k} - \tau_i)$ take the values 1 and –1 with the equal probabilities 1/2. So, in this case the expectation of the statistics $z_i^{(l)}$ is negative. To implement CUSUM procedures it is necessary to provide positive expectations of the statistics $z_i^{(1)}$ and $z_i^{(2)}$ after an increase or decrease in the mean of τ_i correspondingly. So, introducing the notations

$$\begin{aligned} p &= P\{\tau_i \geq \tau_{i-k} | H_{1,2}(t_{i-k}, t_{i-1})\} = P\{\tau_{i-k} \geq \tau_i | H_{2,1}(t_{i-k}, t_{i-1})\}; \\ q &= P\{\tau_i < \tau_{i-k} | H_{1,2}(t_{i-k}, t_{i-1})\} = P\{\tau_{i-k} < \tau_i | H_{2,1}(t_{i-k}, t_{i-1})\}, \end{aligned} \tag{4}$$

where $p > 1/2$, $q < 1/2$, $p + q = 1$ we can obtain the following result.

Theorem 1. *If the parameter δ satisfies the condition*

$$\delta < 2p - 1, \tag{5}$$

then the statistics $z_i^{(j)}$, $j \in \{1, 2\}$ (2) and (3) have the following properties:

$$\begin{aligned} E\left[z_i^{(1)} \middle| H_l(t_{i-k-1}, t_i)\right] &= -m < 0, \quad l = 1, 2; \\ E\left[z_i^{(1)} \middle| H_{2,1}(t_{i-k}, t_{i-1})\right] &= -n(2p - 1) - m < 0; \\ E\left[z_i^{(1)} \middle| H_{1,2}(t_{i-k}, t_{i-1})\right] &= n(2p - 1) - m > 0; \\ E\left[z_i^{(2)} \middle| H_l(t_{i-k-1}, t_i)\right] &= -m < 0, \quad l = 1, 2; \\ E\left[z_i^{(2)} \middle| H_{1,2}(t_{i-k}, t_{i-1})\right] &= -n(2p - 1) - m < 0, \\ E\left[z_i^{(2)} \middle| H_{2,1}(t_{i-k}, t_{i-1})\right] &= n(2p - 1) - m > 0. \end{aligned} \tag{6}$$

Proof. Using (4) and (5) one obtains

$$E\left[z_i^{(1)}\middle| H_l(t_{i-k-1}, t_i)\right] = n\left(E\left[\text{sign}(\tau_i - \tau_{i-k})\middle| H_l(t_{i-k-1}, t_i)\right] - \delta\right)$$
$$= n((1/2 - 1/2) - \delta) = -m < 0;$$
$$E\left[z_i^{(1)}\middle| H_{2,1}(t_{i-k}, t_{i-1})\right] = n\left(E\left[\text{sign}(\tau_i - \tau_{i-k})\middle| H_{2,1}(t_{i-k}, t_{i-1})\right] - \delta\right)$$
$$= n((q - p) - \delta) = -n(2p - 1) - m < 0;$$
$$E\left[z_i^{(1)}\middle| H_{1,2}(t_{i-k}, t_{i-1})\right] = n\left(E\left[\text{sign}(\tau_i - \tau_{i-k})\middle| H_{1,2}(t_{i-k}, t_{i-1})\right] - \delta\right)$$
$$= n((p - q) - \delta) = n(2p - 1) - m > 0;$$
$$E\left[z_i^{(2)}\middle| H_l(t_{i-k-1}, t_i)\right] = n\left(E\left[\text{sign}(\tau_{i-k} - \tau_i)\middle| H_l(t_{i-k-1}, t_i)\right] - \delta\right)$$
$$= n((1/2 - 1/2) - \delta) = -m < 0;$$
$$E\left[z_i^{(2)}\middle| H_{1,2}(t_{i-k}, t_{i-1})\right] = n\left(E\left[\text{sign}(\tau_{i-k} - \tau_i)\middle| H_{1,2}(t_{i-k}, t_{i-1})\right] - \delta\right)$$
$$= n((q - p) - \delta) = -n(2p - 1) - m < 0;$$
$$E\left[z_i^{(2)}\middle| H_{2,1}(t_{i-k}, t_{i-1})\right] = n\left(E\left[\text{sign}(\tau_{i-k} - \tau_i)\middle| H_{2,1}(t_{i-k}, t_{i-1})\right] - \delta\right)$$
$$= n((p - q) - \delta) = n(2p - 1) - m < 0$$

So the average values of statistics (2) and (3) change from a negative value to positive when the intensity of the process changes. Besides, the statistic $z_i^{(1)}$ reacts to decrease the intensity, i.e., to increase the mean length of the interval between events; the statistic $z_i^{(2)}$ reacts to increase the intensity, i.e., to decrease the mean length of the interval between events. These properties determine the construction of the procedures. We introduce positive values h_1 and h_2 as the procedures thresholds and construct the cumulative sums $S_i^{(1)}$ and $S_i^{(2)}$ which are recalculated at the instants t_i. For the first procedure it is defined as follows

$$S_k^{(1)} = m + n;$$
$$S_i^{(1)} = \max\{m + n, S_{i-1}^{(1)} + z_i^{(1)}\}, \quad i > k; \tag{7}$$
$$S_i^{(1)} = m + n, \quad \text{if} \quad S_i^{(1)} \geq h_1.$$

For the second procedure the cumulative sum is defined as follows

$$S_k^{(2)} = m + n;$$
$$S_i^{(2)} = \max\{m + n, S_{i-1}^{(2)} + z_i^{(2)}\}, \quad i > k; \tag{8}$$
$$S_i^{(2)} = m + n, \quad \text{if} \quad S_i^{(2)} \geq h_2.$$

If the cumulative sum $S_i^{(1)}$ reaches the threshold h_1 then we make the decision that the mean time between events increased and hence the intensity of the process decreased, i.e., it changed from λ_1 to λ_2. If the cumulative sum $S_i^{(2)}$ reaches the threshold h_2 then we make the decision that the mean time between events decreased and hence the intensity of the process increased, i.e., it changed from λ_2 to λ_1. Once the sum reaches the threshold it is reset to $m + n$ and the corresponding procedure is restarted.

In connection with sequential change-point detection procedures two types of errors are considered: the false alarm and the skip of the change. A false alarm occurs when one of the cumulative sums reaches the corresponding threshold in the case of the constant intensity of the arrival process. These events can be described as follows:

$$F_1 = \left\{ S_i^{(1)} \geq h_1 \middle| H_l(t_{i-k-1}, t_i) \cup H_{2,1}(t_{i-k}, t_{i-1}) \right\};$$
$$F_2 = \left\{ S_i^{(2)} \geq h_2 \middle| H_l(t_{i-k-1}, t_i) \cup H_{1,2}(t_{i-k}, t_{i-1}) \right\}.$$
(9)

A skip of the change occurs when the change of the parameter occurs but the corresponding cumulative sum does not reach its threshold. These events can be described as follows:

$$G_1 = \left\{ S_i^{(1)} < h_1 \middle| H_{1,2}(t_{i-k}, t_{i-1}) \right\};$$
$$G_2 = \left\{ S_i^{(2)} < h_2 \middle| H_{2,1}(t_{i-k}, t_{i-1}) \right\}.$$
(10)

4 Characteristics of the Algorithm

4.1 Probability p

First, we calculate the probability p (4) as

$$p = P\{\tau_i \geq \tau_{i-k} | H_{1,2}(t_{i-k}, t_{i-1})\}.$$

In the conditions of the hypothesis $H_{1,2}(t_{i-k}, t_{i-1})$ the variable τ_{i-k} is distributed exponentially with the parameter λ_1, and the variable τ_i is distributed exponentially with the parameter λ_2, hence

$$p = \int_0^\infty P\{\tau_i > t\} dP\{\tau_{i-k} < t\} = \int_0^\infty e^{-\lambda_2 t} \lambda_1 e^{-\lambda_1 t} dt = \frac{\lambda_1}{\lambda_1 + \lambda_2}.$$

According to (1), $p > 1/2$, and we obtain

$$p = \frac{\lambda_1}{\lambda_1 + \lambda_2}, \quad q = \frac{\lambda_2}{\lambda_1 + \lambda_2}.$$
(11)

4.2 Average Delay

Then, we investigate the characteristics of the change-point detection procedure if the change occurs more then once. Figure 4 demonstrates an example of the multiple change point.

In the case of the multiple change-point detection problem any parameter change should be detected before the next change occurs. Without loss of generality, we consider the first procedure and suppose that at the instant θ_1 the intensity parameter changes from λ_1 to λ_2, and then at the instant θ_2 it changes from λ_2 to λ_1, so, our procedure should detect the first change. The first change occurs within the interval $[t_{i-1}, t_i]$, and the second change occurs within the

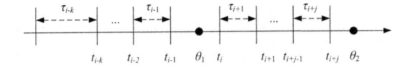

Fig. 2. Multiple change point

interval $[t_{i+j}, t_{i+j+1}]$. It means that the first change should be detected at the interval $[t_i, t_j]$, i.e., the cumulative sum should reach the threshold no later than at the instant t_j. On the other hand, the expectation of the statistics $z_{i+a}^{(1)}$ is positive if and only if $a \leq j$, $a \leq k - 1$, so, the change should be detected not more than in $k-1$ steps. Consequently, the most important characteristics of any multiple change-point detection algorithm are those connected with the delay in the detection (Fig. 2).

Note that in the conditions of the hypothesis $H_{1,2}(t_{i-k}, t_i)$, at the instant t_{i+1}, the cumulative sum can increase by $n - m$ with the probability p, and it can decrease by $n + m$ with the probability q. Let us introduce the following notations

$$n + m = N, \quad n - m = M.$$

To simplify the further calculations, we suppose that $M > 1$.

The cumulative sum is recalculated at the moments when the flow events occur. We will call every such recalculation a step. We denote the mean number of steps for the sum $S_i^{(1)}$ necessary to reach the threshold $h^{(1)}$ if at the moment i the sum is equal to j as $T^{(1)}(j)$. Taking into account that N is the minimal value of the cumulative sum one obtains that the mean delay for the first procedure is

$$T_{delay}^{(1)} = T^{(1)}(N). \tag{12}$$

The values $T^{(1)}(j)$ satisfy the following set of equations [11,12]

$$T^{(1)}(j) = 1 + pT^{(1)}(j + M) + qT^{(1)}(j - N) \tag{13}$$

with the initial conditions

$$T^{(1)}(h^{(1)}) = ... = T^{(1)}(h^{(1)} + M - 1) = 0;$$
$$T^{(1)}(0) = ... = T^{(1)}(N - 1) = T^{(1)}(N). \tag{14}$$

The decision of this system is obtained at [11,12] and can be written as follows

$$T^{(1)}(j) = \frac{j}{qN - pM} + A_1\mu_1^j + ... + A_{N+M}\mu_{N+M}^j, \tag{15}$$

where $\mu_1, ..., \mu_{N+M}$ are the roots of the characteristic polynomial of system (13)

$$P_1(\mu) = p\mu^{N+M} - \mu^N + q. \tag{16}$$

The exact formulas for the decision (15) are given in [11,12]. For the second procedure one can obtain the same result. These imply the following lower bound for the memory depth parameter k

$$k \geq T^{(1)}(N) + 1. \tag{17}$$

Table 1 demonstrates some values of the mean delay $T^{(1)} = T^{(1)}(N)$ subject to the parameters λ_1, λ_2 and $h^{(1)}$. Here $\delta = 1/5$, i.e., $M = 4$, $N = 6$.

Table 1. Average delay

λ_1	λ_2	p	q	h	$T^{(1)}$	h	$T^{(1)}$	h	$T^{(1)}$	h	$T^{(1)}$
2	0.4	0.83	0.17	42	12.12	62	19.01	77	24.12	97	30.9
2	0.6	0.77	0.23	37	13.27	52	20.24	72	29.23	87	35.94
2	0.8	0.71	0.29	32	15.85	47	26.04	67	39.34	82	49.23
2	1	0.67	0.33	32	19.07	47	33.04	62	47.23	77	61.57

Let us calculate now the mean delay not in terms of steps, but in terms of real time between the change point and the instant of its detection. If the change point is detected in b steps, i.e., at the instant t_{i+b} then the time necessary to detect the change can be expressed as follows

$$t_{i+b} - t_i = \sum_{j=1}^{b} \tau_{i+j}.$$

So the mean time of the change-point detection can be written in the form

$$Q_{delay}^{(1)} = E \sum_{j=1}^{b} \tau_{i+j}.$$

If $\theta_1 \in [t_{i-1}, t_i]$, $k > b$ and $\theta_2 > t_{i+b}$ then the random variables τ_{i+j} and b satisfy the following conditions

(a) $\{\tau_{i+j}\}$ are all finite-mean random variables having the same expectations;
(b) $E\tau_{i+j}\mathbf{1}_{b \geq i+j} = E\tau_{i+j}\mathbf{1}_{S_{i+j-1}^{(1)} < h^1} = E\tau_{i+j}P(S_{i+j-1}^{(1)} < h^1);$
(c) b has a finite expectation.

So, using the Wald's identity [13] one can obtain

$$Q_{delay}^{(1)} = E\tau_i Eb = \frac{1}{\lambda_2} T_{delay}^{(1)}. \tag{18}$$

The change point θ_1 should be detected earlier then the next change occurs, so, the necessary condition for this is $t_{i+b} < \theta_2$, or $t_{i+b} - \theta_1 < \theta_2 - \theta_1$. Rewriting the left side of the expression

$$t_{i+b} - \theta_1 = \sum_{j=1}^{b} \tau_{i+j} + (t_i - \theta_1)$$

and taking into account that $t_i - \theta_1$ has exponential distribution with the parameter λ_2, we obtain the following condition

$$E(t_{i+b} - \theta_1) = Q_{delay}^{(1)} + E(t_i - \theta_1) \leq Q_{delay}^{(1)} + \frac{1}{\lambda_2} = \frac{1}{\lambda_2}\left(T_{delay}^{(1)} + 1\right).$$

The variable $\theta_2 - \theta_1$ is distributed exponentially with the parameter α_2. This and (18) implies the condition

$$\frac{1}{\lambda_2}\left(T_{delay}^{(1)} + 1\right) < \frac{1}{\alpha_2} \tag{19}$$

4.3 Average Time Between False Alarms

A false alarm occurs when the cumulative sum reaches the corresponding threshold in the conditions of the hypothesis $H_l(t_{i-k-1}, t_{i+1})$. At the instant t_{i+1}, the cumulative sum can increase by M or decrease by N with the probability $1/2$.

Let us denote the mean number of steps for the sum $S_i^{(1)}$ necessary to reach the threshold $h^{(1)}$ if at the moment i the sum is equal to j as $R^{(1)}(j)$. Taking into account that N is the minimal value of the cumulative sum one obtains that the mean number of steps between false alarms for the first procedure is

$$T_{alarm}^{(1)} = R^{(1)}(N). \tag{20}$$

The values $R^{(1)}(j)$ satisfy the following set of equations [11, 12]

$$R^{(1)}(j) = 1 + \frac{1}{2}R^{(1)}(j + M) + \frac{1}{2}R^{(1)}(j - N) \tag{21}$$

with the initial conditions

$$R^{(1)}(h^{(1)}) = \dots = R^{(1)}(h^{(1)} + M - 1) = 0; \\ R^{(1)}(0) = \dots = R^{(1)}(N - 1) = R^{(1)}(N). \tag{22}$$

The decision of this system is obtained at [11, 12] and can be written as follows

$$R^{(1)}(j) = \frac{2j}{N - M} + C_1\mu_1^j + \dots + C_{N+M}\mu_{N+M}^j, \tag{23}$$

where $\mu_1, \dots, \mu_{N+M}$ are the roots of the characteristic polynomial of system (21)

$$P_1(\mu) = \mu^{N+M} - 2\mu^N + 1. \tag{24}$$

The exact formulas for the decision (23) are given in [11, 12]. For the second procedure one can obtain the same result.

Table 2 demonstrates some values of the mean number of steps between false alarms $R^{(1)} = R^{(1)}(N)$ subject to the parameter $h = h^{(1)}$.

As for the mean delay, we can obtain the formula for the mean time between the false alarms

$$Q_{false}^{(1)} = \frac{1}{\lambda_2}T_{alarm}^{(1)}. \tag{25}$$

Table 2. Average time between false alarms

h	$R^{(1)}$	h	$R^{(1)}$
20	12,31	50	195,17
30	37,99	60	385,13
40	91,86	70	727,26

According to G. Lorden [14], a sequential change-point detection procedure is optimal if both the average delay and the logarithm of the average time between false alarms grow linearly with an increase in the parameter h. Figures 3 and 4 demonstrate this property. Figure 4 indicates that the growth rates of the average delay increase with the decreasing in the probability p.

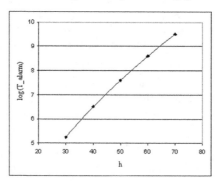

Fig. 3. Logarithm of the average time between false alarms

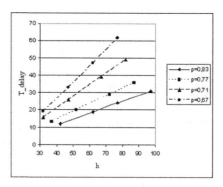

Fig. 4. Average delay

4.4   Memory Depth

Now we consider the event $B_j^{(1)} = \left\{ \sum_{a=1}^{j} \tau_{i+a} + (t_i - \theta_1) < \theta_2 - \theta_1 \right\}$, which means that the next change of the controlling chain state occurs after the instant t_{i+j}.

All the variables τ_a and $t_i - \theta_1$ are independent and identically distributed with the parameter λ_2. Their sum has the gamma distribution $\text{Gamma}(1/\lambda_2, j+1)$:

$$f(x) = \begin{cases} \lambda_2^{j+1} x^j \dfrac{e^{-\lambda_2 x}}{j!}, & \text{if } x \geq 0; \\ 0, & \text{if } x < 0. \end{cases}$$

The difference $\theta_2 - \theta_1$ has the exponential distribution with the parameter α_2; hence, we obtain

$$P\left(\sum_{a=1}^{j} \tau_{i+a} + (t_i - \theta_1) < \theta_2 - \theta_1\right) = \int_0^\infty P(\theta_2 - \theta_1 > x)\, \lambda_2^{j+1} x^j \frac{e^{-\lambda_2 x}}{j!} dx$$

$$= \frac{\lambda_2^{j+1}}{j!} \int_0^\infty e^{-\alpha_2 x} x^j e^{-\lambda_2 x}\, dx = \frac{\lambda_2^{j+1}}{j!} \frac{j!}{(\lambda_2 + \alpha_2)^{j+1}} = \frac{\lambda_2^{j+1}}{(\lambda_2 + \alpha_2)^{j+1}}.$$

For the change of the controlling chain state from the second to the first, we have the same result. Finally, we obtain

$$P\left(B_j^{(1)}\right) = \frac{\lambda_2^{j+1}}{(\lambda_2 + \alpha_2)^{j+1}}, \quad P\left(B_j^{(2)}\right) = \frac{\lambda_1^{j+1}}{(\lambda_1 + \alpha_1)^{j+1}} \tag{26}$$

On the one hand, the next change point should be detected in not more than k steps. On the other hand, the next change should occur later than the previous is detected. So, the probability of the event B_k should exceed the prescribed probability Q which is close to one. This implies the upper bound for the value of the parameter k

$$\frac{\lambda_l^{k+1}}{(\lambda_l + \alpha_l)^{k+1}} \geq Q, \quad l = 1, 2. \tag{27}$$

Before we supposed that $\lambda_l \geq r\alpha_l$, where $r \gg 1$. Table 3 demonstrates the maximum values of k satisfying condition (27) subject to the values of Q (in the rows) and r (in the columns). In practice, the values of $k < 10$ lead to frequent skips of changes; hence, the empty cells correspond to such values of k.

Table 3. Upper bound for the memory depth

$Q \backslash r$	40	60	80	100	120	160	200
0,7	13	22	27	34	41	56	70
0,75	11	16	22	27	33	45	56
0,8		12	16	21	24	34	43
0,85			12	15	18	25	31
0,9					11	16	20

4.5   Error Probabilities

Now we consider the event $A_j = \bigcup_{a=1}^{j} S_{i+a}^{(1)} < h^{(1)}$ which means that the first change point is not detected in j steps. In every step, the cumulative sum can increase by M with the probability p and can decrease by N with the probability q. So, for the event $S_{i+j}^{(1)} < h^{(1)}$, if the condition A_{j-1} holds true, we have

$$P\left(S_{i+j}^{(1)} < h^{(1)}\right) = p\mathbf{1}_{h^{(1)}-M>N}P\left(S_{i+j-1}^{(1)} < h^{(1)} - M\right)$$
$$+q\mathbf{1}_{h^{(1)}>N}P\left(S_{i+j-1}^{(1)} < h^{(1)}\right).$$

By using this formula again for its right side, we obtain

$$P\left(S_{i+j}^{(1)} < h^{(1)}\right) = p\mathbf{1}_{h^{(1)}-M>N}\left[p\mathbf{1}_{h^{(1)}-2M>N}P\left(S_{i+j-2}^{(1)} < h^{(1)} - 2M\right)\right.$$
$$+ q\mathbf{1}_{h^{(1)}-M>N}P\left(S_{i+j-2}^{(1)} < h^{(1)} - M\right)\Big]$$
$$+q\mathbf{1}_{h^{(1)}>N}\left[p\mathbf{1}_{h^{(1)}-M>N}P\left(S_{i+j-2}^{(1)} < h^{(1)} - M\right)\right.$$
$$+ q\mathbf{1}_{h^{(1)}>N}P\left(S_{i+j-2}^{(1)} < h^{(1)}\right)\Big] = p^2\mathbf{1}_{h^{(1)}-2M>N}P\left(S_{i+j-2}^{(1)} < h^{(1)} - 2M\right)$$
$$+2pq\mathbf{1}_{h^{(1)}-M>N}P\left(S_{i+j-2}^{(1)} < h^{(1)} - M\right) + q^2\mathbf{1}_{h^{(1)}>N}P\left(S_{i+j-2}^{(1)} < h^{(1)}\right).$$

By using the mathematical induction method, finally we obtain

$$P(A_j) = \sum_{a=0}^{j} C_a^j p^a q^{j-a} \mathbf{1}_{h^{(1)}-aM>N}P\left(S_i^{(1)} < h^{(1)} - aM\right)$$

If we choose the threshold as $h^{(1)} = N + cM$, where c is an integer, then

$$P(A_j) = \sum_{a=0}^{\min\{j,c-1\}} C_a^j p^a q^{j-a}P\left(S_i^{(1)} < h^{(1)} - aM\right)$$

As the minimum value of the sum $S_i^{(1)}$ is N, we can bound this probability from above

$$P(A_j) \le \sum_{a=0}^{\min\{j,c-1\}} C_a^j p^a q^{j-a}.$$

This formula is used for $c < j$, because the probability should be small enough. So, we obtain

$$P(A_j) \le \sum_{a=0}^{c-1} C_a^j p^a q^{j-a}, \tag{28}$$

where $c = (h^{(1)} - N)/M$.

This formula gives a way to determine the value of the threshold. Let P_1 be the desired value of the probability of the skip of the change, then the value c can be calculated as the maximum integer satisfying the following conditions

$$\sum_{a=0}^{c-1} C_a^j p^a q^{j-a} \le P_1. \tag{29}$$

Note that the probability does not depend on the exact values of the input flow intensities, it includes only the probabilities p and q given by formulas (4). One can use instead of p its minimum desired value p^*, and, instead of q the value $q^* = 1 - p^*$. As a result, the value of the threshold for $p \geq p^*$ will be obtained.

Let $S_i^{(1)} = N$, and the process satisfy the hypothesis $H_l(t_{i-k-1}, t_{i+j})$. Let us consider the event $C_j = \bigcup_{a=1}^{j} S_{i+a}^{(1)} < h^{(1)}$ which means that no false alarms occur in j steps if the initial value of the sum is N. By using the same reasoning as for a skip of the change, we obtain

$$P\left(C_j\right) = \frac{1}{2^j} \sum_{a=0}^{c-1} C_a^j, \tag{30}$$

where $c = (h^{(1)} - N)/M$.

Table 4 gives some results of calculations. Here $P_1 = 0, 1$, $m = 1$, $n = 6$ (so, $M = 5$, $N = 7$); λ_1 and λ_2 are the process parameters, j is the number of steps, p and q are calculated by formulas (4); c is the maximum value satisfying condition (29), $h = N + cM$, $\hat{P}_1$ is the upper bound for the probability of a skip of the change calculated by Eq. (28), $\hat{P}_0$ is the probability of a false alarm calculated by Eq. (30).

The error probabilities depend on the probability p: an increase in p leads to a decrease in the probability of a skip of the change: hence, the value of the threshold can be increased for the fixed value of P_1. In turn, this implies an decrease in the false alarm probability. Both error probabilities are sufficiently small for $p > 0.75$. On the other hand, the change point should be detected in k steps; so, we can decrease the false alarm probability by increasing k and, consequently, the threshold.

Table 5 demonstrates the results of calculations for the values of k taken from Table 3. The parameter h was chosen to provide the false alarm probability less then 0.15. Then the probability of a skip of the change was calculated for different values of λ_2 and, hence, p and q.

Table 4. Error probabilities

λ_1	λ_2	p	q	j	c	h	$\hat{P}_1$	$\hat{P}_0$	j	c	h	$\hat{P}_1$	$\hat{P}_0$
2	0.4	0.83	0.17	10	7	42	0.07	0.172	15	11	62	0.09	0.059
2	0.6	0.77	0.23	10	6	37	0.058	0.377	15	9	52	0.038	0.304
2	0.8	0.71	0.29	10	5	32	0.038	0.623	15	8	47	0.038	0.5
2	1	0.67	0.33	10	5	32	0.077	0.623	15	8	47	0.088	0.5
2	0.4	0.83	0.17	20	14	77	0.037	0.058	25	18	97	0.044	0.022
2	0.6	0.77	0.23	20	13	72	0.069	0.132	25	16	87	0.044	0.115
2	0.8	0.71	0.29	20	12	67	0.087	0.252	25	15	82	0.072	0.212
2	1	0.67	0.33	20	11	62	0.092	0.412	25	14	77	0.092	0.345

Table 5. Error probabilities

λ_1	k	h	$\hat{P}_0$	λ_2	p	q	$\hat{P}_1$	λ_2	p	q	$\hat{P}_1$
2	31	107	0.075	0.4	0.83	0.17	0.0028	0.6	0.77	0.23	0.0037
2	43	137	0.111	0.4	0.83	0.17	0.00008	0.6	0.77	0.23	0.0048
2	56	182	0.07	0.4	0.83	0.17	0.00001	0.6	0.77	0.23	0.002
2	70	212	0.094	0.4	0.83	0.17	0.0000002	0.6	0.77	0.23	0.00019
2	31	107	0.075	0.8	0.71	0.29	0.0016	1	0.67	0.33	0.323
2	43	137	0.111	0.8	0.71	0.29	0.0043	1	0.67	0.33	0.153
2	56	182	0.07	0.8	0.71	0.29	0.03	1	0.67	0.33	0.139
2	70	212	0.094	0.8	0.71	0.29	0.0074	1	0.67	0.33	0.061

Greater values of k provide adequate error probabilities in the most of cases.

Then we compare the results presented in Tables 1 and 4 for the same values of the threshold h and the intensities λ_i. Table 1 demonstrates the mean delay $T^{(1)}$ in the change-point detection, and Table 4 contains the number of steps j used to detect a change point with the probability 0.9 and greater.

For $p = 0.83$, the values $T^{(1)}$ and j differ insignificantly (not more than by 25 %), but the decrease in p leads to the increase of that difference, and for $p = 0.67$ the value $T^{(1)}$ exceeds the value of j twice or more. So, in spite of the big values of the mean delay for small values of p, the change in this case will be detected most probably much earlier than in the mean delay time.

Acknowledgements. Y. Burkatovskaya is supported by The National Research Tomsk State University Academic D.I. Mendeleev Fund Program (NU 8.1.55.2015 L) in 2014–2015 and by RFBR Grant 16-01-00121.

The authors are grateful to Prof. Sergey Vorobeychikov from Tomsk State University for useful comments.

References

1. Fischer, W., Meier-Hellstern, K.: The Markov-modulated Poisson process (MMPP) cookbook. Perform. Eval. **18**(2), 149–171 (1993)
2. Zou, M., Liu, J.: Performance analysis of IEEE 802.16 networks with MMPP arrivals. Perform. Eval. **69**, 492–509 (2012)
3. Nogueira, A., Salvador, P., Valadas, R., Pacheco, A.: Modeling self-similar traffic over multiple time scales based on hierarchical Markovian and L-System models. Comput. Commun. **33**, S3–S10 (2010)
4. Giacomazzi, P.: Closed-form analysis of end-to-end network delay with Markov-modulated Poisson and fluid traffic. Comput. Commun. **32**(4), 640–648 (2009)
5. Choi, D.I., Kim, T.-S., Lee, S.: Analysis of an MMPP/G/1/K queue with queue length dependent arrival rates, and its application to preventive congestion control in telecommunication networks. Eur. J. Oper. Res. **187**(2), 652–659 (2008)

6. Vasil'eva, L.A., Gortsev, A.M.: Estimation of the dead time of an asynchronous double stocastic flow of events under incomplete observability. Autom. Remote Control **64**(12), 1890–1898 (2003)

7. Gortsev, A.M., Nezhelskaya, L.A.: An asynchronous double stochastic flow with initiation of superfluous events. Discrete Math. Appl. **21**(3), 283–290 (2011)

8. Asmussen, S.: Phase-type distributions and related point processes: fitting and recent advances. In: Chakravarthy, S., Alfa, A.S. (eds.) Matrix-Analytic Methods in Stochastic Models. Lecture Notes in Pure and Applied Mathematics, vol. 183, pp. 137–149. Marcel Dekker, New York (1997)

9. Gerhardt, I., Nelson, B.L.: On capturing dependence in point processes: matching moments and other techniques. Technical report, Northwestern University (2009)

10. Burkatovskaya, Y., Kabanova, T., Vorobeychikov, S.: CUSUM algorithms for parameter estimation in queueing systems with jump intensity of the arrival process. In: Dudin, A., Nazarov, A., Yakupov, R. (eds.) ITMM 2015. CCIS, vol. 564, pp. 275–288. Springer, Heidelberg (2015). doi:10.1007/978-3-319-25861-4_24

11. Vorobeychikov, S.E.: On the detection of a change in the mean of a sequence of random variables. Autom. Remote Control **59**(3), 50–56 (1998)

12. Vorobeychikov, S.E., Kabanova, T.V.: Detection of the change point in a sequence of independent random variables (in Russian). J. Commun. Technol. Electron. **47**(10), 1198–1203 (2002)

13. Wald, A.: Sequential Analysis. Wiley, New York (1947)

14. Lorden, G.: Procedures for reacting to a change in distribution. Annals. Math. Statist. **42**, 1897–1971 (1971)

A Mathematical Model of an Insurance Company in the Form of a Queueing System with an Unlimited Number of Servers Considering One-Time Insurance Payments

Diana Dammer[(✉)]

Tomsk State University, Tomsk, Russia
di.dammer@yandex.ru

Abstract. This paper focuses on the research of a model of an insurance company with an unlimited insurance field considering implicit advertising and one-time insurance payments. Using the method of characteristic functions we obtained the probability distribution of a two-dimensional stochastic process of a number of risks that are insured in the company and a number of demands for one-time insurance payments. We also obtain expressions for the expected values and dispersions of components of a two-dimensional process. The total amount of insurance payments has been reviewed and its distribution found.

Keywords: Mathematical model · Insurance company · One-time insurance payments · Queueing system · Characteristic function

1 Introduction

Nowadays mathematical modeling is widely used, both for solving practical tasks and for theoretical research of economic processes. Models of actuarial mathematics, which studies insurance, are getting much attention nowadays. In the main, all the papers devoted to the research of an insurance company's mathematical models have such characteristics of a company's functioning as: expected values of the risk's numbers, capital, bankruptcy possibility and so on. Thus, paper [1] is about a model of an insurance company that takes into account advertising expenses, paper [2] is about a model with the possibility of reassurance of some company's risks. In [3] we got the distribution of number of demands for payments of sums insured with a random variable of the duration of the contract and the stationary Poisson arrival process of risks. In [4] by using the method of asymptotic analysis we have found the probability distribution of a two-dimensional process of a number of demands for insurance payments and the number of a company's risks, given that the arrival process of risks is a stationary Poisson. In paper [5] we researched the two-dimensional process of a number of demands for payments of sums insured and a number of risks, that are insured in the company in the case when the parameter of the arrival process

© Springer International Publishing Switzerland 2016
A. Dudin et al. (Eds.): ITMM 2016, CCIS 638, pp. 34–43, 2016.
DOI: 10.1007/978-3-319-44615-8_3

of insurance risks depends on the number of risks that are already insured in the company, which considers the possibility of implicit advertising. Mathematical models with this arrival process of risks are reviewed in [6], but methods of the model's research are of a different nature and the process of a number of insurance payments is ignored. In this paper we research the mathematical model of an insurance company with one-time insurance payments considering implicit advertising, which is no doubt present in real life.

2   Mathematical Model and Formulation of the Problem

Let us consider the model of an insurance company with an unlimited insurance field [7] in the form of a queueing system with an unlimited number of servers (Fig. 1). The validity of the insurance contract matches the server's duration of request handling. We will assume that risks are flowing into the company, forming an arrival process with an intensity that depends on a number of insured risks. The intensity of that arrival process will be determined by two components: parameter λ, which determines the arrival process of risks that come independently of insured ones, and parameter α, which determines the arrival process of risks that are under the influence of implicit advertising. Each risk located in the company generates a request for a one-time insurance payment with intensity γ for the duration of the insurance contract independently of other risks. These requests also form the stationary Poisson process of events. It is natural to assume that a request for payment is determined by an insured accident. After receiving the insurance payment there is the risk of leaving the company. We will assume that the duration of the insurance contract for each risk located in the company will be a random variable that is distributed by an exponential law with parameter μ.

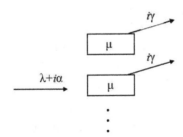

Fig. 1. Model of an insurance company with an unlimited insurance field considering insurance payments and implicit advertising

Let us denote: $n(t)$ – number of requests for payments during the time interval $[0,t]$, $i(t)$ – number of insurance risks located in the company at an instant of time t, $P(i,n,t) = \mathsf{P}\{i(t) = i, n(t) = n\}$ – probability distribution of a two-dimensional process of a number of one-time insurance payments and a number of insurance risks at an instant of time t. The task is to obtain this two-dimensional distribution and also numerical characteristics of a number of insured risks and a number of one-time insurance payments.

3   Probability Distribution of Two-Dimensional Stochastic Process of a Number of Insurance Risks and a Number of One-Time Insurance Payments

Let us write a system of Kolmogorov differential equations [8] for probability distribution $P(i, n, t)$ using the Δt method. First, the prelimit equalities:

$$\begin{aligned}
P(i, n, t + \Delta t) &= P(i, n, t)(1 - (\lambda + i\alpha)\Delta t)(1 - i\gamma\Delta t)(1 - i\mu\Delta t) \\
&+ (\lambda + (i-1)\alpha)\Delta t P(i-1, n, t) \\
&+ (i+1)\gamma\Delta t P(i+1, n-1, t) + (i+1)\mu\Delta t P(i+1, n, t) + o(\Delta t).
\end{aligned}$$

The system of differential equations will have this form:

$$\begin{aligned}
\frac{\partial P(i, n, t)}{\partial t} &= -[\lambda + i(\alpha + \mu + \gamma)]P(i, n, t) + (\lambda + (i-1)\alpha)P(i-1, n, t) \\
&+ (i+1)\mu P(i+1, n, t) + (i+1)\gamma P(i+1, n-1, t).
\end{aligned} \tag{1}$$

Let us introduce a characteristic function to solve system (1):

$$\sum_{i=0}^{\infty} \sum_{n=0}^{\infty} e^{jui} e^{jwn} P(i, n, t) = H(u, w, t),$$

where j is the imaginary unit. We will continue solving the task of determining the form of this function. Then, we form system (1), considering the properties of characteristic functions, we will get a partial differential equation of the first order for the function $H(u, w, t)$:

$$\begin{aligned}
\frac{\partial H(u, w, t)}{\partial t} &= -\lambda H(u, w, t)(1 - e^{ju}) \\
+ j\frac{\partial H(u, w, t)}{\partial u} &(\alpha + \mu + \gamma - \alpha e^{ju} - \mu e^{-ju} - \gamma e^{-ju} e^{jw}).
\end{aligned} \tag{2}$$

The solution for differential Eq. (2) is determined by solving the following system of ordinary differential equations for characteristic curves [9]:

$$\frac{dt}{1} = \frac{du}{-j(\alpha + \mu + \gamma - \alpha e^{ju} - \mu e^{-ju} - \gamma e^{-ju} e^{jw})} = \frac{dH(u, w, t)}{H(u, w, t)\lambda(e^{ju} - 1)}. \tag{3}$$

We will start by finding the two first integrals of this system. First, let us examine this equation:

$$dt = \frac{du}{j(\alpha(e^{ju} - 1) - \mu e^{-ju}(e^{ju} - 1) - \gamma e^{-ju}((e^{ju} - 1) + 1 - e^{jw}))}. \tag{4}$$

We will change variables $e^{ju} - 1 = v$, and, considering

$$u = \frac{\ln(v+1)}{j}, \quad du = \frac{dv}{j(v+1)}, \quad e^{-ju} = \frac{1}{v+1}, \quad j^2 = -1, \quad (5)$$

the Eq. (4) will have this form:

$$dt = \frac{dv}{-(\alpha v^2 + (\alpha - \mu - \gamma)v - \gamma(1 - e^{jw}))}.$$

Let us examine the right part of the last equation. We can write down

$$\alpha v^2 + (\alpha - \mu - \gamma)v - \gamma(1 - e^{jw}) = \alpha(v - v_1)(v - v_2), \quad (6)$$

where v_1 and v_2 are the roots of said quadratic equation. Let us write down expressions for v_1 and v_2:

$$v_1(w) = -\frac{1}{2}\left[\left(1 - \frac{\mu+\gamma}{\alpha}\right) - \sqrt{D(w)}\right],$$

$$ \quad (7)$$

$$v_2(w) = -\frac{1}{2}\left[\left(1 - \frac{\mu+\gamma}{\alpha}\right) + \sqrt{D(w)}\right],$$

where $D(w) = \left(1 - \frac{\mu+\gamma}{\alpha}\right)^2 + 4\frac{\gamma}{\alpha}(1 - e^{jw}) > 0$. Therefore, roots $v_1(w)$ and $v_2(w)$ are real and different, $v_1(w) > 0$ and $v_2(w) \le 0$ for $\alpha < \mu$.

Thus, based on the foregoing, the solution for Eq. (4) could be written in this form:

$$t = \frac{1}{\alpha(v_1(w) - v_2(w))} \ln\left(\frac{v - v_2(w)}{v - v_1(w)}\right) - \ln(\tilde{C}_1),$$

which will determine our first integral. Let us write down expression for $\tilde{C}_1$, we have:

$$\tilde{C}_1 = e^{-t}\left(\frac{v - v_2(w)}{v - v_1(w)}\right)^{\frac{1}{\alpha(v_1(w) - v_2(w))}}.$$

We denote $C_1 = \tilde{C}_1^{\alpha(v_1(w) - v_2(w))}$, then

$$C_1 = e^{-\alpha(v_1(w) - v_2(w))t}\left(\frac{v - v_2(w)}{v - v_1(w)}\right). \quad (8)$$

The other first integral will be found from equation:

$$\frac{dH(u,w,t)}{H(u,w,t)} = \frac{\lambda(e^{ju} - 1)du}{j(\alpha(e^{ju} - 1) - \mu e^{-ju}(e^{ju} - 1) - \gamma e^{-ju}(e^{ju} - e^{jw}))}. \quad (9)$$

Let us make a similar change to variables $e^{ju} - 1 = v$. We will introduce function $G(v,w,t) = H(u,w,t)$. Let us write down Eq. (9) for the function $G(v,w,t)$ while splitting variables considering (5):

$$\frac{dG(v,w,t)}{G(v,w,t)} = \frac{\lambda v\, dv}{-(\alpha v^2 + (\alpha - \mu - \gamma)v - \gamma(1 - e^{jw}))},$$

or considering (6)

$$\frac{dG(v,w,t)}{G(v,w,t)} = \frac{\lambda v dv}{-\alpha(v - v_1(w))(v - v_2(w))}, \tag{10}$$

where v_1 and v_2 are determined by expressions (7). Let us write down the solution to Eq. (10)

$$G(v,w,t) = C_2 \left[\frac{(v - v_2(w))^{v_2(w)}}{(v - v_1(w))^{v_1(w)}} \right]^{\frac{\lambda}{\alpha(v_1(w) - v_2(w))}}. \tag{11}$$

We will introduce an arbitrary differentiable function $\phi(C_1) = C_2$. Then the general solution to Eq. (10) considering (8) will have this form:

$$G(v,w,t) = \phi \left[e^{-\alpha(v_1(w) - v_2(w))t} \left(\frac{v - v_2(w)}{v - v_1(w)} \right) \right]$$
$$\times \left[\frac{(v - v_2(w))^{v_2(w)}}{(v - v_1(w))^{v_1(w)}} \right]^{\frac{\lambda}{\alpha(v_1(w) - v_2(w))}}. \tag{12}$$

We define the particular solution with the help of initial conditions. To do this, we will write down the value of function $H(u,w,t)$ at initial time $t = 0$. Then

$$H(u,w,0) = \sum_{i=0}^{\infty} \sum_{n=0}^{\infty} e^{jui} e^{jwn} P(i,n,0) = \sum_{i=0}^{\infty} e^{jui} P(i), \tag{13}$$

because at the initial time there were no requests for an insurance payment, which means $P(i,n,0) = P(i)$, if $n = 0$, and $P(i,n,0) = 0$, if $n > 0$.

Let us denote $H(u,w,0) = H(u)$, then by using Eq. (2) we can write down the equation for function $H(u)$:

$$j(\mu + \gamma - \alpha e^{ju}) \frac{dH(u)}{du} + \lambda e^{ju} H(u) = 0. \tag{14}$$

The solution will have this form for normality condition $H(0) = 1$:

$$H(u) = \left(\frac{1 - \frac{\alpha}{\mu + \gamma} e^{ju}}{1 - \frac{\alpha}{\mu + \gamma}} \right)^{-\frac{\lambda}{\alpha}}. \tag{15}$$

We note that characteristic function (15) is a discrete analog of gamma distribution. Now we can write down the expression for $H(v,w,t)$ considering $t = 0$:

$$G(v,w,0) = \phi \left(\frac{v - v_2(w)}{v - v_1(w)} \right) \left[\frac{(v - v_2(w))^{v_2(w)}}{(v - v_1(w))^{v_1(w)}} \right]^{\frac{\lambda}{\alpha(v_1(w) - v_2(w))}}, \tag{16}$$

or

$$\left(\frac{\mu + \gamma - \alpha(v + 1)}{\mu + \gamma - \alpha} \right)^{-\frac{\lambda}{\alpha}} = \phi \left(\frac{v - v_2(w)}{v - v_1(w)} \right)$$
$$\times \left(\frac{(v - v_2(w))^{v_2(w)}}{(v - v_1(w))^{v_1(w)}} \right)^{\frac{\lambda}{\alpha(v_1(w) - v_2(w))}}, \tag{17}$$

where

$$v_1(w) = -\frac{1}{2}\left[\left(1 - \frac{\mu+\gamma}{\alpha}\right) - \sqrt{D(w)}\right], v_2(w) = -\frac{1}{2}\left[\left(1 - \frac{\mu+\gamma}{\alpha}\right) + \sqrt{D(w)}\right],$$

$$D(w) = \left(1 - \frac{\mu+\gamma}{\alpha}\right)^2 + 4\frac{\gamma}{\alpha}(1 - e^{jw}).$$

We will define the form of function $\phi(.)$ now. Let us denote

$$f(w) = \frac{v - v_2(w)}{v - v_1(w)}. \tag{18}$$

Then

$$\phi(f(w)) = \left[\frac{\left(1 - \frac{\alpha}{\mu+\gamma}\right)(v_2(w) - v_1(w))}{(1 - f(w)) - \frac{\alpha}{\mu+\gamma}(1 + v_2(w) - (1 + v_1(w))f(w))}\right]^{\frac{\lambda}{\alpha}} \tag{19}$$

$$\times f(w)^{\frac{\lambda v_2(w)}{\alpha(v_2(w) - v_1(w))}},$$

where

$$v = \frac{v_2(w) - f(w)v_1(w)}{1 - f(w)}. \tag{20}$$

Now we can write down the expression for function $\phi(.)$:

$$\phi\left[e^{-\alpha(v_1(w) - v_2(w))t}\left(\frac{v - v_2(w)}{v - v_1(w)}\right)\right] = e^{\lambda v_2(w)t}$$

$$\times [\mu + \gamma - \alpha(v_2(w) - v_1(w))(v - v_1(w))]^{\frac{\lambda}{\alpha}}$$

$$\times \left[((v - v_1(w)) - (v - v_2(w))e^{\alpha(v_2(w) - v_1(w))t})(\mu + \gamma)\right.$$

$$- \alpha\left((v - v_1(w))(1 + v_2(w)) - (v - v_2(w))(1 + v_1(w))e^{\alpha(v_2(w) - v_1(w))t}\right)\right]^{-\frac{\lambda}{\alpha}}$$

$$\times \left(\frac{v - v_2(w)}{v - v_1(w)}\right)^{\frac{\lambda v_2(w)}{\alpha(v_2(w) - v_1(w))}}.$$

Accordingly, we will write down the expression for function $G(v, w, t)$. We have:

$$G(v, w, t) = e^{\lambda v_2(w)t}\left[(\mu + \gamma - \alpha)(v_1(w) - v_2(w))\right]^{\frac{\lambda}{\alpha}}$$

$$\times \left\{(v_1(w) - v)[\mu + \gamma - \alpha(1 + v_2(w))]\right. \tag{21}$$

$$-(v_2(w) - v)e^{-\alpha(v_1(w) - v_2(w))t}\left[\mu + \gamma - \alpha(1 + v_1(w))\right]\right\}^{-\frac{\lambda}{\alpha}}.$$

Let us write down the expression for function $H(u, w, t)$ by passing from variable v to variable u:

$$H(u, w, t) = e^{\lambda v_2(w)t}\left[(\mu + \gamma - \alpha)(v_1(w) - v_2(w))\right]^{\frac{\lambda}{\alpha}}$$

$$\times \left\{(v_1(w) - e^{ju} + 1)[\mu + \gamma - \alpha(1 + v_2(w))]\right. \tag{22}$$

$$-(v_2(w) - e^{ju} + 1)e^{\alpha(v_2(w) - v_1(w))t}\left[\mu + \gamma - \alpha(1 + v_1(w))\right]\right\}^{-\frac{\lambda}{\alpha}}.$$

Resulting function (22) is a characteristic function of a two-dimensional random process of a number of risks that are insured in the company and a number of requests for one-time insurance payments. Knowing this function, we can obtain the functions of marginal distributions of processes $i(t)$ and $n(t)$.

4 Probability Distributions and Numerical Characteristics of a Number of Risks and a Number of Requests for One-Time Insurance Payments

Let us suppose that in (22) $u = 0$, now we can get the marginal characteristic function of process $n(t)$:

$$
\begin{aligned}
H(0, w, t) = H_1(w, t) = e^{\lambda v_2(w)t} & \left[(\mu + \gamma - \alpha)(v_1(w) - v_2(w))\right]^{\frac{\lambda}{\alpha}} \\
& \times \Big\{ v_1(w) \left[\mu + \gamma - \alpha(1 + v_2(w))\right] \\
& - v_2(w) e^{-\alpha(v_1(w) - v_2(w))t} \left[\mu + \gamma - \alpha(1 + v_1(w))\right] \Big\}^{-\frac{\lambda}{\alpha}}.
\end{aligned}
\tag{23}
$$

Let us suppose that in (22) $w = 0$, now we can get the marginal characteristic function of process $i(t)$. Because

$$
v_1(0) = \frac{\mu + \gamma}{\alpha} - 1, \ v_2(0) = 0,
$$

we have:

$$
H(u, 0, t) = H(u) = \left(\frac{1 - \dfrac{\alpha}{\mu + \gamma} e^{ju}}{1 - \dfrac{\alpha}{\mu + \gamma}} \right)^{-\frac{\lambda}{\alpha}}.
\tag{24}
$$

Now we can write down the expected values for a number of risks and a number of requests for one-time insurance payments. We have

$$
\mathsf{E}\{i(t)\} = \frac{1}{j} \frac{dH(u)}{du} \bigg|_{u=0},
$$

$$
\mathsf{E}\{n(t)\} = \frac{1}{j} \frac{\partial H_1(w, t)}{\partial w} \bigg|_{w=0},
$$

then

$$
\mathsf{E}\{i(t)\} = \frac{\lambda}{\mu + \gamma - \alpha},
\tag{25}
$$

and

$$
\mathsf{E}\{n(t)\} = \frac{\lambda \gamma t}{\mu + \gamma - \alpha}.
\tag{26}
$$

Now we can write down the dispersions for a number of risks and a number of requests for one-time insurance payments: We have:

$$
\mathsf{D}\{i(t)\} = \frac{1}{j^2} \frac{\partial^2 H_i(u, t)}{\partial u^2} \bigg|_{u=0} - \mathsf{E}^2\{i(t)\},
$$

and

$$D\{n(t)\} = \frac{1}{j^2} \frac{\partial^2 H_n(w,t)}{\partial w^2}\bigg|_{w=0} - E^2\{n(t)\},$$

then

$$D\{i(t)\} = \frac{\lambda(\mu+\gamma)}{(\mu+\gamma-\alpha)^2}, \tag{27}$$

and

$$D\{n(t)\} = \frac{\lambda\gamma}{\mu+\gamma-\alpha}t + \frac{2\lambda\gamma^2(\mu+\gamma)}{(\mu+\gamma-\alpha)^3}t$$
$$- \frac{2\lambda\gamma^2(\mu+\gamma)}{(\mu+\gamma-\alpha)^4} + \frac{2\lambda\gamma^2(\mu+\gamma)}{(\mu+\gamma-\alpha)^4}e^{-(\mu+\gamma-\alpha)t}. \tag{28}$$

Formulas (25) and (27) match with the result we obtained in work [6] for $\gamma = 0$, where the one-dimensional process of a number of insured risks considering implicit advertising was researched.

Let us review the coefficient of the correlation of processes $i(t)$ and $n(t)$. Knowing function $H(u,w,t)$, we can obtain a joint moment of studied processes. We have:

$$\frac{1}{j^2} \frac{\partial^2 H(u,w,t)}{\partial u \partial w}\bigg|_{u=0,w=0} = E\{i(t)n(t)\}, \tag{29}$$

then, considering the characteristics we obtained earlier, let us write down the expression for the coefficient of correlation:

$$r_{in}(t) = \gamma\sqrt{\lambda(\mu+\gamma)}(1 - e^{-(\mu+\gamma-\alpha)t})$$
$$\times\Big[2\lambda(\mu+\gamma)\gamma^2(\mu+\gamma-\alpha)t + \lambda\gamma(\mu+\gamma-\alpha)^3t$$
$$- 2\lambda(\mu+\gamma)\gamma^2\Big(1 - e^{-(\mu+\gamma-\alpha)t}\Big)\Big]^{-\frac{1}{2}}. \tag{30}$$

The nonzero coefficient of correlation shows the presence of dependence between processes $i(t)$ and $n(t)$.

5   Probability Distributions and Numerical Characteristics of the Value of the Total Amount of One-Time Insurance Payments

We will denote $S(t)$ as a value of the total amount of insurance payments for all insured events during the time interval $[0,t]$, ξ is the value of the payment for one insured event. Let us introduce characteristic function of the value of $S(t)$:

$$\Psi(\eta,t) = E\{e^{-\eta S(t)}\}. \tag{31}$$

In paper [5] it was proved:

$$\Psi(\eta,t) = \sum_{n=0}^{\infty} \theta^n(\eta)P(n,t), \tag{32}$$

where $\theta(\eta) = \mathsf{E}\{e^{-\eta\xi}\}$ is the characteristic function of the value of ξ. With this in mind we can write down

$$\Psi(\eta, t) = \sum_{n=0}^{\infty} \theta^n(\eta) P(n, t) = F(\theta(\eta), t), \qquad (33)$$

where

$$F(e^{j\theta(\eta)}, t) = H_1(\theta(\eta), t).$$

Function $H_1(\theta(\eta), t)$ are determined by expression (23). Let us introduce functions

$$\begin{aligned}
y_1(\eta) = v_1(\theta(\eta)) = -\frac{1}{2}\left[\left(1 - \frac{\mu}{\alpha} - \frac{\gamma}{\alpha}\right) - \sqrt{D(\theta(\eta))}\right], \\
y_2(\eta) = v_2(\theta(\eta)) = -\frac{1}{2}\left[\left(1 - \frac{\mu}{\alpha} - \frac{\gamma}{\alpha}\right) + \sqrt{D(\theta(\eta))}\right],
\end{aligned} \qquad (34)$$

where

$$D(\theta(\eta)) = \left[1 - \frac{\mu}{\alpha} - \frac{\gamma}{\alpha}\right]^2 + 4\frac{\gamma}{\alpha}(1 - \theta(\eta)). \qquad (35)$$

Expressions (34) and (35) are written considering (7). Then, the characteristic function of the total amount of one-time insurance payments will have this form:

$$\begin{aligned}
\Psi(\eta, t) = F(\theta(\eta), t) = e^{\lambda y_2(\eta)t}\left[\left(1 - \frac{\alpha}{\mu + \gamma}\right)(y_1(\eta) - y_2(\eta))\right]^{\frac{\lambda}{\alpha}} \\
\times \left\{y_1(\eta)\left[1 - \frac{\alpha}{\mu + \gamma}(1 + y_2(\eta))\right]\right. \\
\left. - y_2(\eta)e^{\alpha(y_2(\eta) - y_1(\eta))t}\left[1 - \frac{\alpha}{\mu + \gamma}(1 + y_1)(\eta)\right]\right\}^{-\frac{\lambda}{\alpha}}.
\end{aligned} \qquad (36)$$

Now that we know the form of the characteristic function of a value of the total amount of one-time insurance payments, we can obtain the numerical characteristics of value $S(t)$:

$$\mathsf{E}\{S(t)\} = \frac{\lambda\gamma a_1 t}{\mu + \gamma - \alpha} \qquad (37)$$

and

$$\begin{aligned}
\mathsf{D}\{S(t)\} = \frac{\lambda\gamma a_2}{\mu + \gamma - \alpha}t + \frac{2\lambda(\gamma a_1)^2(\mu + \gamma)}{(\mu + \gamma - \alpha)^3}t \\
- \frac{2\lambda(\gamma a_1)^2(\mu + \gamma)}{(\mu + \gamma - \alpha)^4} + \frac{2\lambda(\gamma a_1)^2(\mu + \gamma)}{(\mu + \gamma - \alpha)^4}e^{-(\mu + \gamma - \alpha)t},
\end{aligned} \qquad (38)$$

where $\mathsf{E}\{\xi\} = a_1$, $\mathsf{E}\{\xi^2\} = a_2$.

6   Conclusions

Thus, in this paper we have researched a mathematical model of an insurance company in the form of a queueing system with an unlimited number of servers considering one-time insurance payments. We have found the expression for probability distribution of a number of requests for one-time insurance payments and a number of insurance risks. Also we have found expressions for the numerical characteristics of the said processes. It has been shown that the results are the generalization of particular cases. The characteristic function and numerical characteristics of a value of the total amount of one-time insurance payments have also been found. These results may be used for analysis of indicators of economic activity of insurance companies and other economic systems.

References

1. Akhmedova, D.D., Terpugov, A.F.: Mathevatical model of the insurance company considering advertising expenses. Univerisities News Phys. **1**, 25–29 (2001)
2. Glukhova, E.V., Kapustin, E.V.: Calculation of probability of insurance company's bankruptcy considering reassurance. Universities News Phys. **4**, 3–9 (2000)
3. Nazarov, A.A., Dammer, D.D.: Research of a number of requests for insurance payments in the company with arbitrary length of duration of the contract. Tomsk State University J. **2**(15), 24–32 (2011)
4. Dammer, D.D., Nazarov, A.A.: Research of the mathematical model of the insurance company in form of the infinite queueing system by using method of asymptotic analysis. In: Materials of VII Ferghan Conference Limit Theorems and its Applications, pp. 191–196. Namangan (2015)
5. Dammer, D.: Research of mathematical model of insurance company in the form of queueing system with unlimited number of servers considering implicit advertising. In: Dudin, A., Nazarov, A., Yakupov, R. (eds.) ITMM 2015. CCIS, vol. 564, pp. 163–174. Springer, Heidelberg (2015). doi:10.1007/978-3-319-25861-4_14
6. Glukhova, E.V., Zmeev, O.A., Livshits, K.I.: Mathematical Models of Insurance. Tomsk University, Tomsk (2004)
7. Gafurov, S.R., Gugnin, V.I., Amanov, S.N.: Business Language. Shark, Tashkent (1995)
8. Nazarov, A.A., Terpugov, A.F.: Theory of Queuing. Publishment NTL, Tomsk (2005)
9. Elsgolts, L.E.: Differential Equations and Calculus of Variations. Science, Moscow (1969)

Analysis of the BMAP/SM/1/N Type System with Randomized Choice of Customers Admission Discipline

Alexander Dudin$^{(\boxtimes)}$ and Alexej Shaban

Belarusian State University, 4, Nezavisimosti Ave., 220030 Minsk, Belarus
dudin@bsu.by, a_shaban@mail.ru

Abstract. A single server queue with a finite buffer is analyzed. The input is described by the BMAP (Batch Markovian Arrival Process), and the service process is of Semi-Markovian (SM) type. Due to the batch arrivals, situations may occur when the available capacity of the buffer is not enough to admit to the system a whole arriving batch of customers. A randomized choice of admission disciplines among well-known partial admission, complete admission and complete rejection disciplines is assumed. The stationary queue length distribution at service completion and arbitrary epochs is calculated. The expression for loss probability is given. The problem of optimal randomization between disciplines of complete admission and complete rejection is considered in brief.

Keywords: Batch Markovian Arrival Process · Semi-Markovian service time · Finite buffer · Admission discipline

1 Introduction

The $BMAP$ is a popular mathematical model of bursty correlated flows in modern communication networks, see, e.g., [3, 10, 11]. The Semi-Markovian service process is an essential extension of the renewal process. It enables the taking into account correlation and different distribution of successive service times. The $BMAP/SM/1$ retrial system was considered in [5]. The $BMAP/SM/1$ system with an infinite buffer was considered in [14]. Systems with a finite buffer are important for analysis because the capacity of the buffer in many real life systems is restricted. Systems with a $BMAP$ arrival process and a finite buffer were first considered in [1, 2, 7]. A vacation model of such a type was considered in [16].

It is well known that, due to the capacity limitation and batch arrivals, a situation can occur when the size of the arriving batch exceeds the currently available capacity of the buffer. Three popular admission disciplines for such a situation are known in the literature, see, e.g. [13]:

- partial admission (PA) when only a part of the batch corresponding to the number of free places in the buffer is allowed to join the system;

A. Dudin et al. (Eds.): ITMM 2016, CCIS 638, pp. 44–56, 2016.
DOI: 10.1007/978-3-319-44615-8_4

- complete admission (CA) when the whole batch is allowed to enter the system;
- complete rejection (CR) when the whole batch is rejected.

Only the discipline of PA is well investigated for the models with a general service time distribution. This stems from the fact that the technique of the embedded Markov chains, which is very effective in research, has a difficulty with arrivals accounting between the embedded epochs when the disciplines of CA and CR are applied. However, these disciplines are very important, e. g., in modelling in telecommunications. If the batch is interpreted as a set of packages belonging to one information unit, e.g., message or file, it does not make sense to allow partial admission of the packages. Thus, the discipline of CR or CA type should be chosen in such a situation.

A general $BMAP/SM/1/N$ model with PA discipline was investigated in [6]. The assumption that the input flow is the $BMAP$ allows the use of this result for modelling modern telecommunication networks where the flows of information are correlated and so they can not be well approximated in terms of the stationary Poisson process (even with the use of the Hurst parameter). The assumption that the service process is of SM (Semi-Markovian) type allows to capture a possible correlation of successive service times. The advantage of the paper [6] comparing to earlier papers [2,7], besides consideration of the more general SM service process, consists of the following. In paper [2], direct solving of the finite set of equilibrium equations is performed. So, the existing specifics of the transition probability matrix is practically ignored. In paper [7], such specifics are taken into account effectively. However, the algorithm developed for computing the stationary distribution of the system states is not numerically stable for large values of the buffer capacity N due to the presence of a subtraction operation over the matrices involved to the recursive computations. The algorithm presented in [6] takes into account the special structure of the transition probability matrix of the embedded Markov chain and simultaneously is very stable numerically.

In paper [16], the system $BMAP/G/1/N$ was analysed by means of the method of supplementary variables for PA and CR disciplines. Disciplines CA and CR in the $BMAP/G/1/N$ system were considered in [8]. In this paper we extend the results of [8] where the service times of successive customers are independent identically arbitrarily distributed random variables to the case of SM service and consider more general customers admission discipline. Namely, we assume that at a batch arrival moment in a situation when the buffer is not full, but its capacity is not enough to admit a whole batch, a randomized decision about the fate of the arrived batch is made. Some probabilities, p_{cr}, p_{pa} and p_{ca} such as

$$p_{cr} + p_{pa} + p_{ca} = 1$$

are fixed. With probability p_{pa} we admit a part of the batch that matches the available capacity. With probability p_{ca} we admit the whole batch. With probability p_{cr} we reject the whole batch.

Note that in different real life applications some disciplines may not be applicable. E.g., if the customers of a batch constitute one entity, it does not makes sense to apply PA discipline. If there is no additional buffer along with the main buffer of the system, the discipline CA is not applicable.

If the discipline PA does not makes sense, but both other disciplines may be applied, one can have a difficult choice which discipline, CR or CA, should be applied. CR discipline is easy realised in any system. However, e.g., if the size of the batch may be of the same order as the capacity of the buffer, CR discipline may lead to high probability of customer loss and essential degradation of the system. Many customers are lost while the server is often idle and the throughput of the system is low. CA discipline is more difficult to realise in a real world system. It is necessary to have additional storage to temporarily keep customers from the batch who did not find a place in the main buffer. The use of additional storage may be charged explicitly or implicitly (because the use of additional storage, e.g. external memory, may lead to more slow processing of customers). However, it can be verified that CA discipline can provide, among the three considered disciplines, the lowest value of loss probability and highest value of the system throughput. Because the mentioned disciplines have their own advantages and shortcomings, it seems reasonable to use a combination of these disciplines. The results presented in our paper show that a randomized choice between the available classical disciplines may be profitable for the system.

2   Mathematical Model

We consider a single server queue with a finite buffer. The capacity of the buffer is N. Customers arrive into the system according to the $BMAP$. The behavior of the $BMAP$ is defined by the underlying process ν_t, which is a continuous-time irreducible Markov chain with the state space $\{0, 1, \ldots, W\}$. The customers arrive at the epochs when the chain ν_t makes transitions. The matrix D_k defines the intensities of transitions of the chain ν_t, which are accompanied by arrival of a batch consisting of k customers, $k \geq 0$. Denote $D(z) = \sum_{k=0}^{\infty} D_k z^k$, $|z| \leq 1$. The matrix $D(1)$ is the infinitesimal generator of the process ν_t. The vector $\boldsymbol{\theta}$ of steady state distribution of the chain ν_t, satisfies the system

$$\boldsymbol{\theta} D(1) = 0, \ \boldsymbol{\theta}\mathbf{e} = 1.$$

Here $\mathbf{e}$ is a column vector consisting of all ones. The average intensity λ of the $BMAP$ (fundamental rate) is calculated as $\lambda = \boldsymbol{\theta} D'(1)\mathbf{e}$, and the intensity λ_g of group arrivals is defined as

$$\lambda_g = \boldsymbol{\theta}(-D_0)\mathbf{e}.$$

The variance v of intervals between group arrivals is calculated as

$$v = 2\lambda_g^{-1}\boldsymbol{\theta}(-D_0)^{-1}\mathbf{e} - \lambda_g^{-2},$$

while the correlation coefficient c_{cor} of intervals between successive group arrivals is given by

$$c_{cor} = (\lambda_g^{-1}\boldsymbol{\theta}(-D_0)(D(1) - D_0)(-D_0)^{-1}\mathbf{e} - \lambda_g^{-2})/v.$$

The $BMAP$ is a popular descriptor of flows in modern telecommunication networks. It allows to be bursty their captured correlated nature. So, the models of queues with the $BMAP$ are investigated intensively. For more details about the $BMAP$ and related research see [3,15].

It was assumed in [8] that service times of the customers are independent identically distributed random variables having distribution function $B(t)$. Such a suggestion does not allow us to consider the systems where the successive service times may be dependent. So, here we assume that the service of customers is governed by the semi-Markovian process m_t. It is characterized by the state space $\{1, ..., M\}$ and the semi-Markovian kernel $B(x)$. This kernel is a matrix with entries $B_{m,m'}(x)$, $m, m' = \overline{1, M}$. The successive service times of customers are defined as the sojourn times of the process m_t in its states. The average service time is given by the formula

$$b_1 = \mathbf{b}\boldsymbol{\delta},$$

where $\mathbf{b}$ is the invariant probability vector of the stochastic matrix $B(+\infty)$ (i.e., it is defined as a solution to the system and $\mathbf{b} = \mathbf{b}B(+\infty)$, $\mathbf{be} = 1$ and $\boldsymbol{\delta}$ is the column vector defined by formula

$$\boldsymbol{\delta} = \int\limits_0^\infty x dB(x)\mathbf{e}.$$

Without essential loss of generality, we will assume that the kernel $B(x)$ has the following form:

$$B(x) = diag\{B_1(x), \ldots, B_M(x)\}Q$$

where $B_m(x)$ is a distribution function of the sojourn time of the process m_t in the state m and $Q = B(\infty)$ is a transition probability matrix of the Markov chain embedded at the moments of the jumps of the process m_t.

3   The Process of the System States and the Embedded Markov Chain

Let us consider the process i_t which is the number of customers in the system at the moment t, $t > 0$. In general, because CA discipline can be applied, the state space of this process is infinite. This process, in general, is not Markovian. So, let us first consider the embedded process i_{t_n+0} where t_n is the n-th service completion epoch, $n \geq 1$. In the sequel, we use notation $i_n = i_{t_n+0}$, $n \geq 1$. The process i_n, $n \geq 1$, is also non-Markovian. But the three-dimensional process $\xi_n = \{i_n, \nu_n, m_n\}$, $n \geq 1$, where $\nu_n = \nu_{t_n}$, $m_n = m_{t_n+0}$ is a Markov chain. To make this clear, we have to compute the one-step transition probabilities of this process.

Let $P_{i,l}$ be the matrices with entries

$$P\{i_{n+1} = l, \ \nu_{n+1} = \nu', \ m_{n+1} = m'|i_n = i, \ \nu_n = \nu, \ m_n = m\},$$

$$\nu, \nu' = \overline{0, W}, \ m, m' = \overline{1, M}.$$

To calculate the matrices $P_{i,l}$, which define the probability of transitions of the Markov chain $\xi_n, n \geq 0$, between two successive service completion epochs, we should first calculate the matrices $P^{(j)}(n,t)$, which define the following conditional probabilities. The (ν, ν')-th entry of the matrix $P^{(j)}(n,t)$ is the probability to admit n customers during the time interval $(0,t]$ and to have the state ν' of the underlying process of the $BMAP$ at the epoch t conditional that the state of this process was ν at the epoch 0 and at most j customers can be admitted during the interval $(0,t]$, $n = \overline{0,j}$.

In the case of PA discipline, the matrices $P^{(j)}(n,t)$ are easily calculated as:

$$
P^{(j)}(n,t) = \begin{cases} P(n,t), & n < j, \\ \sum\limits_{l=j}^{\infty} P(l,t), & n = j \end{cases}
$$

where the matrices $P(n,t)$ are defined (see, e.g., [15]) as the coefficients in the following matrix expansion:

$$
e^{D(z)t} = \sum_{n=0}^{\infty} P(n,t) z^n.
$$

Unfortunately, it is not possible to compute the matrices $P(n,t)$ directly from this matrix expansion and D. Lucantoni (see [15]) has offered the following procedure for computing these matrices.

Let ψ be defined as $\psi = \max(-D_0)_{\nu,\nu}$, $\nu = \overline{0,W}$. Then

$$
P(n,t) = e^{-\psi t} \sum_{i=0}^{\infty} \frac{(\psi t)^i}{i!} U_n^{(i)},
$$

where the matrices $U_n^{(i)}$ are computed from recursions

$$
U_n^{(0)} = \begin{cases} I, & n = 0, \\ 0, & n > 0, \end{cases}
$$

$$
U_n^{(i+1)} = U_n^{(i)}(I + \psi^{-1}D_0) + \psi^{-1} \sum_{l=0}^{n-1} U_l^{(i)} D_{n-l}, i \geq 0, n \geq 0.
$$

These formulas are derived using the notion of uniformization of the Markov chain based on the well known system of differential equations

$$
\dot{P}(n,t) = \sum_{k=0}^{n} P(k,t) D_{n-k}, \ n \geq 0,
$$

for matrices $P(n,t)$, $n \geq 0$.

Using this approach for computation of the matrices $P^{(j)}(n,t)$ for the system under study, it is possible to verify that the following statement is true.

Lemma 1. The matrices $P^{(j)}(n,t)$ are computed by formulas

$$P^{(j)}(n,t) = e^{-\psi t} \sum_{i=0}^{\infty} \frac{(\psi t)^i}{i!} U_n^{(i)}(j)$$

where the matrices $U_n^{(i)}(j)$ are computed by recursions

$$U_n^{(0)}(j) = \begin{cases} I, & n = 0, \\ 0, & n > 0, \end{cases}$$

$$U_n^{(i+1)}(j) = \psi^{-1} \sum_{l=0}^{n-1} U_l^{(i)}(j) D_{n-l} + U_n^{(i)}(j) \left(I + \psi^{-1}(D_0 + p_{cr} \sum_{m=j+1-n}^{\infty} D_m) \right),$$

$$i \geq 0, \ j = \overline{0, N+1}, \ n = \overline{0, j-1},$$

$$U_j^{(i+1)}(j) = U_j^{(i)}(j)(I + \psi^{-1}D(1)) + \psi^{-1} \sum_{l=0}^{j-1} U_l^{(i)}(j)$$

$$\times \left((p_{cr} + p_{ca}) D_{n-l} + p_{pa} \sum_{m=n-l}^{\infty} D_m \right), \ i \geq 0, \ j = \overline{0, N+1},$$

$$U_n^{(i+1)}(j) = U_n^{(i)}(j) \left(I + p_{ca} \psi^{-1} D(1) \right) + p_{ca} \psi^{-1} \sum_{l=0}^{j-1} U_l^{(i)}(j) D_{n-l},$$

$$i \geq 0, \ j = \overline{0, N+1}, \ n > j.$$

These recursions are derived from the differential equations for matrices $P^{(j)}(n,t)$ having the following form:

$$\dot{P}^{(j)}(n,t) = P^{(j)}(n,t) \left(D_0 + p_{cr} \sum_{m=j+1-n}^{\infty} D_m \right)$$

$$+ \sum_{l=0}^{n-1} P^{(j)}(l,t) D_{n-l}, j = \overline{0, N+1}, \ n = \overline{0, j-1},$$

$$\dot{P}^{(j)}(j,t) = P^{(j)}(j,t) D(1)$$

$$+ \sum_{l=0}^{j-1} P^{(j)}(l,t) \left((p_{cr} + p_{ca}) D_{n-l} + p_{pa} \sum_{m=n-l}^{\infty} D_m \right), \ j = \overline{0, N+1},$$

$$\dot{P}^{(j)}(n,t) = p_{ca}(P^{(j)}(n,t) D(1)$$

$$+ \sum_{l=0}^{j-1} P^{(j)}(l,t) D_{n-l}), \ j = \overline{0, N+1}, \ n > j.$$

Now we are able to derive expressions for matrices $P_{i,l}$.

Lemma 2. Transition probability matrices $P_{i,l}$ are computed by formulas

$$P_{i,l} = \int_0^\infty P^{(N+1-i)}(l+1-i,t) \otimes dB(t), i = \overline{1,N}, \, l \geq i-1,$$

$$P_{i,l} = 0, \, i > N, \, l \neq i-1 \, i > 0, \, l < i-1,$$
$$P_{i,l} = G, \, i > N, \, l = i-1,$$

$$P_{0,l} = \tilde{D}_p \sum_{k=1}^{l+1} \tilde{D}_k P_{k,l}, \, l = \overline{0,N-1},$$

$$P_{0,N} = \tilde{D}_p \sum_{k=1}^{N+1} \tilde{D}_k P_{k,N} + p_{pa}(-\tilde{D}_0)^{-1} \sum_{m=N+2}^\infty \tilde{D}_m G,$$

$$P_{0,l} = p_{ca}(-\tilde{D}_0)^{-1} \left(\tilde{D}_{l+1}G + \sum_{k=1}^N \tilde{D}_k P_{k,l} \right), \, l > N,$$

$$\tilde{D}_p = - \left(p_{cr} \left(\tilde{D}_0 + \sum_{m=N+2}^\infty \tilde{D}_m \right)^{-1} + (p_{pa} + p_{ca})\tilde{D}_0^{-1} \right),$$

where

$$G = \int_0^\infty e^{D(1)t} \otimes dB(t) = \int_0^\infty P^{(0)}(0,t) \otimes dB(t),$$

$$\tilde{D}_l = D_l \otimes I_M, \, l \geq 0,$$

$\otimes$ is the symbol of the Kronecker product of matrices, see [9].

Having proved two lemmas and accounting the special form of the kernel $B(t)$ we get the following formula useful for computation of transition probability matrices $P_{i,l}$.

$$\int_0^\infty P^{(j)}(l,t) \otimes dB(t) = \sum_{i=0}^\infty U_n^{(i)}(j) \int_0^\infty e^{-\psi t} \frac{(\psi t)^i}{i!} \otimes dB(t)$$
$$= \sum_{i=0}^\infty U_n^{(i)}(j) \otimes diag\{\gamma_i^{(1)}, \ldots, \gamma_i^{(M)}\}Q,$$

where

$$\gamma_i^{(m)} = \int_0^\infty e^{-\psi t} \frac{(\psi t)^i}{i!} dB_m(t), \, m = \overline{1,M}.$$

Integrals like $\gamma_i = \int_0^\infty e^{-\psi t} \frac{(\psi t)^i}{i!} dB_m(t)(t), \, i \geq 0$, in the case of arbitrary distribution can be computed numerically. But for distributions many popular in queueing theory these integrals are computed explicitly.

If service time is deterministic and equals b_1 then

$$\gamma_i = \frac{\rho^i}{i!} e^{-\rho}, \, \rho = \psi b_1.$$

If service time has gamma-distribution with parameters (μ, δ) then $B'(t) = \frac{\mu^\delta}{\Gamma(\delta)} t^{\delta-1} e^{-\mu t}$:

$$\gamma_i = \frac{\prod\limits_{l=0}^{i-1}(\delta + l)}{i!} \frac{\rho^i}{(1+\rho)^{i+\delta}}, \; \rho = \psi b_1, \; b_1 = \frac{\delta}{\mu}.$$

In the particular case of Erlang's distribution with parameters (k, μ) we have

$$\gamma_i = C^i_{k+1} \frac{\rho^i}{(1+\rho)^{i+k}}$$

where $\rho = \psi b_1$, $b_1 = \frac{k}{\mu}$.

If service time has uniform distribution in interval $[t_1, t_2]$, then

$$\gamma_i = \frac{1}{\psi(t_2 - t_1)}\left(e^{-\psi t_1}\sum_{m=0}^i \frac{(\psi t_1)^m}{m!} - e^{-\psi t_2}\sum_{m=0}^i \frac{(\psi t_2)^m}{m!}\right).$$

4 Stationary Distribution of the Embedded Markov Chain

Using the results from [12], it is possible to prove that the considered Markov chain is ergodic if and only if the inequality

$$\lambda b_1 < 1$$

holds good.

Let this condition be fulfilled. Then the following limits exist:

$$\pi(i, \nu, m) = \lim_{n\to\infty} P\{i_n = i, \nu_n = \nu, m_n = m\}, \; \nu = \overline{0, W}, \; m = \overline{1, M}.$$

Let π_i be the vector consisting of probabilities $\pi(i, \nu, m)$, enumerated in lexicographic order, $i \geq 0$.

Using the results from [12], it is also possible to prove the following statement.

Theorem 1. Vectors π_i can be computed by formalas

$$\pi_i = \pi_0 \Phi_i, \; i \geq 0,$$

where the matrices Φ_i are computed from recursion

$$\Phi_0 = I, \; \Phi_l = \sum_{i=0}^{l-1} \Phi_i \bar{P}_{i,l}(I - \bar{P}_{l,l})^{-1}, \; l > 0,$$

and the vector π_0 is the unique solution to the system

$$\pi_0(I - \bar{P}_{0,0}) = 0, \; \pi_0 \sum_{l=0}^\infty \Phi_l e = 1.$$

Here the matrices $\bar{P}_{i,l}$ are computed from recursion

$$\bar{P}_{i,l} = P_{i,l} + \bar{P}_{i,l+1}G_l, \ i \geq 0, \ l \geq i,$$

matrices G_i are computed from backward recursion

$$G_i = G, \ i \geq N,$$

$$G_i = (I - \sum_{l=i+1}^{\infty} P_{i+1,l}G_{l-1}G_{l-2} \cdot \ldots \cdot G_{i+1})^{-1}P_{i+1,i},$$

$$i = \overline{0, N-1}.$$

The procedure for computation of probability vectors $\boldsymbol{\pi}_i$ defined by this theorem is numerically stable due to avoiding operations with matrices having the negative entries.

5   Stationary Distribution of the System States at Arbitrary Time

Let us consider the stationary probabilities

$$p(i, \nu, m) = \lim_{t \to \infty} P\{i_t = i, \ \nu_t = \nu, \ m_t = m\}, \ \nu = \overline{0, W}, \ m = \overline{1, M},$$

and let $\mathbf{p}_i$ be the vector consisting of these probabilities enumerated in lexico-graphic order, $i \geq 0$.

The necessary and sufficient condition for the existence of these limits also is the fulfillment of inequality $\lambda b_1 < 1$.

Theorem 2. Vectors $\mathbf{p}_i$, $i \geq 0$, are defined by

$$\mathbf{p}_0 = \tau^{-1}\boldsymbol{\pi}_0(-D_0)^{-1},$$

$$\mathbf{p}_i = \sum_{k=1}^{i} \left(\mathbf{p}_0\tilde{D}_k + \tau^{-1}\boldsymbol{\pi}_k\right)$$

$$\times \int_0^{\infty} P^{(N+1-k)}(i-k,t) \otimes \left(I_M - \tilde{B}(t)\right) dt, \ i = \overline{1, N+1},$$

$$\mathbf{p}_i = \sum_{k=1}^{N+1} \left(p_{ca}\mathbf{p}_0\tilde{D}_k + (1 - \delta_{k,N+1})\tau^{-1}\boldsymbol{\pi}_k\right)\int_0^{\infty} P^{(N+1-k)}(i-k,t) \otimes \left(I_M - \tilde{B}(t)\right) dt$$

$$+ \left(p_{ca}\mathbf{p}_0\tilde{D}_i + \tau^{-1}\boldsymbol{\pi}_i\right)\int_0^{\infty} e^{D(1)t} \otimes \left(I_M - \tilde{B}(t)\right) dt, \ i > N+1,$$

where the average inter-departure time τ is computed by

$$\tau = b_1 + \boldsymbol{\pi}_0\tilde{D}_p\mathbf{e}$$

where $\tilde{B}(t)$ is the diagonal matrix with entries $B(t)\mathbf{e}$.

The proof of the theorem is performed based on the known results for Markov renewal processes, see [4]. Here the matrices

$$S_n^{(j)} = \int_0^\infty P^{(j)}(n,t) \otimes \left(I_M - \tilde{B}(t)\right) dt$$

can be computed using the modified algorithm from [15] having the form:

$$S_n^{(j)} = \sum_{i=0}^\infty U_n^{(i)}(j) \otimes diag\{\hat{\gamma}_i^{(1)}, \ldots, \hat{\gamma}_i^{(M)}\}Q,$$

where

$$\hat{\gamma}_i^{(m)} = \psi^{-1}\left(1 - \sum_{k=0}^i \gamma_k^{(m)}\right), \; i \geq 0, \; m = \overline{1,M}.$$

A numerically stable procedure for computation of values $\hat{\gamma}_i^{(m)}$ is given by:

$$\hat{\gamma}_i^{(m)} = \begin{cases} \psi^{-1} \sum_{k=i+1}^{N_\gamma} \gamma_k^{(m)}, & i = \overline{0, N_\gamma - 1}, \; m = \overline{1,M}, \\ 0, & i \geq N_\gamma, \; m = \overline{1,M}, \end{cases}$$

where, for an arbitrary chosen small number ε_γ, the integer N_γ is chosen in such a way as

$$\sum_{j=0}^{N_\gamma} \gamma_j^{(m)} > 1 - \varepsilon_\gamma.$$

6   Loss Probability

By loss probability of an arbitrary customer in the system we mean the value

$$P_{loss} = 1 - \lim_{t\to\infty} \frac{N_t^{(s)}}{N_t},$$

where N_t is the number of customers arriving into the system during the interval $(0,t)$ and $N_t^{(s)}$ is the number of customers being served in the system during this interval.

Theorem 3. Probability of arbitrary customer loss P_{loss} is given by the formula

$$P_{loss} = 1 - \frac{1}{\tau\lambda}.$$

The theorem is proved using the ergodic theorems for functionals defined on the trajectories of Markov chains.

Alternative formulas for computation of P_{loss} can be written down as straightforward extension of the corresponding formulas presented in [6] (for PA discipline) and [8] (for CA and CR disciplines). The existence of different formulas is helpful for control at the stage of computer implementation.

7   Optimization Problem

The set of probabilities p_{cr}, p_{pa} and p_{cr} might be chosen to provide better quality of system operation. To evaluate the quality of the system operation, we consider the charge c_{loss} for a loss of one customer and the charge c_{extra} for occupation of one place beyond the main buffer of the system per unit of time.

The average number N_{extra} of additionally occupied places is computed by

$$N_{extra} = \sum_{i=N+2}^{\infty} (i - N - 1)\mathbf{p}_i\mathbf{e}.$$

Then the total charge of the system paid during a unit of time is given by the formula

$$F = c_{loss}\lambda P_{loss} + c_{extra}N_{extra}.$$

The optimal set of probabilities p_{cr}, p_{pa} and p_{cr} can be found by means of minimization of the value F as a function of parameters p_{cr}, p_{pa} and p_{cr} with the help of a computer.

8   Numerical Result

Let the arrival flow of customers be a group stationary Poisson process given by

$$D_0 = (-5), \; D_k = (1), \; k = \overline{5, 9}.$$

Service time distribution is exponential with parameter 100, buffer capacity is $N = 7$, charges for customers loss and the rent of additional buffer space are defined by $c_{loss} = 1$ and $c_{extra} = 100$.

We suggest that the application of the PA discipline is not appropriate in the system under study because the customers are not independent and partial

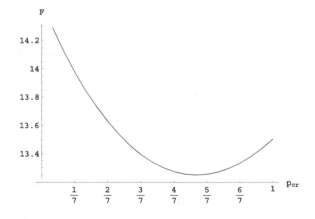

Fig. 1. Cost criterion as function of probability p_{cr}

loss is not allowed. The CR discipline is applied with probability p_{cr} while the CA discipline is applied with complementary probability $p_{ca} = 1 - p_{cr}$. We vary the value of p_{cr} from 0 to 1 with step 0.05 and compute the value of the cost criterion F. The results are presented the Fig. 1.

It is evidently seen that the optimal value of probability p_{cr} is 0.65. Thus, in the situation when the number of customers arriving in a batch to a non-full system exceeds the available capacity of the system, in 65 % of cases we have to admit this batch and we have to reject this batch in 35 % of cases.

9 Conclusion

We analysed a single server queue with a finite buffer under very general assumptions about the arrival and service processes. We introduced customers admission discipline which includes all previously studied disciplines as partial cases. Stationary distributions of the embedded at service completion moments multi-dimensional Markov chain and the system states at an arbitrary moment are calculated. A formula for the probability of arbitrary customer loss is presented. The brief numerical result shows that the introduced admission discipline may have an advantage over the previously applied disciplines.

Acknowledgments. This work has been financially supported by the Russian Science Foundation and the Department of Science and Technology (India) via grant #16-49-02021 for the joint research project by the V.A. Trapeznikov Institute of control Sciences and the CMS College Kottayam.

References

1. Baiocchi, A., Blefari-Melezzi, N.: Analysis of the loss probability of the $MAP/G/1/K$ queue. Commun. Stat. Stoch. Models **10**, 867–925 (1994)
2. Blondia, C.: The $N/G/1$ finite capacity queue. Commun. Stat. Stoch. Models **5**, 273–294 (1989)
3. Chakravarthy, S.R.: The batch Markovian arrival process: a review and future work advances. In: Krishnamoorthy, A., et al. (eds.) Probability Theory and Stochastic Processes, pp. 21–49. Notable Publications, NJ (2001)
4. Cinlar, E.: Introduction to Stochastic Processes. Prentice-Hall, New Jersey (1971)
5. Dudin, A.N., Klimenok, V.I.: A retrial $BMAP/SM/1$ system with linear repeated requests. Queueing Syst. **34**, 47–66 (2000)
6. Dudin, A.N., Klimenok, V.I., Tsarenkov, G.V.: Characteristics calculation for a single server queue with the batch Markovian arrival process, semi-Markovian service and finite buffer. Autom. Remote Control **63**(8), 87–100 (2002)
7. Dudin, A.N., Nishimura, S.: Optimal hysteretic control for a $BMAP/SM/1/N$ queue with two operation modes. Math. Probl. Eng. **5**, 397–419 (2000)
8. Dudin, A.N., Shaban, A.A., Klimenok, V.I.: Analysis of a queue in the $BMAP/G/1/N$ system. Int. J. Simul. **6**, 13–23 (2005)
9. Graham, A.: Kronecker Products and Matrix Calculus with Applications. Ellis Horwood, Cichester (1981)

10. Heyman, D., Lucantoni, D.: Modelling multiple IP traffic streams with rate limits. IEEE ACM Trans. Netw. **11**, 948–958 (2010)
11. Klemm, A., Lindermann, C.: Modelling IP traffic using the batch Marcovian arrival process. Perform. Eval. **54**, 149–173 (2003)
12. Klimenok, V.I., Dudin, A.N.: Multi-dimensional asymptotically quasi-Toeplitz Markov chains and their application in queueing theory. Queueing Syst. **54**, 245–259 (2006)
13. Klimenok, V.I., Kim, C.S., Orlovsky, D.S., Dudin, A.N.: Lack of invariant property of Erlang $BMAP/PH/N/0$ model. Queueing Syst. **49**, 187–213 (2005)
14. Lucantoni, D., Neuts, M.: Some steady-state distributions for the $MAP/SM/1$ queue. Commun. Stat. Stoch. Models **10**, 575–578 (1994)
15. Lucantoni, D.: New results on the single server queue with a batch Markovian arrival process. Commun. Stat. Stoch. Models **7**, 1–46 (1991)
16. Niu, Z., Shu, T., Takahashi, Y.: A vacation queue with setup and close-down times and batch Markovian arrival process. Perform. Eval. **54**, 225–248 (2003)

Optimal Control of a Two-Server Heterogeneous Queueing System with Breakdowns and Constant Retrials

Dmitry Efrosinin[1,2](✉) and Janos Sztrik[3]

[1] Johannes Kepler University Linz, Altenbergerstrasse 69, 4040 Linz, Austria
dmitry.efrosinin@jku.at
[2] Institute of Control Sciences, RAS, Profsoyuznaya street 65,
117997 Moscow, Russia
[3] University of Debrecen, Egyetem tér 1, Debrecen 4032, Hungary
jsztrik@inf.unideb.hu
http://www.jku.at, http://www.ipu.ru, http://www.unideb.hu

Abstract. Heterogeneous servers which can differ in service speed and reliability are becoming more popular in the modelling of modern communication systems. For a two-server queueing system with one non-reliable server and constant retrial discipline we formulate an optimal allocation problem for minimizing a long-run average cost per unit of time. Using a Markov decision process formulation we prove a number of monotone properties for the increments of the dynamic-programming value function. Such properties imply the optimality of a two-level threshold control policy. This policy prescribes the usage of a less productive server if the number of customers in the queue becomes higher than a predefined level which depends on the state of a non-reliable more powerful server. We provide also a heuristic solution for the optimal threshold levels in explicit form as a function of system parameters.

Keywords: Optimal allocation · Markov decision process · Monotonicity properties · Threshold policy · Heterogeneous servers · Long-run average cost

1 Introduction

In modern communication systems the speed of data transmission in a link can interact with its reliability. In many cases this interaction is differently directed. The complementary properties of different links lead engineers to an idea to combine them in such a way that the advantages of high speed links could guarantee acceptable values of the cost and reliability characteristics. As example of a system where the data transmission links differ in speed and reliability is a Radio Frequency/Free Space Optic (RF/FSO) channel [33]. The capacity of the

D. Efrosinin—This work was funded by the Russian Foundation for Basic Research, Project No. 16-37-60072 mol_a_dk.

© Springer International Publishing Switzerland 2016
A. Dudin et al. (Eds.): ITMM 2016, CCIS 638, pp. 57–72, 2016.
DOI: 10.1007/978-3-319-44615-8_5

RF link is limited to link throughputs in the order of tens of Mbps over distances of tens of meters. However, link availability can be maintained in most weather conditions. In contrast, the commercial FSO currently provides throughputs of several Gbps with link distances of a kilometre or more. One of the major limitations of FSO technology is the need for optical links to maintain line-of-sight and sensitivity to adverse weather conditions such as fog or heavy snowfalls. Another example of a system with combined technologies is a modern call-centre where the human operators are working together with a self-service facility which serves the calls on the basis of speech recognition methods [7].

Queueing systems with heterogeneous servers are appropriate models to describe the behaviour of communication systems with the properties mentioned above. In [14] it was proved that heterogeneous multi-server systems are superior in performance to homogeneous ones with the same total service time. This is also established in [24] in the context of manufacturing cells and systems, and it was confirmed in [30] using simulation results. In [5] a non-trivial application of a queue with heterogeneous servers to the performance evaluation of a wireless communication system is presented. The question of how to allocate the customers between heterogeneous servers in order to minimize the mean number of customers in the system has been studied by many authors. In [13] the optimality of a threshold policy was conjectured that prescribes the usage of the faster server whenever it is idle and there are customers in the system and the slower server must be used only if the queue length exceeds some prespecified threshold level. Based on a dynamic programming approach, the authors in [16] considered a similar problem and proved the optimality of a threshold policy. Some alternative proofs of this result were later given in [11,18,34]. The problem of optimal allocation in the system with more than two servers was studied in [2,21,22,32]. The allocation problem between the servers with failures was recently studied in [25], where the optimality of threshold policy and dependence of a state of an unreliable server was proved.

Analysis control procedures for queueing systems with additional cost structure with the aim of minimizing the long-run average cost per unit of time is notoriously more difficult. Some progress has been made in [20], where a model with set-up costs and hysteresis control policy was studied. The paper [23] generalizes results to the case of a multi-server system with a cost structure. A recent paper [17] introduces the optimal routing problem in two-server systems which differ in speed and quality of resolution. Multi-server retrial queueing systems have been extensively studied only for homogeneous servers. Queueing systems that combine heterogeneous servers and retrial effect were only rarely examined in research, especially in queueing theory. The authors in [28] evaluated the mean performance measures for the fastest free server allocation policy using the MOSEL performance modelling tool in the case of a finite-population multi-server retrial queueing system. In [9] the optimality of a threshold policy was proved in a queue with a constant retrial discipline, where only the customer at the head of the queue repeats an attempt to occupy the server.

The results presented here differ from those obtained in previous papers. The paper studies the structural properties of the optimal control policy for a two-server heterogeneous queueing system where we combine the classical model with an additional cost structure, unreliability of the server and constant retrial discipline. Our primary contribution includes a rigorous proving of the monotonicity properties of the dynamic programming value function for optimization problems with a constraint, of the threshold structure of the optimal allocation policy, evaluation of the stationary state probabilities in matrix form for the corresponding quasi-birth-and-death (QBD) process and derivation of the heuristic solution for the optimal threshold levels as a function of system parameters.

The remainder of the paper is organized as follows. Section 2 describes the Markov decision process. In Sect. 3 the optimization problem is formulated and the optimality equations for the dynamic-programming value function together with the relationship to the control policy is derived. Section 4 presents the monotonicity properties of the value function needed to establish the optimality of threshold policies. Section 5 is devoted to stationary probabilistic analysis under the given control policy. Section 6 presents the heuristic results for explicit evaluation of optimal thresholds.

2   Mathematical Model

Let us consider a controllable heterogeneous queueing system $M/M/2$ with breakdowns and a constant retrial rate. Customers arrive at the system according to a Poisson process with intensity $\lambda, 0 < \lambda < \infty$. The service facility consists of two heterogeneous exponential servers with intensities μ_1 and μ_2 with $0 < \mu_2 \leq \mu_1 < \infty$. The customer, who is rejected to occupy one of the servers, joins a queue of orbiting customers. According to a constant retrial discipline the customer at the head of the queue retries for service after an exponential distributed time with intensity $\tau, 0 < \tau < \infty$. Server 1 is assumed to be nonreliable with exponential distributed life times with intensity α. A failure can occur in both cases, if a server is idle or busy. If the server fails, the repair starts immediately and a customer leaves this server if it was busy and joins the head of the queue. The repair time is exponential with intensity $\beta, 0 < \beta < \infty$. Let us assume that a customer who gets service at a certain server, cannot change it up to the moment of leaving the system, e.g. after service completion and at a failure time. All interarrival times, intervals of successive retrials and service times, times to failure and repair times are assumed to be mutually independent. The system is supplied with a controller who gets full information about system states and can allocate customers between two servers or put it to the queue of the orbiting customers at each arrival, failure of the busy server or retrial epoch. The objective is to minimize the long-run average cost per unit of time. We propose to formulate the allocation problem as a Markov decision process and then to use the value iteration technique to prove some structural properties, e.g. threshold structure, of the optimal control policy.

Let $Q(t)$ denote the number of customers in the queue at time t, and $D_j(t)$ denote the state of server $j \in \{1, 2\}$ at time t, where $D_j(t) = 0$ if Server j is idle,

$D_j(t) = 1$ if Server j is busy and $D_1(t) = 2$ if Server 1 is failed. The system states at time t are described by a Markov process

$$\{X(t)\}_{t \geq 0} = \{Q(t), D_1(t), D_2(t)\}_{t \geq 0}. \tag{1}$$

The controllable model associated with a Markov process $\{X(t)\}$, which is referred to as a Markov decision process (MDP), is a five-tuple

$$\{E, A, \{A_0(x), A_{b,1}(x), A_s(x), x \in E, j \in \{1,2\}\}, \lambda_{xy}(a), c(x)\}. \tag{2}$$

– E is a *state space*,

$$E = \{x = (q(x), d_1(x), d_2(x)); q(x) \geq 0, d_1(x) \in \{0,1,2\}, d_2(x) \in \{0,1\}\}. \tag{3}$$

Further in the paper the notations $q(x)$ and $d_j(x)$ will be used to specify the certain components of the vector state $x \in E$. For every state x we denote by $J_0(x), J_1(x)$ and $J_2(x)$ the sets of indices of idle, busy and failed servers,

$$J_0(x) = \{j; d_j(x) = 0\}, \ J_1(x) = \{j; d_j(x) = 1\}, \ J_2(x) = \{1; d_1(x) = 2\}. \tag{4}$$

– $A = \{0,1,2\}$ is an *action space* with elements $a \in A$, where $a = 0$ means "to send a customer to the queue" and $a = 1,2$ - "to send a customer to Server 1 or 2". $A_0(x), A_{b,1}, A_s \subseteq A$ denote *subsets of control actions in state* x, respectively in case of a new arrival, failure of busy Server 1 and a retrial arrival, where $A_0(x) = J_0(x) \cup \{0\}$. All other subsets can be expressed through the subset $A_0(x)$,

$$A_{b,1}(x) = A_0(S_1 x), \ q(x) \geq 0, \ 1 \in J_1(x), \tag{5}$$
$$A_s(x) = A_0(S_0^{-1} x), \ q(x) > 0,$$

where $S_0, S_0^{-1}, S_j, S_j^{-1}, j \in \{1,2\}$ stand for the shift operators defined as

$$S_0 x = x + \mathbf{e}_0, \ S_0^{-1} x = x - \mathbf{e}_0, \ q(x) > 0, \tag{6}$$
$$S_j x = x + \mathbf{e}_j, \ j \in J_0(x) \cup J_1(x), \ S_j^{-1} x = x - \mathbf{e}_j, \ j \in J_1(x) \cup J_2(x),$$

where $\mathbf{e}_j$ is used for the vector of dimension 3 with 1 in the jth position (beginning from 0th) and 0 elsewhere.

– $\lambda_{xy}(a)$ is a *transition intensity* to go from state x to state y under a control action a. It is assumed that the model is *conservative*, i.e.

$$\lambda_{xy}(a) \geq 0, \ y \neq x, \ \lambda_{xx}(a) = -\lambda_x(a) = -\sum_{y \neq x} \lambda_{xy}(a), \ \lambda_x(a) < \infty.$$

– $c(x)$ is an *immediate cost*, in state $x \in E$,

$$c(x) = c_0 q(x) + c_{11} 1_{\{1 \in J_1(x)\}} + c_2 1_{\{2 \in J_1(x)\}} + c_{12} 1_{\{1 \in J_2(x)\}}, \tag{7}$$

where c_0, c_{11}, c_{12} and c_2 is respectively the waiting cost per unit of time for each customer in the system, the usage and repair costs for Server 1 and usage cost for Server 2.

We will next explain how the controller chooses its actions. First the concept of a *stationary policy* must be introduced.

Definition 1. *A stationary policy f for the given model is a vector of functions*

$$f = (f_0, f_{b,0}, f_s), \quad where \ f_\bullet : E \to A_\bullet(x). \tag{8}$$

The policy f specifies the control actions that must be chosen at different decision epochs whenever they occur just after an event at state $x \in E$:

- just after an arrival, if the queue is not empty, a customer can be sent to one of the idle servers or sent to the queue;
- just after a failure of a busy server a customer from the server can be sent to the queue or to another idle server;
- just after a retrial arrival a customer can be sent from the head of the queue, if it is not empty, to an idle server.

We assume that just after any other transition no control action has to be chosen.

3   Optimization Problem for Performance Characteristics

For every fixed stationary policy f we wish to guarantee that the process $\{X(t)\}_{t \geq 0}$ with a state-space E is an irreducible, positive recurrent Markov process defined through its infinitesimal matrix $\Lambda = [\lambda_{xy}(f(x))]$. As is known [29], for an ergodic Markov process with costs the long-run average cost per unit of time (also referred to as *gain*) for the policy f coincides with the corresponding assemble average,

$$g^f = \lim_{t \to \infty} \frac{1}{t} V^f(x, t) = \sum_{y \in E} c(y) \pi_y^f, \tag{9}$$

where $V^f(x, t) = \int_0^t \sum_{y \in E} \mathbb{P}^f[X(u) = y | X(0) = x] c(y) du$ denotes the *total average cost up to time* t when the process starts in state x and $\pi_y^f = \mathbb{P}^f[X(t) = y]$ denotes stationary probability of the process given policy f. The policy f^* is said to be optimal when for any admissible policy f

$$g^{f^*} = \min_f g^f. \tag{10}$$

In many applications it is often necessary to find a policy f^* which minimizes the long-run average cost per unit of time under the constraint on the sojourn time or the number of customers in the system (due to Little's Law), namely

$$g^{f^*} = \min_f g^f \quad subject \ to \quad \bar{N} \leq \gamma, \tag{11}$$

where $\bar{N}$ is a mean number of customers in the system. The constrained Markov decision problem can be rewritten as an unconstrained one using Lagrange multipliers, see e.g. [1,4]. The application of MDP with constraints to the problem of optimal allocation in queueing systems was illustrated e.g. in [35].

The optimal policy f^* can be evaluated by means of a *Howard iteration algorithm* [10], which constructs a sequence of improved policies until the average cost optimal is reached. The key role in this algorithm is played by the *dynamic programming value function* $v : E \to \mathbb{R}_+$ which indicates a transition effect of an initial state x to the total average cost and satisfies an asymptotic relation,

$$V^f(x,t) = g^f t + v^f(x) + o(1), \ x \in E, \ t \to \infty. \tag{12}$$

The functions V^f, v^f and g^f further in the paper will be denoted by V, v and g without upper index f. The system will be uniformized as in [26] with

$$\lambda + \mu_1 + \mu_2 + \alpha + \beta + \tau = 1,$$

which can be obtained by time scaling. As is well known, the optimal policy f and the optimal average cost g are solutions of the optimality equation

$$Bv(x) = v(x) + g, \tag{13}$$

where B is the *dynamic programming operator* acting on value function v.

Theorem 1. *The dynamic programming operator B is defined as follows*

$$Bv(x) = c(x) + \eta l(x) + \lambda T_0 v(x) + \sum_{j \in J_1(x)} \mu_j T_j v(x) \tag{14}$$

$$+ \alpha T_{b,0} v(x) + \alpha T_{b,1} v(x) + \beta T_r v(x) + \tau T_s v(x)$$

$$+ \left(\sum_{j \notin J_1(x)} \mu_j + \beta 1_{\{1 \notin J_2(x)\}} \right) T_d v(x),$$

where $T_0, T_j, j \in \{1,2\}, T_{b,0}, T_{b,1}, T_r, T_s$ and T_d – event operators, respectively, for a new arrival, for service completion on server j, failure occurrence in idle or busy state, repair completion of Server 1 and retrial arrival, the last operator stands for dummy transitions,

$$T_0 v(x) = \min_{a \in A_0(x)} v(x + e_a), \ T_j v(x) = v(S_j^{-1} x), \ j \in J_1(x),$$

$$T_{b,0} v(x) = v(S_1^2 x), \ 1 \in J_0(x), \ T_{b,1} v(x) = T_0 v(S_1 x), \ 1 \in J_1(x),$$

$$T_r v(x) = v(S_1^{-2} x), \ 1 \in J_2(x), \ T_s v(x) = \begin{cases} T_0 v(S_0^{-1} x) & q(x) > 0, \\ v(x) & q(x) = 0. \end{cases}, \ T_d v(x) = v(x).$$

Proof. The optimality equation is obtained by analysing the function $V(x,t)$ in some infinitesimal interval $[t, t+dt]$. It leads to a differential equation. Applying further the limit expression for $dt \to 0$ and due to the Markov property of $\{X(t)\}_{t \geq 0}$ with asymptotic relation (12) ones get (14). Here $\eta > 0$ is the Lagrange multiplier, $l(y)$ stands for the number of customers in state y. When η increases, then the value of $\bar{N}$ decreases. Therefore, there exist values η' and η'', such that $\bar{N}(\eta') > \gamma$ and $\bar{N}(\eta'') \leq \gamma$, where $\eta'' = \eta' + \varepsilon$ for a small $\varepsilon \geq 0$.

The structure of the system (14) implies for each $x \in E$ the following statement.

Corollary 1. *The optimal policy $f = (f_0, f_{b,1}, f_s)$ is defined through the value function $v(x)$ and depends only on its component f_0 for the shifted states,*

$$f_0(x) = \arg\min_{a \in A_0(x)}\{v(S_a x)\}, \tag{15}$$

$$f_{b,1}(x) = \arg\min_{a \in A_{b,1}(x)}\{v(S_a S_1 x)\} = f_0(S_1 x), \ 1 \in J_1(x),$$

$$f_s(x) = \arg\min_{a \in A_s(x)}\{v(S_a S_0^{-1} x)\} = f_0(S_0^{-1} x), \ q(x) > 0.$$

Therefore, the optimal component f_0 completely defines the optimal policy f. To show the structural properties of the optimal policy f some monotonicity properties of the increments of the value function $v(x)$ must be proved.

The relations (15) show that the structural and monotone properties of the optimal control policy f can be derived by analysing the monotonicity properties of the value function v. Such properties for other types of controlled queues in a tandem were studied also in [12, 15, 31]. It was shown that the value function has some monotonicity properties like non-decreasing and superconvexity. To prove such inequalities it is necessary to solve (13). Since the solution of the optimality equation in analytic form is hardly available, it can be solved recursively defining $v_{n+1} = Bv_n$ for some arbitrary initial v_0. Due to the limit relation

$$\lim_{n \to \infty} B^n v_0(x) = v(x) \tag{16}$$

we get an optimal solution for the value function. For existence and convergence solutions and optimal policies we refer to [3, 25–27].

4   Optimality of the Threshold Policy

Now some monotonicity properties of the value function for the system under study will be presented and proved, but first we have to make a statement.

Theorem 2. *The value function v satisfies the conditions for any $x \in E$:*
C1. Non-decreasing condition

$$\begin{aligned}
&(a) \quad v(x) - v(S_0 x) \leq 0, \\
&(b) \quad v(x) - v(S_j x) \leq 0, \ j \in \{1, 2\}, \ d_j(x) = 0, \\
&(c) \quad v(S_1 x) - v(S_2 x) \leq 0, \ d_1(x) = d_2(x) = 0, \\
&(d) \quad v(S_1 x) - v(S_0 x) \leq 0, \ d_1(x) = 0.
\end{aligned}$$

C2. Superconvexity condition

$$\begin{aligned}
&(a) \quad v(S_0 x) - v(S_2 x) \leq v(S_0^2 x) - v(S_0 S_2 x), \ d_1(x) \in \{0, 1, 2\}, \ d_2(x) = 0, \\
&(b) \quad v(S_1 x) - v(S_2 x) \leq v(S_0 S_1 x) - v(S_1 S_2 x), \ q(x) = 0, \ d_1(x) = d_2(x) = 0.
\end{aligned}$$

C3. Supermodularity condition

    (a) $v(S_0 x) - v(x) \leq v(S_0 S_2 x) - v(S_2 x)$, $d_1(x) \in \{0, 1, 2\}$, $d_2(x) = 0$,

    (b) $v(S_1 x) - v(x) \leq v(S_1 S_2 x) - v(S_2 x)$, $q(x) = 0$, $d_1(x) = 0$, $d_2(x) = 0$.

C4. Convexity condition

    (a) $2v(S_0 x) - v(x) - v(S_0^2 x) \leq 0$, $d_1(x) \in \{1, 2\}$, $d_2(x) = 0$,

    (b) $2v(S_0 S_2 x) - v(S_2 x) - v(S_0^2 S_2 x) \leq 0$, $d_1(x) \in \{1, 2\}$, $d_2(x) = 0$,

    (c) $2v(S_1 x) - v(x) - v(S_0 S_1 x) \leq 0$, $d_1(x) = 0$, $d_2(x) = 0$.

Statement 1. The inequality (C4-a) directly follows by summing up the conditions (C2-a) and (C3-a), the inequality (C4-b) – by summing up the conditions (C2-a) and (C3-a) in state x and $S_0 x$ respectively and (C4-c) – by summing up the conditions (C2-b) and (C3-b), which are the boundary conditions for the state $x = (0, 0, 0)$. The name of condition (C3) is borrowed from [6].

Statement 2. (C1-c) can be explicitly proved only under assumption $c_{11} - c_2 \leq 0$ and (C1-d) – under assumption $c_{11} - c_0 \leq 0$, which are of course too strong. We assume that for conditions (C1)–(C4) the expression (26) must be positive.

    Due to lack of space we demonstrate the proof only for the properties (C1-c) and (C2-a). The rest of the inequalities can be proved in a similar way.

Proof (Condition C1). The proof is by induction on n in v_n. Let us define $v_0(x) = 0$ for all states $x \in E$. This function obviously satisfies the conditions (C1)–(C3). Now, we assume (C1)–(C3) for the function $v_n(x), x \in E$, and some $n \in \mathbb{N}$. One has to prove that $v_{n+1}(x)$ satisfies the proposed conditions as well.

(C1-c). Let us consider the inequality (C1-c). For this increment we get,

$$
\begin{aligned}
v_{n+1}(S_1 x) - v_{n+1}(S_2 x) &= c(S_1 x) - c(S_2 x) + \eta(l(S_1 x) - l(S_2 x)) & (I) \\
&+ \lambda[T_0 v_n(S_1 x) - T_0 v_n(S_2 x)] & (II) \\
&+ \mu_1[T_1 v_n(S_1 x) - T_d v_n(S_2 x)] & (III) \\
&+ \mu_2[T_d v_n(S_1 x) - T_2 v_n(S_2 x)] & (IV) \\
&+ \alpha[T_{b,1} v_n(S_1 x) - T_{b,0} v_n(S_2 x)] & (V) \\
&+ \tau[T_s v_n(S_1 x) - T_s v_n(S_2 x)] & (VI) \\
&+ \beta[T_d v_n(S_1 x) - T_d v_n(S_2 x)] \leq 0. & (VII)
\end{aligned}
$$

The term (I) is equal to $c_{11} - c_2 \leq 0$, which is true by assumption. The term (II)

$$(II) = T_0 v_n(S_1 x) - v_n(S_1 S_2 x) \leq v_n(S_1 S_2 x) - v_n(S_1 S_2 x) = 0$$

by virtue of (C1-c) and (C1-d). Now we join the terms (III) and (IV),

$$
\begin{aligned}
(III) + (IV) &= \mu_2[T_d v_n(S_1 x) - T_2 v_n(S_2 x)] - \mu_1[T_d v_n(S_2 x) - T_1 v_n(S_1 x)] \\
&= \mu_2[v_n(S_1 x) - v_n(x)] - \mu_1[v_n(S_2 x) - v_n(x)] \\
&= \mu_1 \mu_2 \left[\frac{v_n(S_1 x) - v_n(x)}{\mu_1} - \frac{v_n(S_2 x) - v_n(x)}{\mu_2} \right] \leq 0,
\end{aligned}
$$

for $q(x) > 0$ due to $\mu_1 \geq \mu_2$ and (C1-b). The term (V) is non-positive,

$$(V) = T_0 v_n(S_1^2 x) - v_n(S_1^2 S_2 x) \leq v_n(S_1^2 S_2 x) - v_n(S_1^2 S_2 x) = 0.$$

For the term (VI) we have

$$(VI) = T_0 v_n(S_0^{-1} S_1 x) - T_0 v_n(S_0^{-1} S_2 x) = T_0 v_n(S_0^{-1} S_1 x) - v_n(S_0^{-1} S_1 S_2 x)$$
$$\leq v_n(S_0^{-1} S_1 S_2 x) - v_n(S_0^{-1} S_1 S_2 x) = 0, \quad \text{for } q(x) > 0,$$
$$(VI) = v_n(S_1 x) - v_n(S_2 x) \leq 0, \quad \text{for } q(x) = 0,$$

by (C1-c). The term (VII) is non-positive according to (C1-c). If (C1) holds for $v_n(x)$ then it holds by induction for any n and by (16) for the function $v(x)$.

Proof (Condition C2). The proof is made by induction on n in v_n. Let us define $v_0(x) = 0$ for all states $x \in E$. It is clear that this function satisfies condition (C2). Now suppose that properties (C1)–(C4) hold for $v_n, n \in \mathbb{N}$. Now we prove that it holds for $n + 1$ as well.

(C2-a). In this case for the function $v_n(x)$ holds the following inequality

$$
\begin{aligned}
&v_{n+1}(S_0 x) - v_{n+1}(S_2 x) - v_{n+1}(S_0^2 x) + v_{n+1}(S_0 S_2 x) \\
&= c(S_0 x) - c(S_2 x) - c(S_0^2 x) + c(S_0 S_2 x) && (I) \\
&+ \eta(l(S_0 x) - l(S_2 x) - l(S_0^2 x) + l(S_0 S_2 x)) \\
&+ \lambda[T_0 v_n(S_0 x) - T_0 v_n(S_2 x) - T_0 v_n(S_0^2 x) + T_0 v_n(S_0 S_2 x)] && (II) \\
&+ \mu_1[T_1 v_n(S_0 x) - T_1 v_n(S_2 x) - T_1 v_n(S_0^2 x) + T_1 v_n(S_0 S_2 x)] && (III) \\
&+ \alpha[T_{b,1} v_n(S_0 x) - T_{b,1} v_n(S_2 x) - T_{b,1} v_n(S_0^2 x) + T_{b,1} v_n(S_0 S_2 x)] && (IV) \\
&+ \beta[T_r v_n(S_0 x) - T_r v_n(S_2 x) - T_r v_n(S_0^2 x) + T_r v_n(S_0 S_2 x)] && (V) \\
&+ \mu_2[T_d v_n(S_0 x) - T_2 v_n(S_2 x) - T_d v_n(S_0^2 x) + T_2 v_n(S_0 S_2 x)] && (VI) \\
&+ \tau[T_s v_n(S_0 x) - T_s v_n(S_2 x) - T_s v_n(S_0^2 x) + T_s v_n(S_0 S_2 x)] && (VII) \\
&+ ((\alpha + \mu_1) 1_{\{d_1(x)=2\}} + \beta 1_{\{d_1(x)=1\}}) \\
&\times [T_d v_n(S_0 x) - T_d v_n(S_2 x) - T_d v_n(S_0^2 x) + T_d v_n(S_0 S_2 x)] \leq 0. && (VIII)
\end{aligned}
$$

The term (I) is obviously 0. We show that the term (II) is non-positive. To perform it let us consider the following two subcases. If $f_0(S_2 x) = f(S_0^2 x) = 0$, then

$$(II) = T_0 v_n(S_0 x) - v_n(S_0 S_2 x) - v_n(S_0^3 x) + T_0 v_n(S_0 S_2 x)$$
$$\leq v_n(S_0^2 x) - v_n(S_0 S_2 x) - v_n(S_0^3 x) + v_n(S_0^2 S_2 x) \leq 0$$

due to (C2-a) in state $S_0 x$. If $f_0(S_2 x) = 0$ and $f(S_0^2 x) = 2$, then

$$(II) = T_0 v_n(S_0 x) - v_n(S_0 S_2 x) - v_n(S_0^2 S_2 x) + T_0 v_n(S_0 S_2 x)$$
$$\leq v_n(S_0 S_2 x) - v_n(S_0 S_2 x) - v_n(S_0^2 S_2 x) + v_n(S_0^2 S_2 x) = 0.$$

The term (III) satisfies

$$(III) = v_n(S_0 S_1^{-1} x) - v_n(S_1^{-1} S_2 x) - v_n(S_0^2 S_1^{-1} x) + v_n(S_0 S_1^{-1} S_2 x) \leq 0,$$

which follows from the inequalities (C2-a). Let us consider the term (IV). We have

$$(IV) = T_0 v_n(S_0 S_1^2 x) - T_0 v_n(S_1^2 S_2 x) - T_0 v_n(S_0^2 S_1^2 x) + T_0 v_n(S_0 S_1^2 S_2 x) \leq 0,$$

with respect to the property of the operator T_0 applied to (C2-a) in state $S_1^2 x$. The term (V) is of the form

$$(V) = v_n(S_0 S_1^{-2} x) - v_n(S_1^{-2} S_2 x) - v_n(S_0^2 S_1^{-2} x) + v_n(S_0 S_1^{-2} S_2 x) \leq 0,$$

which follows from the inequality (C2-a) in state $S_1^{-2} x$ for $d_1(x) = 2$. For the term (VI) we have

$$(VI) = v_n(S_0 x) - v_n(x) - v_n(S_0^2 x) + v_n(S_0 x) = 2 v_n(S_0 x) - v_n(x) - v_n(S_0^2 x) \leq 0,$$

by (C4-a). The term (VII) is non-positive for $q(x) > 0$ due to the property of the operator T_s and T_0 applied to the state $S_0^{-1} x$ in (C2-a),

$$(VII) = T_0 v_n(x) - T_0 v_n(S_0^{-1} S_2 x) - T_0 v_n(S_0 x) + T_0 v_n(S_2 x) \leq 0.$$

For $q(x) = 0$ we get

$$(VII) = T_0 v_n(x) - v_n(S_2 x) - T_0 v_n(S_0 x) + T_0 v_n(S_2 x).$$

If $f_0(S_0 x) = 0$, then we have

$$(VII) = T_0 v_n(x) - v_n(S_2 x) - v_n(S_0^2 x) + T_0 v_n(S_2 x)$$
$$\leq v_n(S_0 x) - v_n(S_2 x) - v_n(S_0^2 x) + v_n(S_0 S_2 x) \leq 0$$

as in (C2-a). If $f_0(S_0 x) = 2$, then

$$(VII) = T_0 v_n(x) - v_n(S_2 x) - v_n(S_0 S_2 x) + T_0 v_n(S_2 x)$$
$$\leq v_n(S_2 x) - v_n(S_2 x) - v_n(S_0 S_2 x) + v_n(S_0 S_2 x) = 0.$$

The last term (VIII) is non-positive by (C2-a).

Hence, we conclude, by taking the limit $n \to \infty$, that the value function $v(x)$ preserves the conditions (C2).

Corollary 2. *The optimal allocation policy is of threshold type. There exists a two-level threshold policy* $f^* = (q_1^*, q_2^*)$ *such that*

1. *Server 1 must be used whenever it is free while a customer tries to get service at a service facility.*
2. *Server 2 must be used whenever it is free, Server 1 is busy or failed, a customer tries to get service and the queue length $q(x)$ upon arrival in state x satisfies $q(x) \geq q_k^* - 1$ if $d_1(x) = k, k \in \{1, 2\}$.*

Proof. The statement follows directly from the definition (15) of the control policy $f(x)$ and monotonicity properties of the value function (C1-c), (C1-d) and (C2-a) given in Theorem 2.

5 Stationary State Probabilities

Now we fix the threshold policy $f = (q_1, q_2)$. The possible states of the system at any time are presented by the vector state $x = (q, d_1, d_2) \in E$. Let us define the levels $\mathbf{0, 1, 2}, \ldots$ as a set (block) of states,

$$\mathbf{q} = \{(q, 0, 0), (q, 1, 0), (q, 2, 0), (q, 0, 1), (q, 1, 1), (q, 2, 1)\}$$

for $q \geq 0$. Let the row-vector of the stationary state probabilities $\boldsymbol{\pi}$ be partitioned as $\boldsymbol{\pi} = (\boldsymbol{\pi}_0, \boldsymbol{\pi}_1, \boldsymbol{\pi}_2, \ldots)$, where $\boldsymbol{\pi}_q = \{\pi_{(q,0,0)}, \pi_{(q,1,0)}, \pi_{(q,2,0)}, \pi_{(q,0,1)}, \pi_{(q,1,1)}, \pi_{(q,2,1)}\}$. Let us denote by $\mathbf{e}_j$ the vector of dimension 6 with 1 in the jth position (beginning from 0th) and 0 elsewhere and by $\mathbf{e}$ the unit vector of the same dimension.

Theorem 3. *If the stability condition holds, the vector $\boldsymbol{\pi}$ exists and is a unique solution of the system $\boldsymbol{\pi}\Lambda = \mathbf{0}, \boldsymbol{\pi}\mathbf{e} = 1$, where $\Lambda := \Lambda(q_1, q_2)$ specifies a three diagonal block infinitesimal matrix of $\{X(t)\}_{t \geq 0}$, which is of the QBD type:*

$$\Lambda(q_1, q_2) = diag(Q_{1,0}, \underbrace{Q_{1,1}, \ldots, Q_{1,1}}_{q_2 - 2}, Q_{1,2}, \underbrace{Q_{1,3}, \ldots, Q_{1,3}}_{q_1 - q_2 - 1}, Q_{1,4}, Q_{1,5}, \ldots)$$

$$+ diag^+ (\underbrace{Q_{0,1}, \ldots, Q_{0,1}}_{q_2 - 2}, \underbrace{Q_{0,2}, \ldots, Q_{0,2}}_{q_1 - q_2 - 1}, Q_{0,3}, \ldots) \tag{17}$$

$$+ diag^- (\underbrace{Q_{2,1}, \ldots, Q_{2,1}}_{q_2 - 2}, \underbrace{Q_{2,2}, \ldots, Q_{2,4}}_{q_1 - q_2 - 1}, Q_{2,3}, \ldots).$$

with $(Q_{1,0} + Q_{0,1})\mathbf{e} = (Q_{2,1} + Q_{1,1} + Q_{0,1})\mathbf{e} = (Q_{2,1} + Q_{1,2} + Q_{0,2})\mathbf{e} = (Q_{2,2} + Q_{1,3} + Q_{0,2})\mathbf{e} = (Q_{2,2} + Q_{1,4} + Q_{0,3})\mathbf{e} = (Q_{2,3} + Q_{1,5} + Q_{0,3})\mathbf{e} = \mathbf{0}$.

Blocks $Q_{1,n}, n = \overline{0,5}$, *include outgoing intensities for the certain block state,*

$$Q_{1,0} = \begin{pmatrix} -(\lambda + \alpha) & \lambda & \alpha & 0 & 0 & 0 \\ \mu_1 & -(\lambda + \mu_1 + \alpha) & 0 & 0 & 0 & 0 \\ \beta & 0 & -(\lambda + \beta) & 0 & 0 & 0 \\ \mu_2 & 0 & 0 & -(\lambda + \mu_2 + \alpha) & \lambda & \alpha \\ 0 & \mu_2 & 0 & \mu_1 & -(\lambda + \mu_1 + \mu_2 + \alpha) & 0 \\ 0 & 0 & \mu_2 & \beta & 0 & -(\lambda + \mu_2 + \beta) \end{pmatrix},$$

$Q_{1,1} = Q_{1,0} + \tau(\mathbf{e}_0 \otimes \mathbf{e}_0' + \mathbf{e}_3 \otimes \mathbf{e}_3')$, $Q_{1,2} = Q_{1,1} + \lambda \mathbf{e}_2 \otimes \mathbf{e}_5' + \alpha \mathbf{e}_1 \otimes \mathbf{e}_5'$,
$Q_{1,3} = Q_{1,2} + \tau \mathbf{e}_2 \otimes \mathbf{e}_2'$, $Q_{1,4} = Q_{1,3} + \lambda \mathbf{e}_1 \otimes \mathbf{e}_4'$, $Q_{1,5} = Q_{1,4} + \tau \mathbf{e}_1 \otimes \mathbf{e}_1'$.

Blocks $Q_{0,n}, n = \overline{1,3}$, *include the incoming intensities from the lower level,*

$$Q_{0,1} = diag(0, \lambda, \lambda, 0, \lambda, \lambda) + diag^+ (0, \alpha, 0, 0, \alpha),$$
$$Q_{0,2} = Q_{0,1} - \lambda \mathbf{e}_2 \otimes \mathbf{e}_2' - \alpha \mathbf{e}_1 \otimes \mathbf{e}_2', \quad Q_{0,3} = Q_{0,2} - \lambda \mathbf{e}_1 \otimes \mathbf{e}_1'.$$

Blocks $Q_{2,n}, n = \overline{1,3}$, *include the incoming intensities from the upper level,*

$$Q_{2,1} = diag^+ (\tau, 0, 0, \tau, 0), \quad Q_{2,2} = Q_{2,1} + \tau \mathbf{e}_2 \otimes \mathbf{e}_5', \quad Q_{2,3} = Q_{2,2} + \tau \mathbf{e}_1 \otimes \mathbf{e}_4'.$$

For the given QBD process to be stable, it is necessary ([19], Theorem 3.1.1, p.82) that $\mathbf{p}Q_{0,3}\mathbf{e} < \mathbf{p}Q_{2,3}\mathbf{e}$, where $\mathbf{p}$ is an invariant probability, which can be evaluated from $\mathbf{p}(Q_{0,3} + Q_{1,5} + Q_{2,3}) = \mathbf{0}$ and $\mathbf{p}\mathbf{e} = 1$.

Theorem 4. *The macro-vectors* $\boldsymbol{\pi}_q, q \geq 0$, *satisfy the relations*

$$\boldsymbol{\pi}_q = \boldsymbol{\pi}_{q_1} \prod_{j=1}^{q_1-q} M_{q_1-j}, \, q = \overline{0, q_1 - 1}, \tag{18}$$

$$\boldsymbol{\pi}_q = \boldsymbol{\pi}_{q_1} R^{q-q_1}, \, q \geq q_1,$$

where matrices M_q *are of the form,*

$$M_0 = -Q_{2,1}Q_{1,0}^{-1}, \, M_q = -Q_{2,1}(M_{q-1}Q_{0,1} + Q_{1,1})^{-1}, \, q = \overline{1, q_2 - 2}, \tag{19}$$

$$M_{q_2-1} = -Q_{2,2}(M_{q_2-2}Q_{0,1} + Q_{1,2})^{-1}, \, M_{q_1-1} = -Q_{2,3}(M_{q_1-2}Q_{0,2} + Q_{1,4})^{-1},$$

$$M_q = -Q_{2,2}(M_{q-1}Q_{0,2} + Q_{1,3})^{-1}, \, q = \overline{q_2, q_1 - 2}.$$

The vector $\boldsymbol{\pi}_{q_1}$ *is a unique solution of the system of equations*

$$\boldsymbol{\pi}_{q_1}\left[\sum_{q=0}^{q_1-1}\prod_{j=1}^{q_1-q} M_{q_1-j} + (I - R)^{-1}\right]\mathbf{e} = 1, \tag{20}$$

$$\boldsymbol{\pi}_{q_1}(M_{q_1-1}Q_{0,3} + Q_{1,5} + RQ_{2,3}) = \mathbf{0}.$$

Matrix R *is a minimum non-negative solution of the matrix quadratic equation,*

$$R^2 Q_{2,3} + RQ_{1,5} + Q_{0,3} = 0. \tag{21}$$

Proof. The results for boundary blocks, if $q < q_1$, follows by recursive solution of the balance equation after routine block identification. For $q \geq q_1$ we have a matrix geometric solution as shown in [19].

Corollary 3. *Optimization problem consists in minimizing of the function*

$$g = c_0\bar{Q} + c_{11}\bar{U}_1 + c_2\bar{U}_2 + c_{12}\bar{B} \tag{22}$$

subject to $\bar{N} \leq \gamma$. *Here* $\bar{Q}$ *is the mean number of customers in the queue,*

$$\bar{Q} = \left[\sum_{q=0}^{q_1-1} q\boldsymbol{\pi}_q + \boldsymbol{\pi}_{q_1}(R + q_1(I - R))(I - R)^{-2}\right]\mathbf{e},$$

$\bar{U}_j = \mathbb{P}[D_j(t) = 1]$ *stands for the utilization of the Server* j,

$$\bar{U}_j = \left[\sum_{q=0}^{q_1-1}\boldsymbol{\pi}_q + \boldsymbol{\pi}_{q_1}(I - R)^{-1}\right](\mathbf{e}_1 1_{\{j=1\}} + (\mathbf{e}_3 + \mathbf{e}_5)1_{\{j=2\}} + \mathbf{e}_4),$$

$\bar{B} = \mathbb{P}[D_1(t) = 2]$ *is a failure index of the Server* 1,

$$\bar{B} = \left[\sum_{q=0}^{q_1-1}\boldsymbol{\pi}_q + \boldsymbol{\pi}_{q_1}(I - R)^{-1}\right](\mathbf{e}_2 + \mathbf{e}_5),$$

and $\bar{N}$ *is the mean number of customers in the system,* $\bar{N} = \bar{Q} + \bar{U}_1 + \bar{U}_2$.

6   Heuristic Solution for Optimal Thresholds

The Howard iteration algorithm allows us to evaluate numerically the optimal threshold levels. This method has weak spots connected with a necessity to solve the system of $6N$ linear equations for any fixed set of values of the system parameters. Here N specifies a truncation parameter to transform an infinite buffer system to a finite equivalent. A direct analytic minimization of the function $g(q_1, q_2)$ is infeasible and numerical evaluation requires evaluation of the corresponding matrices $M_q, q = \overline{0, q_1 - 1}$ for each pair (q_1, q_2).

Let us consider first the equivalent scheduling problem where is assumed that there are customers in the system and the problem consists in allocation between the servers until the system becomes empty with the aim of the total expected cost minimization. That means that in the original system the arrival rate λ must be set to be 0.

Theorem 5. *The optimal thresholds* $q_k^*, k \in \{1, 2\}$, *for the scheduling problem satisfy the relations,*

$$q_k^* = \max\left\{1, \left\lfloor \frac{1}{c_0 A_k}\left[\frac{c_2}{\mu_2} - \frac{c_{11}}{\mu_1} - c_0 B_k - c_{12} C_k\right]\right\rfloor\right\}, \text{ where} \tag{23}$$

$$A_1 = \frac{(\alpha + \beta)(\alpha + \mu_1 + \tau)}{\mu_1 \beta \tau}, \quad A_2 = A_1 + \frac{\alpha + \beta + \tau}{\beta \tau}, \tag{24}$$

$$B_1 = B_2 = \frac{\alpha(\alpha + \beta + \tau)}{\mu_1 \beta \tau}, \quad C_1 = \frac{\alpha(\alpha + \mu_1 + \tau)}{\mu_1 \beta \tau}, \quad C_2 = C_1 + \frac{\alpha + \tau}{\beta \tau}.$$

Proof. Let us consider the case $\lambda = 0$. Due to the threshold structure of the control policy f, we can calculate the total average cost $V(x)$ incurred by the customers presented in the system until it becomes empty given initial state x. Assuming the known values of (q_1, q_2) after some algebra we get for the state $x = (0, 0, 0)$,

$$V(S_0^q S_1 x) = \frac{c_{11}}{\mu_1} + c_0 q \frac{(\alpha + \beta)(\alpha + \mu_1 + \tau)}{\mu_1 \beta \tau} + c_0 \frac{\alpha(\alpha + \beta + \tau)}{\mu_1 \beta \tau}$$
$$+ c_{12} \frac{\alpha(\alpha + \mu_1 + \tau)}{\mu_1 \beta \tau} + V(S_0^{q-1} S_1 x),$$

$$V(S_0^{q+1} S_1^2 x) = \frac{c_{11}}{\mu_1} + c_0 q \left[\frac{(\alpha + \beta)(\alpha + \mu_1 + \tau)}{\mu_1 \beta \tau} + \frac{\alpha + \beta + \tau}{\beta \tau}\right] + c_0 \frac{\alpha(\alpha + \beta + \tau)}{\mu_1 \beta \tau}$$
$$+ c_{12} \left[\frac{\alpha(\alpha + \mu_1 + \tau)}{\mu_1 \beta \tau} + \frac{\alpha + \tau}{\tau \beta}\right] + V(S_0^{q-1} S_1 x).$$

By successive substitution of expressions for $v(S_0^q S_1 x)$ we obtain

$$V(S_0^q S_1 x) = \frac{c_{11}}{\mu_1} + c_0 \frac{q(q+1)}{2} \frac{(\alpha + \beta)(\alpha + \mu_1 + \tau)}{\mu_1 \beta \tau} \tag{25}$$
$$+ (q+1)\left[c_0 \frac{\alpha(\alpha + \beta + \tau)}{\mu_1 \beta \tau} + c_{12} \frac{\alpha(\alpha + \mu_1 + \tau)}{\mu_1 \beta \tau}\right]$$
$$= \frac{c_{11}}{\mu_1} + c_0 \frac{q(q+1)}{2} A_1 + (q+1)[c_0 B_1 + c_{12} C_1]$$

$$V(S_0^q S_1^2 x) = \frac{c_{11}}{\mu_1} + c_0 \frac{q(q+1)}{2} \left[\frac{(\alpha+\beta)(\alpha+\mu_1+\tau)}{\mu_1 \beta \tau} + \frac{\alpha+\beta+\tau}{\beta\tau} \right]$$
$$+ (q+1) \left[c_0 \frac{\alpha(\alpha+\beta+\tau)}{\mu_1 \beta \tau} + c_{12} \left[\frac{\alpha(\alpha+\mu_1+\tau)}{\mu_1 \beta \tau} + \frac{\alpha+\tau}{\tau\beta} \right] \right]$$
$$= \frac{c_{11}}{\mu_1} + c_0 \frac{q(q+1)}{2} A_2 + (q+1)[c_0 B_2 + c_{12} C_2].$$

When the orbit has reached the level q_1 or q_2, it becomes optimal to use the second server. Then the following inequalities must hold,

$$V(S_0^{q-1} S_1 S_2 x) = \frac{c_2}{\mu_2} + V(S_0^{q-1} S_1 x) \le V(S_0^q S_1 x)$$
$$V(S_0^{q-1} S_1^2 S_2 x) = \frac{c_2}{\mu_2} + V(S_0^{q-1} S_1 x) \le V(S_0^q S_1^2 x).$$

Solving the last two inequalities using the notations (24) for the expressions (25) we get the relations (23) for the optimal thresholds (q_1^*, q_2^*).

In [8] the heuristic solution was obtained for the ordinary $M/M/2$ queue. Substituting in this relation the mean service $\frac{1}{\mu_1}$ by the effective mean service time of the model under study A_k, taking into account the cost structure and form of the scheduling threshold levels, we can make the following conjecture.

Conjecture 1. The optimal threshold $q_k^*, k \in \{1, 2\}$, for the case $\lambda > 0$ can be calculated approximately by

$$q_k^* \approx \hat{q}_k^* = \max\left\{1, \left\lfloor \frac{1}{c_0 A_k} \left[\frac{c_2}{\mu_2} F_k - \frac{c_{11}}{\mu_1} - c_0 B_k - c_{12} C_k \right] \right\rfloor \right\}, \quad \text{where} \quad (26)$$
$$F_k = \frac{1 - \lambda A_k + \sqrt{(1 - \lambda A_k)^2 + 4\lambda \mu_2 A_k^2}}{2} \tag{27}$$

Statement 3. The threshold levels defined by (23) satisfy the inequalities

$$\frac{1}{c_0 A_k} \left[\frac{c_2}{\mu_2} (1 - \lambda A_k) - \frac{c_{11}}{\mu_1} - c_0 B_k - c_{12} C_k \right] \le \hat{q}_k \tag{28}$$
$$\le \frac{1}{c_0 A_k} \left[\frac{c_2}{\mu_2} - \frac{c_{11}}{\mu_1} - c_0 B_k - c_{12} C_k \right], \quad \text{for } \mu_2 \le \frac{1}{A_k}$$

Proof. The left inequality of (28) follows directly from

$$F_k \ge \frac{1 - \lambda A_k + \sqrt{(1 - \lambda A_k)^2}}{2} = 1 - \lambda A_k.$$

To prove the inequality at the right hand side we show that $F_k \le 1$. By solving this inequality using simple algebraic manipulations we get $\mu_2 \le A_k^{-1}$, where A_k represents the mean effective service time of the customer.

Statement 4. Because of the restricted volume space we have skipped numerical examples and make here only two general observations:

1. The optimal threshold policy can be up to 25 % superior in performance compared with policies like Fastest Free Server or Random Server Selection.
2. The difference between the real and heuristic policies does not exceed 1.5 % of the performance value.

7   Conclusion

In this paper we have studied a dynamic allocation problem for a two-server heterogeneous queueing system. The more productive server is unreliable and subject to breakdowns whereas the less productive is absolutely reliable. Under some assumption about the costs it was proved that the optimality of a threshold control policy depends on the state of the unreliable server. We expect that this result can be generalized to the multi-server retrial queue where all servers are unreliable and to the limited case where retrial intensity tends to infinity. But in this case, as for the ordinary $M/M/K$ queue, the optimal threshold levels may depend also on the states of slower servers although this influence is very negligible.

References

1. Altman, E.: Constrained Markov Decision Processes. Chapman and Hall, London (1999)
2. Armony, M., Ward, A.R.: Fair dynamic routing in large-scale heterogeneous-server systems. Oper. Res. **58**(3), 624–637 (2010)
3. Aviv, Y., Federgruen, A.: The value-iteration method for countable state Markov decision processes. Oper. Res. Lett. **24**(5), 223–234 (1999)
4. Beutler, F.J., Ross, K.W.: Optimal policies for controlled Markov chains with a constraint. J. Math. Anal. Appl. **112**, 236–252 (1985)
5. Chakka, R., Do, T.V.: The $MM \sum_{k=1}^{K} CPP_k/GE/c/L$ G-queue with heterogeneous servers: steady state solution and application to performance evaluation. Perform. Eval. **64**, 191–209 (2007)
6. Ghoneim, H.A., Stidham, S.: Control of arrivals to two queues in series. Eur. J. Oper. Res. **21**, 399–409 (1985)
7. Efrosinin, D., Farhadov, M.: Performance analysis and monotone control of a tandem queueing system. DCCN Commun. Comput. Inform. Sci. **279**, 241–255 (2014)
8. Efrosinin, D., Rykov, V.: Heuristic solution for the optimal thresholds in a controllable multi-server heterogeneous queueing system without preemption. DCCN Commun. Comput. Inform. Sci. **601**, 238–252 (2016)
9. Efrosinin, D., Sztrik, J.: Performance analysis of a two-server heterogeneous retrial queue with threshold policy. Qual. Technol. Quant. Manage. **8**(3), 211–236 (2011)
10. Howard, R.: Dynamic Programming and Markov Processes. Wiley, New York (1960)
11. Koole, G.: A simple proof of the optimality of a threshold policy in a two-server queueing system. Syst. Control Lett. **26**, 301–303 (1995)
12. Koole, G.: Convexity in tandem queues. Prob. Eng. Inf. Sci. **18**(1), 13–31 (2004)
13. Larsen, R.L., Agrawala, A.K.: Control of a heterogeneous two-server exponential queueing system. IEEE Trans. Software Eng. **9**(4), 522–526 (1983)

14. Lehtonen, T.: Stochastic comparisons for many server queues with non-homogeneous exponential servers. Opsearch **20**(1), 1–15 (1983)
15. Liang, H.M., Kulkarni, V.G.: Optimal routing control in retrial queues. In: Shanthikumar, J.G., et al. (eds.) Applied Probability and Stochastic Processes, pp. 203–218. Kluwer Academic Publishers, Boston (1999). Chap. 14
16. Lin, W., Kumar, P.R.: Optimal control of a queueing system with two heterogeneous servers. IEEE Trans. Autom. Control **29**, 696–703 (1984)
17. Legros, B., Jouini O.: Routing in a queueing system with two heterogeneous servers in speed and in quality of resolution (2016). http://www.lgi.ecp.fr/~jouini
18. Luh, H.P., Viniotis, I.: Threshold control policies for heterogeneous server systems. Math. Methods Oper. Res. **55**, 121–142 (2002)
19. Neuts, M.F.: Matrix-geometric solutions in stochastic models. The John Hopkins University Press, Baltimore (1981)
20. Nobel, R., Tijms, H.C.: Optimal control of a queueing system with heterogeneous servers and set-up costs. IEEE Trans. Autom. Control **45**(4), 780–784 (2000)
21. Rosberg, Z., Makowski, A.M.: Optimal routing to parallel heterogeneous servers - small arrival rates. Trans. Autom. Control **35**(7), 789–796 (1990)
22. Rykov, V.: Monotone control of queueing systems with heterogeneous servers. QUESTA **37**, 391–403 (2001)
23. Rykov, V., Efrosinin, D.: On the slow server problem. Autom. Remote Control **70**(12), 2013–2023 (2009)
24. Stecke, K., Kim, I.: Performance evaluation for systems of pooled machines of unequal sizes: Unbalancing vs. blocking. European Journal of. Oper. Res. **42**, 22–38 (1989)
25. Özkan, E., Kharoufeh, J.P.: Optimal control of a two-server queueing system with failures. Probab. Eng. Inform. Sci. **28**(4), 489–527 (2014)
26. Puterman, M.L.: Markov Decision Process: Discrete Stochastic Dynamic Programming. Wiley, New York (1994)
27. Sennott, L.I.: Stochastic Dynamic Programming and the Control of Queueing Systems. Wiley, New York (1999)
28. Sztrik, J., Roszik, J.: Performance analysis of finite-source retrial queueing systems with nonreliable heterogeneous servers. J. Math. Sci. **146**(4), 6033–6038 (2007)
29. Tijms, H.C.: Stochastic Models. An Algorithmic Approach. Wiley, New York (1994)
30. Trancoso, P.: One size does not fit all: a case for heterogeneous multiprocessor systems. In: Proceedings of the IADIS International Conference Applied Computing, Algarve, Portugal (2005)
31. Veatch, M.N., Wein, L.M.: Monotone control of queueing networks. Queueing Syst. **12**, 391–408 (1992)
32. Viniotis, I., Ephremides, A.: Extension of the optimality of a threshold policy in heterogeneous multi-server queueing systems. IEEE Trans. Autom. Control **33**, 104–109 (1988)
33. Wang, W., Abouzeid, A.A.: Throughput of hybrid radio-frequency and free-space-optical (RF/FSO) multi-hop networks. In: Information Theory and Applications Workshop, USA, pp. 1–8, (2007)
34. Weber, R.: On a conjecture about assigning jobs to processors of different speeds. IEEE Trans. Autom. Control **38**, 166–170 (1993)
35. Yang, R., Bhulai, S.: Mei., R.: Structural Properties of the Optimal Resource Allocation Policy for Single-Queue Systems. Ann. Oper. Res. **202**, 211–233 (2013)

How Does a Queuing Network React to a Change of Different Flow Control Parameters?

Mais Farkhadov[(✉)], Nina Petukhova, Alexander Abramenkov,
and Olga Blinova

V.A. Trapeznikov Institute of Control Sciences of RAS, Moscow, Russia
mais.farhadov@gmail.com

Abstract. We investigate the simulation methods for a multiphase queuing network with non-Poisson flows. Specifically, we study how the network internal flows react to a change of the network flow control parameters. The parameters are: the number of phases, system load, queue capacity, time interval distribution between applications. Also, we investigate how the internal flows react on different routing rules in the system. All these simulations are compared with the analytical results, obtained under the assumption that the flows are either Poisson or Erlang. Finally, we demonstrate when it is justifiable to use Poisson approximation for the input flows for these kind of systems.

Keywords: Networks with non-poisson flows · Queuing networks with restrictions · Serial connection · Simulations · The reliability characteristics of systems · Flow control

1 Introduction

As information networks (IN) become widespread and complex, society demands more efficient designs and more useful functionality indicators for the networks. Queuing Theory (QT) is often used to describe how information networks work. Modern networks require new, better, and more adequate models of real-life objects. Classic QT formulas were developed for simplest Poisson input flows, and it appears that for many problems this approach works. However, a lot of research works indicate that in many modern real-life systems, a model behaves very differently from its real life counterpart as a Poisson flow is substituted for a real one.

In 1993 the group of researchers W. Leland, M. Taqqu, W. Willinger, and D. Wilson published the paper "On the SelfSimilar Nature of Ethernet Traffic" [1]. It was discovered that conventional approximation for elementary flows in IN leads to incorrect results. It appears that on a large scale the network traffic is self-similar. In other words, these flows are already of a completely different structure than in the classical Teletraffic theory. Self-similar telecommunication

© Springer International Publishing Switzerland 2016
A. Dudin et al. (Eds.): ITMM 2016, CCIS 638, pp. 73–82, 2016.
DOI: 10.1007/978-3-319-44615-8_6

network traffic is associated with a widespread use of packet switching rather than circuit-switching, is often used in telephone networks [2].

In addition to the type of switching, there is a number of parameters that affect the flow distribution shape. These parameters include, for example, a loss of orders due to queue overflow, the rules of request routing, and so on.

A large number of works has been devoted to construct and study queuing models with non-Poisson input flow. In those works, a number of methods that beget adequate results has been proposed [3–5]. However, there are problems that require a systematic approach:

- different models are built by different methods to obtain adequate results for specific tasks. There is no algorithm that allows to the choice of the optimal method. Existing models are diverse, each new model requires a lot of computation work. There is no general classification and structuring of models and methods to study IN;
- practically, there is no sufficiently developed general theoretical framework to model systems with traffic other than a simple flow. QT requires generalization to a larger number of tasks that meet modern needs [6]. There is no single algorithm to calculate the IN quality indicators; also, a unified method to assess occurring errors is missing;
- there is no unified framework to study the mutual influence of flow distributions and system operational parameters. The emergence of self-similar traffic and the deviation of the real application flow from the simplest one are often due to the structure of the network and the rules of the network's construction. Even as a simple Poisson application flow enters the system, the flow shape is gradually distorted under the influence of various parameters of the network.

It becomes obvious that we need a comprehensive study to systematize the already obtained results in how the application flow and the network performance parameters depend on the network setup. Our report presents an overview of the works on the subject. We demonstrate the results of a simulation model of a queuing system; the simulation was conducted by means of a software package we developed specifically for this project. Also, we show that our model can be flexibly adjusted for different network setups to study how the network parameters and the flow distributions are related; our software is also capable of comparing simulation results with analytical ones.

Queuing networks are often taken as models of real life processes. However, it is important that those models adequately represent the original processes. For many problems it is sufficient to select a Poisson-distributed input flow. Internal flows circulating inside queuing networks are of particular interest in case of serially connected queuing systems (QS), that is when the output flow of one QS is the input flow of another QS, or when applications are rerouted. It is known that a QS network with exponential service time without restriction on queue length has all its internal flows to remain Poisson. For a QS with restrictions, the output flow is distorted, and the next QS in line receives a modified flow [7–9].

Relatively few works are devoted to QSs with losses or with different application flows. However, there are some papers that are of interest for this study [4,10–12].

In [10] a multiphase model of a QS with restrictions on queue length is investigated. The work is dedicated to find the average number of lost applications for each phase of service. The formulas to calculate that number are obtained on the assumption of Poisson flow. In [11] a model of á two-phase QS is constructed; the formulas for the probability distribution of system states and the probability of loosing an application are found for that model.

2   Multiphase Systems with Constraints

To study flows in a multiphase queuing network, let us take the system that was investigated in the work [10] as an example. That system was a 5-phase queuing system that has a single channel in each phase. The time it takes the system to service a request in the i-th phase is a random variable, distributed exponentially with parameter μ_i. The system is fed a stationary Poisson application flow with parameter λ. If an application finds the system busy during a particular phase, the application is considered lost in that phase.

For this study we take a 5-phase queuing network with a limited queue. Analytical calculations of the number of lost applications were made with the assumption that the internal flows are Poisson. We have built a simulation model to compare the results. The table shows the results obtained analytically and by means of a simulation for each phase and varying intensity of the input stream. The difference in the results obtained is noticeable - so internal flows are significantly different from the Poisson distribution.

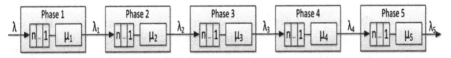

The authors of [10] built a simulation model that allowed them to compute the average number of applications lost during every phase. Table 1 shows the results. The column "Analytics" indicates the results obtained from the analytical formulas derived in [10]:

$$z_1 = \frac{\rho}{1+\rho}, \quad z_2 = \frac{\rho}{2(1+\rho)^2}, \quad z_3 = \frac{\rho(3+\rho)}{8(1+\rho)^3}$$
$$z_4 = \frac{\rho(\rho^2+4\rho+5)}{16(1+\rho)^4}, \quad z_5 = \frac{\rho(5\rho^3+25\rho^2+47\rho+35)}{128(1+\rho)^5}$$

where $z_1, z_2,..., z_n$ are the average fractions of the applications that are lost during the 1-, 2 - ,..., n-th phase accordingly. $\rho = \frac{\mu}{\lambda}$, $\mu = \mu_1 = \mu_2 = \mu_3 = \mu_4 = \mu_5 = 1$.

The difference is noticeable even at low intensity of the input flow; the greater the intensity, the greater the difference. As the input flow intensity in the phases 4–5 becomes high, the results differ by more than a factor of 2.

Table 1. The fraction of lost applications in each phase vs. incoming flow

Incoming flow	Phase number (analytics)					Phase number (simulation)				
	1	2	3	4	5	1	2	3	4	5
0.1	0.091	0.041	0.029	0.023	0.019	0.092	0.046	0.033	0.027	0.023
0.2	0.167	0.069	0.046	0.035	0.029	0.166	0.083	0.061	0.049	0.042
0.3	0.231	0.089	0.056	0.041	0.032	0.231	0.115	0.083	0.066	0.056
0.4	0.286	0.102	0.062	0.044	0.034	0.286	0.143	0.101	0.08	0.067
0.5	0.333	0.111	0.065	0.045	0.034	0.333	0.166	0.117	0.091	0.076
0.6	0.375	0.117	0.066	0.044	0.033	0.375	0.188	0.13	0.1	0.083
0.7	0.412	0.121	0.066	0.043	0.032	0.412	0.206	0.141	0.108	0.088
0.8	0.444	0.123	0.065	0.042	0.03	0.445	0.222	0.151	0.115	0.93
0.9	0.474	0.125	0.064	0.041	0.029	0.474	0.237	0.159	0.12	0.097

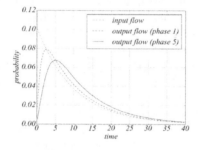

Fig. 1. Internal flow distributions in the system under a moderate load

Obviously, the analytical expressions derived under the assumption that the input flow is Poisson are of limited accuracy for multiphase QSs with losses.

Let us compare how the intervals between the output flow of 1st and 5th phase and the Poisson-distributed input flow are distributed (Figs. 1 and 2). The service intensity in both cases is equal to 1, the restriction on the queue length is 0. This means that if the device is busy, then an incoming application is lost. Figure 1 describes the flow circulating in the network with a low load for the input stream 0.1. Figure 2 does the same for the input flow intensity increased to 0.9.

It is clear that the profile of the flow probability distribution differs significantly from a Poisson one. Moreover, the difference increases as the number of phases the system passes increases; also the difference increases and the load increases.

Let us see what happens when we increase the queue's capacity. Figures 3 and 4 show how the circulating internal flows react to one more element in the queue.

Under a moderate system load, a small growth in the queue causes significant reduction of application losses. Also in this regime, the internal flows approach

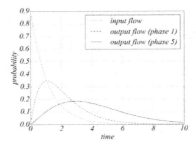

Fig. 2. Internal flow distributions in the system as the load increases

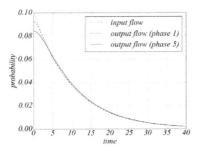

Fig. 3. Time interval distribution function between applications, the internal flow is 0.1

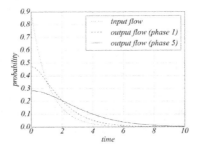

Fig. 4. Time interval distribution function between applications, the internal flow is 0.9

exponential distribution. As the system load increases, the difference between analytical and simulation results remains very significant. The former flow can be approximated by a Poisson flow, the latter flow bears similarity neither with Poisson nor with Erlang flows.

3   Flow Approximations

We demonstrated that the flows may differ significantly from Poisson. Let us see how significant the difference is for our system's performance. For example, we take the average residence time of an application in a queue with exponential service and maintenance intensity $\mu = 1.0$. Next, we replace the flow obtained from the simulation with the Poisson flow that has the same expectation value (Fig. 5).

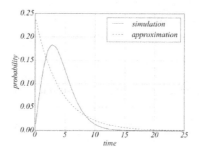

Fig. 5. Approximation of the flow by a Poisson one

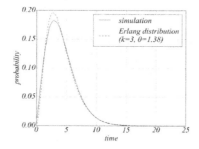

Fig. 6. Approximation of the simulated flow by a 3rd order Erlang flow

The mean value of the time between applications of the input flow is 4.14. Our simulation gave the following value of the mean residence time of an application in our system: $t = 1.098$.

The analytical expression for the mean residence time of an application in a $M/M/1$ system is $t = \frac{1}{\mu-\lambda}$. The input flow is $\lambda = 1/4.14 = 0.242$. Thus, the mean residence time of an application in our system under Poisson approximation is $t = 1.318$.

It follows that making the assumption that the input flow is Poisson to solve applied problems is not always justified, as the error in this case may be very significant. However, the analytical results obtained in this paper can be used to estimate the upper limit of performance of a QS. A real life system will perform better, because a Poisson flow gives maximum dispersion of intervals between applications. Obviously, the Poisson approximation of a flow results in far too high a value of the average residence time of an application in the system.

Perhaps for some special cases it will be possible to select some of the well-studied distributions, which allows us to calculate the properties of a QS with minimal error. For example, the distribution obtained in the first experiment is similar to an Erlang distribution. Figure 6 shows how the Erlang distribution works to approximate the output flow from the fifth phase of the previously considered case (the intensity of the input is 0.9).

4   Role of Routing

Another commonly used method of flow control inside a queuing network is routing of applications. Let us see how the internal flows of a QS change, if different routing rules are applied to the network. We take a simple QS that consists of two parallel-connected QS with exponentially distributed service time and intensity of service $\mu_1 = \mu_2 = 1.0$.

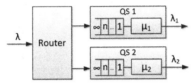

If a request is routed at random (with a given probability) among the queues, the output flow of every QS is Poisson; this fact allows us to use standard analytical formulas to analyze the system. In the case of a sequential routing algorithm, the output flow takes the form shown in Fig. 7 (the intensity of the incoming application flow is 0.1) and Fig. 8 (the intensity of the incoming application flow is 0.9).

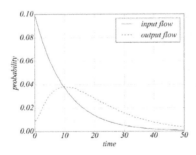

Fig. 7. Residence time distribution in case of sequential routing (the intensity of the input flow is 0.1)

The following routing rule is intended to minimize queues and load the devices as uniformly as possible; an application is assigned to the queue with minimal length. If any two queues are the same, then the application is assigned to the first one. Our simulation was carried out for two values of the flow intensities - 0.1 (Fig. 9) and 0.9 (Fig. 10).

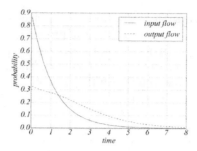

Fig. 8. Residence time distribution in case of sequential routing(the intensity of the input flow is 0.9)

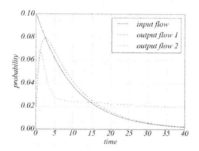

Fig. 9. Residence time distribution in case of minimal queue routing (the input flow intensity is 0.1)

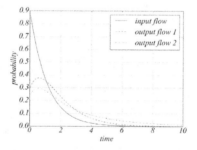

Fig. 10. Residence time distribution in case of minimal queue routing (the input flow intensity is 0.9)

5   Conclusion

This work shows that for a QS with an incoming Poisson flow, the internal flows remain the same only if the queues are infinite and the flows are not routed. If there is a queue with restrictions or a flow is divided between multiple devices, the flows actually coming to each QS differ from Poisson. If this fact is ignored and analytical formulas for the elementary streams are used without caution, significant errors are inevitable as one tries to predict how the QS performs.

Sometimes it is easy to guess a distribution to approximate and compute the characteristics of a QS: some real life flows are well approximated by an Erlang flow (Fig. 6); in some cases, approximation by a Poisson flow gives a small error. But there are times when it is difficult to choose a distribution to approximate a flow, and analytical formulas for a QS exist only for some well-studied distributions. There are no analytical methods to determine the shape and characteristics of an output flow depending on the characteristics of an incoming flow. It is time-consuming and difficult to build a simulation model for each specific problem. Perhaps statistical information collected for further processing will allow us to learn more about the laws governing internal flows in QSs.

A piece of software, written while working on this paper, allows us to quickly build simulations of different QSs, including QSs with complex topology and routing rules. These features allow us to use the program as a tool to study the distributions of flows depending on input data and error estimations, which are obtained analytically. The ability to build a large number of different simulation models allows us to collect statistics to study flows, that are different from a Poisson one, circulating in QSs. It is possible to trace the dynamics of a flow by varying a particular parameter to estimate the influence of the parameter on the shape of the distribution of the flow. These data, combined with a modern research methodology, could help us to identify laws governing QS's flows and their characteristics.

References

1. Walter, W., et al.: Self-similarity in high-speed packet traffic: analysis and modeling of ethernet traffic measurements. Stat. Sci. **10**(1), 6785 (1995)
2. Kartashevskii, V.G.: Analysis of Probability-Time Characteristics of Nodes Poisson Processing Multimedia Traffic Multiservice Networks. Kand. diss, Samara (2014)
3. Petrov, V.V.: Teletraffic structure and algorithm to ensure service quality under the influence of self-similarity effect: dis. Moscow: [Mosk. energy resou. Inst (Tech. Univ)] (2005)
4. Abramov, P.B., Lenshin, A.V.: Estimation of the parameters of a queuing system as the service discipline is approximated with Erlang flows. Russ. Inter. Ministry Rep. **2**, 13–18 (2012)
5. Ageev, D.V.: Method for determining the parameters of flow in different parts of the multiservice telecommunication networks, taking into account the effect of self-similarity. Probl. Telecommun. **3**(5), 18–37 (2011). Ageev, D.V., Ignatenko, A.A., Kopylov, A.N
6. Zadorozhnyi, V.N.: Simulation modeling of fractal queues. In: Dynamics of Systems, Mechanisms and Machines (Dynamics), pp. 1–4, December 2014. doi:10. 1109/Dynamics.2014.7005703
7. Nazarov, A.A., Nosova, M.G.: Multiphase autonomous queue system and its application to demographics problems. Tomsk Polytech. Univ. Rep. **315**(5), 183–186 (2009)
8. Popov, A.V.: Probabilistic modeling of logistics. Radio-Electron. Comput. Syst. **1**, 144–151 (2012). Popov, A.V., Obresanova, E.R., Sinebriukova, E.Y
9. Bromberg, M.A.: Multi-phase systems with losses with exponential servicing. Avtomatika i Telemekhanika (10), 27–31 (1979)

10. Efanova, T.L., Kochetkov, E.S.: Multiphase queue systems with losses. Autom. Telemech. **35**(10), 32–34 (1974)
11. Viskova, E.V.: Two phase queue system with a Markov flow serviced in discrete time. Russ. Univ. Friendship Nat. **5**(3), 247–257 (2005)
12. Abramov, P.B., Lenshin, A.V.: Estimation of the parameters of multichannel queue systems with memory about serviced application taken into account. Russ. Inter. Ministry Rep. **2**, 130–135 (2013)

Quasi-Geometric, Gamma and Gaussian Approximations for Multiserver Retrial Queueing Systems

Ekaterina Fedorova[(✉)]

Tomsk State University, Tomsk, Russia
moiskate@mail.ru

Abstract. In the paper, methods of quasi-geometric, gamma and Gaussian approximation of the probability distribution of the calls number in the orbit for multiserver retrial queueing systems are proposed. A description and analysis of the application area of each method for retrial queueing system $M|M|N$ are given. In addition, the results of approximations are compared and a table of decision making on the choice of the approximation type has been composed.

Keywords: Multiserver retrial queue · Number of calls in the orbit · Gamma approximation · Quasi-geometric approximation · Gaussian approximation

1 Introduction

In queueing theory, there are two classes of queueing systems: systems with queue and loss systems. In real systems, there are situations when a queue cannot be explicitly identified, but also we cannot say that calls are lost if they come when the service device is unavailable. Usually, a primary call does not refuse service and performs repeated calls to get the desired service in random time intervals. Examples of these situations are telecommunication systems. Thus a new class of queueing systems has appeared: systems with a source of repeated calls or retrial queueing systems.

The retrial queueing systems model is a class of queueing systems which appeared because of practical problems in telecommunication systems [1–6]. The main feature of this model is repeated calls which perform attempts to get the service in random time intervals.

The most comprehensive description, detailed comparison of classical queueing systems and retrial queues and main results in this scientific area are found in books by J.R. Artalejo, A. Gomez-Corral, G.I. Falin and J.G.C. Templeton [7–9]. But analytical results are obtained only in cases of simple input and service processes (e.g., Poisson arrivals or the exponential distribution of service law and only for $N = 2$ servers) [8]. The majority of studies of retrial queues are performed numerically or via computer simulation [10, 11] and matrix

© Springer International Publishing Switzerland 2016
A. Dudin et al. (Eds.): ITMM 2016, CCIS 638, pp. 83–93, 2016.
DOI: 10.1007/978-3-319-44615-8_7

methods [12,13]. Asymptotic and approximate methods were developed by G.I. Falin [14], J.R. Artalejo [15], V.V. Anisimov [16] and others [17,18].

In a number of our previous papers devoted to the study of various single-server retrial queues [19,20], we proposed the asymptotic analysis method for retrial queueing systems under a heavy load condition and quasi-geometric and gamma approximation methods.

In this paper we generalize our results for multiserver retrial queueing systems. Furthermore, we add the method of Gaussian approximation which allows to extend the range of applicability of the results.

2   Mathematical Model

Let us consider a retrial queueing system of the $M|M|N$ type. The structure of the system is presented in Fig. 1.

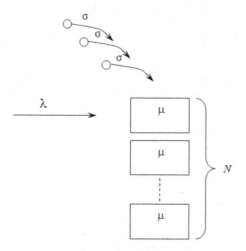

Fig. 1. Retrial queueing system $M|M|N$

The input process is a Poisson arrival process with rate λ. There are N service devices (servers). The service time of each call is exponentially distributed with rate μ. If a call arrives when there is a free server, the call occupies it for the service. If all servers are busy, the call goes to the orbit. There, the call stays during a random time distributed by exponential law with parameter σ. Then the call from the orbit makes an attempt to reach the server again. If there is a free server, the call occupies it, otherwise the call instantly returns to the orbit.

Let $i(t)$ be the random process described by the number of calls in the orbit and $k(t)$ be the random process which defines the server state as follows:

$$
k(t) = \begin{cases}
0, & \text{if all servers are free at the moment } t, \\
1, & \text{if 1 server is busy at the moment } t, \\
2, & \text{if 2 servers are busy at the moment } t, \\
\ldots \\
N, & \text{if all servers are busy at the moment } t.
\end{cases}
$$

The problem is to find the probability distribution of the number of calls in the orbit.

The process $i(t)$ is not Markovian, so we consider the two-dimensional process $\{k(t), i(t)\}$ which is a continuous time Markov chain.

We denote the probability that the device is in state k and there are i calls in the orbit at the time moment t by $P(k, i, t) = P\{k(t) = k, i(t) = i\}$. So, the following direct system of Kolmogorov differential equations for the system states probability distribution $P(k, i, t)$ can be written for $i \geq 0$:

$$
\begin{cases}
\dfrac{\partial P(0, i, t)}{\partial t} = -(\lambda + i\sigma)P(0, i, t) + \mu P(1, i, t), \\
\dfrac{\partial P(k, i, t)}{\partial t} = -(\lambda + k\mu + i\sigma)P(k, i, t) + \lambda P(k - 1, i, t) \\
+(i + 1)\sigma P(k - 1, i + 1, t) + (k + 1)\mu P(k + 1, i, t), \text{ for } k = \overline{1, N - 1}, \\
\dfrac{\partial P(N, i, t)}{\partial t} = -(\lambda + N\mu)P(N, i, t) + \lambda P(N - 1, i, t) \\
+(i + 1)\sigma P(N - 1, i + 1, t) + \lambda P(N, i - 1, t).
\end{cases} \tag{1}
$$

We denote $P(k, i) = \lim\limits_{t \to \infty} P(k, i, t)$. Then in stationary state, the system (1) has the following form:

$$
\begin{cases}
-(\lambda + i\sigma)P(0, i) + \mu P(1, i) = 0, \\
-(\lambda + k\mu + i\sigma)P(k, i) + \lambda P(k - 1, i, t) + (i + 1)\sigma P(k - 1, i + 1) \\
+(k + 1)\mu P(k + 1, i) = 0, \text{ for } k = \overline{1, N - 1}, \\
-(\lambda + N\mu)P(N, i) + \lambda P(N - 1, i) + (i + 1)\sigma P(N - 1, i + 1) \\
+\lambda P(N, i - 1) = 0.
\end{cases} \tag{2}
$$

Analytical results for considered model are not known in the literature. But the system (2) can be easily solved by a numerical algorithm.

Obviously, the results of any numerical algorithm depend on the dimension of solved systems and values of parameters. Furthermore, numerical algorithms are applied for truncated model (for $i \leq I_{max}$). Moreover, for more complex retrial queues (e.g., not with Poisson input process) the use of numerical algorithms is more difficult or impossible.

In this regard, in the paper, we propose methods of approximation of the probability distribution of the calls number in the orbit (gamma, quasi-geometric and Gaussian approximations).

3   Gamma Approximation

In this section, we offer to approximate the probability distribution of the number of calls in the orbit $P_i = \sum\limits_{k=0}^{\infty} P(k, i)$ by discrete analogue (defined below) of the gamma distribution G_i with a shape parameter α and an inverse scale parameter β.

The method of gamma approximation consists in approximation of the probability distribution P_i by the discrete analogue of the gamma distribution G_i whose parameters are calculated via equating means and variances of distributions P_i and G_i. So parameters α and β are the following

$$\alpha = \frac{E\{i(t)\}}{\text{var}\{i(t)\}}, \quad \beta = \frac{E^2\{i(t)\}}{\text{var}\{i(t)\}}$$

where $E\{i(t)\}$ is a mean and $\text{var}\{i(t)\}$ is a variance of the probability distribution P_i of the number of calls in the orbit or their estimates obtained statistically in practical problems.

Many ways can be offered for getting discrete distributions from the gamma distribution. In particular, they are the following:

1. $G_1(i) = c_1 f(i)$ where c_1 is a normalizing constant, and $f(i)$ is the density of the gamma distribution at point i.
2. $G_2(i) = F(i+1) - F(i)$ where $F(i)$ is function of the gamma distribution at point i.

We choose the second way for the calculation of G_i.

Let us compare the probability distribution of the number of calls in the orbit P_i calculated by a numerical algorithm and its gamma approximation G_i for different values of the system parameters.

Lets there be $N = 3$ servers in the system and the service rate $\mu = 1$. The rate of input process is equal to $\lambda = \rho N \mu$ where variable ρ is the system load. We analyse the results of approximation depending on the parameter of load ρ and delay rate σ.

In Table 1, we show the Kolmogorov distance [21] between approximate and exact distributions calculated as follows:

$$\Delta = \max_{0 \leq i \leq I_{max}} \left| \sum_{n=0}^{i} G_n - \sum_{n=0}^{i} P_n \right|.$$

In Table 1, the note "–" means that the numerical algorithm does not work.

Comparing the values in Table 1 with the results of the approximation in the singleserver system [20], we expect that the approximation accuracy increases with number of servers N.

In Table 2 we show the Kolmogorov distance [21] between approximate and exact distributions for $N = 10$:

In Table 2 the note 0 means that $\Delta \leq 10^{-4}$.

It is clear that the results of the approximation in the system with $N = 10$ servers is better than in the previous example for most values of system parameters, so we can conclude that the approximation methods became more accurate when the number of servers N grows.

Table 1. Kolmogorov distance between the gamma approximation and the exact distribution for $N = 3$

Values of the system rates	$\sigma = 0.01$	$\sigma = 0.1$	$\sigma = 0.5$	$\sigma = 1$	$\sigma = 2$	$\sigma = 10$
$\rho = 0.1$	0.066	0.007	0.002	0.001	0.006	0.001
$\rho = 0.3$	0.096	0.188	0.066	0.107	0.028	0.016
$\rho = 0.5$	0.031	0.100	0.162	0.129	0.098	0.063
$\rho = 0.7$	0.012	0.038	0.075	0.088	0.094	0.077
$\rho = 0.9$	0.064	0.010	0.021	0.027	0.032	0.039
$\rho = 0.95$	–	0.016	0.013	0.014	0.019	0.035

Table 2. Kolmogorov distance between the gamma approximation and the exact distribution for $N = 10$

Values of the system rates	$\sigma = 0.01$	$\sigma = 0.1$	$\sigma = 0.5$	$\sigma = 1$	$\sigma = 2$	$\sigma = 10$
$\rho = 0.1$	0	0	0	0	0	0
$\rho = 0.3$	0.011	0.014	0.003	0.001	0.001	0
$\rho = 0.5$	0.055	0.117	0.074	0.045	0.028	0.013
$\rho = 0.7$	0.015	0.046	0.097	0.102	0.085	0.049
$\rho = 0.9$	–	0.037	0.023	0.030	0.035	0.046
$\rho = 0.95$	–	0.048	0.012	0.022	0.025	0.060

Table 3. The range of the applicability of the gamma approximation method for $N = 10$

Values of the retrial rate σ	Values of the load rate ρ
$\sigma = 0.01$	$\rho \leq 0.7$
$\sigma = 0.1$	$\rho < 0.4$ or $\rho \geq 0.7$
$\sigma = 0.5$	$\rho < 0.5$ or $\rho \geq 0.8$
$\sigma = 1$	$\rho \leq 0.5$ or $\rho \geq 0.9$
$\sigma = 2$	$\rho \leq 0.6$ or $\rho \geq 0.9$
$\sigma = 10$	$\rho \leq 0.9$

We assume that the criterion of applicability of methods is the following inequality holding: $\Delta \leq 0.05$. Having considered more numerical examples we defined the range of the method applicability (Table 3).

In some numerical examples, the probability distribution of the number of calls in the orbit was similar to geometric or Gaussian ones (Fig. 2).

So, in the following sections we suggest applying quasi-geometric and Gaussian approximations of P_i for results improving.

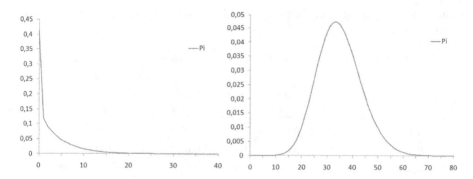

Fig. 2. The probability distribution of the number of calls in the orbit is geometric or Gaussian

4   Quasi-Geometric Approximation

The method of quasi-geometric approximation of the probability distribution of the number of calls in the orbit consists in the approximation of the probability distribution P_i by the quasi-geometric distribution Qg_i [22] whose parameters p_0 and δ are calculated through equating means and variances of distributions P_i and Qg_i as follows

$$\delta = \frac{\mathrm{var}\{i(t)\}}{2\mathrm{E}\{i(t)\} + \mathrm{var}\{i(t)\}} \text{ and } p_0 = 1 - (1-\delta)\mathrm{E}\{i(t)\}. \tag{3}$$

Let us note that the value of p_0, defined by the formula (3) may be negative. In this case we assume $p_0 = 0$. Then quasi-geometric distribution is shifted.

We compare the probability distribution of the number of calls in the orbit P_i obtained numerically and quasi-geometric approximation Qg_i for different values of the system parameters. The results of comparison are presented in Tables 4 and 5.

In Table 6, the range of the method's applicability is demonstrated.

Table 4. Kolmogorov distance between the quasi-geometric approximation and the exact distribution for $N = 3$

Values of the system rates	$\sigma = 0.01$	$\sigma = 0.1$	$\sigma = 0.5$	$\sigma = 1$	$\sigma = 2$	$\sigma = 10$
$\rho = 0.1$	0.026	0.003	0.001	0	0	0
$\rho = 0.3$	0.678	0.044	0.027	0.017	0.012	0.007
$\rho = 0.5$	0.999	0.337	0.008	0.034	0.033	0.023
$\rho = 0.7$	0.999	0.845	0.175	0.208	0.078	0.015
$\rho = 0.9$	0.999	0.988	0.617	0.400	0.239	0.210
$\rho = 0.95$	–	0.996	0.701	0.514	0.371	0.185

Table 5. Kolmogorov distance between the quasi-geometric approximation and the exact distribution for $N = 10$

Values of the system rates	$\sigma = 0.01$	$\sigma = 0.1$	$\sigma = 0.5$	$\sigma = 1$	$\sigma = 2$	$\sigma = 10$
$\rho = 0.1$	0	0	0	0	0	0
$\rho = 0.3$	0.041	0.006	0.001	0	0	0
$\rho = 0.5$	0.841	0.008	0.031	0.020	0.013	0.006
$\rho = 0.7$	0.999	0.636	0.161	0.023	0.016	0.017
$\rho = 0.9$	–	0.986	0.549	0.319	0.161	0.274
$\rho = 0.95$	–	0.995	0.692	0.505	0.308	0.153

Table 6. The range of the applicability of the quasi-geometric approximation method

Values of the retrial rate σ	Values of the load rate ρ
$\sigma = 0.01$	$\rho < 0.4$
$\sigma = 0.1$	$\rho \leq 0.5$
$\sigma = 0.5$	$\rho < 0.7$
$\sigma = 1$	$\rho \leq 0.7$
$\sigma = 2$	$\rho < 0.8$
$\sigma = 10$	$\rho \leq 0.8$

5 Gaussian Approximation

In the literature (e.g., [23]), it was shown that the probability distribution of the number of calls in the orbit is asymptotically Gaussian under the long delay condition. Thus we consider a Gaussian approximation.

The method of Gaussian approximation consists in approximation the probability distribution P_i by the discrete analogue of the normal distribution Gs_i whose parameters are calculated through equating means and variances of distributions P_i and Gs_i.

Table 7. Kolmogorov distance between the Gaussian approximation and the exact distribution for $N = 3$

Values of the system rates	$\sigma = 0.001$	$\sigma = 0.01$	$\sigma = 0.1$
$\rho = 0.1$	0.128	0.062	0.010
$\rho = 0.3$	0.014	0.055	0.229
$\rho = 0.5$	0.009	0.025	0.117
$\rho = 0.7$	–	0.017	0.059
$\rho = 0.9$	–	0.009	0.039
$\rho = 0.95$	–	–	0.018

Table 8. Kolmogorov distance between the Gaussian approximation and the exact distribution for $N = 10$

Values of the system rates	$\sigma = 0.001$	$\sigma = 0.01$	$\sigma = 0.1$
$\rho = 0.1$	0	0	0
$\rho = 0.3$	0.099	0.155	0.020
$\rho = 0.5$	0.016	0.050	0.028
$\rho = 0.7$	–	0.022	0.094
$\rho = 0.9$	–	–	0.042
$\rho = 0.95$	–	–	0.017

Table 9. The range of the applicability of the Gaussian approximation method

Values of the retrial rate σ	Values of the load rate ρ
$\sigma = 0.001$	$\rho < 0.2$ or $\rho \geq 0.5$
$\sigma = 0.01$	$\rho \leq 0.1$ or $\rho \geq 0.5$
$\sigma = 0.1$	$\rho \leq 0.6$ or $\rho \geq 0.9$

For getting the discrete analogue of Gaussian distribution we use the first way of transformations described in Sect. 3.

In Tables 7 and 8, we show the Kolmogorov distance between the approximate and exact distributions for the delay rate $0.001 \leq \sigma \leq 0.1$.

In the Table 9, the range of the method applicability for $N = 10$ is demonstrated.

6   Decision Making About the Type of Applying Approximation

In Fig. 3, we demonstrate the comparison of the results of proposed methods: quasi-geometric and gamma approximations for different values of system parameters.

It is obvious that in case A in Fig. 3, the exact distribution is similar to a geometric one, so the quasi-geometric approximation can be applied and in case B it is better to apply gamma approximation.

From Tables 3, 6 and 9, the overall table of decision making about the type of applying approximation in multiserver retrial queues is composed (Table 10).

By comparing Table 10 with the same table for the singleserver system [20] we note that the accuracy of proposed approximation methods increases with the number of servers N.

Also it was shown in [20] that the proposed methods can be applied to singleserver systems with more complex models and the table of decision making about the type of applying approximation is the same.

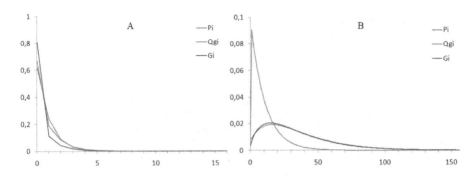

Fig. 3. Comparison of approximation methods

Table 10. Decision making about the type of approximation in use

Values of the load rate	$\sigma = 0.001$	$\sigma = 0.01$	$\sigma = 0.1$	$\sigma = 0.5$	$\sigma = 1$	$\sigma = 2$	$\sigma = 10$
$\rho = 0.1$	Qg_i	Qg_i	Qg_i	Qg_i	Qg_i	Qg_i	Qg_i
$\rho = 0.3$	Gs_i	G_i	Qg_i	Qg_i	Qg_i	Qg_i	Qg_i
$\rho = 0.5$	Gs_i	Gs_i	G_i	Qg_i	Qg_i	Qg_i	Qg_i
$\rho = 0.7$	Gs_i	Gs_i	G_i		Qg_i	Qg_i	Qg_i
$\rho = 0.9$	Gs_i	Gs_i	G_i	G_i	G_i	G_i	G_i
$\rho = 0.95$	Gs_i	Gs_i	Gs_i	G_i	G_i	G_i	G_i

Thus we can conclude that Table 10 can be used for any multiserver retrial queues: with Markovian arrival process (MAP), Batch Markovian arrival process (BMAP) or Semi-Markov arrivals, etc.

7 Conclusions

In the paper, methods of quasi-geometric, gamma and Gaussian approximations of the probability distribution of the calls number in the orbit for multiserver retrial queueing systems are proposed. Based on a study of the $M|M|N$ system, methods are described and obtained distributions are compared with exact ones. In this way conclusions about methods applicability are made. Finally table for decision making about the type of using approximation was composed. From Table 10, we conclude that the proposed approximations can be applied to almost all values of system parameters.

In addition, methods of quasi-geometric, gamma and Gaussian approximations can be applied to more complex multiservers retrial queueing systems (e.g., with MAP arrivals or general distribution function of service time). The main advantage of methods is that for their application we need to know only the mean and variance or their estimates statistically obtained in practical problems.

Acknowledgments. The reported study was funded by RFBR according to the research project No. 16-31-00292 mol-a.

References

1. Wilkinson, R.I.: Theories for toll traffic engineering in the USA. Bell Syst. Techn. J. **35**(2), 421–507 (1956)
2. Cohen, J.W.: Basic problems of telephone trafic and the influence of repeated calls. Philips Telecommun. Rev. **18**(2), 49–100 (1957)
3. Elldin, A., Lind, G.: Elementary Telephone Trafic Theory. Ericsson Public Telecommunications, Stockholm (1971)
4. Gosztony, G.: Repeated call attempts and their efect on trafic engineering. Budavox Telecommun. Rev. **2**, 16–26 (1976)
5. Kuznetsov, D.Y., Nazarov, A.A.: Analysis of non-Markovian models of communication networks with adaptive protocols of multiple random access. Avtomatika i Telemekhanika **5**, 124–146 (2001)
6. Nazarov, A.A., Tsoj, S.A.: Common approach to studies of Markov models for data transmission networks controlled by the static random multiple access protocols. Avtomatika i Vychislitel'naya Tekhnika **4**, 73–85 (2004)
7. Artalejo, J.R., Gómez-Corral, A.: Retrial Queueing Systems. A Computational Approach. Springer, New York (2008)
8. Falin, G.I., Templeton, J.G.C.: Retrial Queues. Chapman & Hall, London (1997)
9. Artalejo, J.R., Falin, G.I.: Standard and retrial queueing systems: a comparative analysis. Revista Matematica Complutense **15**, 101–129 (2002)
10. Neuts, M.F., Rao, B.M.: Numerical investigation of a multiserver retrial model. Queueing Syst. **7**(2), 169–189 (2002)
11. Ridder, A.: Fast simulation of retrial queues. In: Third Workshop on Rare Event Simulation and Related Combinatorial Optimization Problems, pp. 1–5, Pisa (2000)
12. Kim, C.S., Mushko, V.V., Dudin, A.: Computation of the steady state distribution for multi-server retrial queues with phase type service process. Ann. Oper. Res. **201**(1), 307–323 (2012)
13. Gómez-Corral, A.: A bibliographical guide to the analysis of retrial queues through matrix analytic techniques. Ann. Oper. Res. **141**, 163–191 (2006)
14. Falin, G.I.: Asymptotic properties of probability distribution of the number of request in system M/G/1/1 with repeated calls. VINITI, pp. 5418–5483 (In Russian) (1983)
15. Artalejo, J.R.: Information theoretic approximations for retrial queueing systems. In: Transactions of the 11th Prague Conference on Information Theory, Statistical Decision Functions and Random Processes, pp. 263–270. Kluwer Academic Publishers, Dordrecht (1992)
16. Anisimov, V.V.: Asymptotic analysis of highly reliable retrial systems with finite capacity. In: Queues, Flows, Systems, Networks: Proceedings of the International Conference Modern Mathematical Methods of Investigating the Telecommunication Networks, pp. 7–12, Minsk (1999)
17. Yang, T., Posner, M.J.M., Templeton, J.G.C., Li, H.: An approximation method for the M/G/1 retrial queue with general retrial times. Eur. J. Oper. Res. **76**, 552–562 (1994)
18. Diamond, J.E., Alfa, A.S.: Approximation method for M/PH/1 retrial queues with phase type inter-retrial times. Eur. J. Oper. Res. **113**, 620–631 (1999)

19. Moiseeva, E., Nazarov, A.: Asymptotic Analysis of RQ-systems M/M/1 on heavy load condition. In: Proceedings of the IV International Conference Problems of Cybernetics and Informatics, pp. 164–166, Baku, Azerbaijan (2012)

20. Fedorova, E.: Quasi-geometric and gamma approximation for retrial queueing systems. In: Dudin, A., Nazarov, A., Yakupov, R., Gortsev, A. (eds.) ITMM 2014. CCIS, vol. 487, pp. 123–136. Springer, Heidelberg (2014)

21. Kovalenko, I.N., Filippova, A.A.: Probability Theory and Mathematical Statistics. A Textbook. Vyschaya shkola, Moscow (1982). (In Russian)

22. Nazarov, A.A., Lyubina, T.V.: The non-Markov dynamic RQ system with the incoming MMP flow of requests. Autom. Remote Control **74**(7), 1132–1143 (2013)

23. Nazarov, A., Chernikova, Y.: Gaussian approximations of probabilities distribution of states of the retrial queueing system with r-persistent exclusion of alternative customers. In: Dudin, A., Nazarov, A., Yakupov, R. (eds.) Information Technologies and Mathematical Modelling - Queueing Theory and Applications. CCIS, vol. 564, pp. 200–208. Springer, Switzerland (2015)

A Methodology for the Identification of Extremal Loading in Data Flows in Information Systems

Andrey Gorshenin[1,2]([✉]) and Victor Korolev[1,3]

[1] Institute of Informatics Problems, Federal Research Center "Computer Science and Control" of the Russian Academy of Sciences, Moscow, Russia
agorshenin@frccsc.ru
[2] Moscow Technological University (MIREA), Moscow, Russia
[3] Faculty of Computational Mathematics and Cybernetics,
Lomonosov Moscow State University, Moscow, Russia
victoryukorolev@yandex.ru

Abstract. The paper presents two techniques for the identification of extremal loading via determination of special thresholds in order to distinguish between "normal" and "extreme" values in information data flows. Both algorithms are based on the Rényi limit theorem on rarefaction of renewal processes flows and the Pickands–Balkema–de Haan theorem on the asymptotic distribution for peaks over large thresholds. The methodology can be applied to various information systems. The two methods differ in the direction of threshold moving. The ascending algorithm increases the value of the threshold, while the descending one decreases it step-by-step. In addition, for the descending method we suggest a way to process the cumulative data. The key stages of both methods are represented by flowcharts. Some graphical results are demonstrated for test data generated by a special information system.

Keywords: Extreme values · Threshold · Peak over Threshold · Statistical data analysis · Pickands–Balkema–de Haan theorem · Rényi theorem · Probabilistic models

1 Introduction

The problem of finding an appropriate threshold is very important in many applied problems. First of all, this is so in the problems of identification of extremal loading in information data flows. These thresholds can be used for the classification of observations into the standard and dangerous elements, often with the help of the Peak over Threshold (POT) method [1].

Let us consider a marked point process $\{\tau_i, X_i;\ i = 1, 2, \ldots\}$, where $\tau_0 = 0$ and τ_i are random variables such that $\tau_i < \tau_{i+1}$, and X_i are nonnegative random variables, $i \geq 0$. This process can be interpreted as follows: τ_i is the time moment at which the value X_i was observed. Let us assume that the random

© Springer International Publishing Switzerland 2016
A. Dudin et al. (Eds.): ITMM 2016, CCIS 638, pp. 94–103, 2016.
DOI: 10.1007/978-3-319-44615-8_8

variables $X_1, X_2, \ldots$ are identically distributed with common distribution function $F(x) = \mathsf{P}(X_i < x)$, $x \geq 0$.

Traditionally the threshold for the identification of "extremal" observations among the X_i's is determined as some high-order quantile of the distribution function $F(x)$. Within such an approach the determination of the threshold is preceded by the convention that the fraction of "extremal" values among the X_i's should be equal to some given number γ. Hence, the threshold is determined as the $(1 - \gamma)$-quantile of the distribution function $F(x)$. However, the assignment of the particular value of γ is rather subjective. It is extremely desirable to have some more or less objective reasonable algorithm free from subjective conventions. An approach to the construction of such objective algorithms is proposed in this paper.

The algorithms described in the paper differ in the direction of threshold moving. The ascending algorithm increases the value of the threshold, while the descending one decreases it step-by-step. In addition, for the descending method we suggest a way to process the cumulative data. The proposed methodology for finding thresholds is based on the Rényi limit theorem on rarefaction of renewal processes flows and the Pickands–Balkema–de Haan theorem on the asymptotic distribution for peaks over large thresholds. It does not depend on any assumptions concerning the data structure, probability distributions, etc. The universality of methodology leads to the possibility of its correct application to various information systems. The theoretical backgrounds of this approach were described in [2–7].

2 Ascending Method

Figure 1 demonstrates the flowchart of the algorithm for the ascending method to find the threshold. First of all, the parameters

- `Data` (the initial data);
- `Name` (the part of title or filename for the graphical output);
- `step` (the value to modify current threshold level consecutively);
- `alpha` (the significance level of the Pearson's χ^2-test).

should be input by the user. The initial value of the parameter `lvl`, which is used for finding the threshold, equals 0. The auxiliary sample contains the differences between the time moments τ_i when `Data` exceeds the current value of `lvl`. Then, the hypothesis about the exponentiality of the auxiliary sample is tested with the help of Pearson's χ^2-test at the significance level `alpha`. The method starts with the zero level and, if necessary, `lvl` is increased by the `step`.

The form of the null hypothesis is due to the Rényi theorem on the rarefaction of renewal processes [8] which establishes that a stationary point process converges to the Poisson process under ordinary rarefaction when each point is deleted with probability $1 - p$ and left as it is with probability $p \to 0$ accompanied by an appropriate change of scale to provide the non-degeneratedness of the limit process. As is known, the Poisson process is characterized by the intervals between successive points being independent identically distributed random variables with exponential distribution.

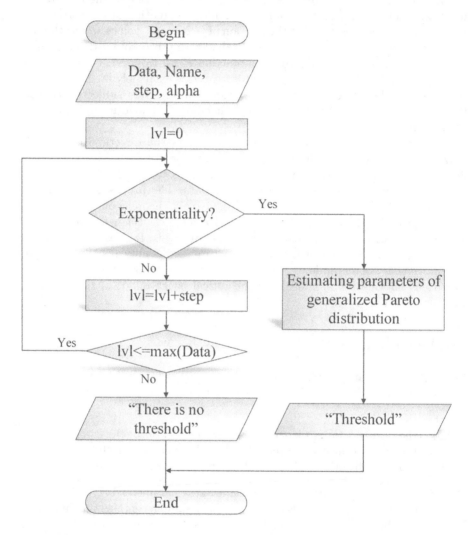

Fig. 1. The algorithm for the ascending method to find the threshold

If the value of the parameter lvl exceeds the maximum of Data, the notification "*There is no threshold*" should be printed. Otherwise, in the case of exponentiality of the auxiliary sample, the value of lvl should be subtracted from the data. The empirical distribution function of a new sample could be well approximated by the generalized Pareto distribution according to the Pickands–Balkema–de Haan theorem [9,10]: for a large class of distribution functions of a random variable X

$$F(X - u \leqslant y \mid X > u) \to F_{\xi,\sigma,\mu}(y), \quad \text{as } u \to \infty,$$

where $F_{\xi,\sigma,\mu}(\cdot)$ denotes a cumulative density function of the generalized Pareto distribution:

$$F_{\xi,\sigma,\mu}(x) = \begin{cases} 1 - \left(1 + \frac{\xi(x-\mu)}{\sigma}\right)^{-1/\xi}, & \text{if } \xi \neq 0, \\ 1 - exp\left(-\frac{x-\mu}{\sigma}\right), & \text{otherwise.} \end{cases}$$

Using Pearson's χ^2-test at the significance level `alpha`, the goodness of fit with the generalized Pareto distribution is tested. The histograms of the two samples mentioned above are plotted together with the approximating distributions. An example of application of an ascending algorithm for the test data can be found in [11].

3   Descending Method

As an alternative way for finding the threshold, the descending method can be used. Unlike the ascending algorithm, in this case the parameter `lvl` should be initialized by the maximum of `Data` and then it should be decreased by the parameter `step`. For the purpose of correct automatization of testing, the initial value of `lvl` could be specified by the parameter corresponding to the minimal number of elements that exceed the current level. Moreover, testing a hypothesis is often based on the P-value, so the required value should be determined beforehand. Within the framework of a descending algorithm, the P-values are evaluated for each level. The relationship between the P-value and the current level can be very informative.

Figure 2 demonstrates the flowchart of a descending algorithm for cumulative data. The initial sample is divided into consecutive non-zero domains while zero domains are located between them. Each element represents the sum of all the previous observations and a current one inside the domain. The details of domain forming are presented in Fig. 2 with the help of the variable `TotalVol`.

The procedure for finding locations when data in the domain exceed the threshold, is implemented in the block `Moments`. The first locations from non-zero domains form a new sample while the others are ignored. Moreover, the user can specify the minimum number of elements above the threshold at this stage. It leads to changing the start value of `lvl`.

The vectors `p(k)` (the P-value for each level) and `LVL(k)` (the current level) are the results of the descending method. Plotting graphs `p(LVL)` (see Fig. 3) and `LVL(p)` (see Fig. 4), the correct value for the threshold can be obtained. The solid lines represent smoothed discontinuous data by the moving average, whereas the corresponding relationships `p(LVL)` and `LVL(p)` are plotted by dots.

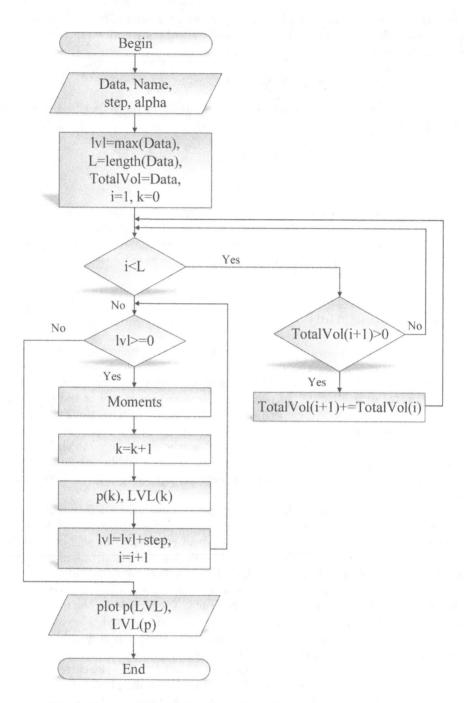

Fig. 2. The algorithm for the descending method to find the threshold

The graph in Fig. 4 is rotated by 90° in Fig. 3. These figures are intended for the evaluation of the P-value which is enough for the acceptance of the null hypothesis for the real data. The level is moved down in Fig. 3 or right in Fig. 4 step-by-step.

Figure 5 demonstrates the cumulative data with the evaluated threshold (a solid horizontal line).

Suppose that the observations can be represented as $X_{n,j}$ where n is the number of the current non-zero domain, j is the number of observation in the domain. If t is a start time of the domain, the cumulative data equals:

- $X_{n,1}$ at time t;
- $X_{n,1} + X_{n,2}$ at time $t + 1$;

  ...

- $X_{n,1} + \ldots + X_{n,k_n}$ at time $t + k_n$ where k_n is a length of the current domain.

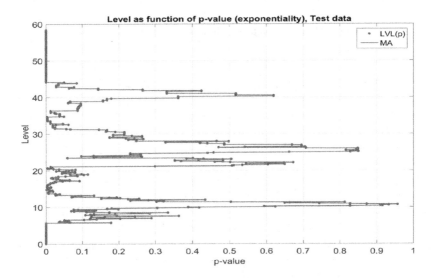

Fig. 3. Levels as a function of p-values for test data

The threshold is exceeded at time $t + j$ (for each non-zero domain), where the value j is determined by the following inequalities:

$$X_{n,1} + \ldots + X_{n,j-1} \leqslant u,$$
$$X_{n,1} + \ldots + X_{n,j} > u.$$

Other moments $(t + j + 1, t + j + 2,$ etc.) should be ignored.

The examples of fitting empirical distribution functions with the exponential (Fig. 6) for the differences of time moments and the generalized Pareto (Fig. 6)

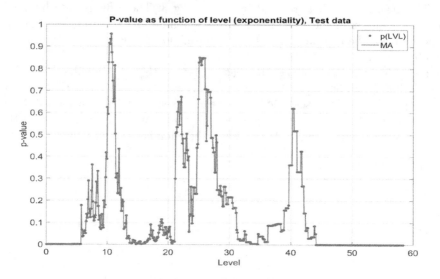

Fig. 4. P-values as a function of levels for test data

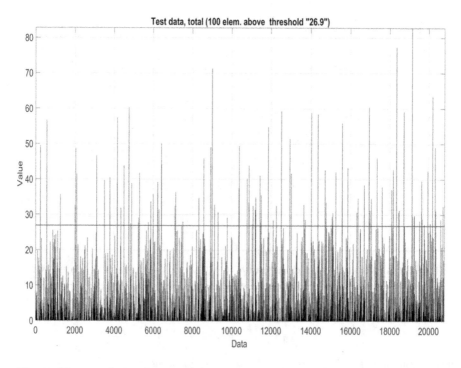

Fig. 5. The cumulative data with the evaluated threshold (a solid horizontal line)

distributions for the values above the evaluated threshold with the corresponding P-values are demonstrated. The histograms in Figs. 6 and 7 are normalized to estimate the probability density function.

The height of each bar is determined by the expression

$$\frac{N_{bin}}{N_{obs} \cdot W_{bin}},$$

where N_{bin} is the number of observations in the bin, N_{obs} is the total number of observations and W_{bin} is the width of a bin. The area of each bar is the relative number of observations. The sum of the bar areas is 1.

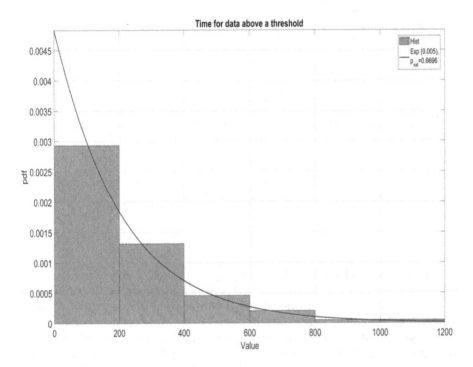

Fig. 6. An example of fitting the empirical distribution function with the exponential distribution (the P-value is shown in the legend)

For the test data the value of the threshold equals 26.9. The number of elements that exceed the threshold (in the sense mentioned above) is 100. The maximum likelihood estimator for the parameter of the exponential distribution (a solid line) is 0.005 (the P-value of the Pearson's χ^2-test equals 0.6696, so the null hypothesis is not rejected). The maximum likelihood estimators for the parameters of the generalized Pareto distribution (a solid line) are 0.165, 6.378 (the P-value of the Pearson's χ^2-test equals 0.813, so the null hypothesis is not rejected).

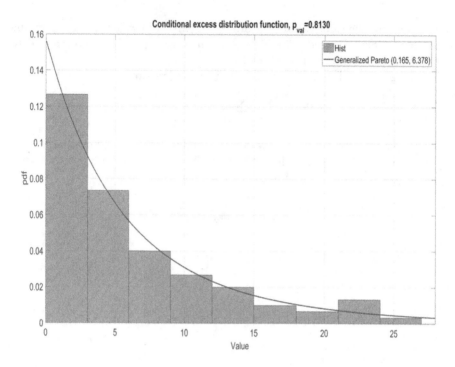

Fig. 7. An example of fitting empirical distribution function with the generalized Pareto distribution (the P-value is shown in the title)

4   Conclusions

The representation of data in the form $\sum_{j=1}^{N_n} X_{n,j}$ (N_n is a random variable) leads to interesting ideas for further research. The probability distribution of the random variables $X_{n,j}$ in the described information system can be successfully fitted by gamma or Pareto distributions, while N_n is the negative binomial random variable. It is known that these types of probability distributions can be represented as mixtures [12] of some distributions (Poisson, exponential mixture models). So, it is possible to clarify the nature of processes in data flows.

Acknowledgments. The research is supported by the Russian Foundation for Basic Research (projects 15-37-20851, 15-07-04040 and 16-07-00736).

References

1. Leadbetter, M.R.: On a basis for "Peaks over Threshold" modeling. Stat. Probab. Lett. **12**(4), 357–362 (1991)
2. Korolev, V.Yu, Sokolov, I.A.: Some problems of the analysis of catastrophic risks related to non-homogeneous flows of extremal events. Syst. Means Inform., 109–125 (2005). Special Issue

3. Korolev, V.Yu, Sokolov, I.A., Gordeev, A.S., Grigoryeva, M.E., Popov, S.V., Chebonenko, N.A.: Some methods for the analysis of temporal characteristics of catastrophes in non-homogeneous flows of extremal events. Syst. Means Inform., 5–23 (2006). Special Issue

4. Korolev, V.Yu., Sokolov, I.A., Gordeev, A.S., Grigoryeva, M.E., Popov, S.V., Chebonenko, N.A.: Some methods for prediction of characteristics of risks related to catastrophic events. Actuary **1**, 34–40 (2007)

5. Korolev, V.Yu., Sokolov, I.A.: Mathematical Models of Non-Homogeneous Flows of Extremal Events. Torus Press, Moscow (2008)

6. Korolev, V.Yu., Shorgin, S.Ya.: Mathematical Methods for the Analysis of the Stochastic Structure of Information Flows. IPI RAS, Moscow (2011)

7. Grigoryeva, M.E., Korolev, V.Yu., Sokolov, I.A.: A limit theorem for geometric sums of independent nonidentically distributed random variables and its application to the prediction of the probabilities of catastrophes in nonhomogeneous flows of extremal events. Inform. Appl. **7**(4), 11–19 (2013)

8. Gnedenko, B.V., Korolev, V.Yu.: Random Summation: Limit Theorems and Applications. CRC Press, Boca Raton (1996)

9. Balkema, A., de Haan, L.: Residual life time at great age. Ann. Probab. **2**, 792–804 (1974)

10. Pickands, J.: Statistical inference using extreme order statistics. Ann. Stat. **3**, 119–131 (1975)

11. Gorshenin, A.K., Korolev, V.Yu.: Statistical approach to determine thresholds. In: Proceedings of ITTMM-2016, pp. 90–92. Peoples Friendship University of Russia, Moscow, Russia (2016)

12. Korolev, V.Yu.: Probabilistic and statistical methods of decomposition of volatility of chaotic processes. Moscow University Publishing House, Moscow (2011)

Maximum Likelihood Estimation of the Dead Time Period Duration of a Modulated Synchronous Flow of Events

Alexander Gortsev and Mariya Sirotina[✉]

National Research Tomsk State University, Tomsk, Russia
mashuliagol@mail.ru

Abstract. A modulated synchronous doubly stochastic flow is considered. The flow under study is considered in conditions of a fixed dead time. It means that after each registered event there is a time of the fixed duration T (dead time), during which other flow events are inaccessible for observation. When duration of the dead time period finishes, the first event to occur creates the dead time period of duration T again and etc. It is supposed that the dead time period duration is an unknown variable. Using the maximum likelihood method and a moments of observed events occurrence the problem of dead time period estimation is solved.

Keywords: Modulated synchronous doubly stochastic flow · Fixed dead time · Likelihood function · Maximum likelihood estimation · Dead time period duration

1 Introduction

This paper is a continuation of the modulated synchronous flow investigation which was started in papers [1,2].

Mathematical models of queueing theory are widely used when describing real physical, technological and other processes and systems. In connection with the rapid development of computer equipment and information technologies an important sphere of queueing theory applications appeared. This sphere was called design and creation of data-processing networks, computer communication networks, satellite networks and telecommunication networks [3,4].

In practice, the intensity of input flow varies along with time. Moreover, these variations are often of a random nature. This leads to consideration of a doubly stochastic flow of events [5–10]. An example of such a flow is a modulated synchronous doubly stochastic flow [11,12].

When considering doubly stochastic flow of events there are problems of flow state estimation [13–18] and flow parameter estimation [19–27]. One of the confounding factors when estimating flow states and parameters is the dead time of registering devices, which is initiated by registered flow event. All other events occurring during the dead time period are inaccessible for observation [28–37].

© Springer International Publishing Switzerland 2016
A. Dudin et al. (Eds.): ITMM 2016, CCIS 638, pp. 104–119, 2016.
DOI: 10.1007/978-3-319-44615-8_9

The modulated synchronous doubly stochastic flow of events was introduced into consideration in papers [1, 2]. This flow is a generalization of the synchronous flow and it is related to the class of MAP-flows of the second order. In this paper, which is a continuation of investigations [1, 2], the maximum likelihood method is applied for solving the problem of dead time period estimation, because the estimation results obtained when using this method usually have interesting characteristics.

2   Problem Statement

Let us consider a modulated synchronous doubly stochastic flow of events, whose rate is a piecewise constant random process $\lambda(t)$ with two states: λ_1, λ_2 ($\lambda_1 > \lambda_2 \geqslant 0$). The sojourn time of the process $\lambda(t)$ in the state λ_i has an exponential probability distribution function with the parameter α_i, $i = 1, 2$. If at the moment t the process $\lambda(t)$ sojourns in the state λ_i than in the small half-interval $[t, t + \Delta t)$, with the probability $\alpha_i \Delta t + o(\Delta t)$ the process finishes its stay in state λ_i and moves to state λ_j with probability of one ($i, j = 1, 2$, $i \neq j$). During the time random interval when $\lambda(t) = \lambda_i$ a Poisson flow with rate λ_i, $i = 1, 2$ arrives. A state transition of the process $\lambda(t)$ may also occur at the moment of the Poisson flow event arrival. Moreover, transition from state λ_1 to state λ_2 is realized only at the moment of event occurrence with the probability p, $(0 \leqslant p \leqslant 1)$. With the complementary probability $1 - p$ the process remains at state λ_1. Transition from state λ_2 to state λ_1 is also realized only at the moment of event occurrence with the probability q, $(0 \leqslant q \leqslant 1)$. With the complementary probability $1 - q$ the process remains at the state λ_2. In the described conditions $\lambda(t)$ is the Markovian process.

Block matrixes of infinitesimal coefficients are of the form:

$$D_1 = \begin{vmatrix} (1 - p)\lambda_1 & p\lambda_1 \\ q\lambda_2 & (1 - q)\lambda_2 \end{vmatrix}, \ D_0 = \begin{vmatrix} -(\lambda_1 + \alpha_1) & \alpha_1 \\ \alpha_2 & -(\lambda_2 + \alpha_2) \end{vmatrix}.$$

The elements of the matrix D_1 are the intensities of the process $\lambda(t)$ transition from the state to the state with an event occurrence. Off-diagonal elements of the matrix D_0 are the intensities of the process $\lambda(t)$ transition from the state to the state without an event occurrence. Diagonal elements of the matrix D_0 are the intensities of the process $\lambda(t)$ leaving its states, which are taken with the opposite sign. We should note that if $\alpha_i = 0$, $i = 1, 2$ there is a usual synchronous flow of events [10].

After each registered event there is a period of fixed duration T (dead time later) during which other flow events are inaccessible for observation. It is considered a fixed dead time, which means that events occurring during the dead time interval do not initiate its prolongation. When the dead time period finishes, the first event creates a dead time period of duration T again, etc.

An example of this situation is shown in the Fig. 1, where λ_1, λ_2 are the states of the process $\lambda(t)$, $t_1, t_2, \ldots$ are the moments of the observable flow events occurrence, crosshatching lines are the dead time periods of T duration, axis

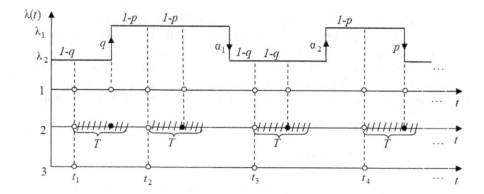

Fig. 1. Forming the flow of observed events

number 1 shows the initial modulated synchronous flow of events, axis number 2 shows the scheme of dead time creation, axis number 3 shows an observable events of the modulated synchronous flow.

Let us emphasize that there is an assumed priority of the event occurrence in the problem statement. An event occurs and after that the process $\lambda(t)$ makes a transition from the state to the state. This circumstance is irrelevant when obtaining analytical results because event occurrence and the process $\lambda(t)$ state transition happen immediately. When obtaining numerical results using simulation modeling, it is necessary to have the definiteness what is the first: the event occurrence or the state transition.

The process $\lambda(t)$ and event types (Poisson flow events of λ_1 and λ_2 intensities) are not observable in principle (in prepositions made $\lambda(t)$ is a latent Markovian process). Only the moments of observable events occurrence $t_1, t_2, \ldots t_k, \ldots$ are accessible for observation. A stationary mode of flow functioning is considered. A sequence of time moments of event occurrence $t_1, t_2, \ldots t_k, \ldots$ is an imbedded Markov chain $\{\lambda(t_k)\}$. So the flow has a Markovian chain characteristic when its evolution is considered from the moment t_k, $k = 1, 2, \ldots$ (the moment of flow event occurrence).

The main purpose of the research is the creation of a dead time period estimation $\hat{T}$ using the maximum likelihood method (assuming that all other flow parameters $\lambda_1, \lambda_2, \alpha_1, \alpha_2, p, q$ are known).

3   Likelihood Function Creation

Let $\tau_k = t_{k+1} - t_k$, $k = 1, 2, \ldots$, is a value of the k interval duration between the moments of the adjacent observable flow events occurrence.

Since the flow functions in stationary mode then a probability density function of interval duration between the moments of adjacent observable flow events occurence is $p_T(\tau_k) = p_T(\tau)$, $\tau \geqslant 0$, for any k (index T gives an accent that the probability density function depends on a dead time period). So without the loss of generality the moment t_k of the event occurrence can be assumed equal to null, which means $\tau = 0$.

Then the probability density function $p_T(\tau)$ of the interval duration between the moments of modulated synchronous flow adjacent event occurrence is written in the form of [2]:

$$
p_T(\tau) = \begin{cases} 0, \ 0 \leqslant \tau < T, \\ \gamma(T) z_1 e^{-z_1(\tau - T)} + (1 - \gamma(T)) z_2 e^{-z_2(\tau - T)}, \ \tau \geqslant T, \end{cases}
$$

$$
\gamma(T) = \frac{1}{z_2 - z_1}(z_2 - \pi_1(T)\lambda_1 - \pi_2(T)\lambda_2),
$$

$$
z_1 = (\lambda_1 + \alpha_1 + \lambda_2 + \alpha_2) - \sqrt{(\lambda_1 + \alpha_1 - \lambda_2 - \alpha_2)^2 + 4\alpha_1\alpha_2},
$$

$$
z_2 = (\lambda_1 + \alpha_1 + \lambda_2 + \alpha_2) + \sqrt{(\lambda_1 + \alpha_1 - \lambda_2 - \alpha_2)^2 + 4\alpha_1\alpha_2},
$$

$$
\pi_1(T) = \pi_1 - (\pi_1 - \pi_1(0|T))e^{-(\alpha_1 + p\lambda_1 + \alpha_2 + q\lambda_2)T},
$$

$$
\pi_2(T) = \pi_2 - (\pi_2 - \pi_2(0|T))e^{-(\alpha_1 + p\lambda_1 + \alpha_2 + q\lambda_2)T},
$$

$$
\pi_1(0|T) = \Big[q\lambda_2\alpha_1 + (1 - p)\lambda_1(\lambda_2 + \alpha_2) - \lambda_1\lambda_2(1 - p - q) \tag{1}
$$
$$
\times (\pi_2 + \pi_1 e^{-(\alpha_1 + p\lambda_1 + \alpha_2 + q\lambda_2)T}) \Big] \Big\{ z_1 z_2 - \lambda_1\lambda_2(1 - p - q)
$$
$$
\times (\pi_2 + \pi_1 e^{-(\alpha_1 + p\lambda_1 + \alpha_2 + q\lambda_2)T}) \Big\}^{-1},
$$

$$
\pi_2(0|T) = \Big[(1 - q)\lambda_2(\lambda_1 + \alpha_1) + p\lambda_1\alpha_2 - \lambda_1\lambda_2(1 - p - q)
$$
$$
\times (\pi_2 + \pi_1 e^{-(\alpha_1 + p\lambda_1 + \alpha_2 + q\lambda_2)T}) \Big] \Big\{ z_1 z_2 - \lambda_1\lambda_2(1 - p - q)
$$
$$
\times (\pi_2 + \pi_1 e^{-(\alpha_1 + p\lambda_1 + \alpha_2 + q\lambda_2)T}) \Big\}^{-1},
$$

$$
\pi_1 = \frac{\alpha_2 + q\lambda_2}{\alpha_1 + p\lambda_1 + \alpha_2 + q\lambda_2}, \ \pi_2 = \frac{\alpha_1 + p\lambda_1}{\alpha_1 + p\lambda_1 + \alpha_2 + q\lambda_2}.
$$

Let $\tau_1 = t_2 - t_1, \ \tau_2 = t_3 - t_2, \ \ldots, \ \tau_k = t_{k+1} - t_k, \ k = 1, 2, \ldots$ is a sequence of interval durations between the adjacent events of observable modulated synchronous flow after its observation on the interval $(0, t)$. Let us put in order the values $\tau_1, \tau_2, \ldots \tau_k$ by increasing: $\tau^{(1)} < \tau^{(2)} < \ldots \tau^{(k)}$. Then the likelihood function with regard to (1) can be written as follows:

$$
L(\lambda_i, \alpha_i, p, q, T | \tau^{(1)}, \tau^{(2)}, \ldots, \tau^{(k)}) = 0, \ 0 < \tau_{min} < T,
$$

$$
L(\lambda_i, \alpha_i, p, q, T | \tau^{(1)}, \tau^{(2)}, \ldots, \tau^{(k)}) = \prod_{j=1}^{k} p_T(\tau^{(j)}), \ T \leqslant \tau_{min}.
$$

Because the main purpose of the research is the creation of a dead time period estimation $\hat{T}$, the solution of the problem under study is obtaining the maximum of the likelihood function by the unknown variable T:

$$L(T|\tau^{(1)}, \tau^{(2)}, \ldots, \tau^{(k)}) = \prod_{j=1}^{k} (p_T(\tau^{(j)})) = \prod_{j=1}^{k} \left(\gamma(T) z_1 e^{-z_1(\tau^{(j)}-T)} \right.$$

$$\left. + (1 - \gamma(T)) z_2 e^{-z_2(\tau^{(j)}-T)} \right) \Rightarrow \max_{T}, \ 0 < \tau_{min} < T, \tag{2}$$

where $z_1, z_2, \gamma(T)$ are defined in (1).

The value of T wherein (2) reaches its global maximum is a dead time period estimation $\hat{T}$.

4   Solution of the Optimization Problem

Let us make a redefinition: $\tau_{min} = \tau_m$. Because the likelihood function (2) is not equal to zero when $0 \leqslant T \leqslant \tau_m$ then let $p_T(\tau^{(j)}) = 0$, $j = \overline{2,k}$ when $T > \tau_m$, ($\tau_m > 0$). Let us treat the probability density $p_T(\tau_m)$, $0 \leqslant T \leqslant \tau_m$ as a function of the variable T. In future investigation the situation when $\tau_m = 0$ means determination of functions under study in a boundary point. Let us consider a derivative function $p_T'(\tau_m)$ of $p_T(\tau_m)$ function by the variable T. The differential equation is of the form:

$$p_T'(\tau_m) = \frac{F_1(T)e^{-z_1(\tau_m-T)} - F_2(T)e^{-z_2(\tau_m-T)}}{(z_2 - z_1)(\beta_1 + \beta_2)},$$

$$F_1(T) = z_1 \left[(\beta_1 + \beta_2) z_1 z_2 - z_1 f(T) - f'(T) \right],$$

$$F_2(T) = z_2 \left[(\beta_1 + \beta_2) z_1 z_2 - z_2 f(T) - f'(T) \right],$$

$$f(T) = P_1 + P_2 + (\lambda_1 - \lambda_2) \frac{(\beta_1 P_1 - \beta_2 P_2)e^{-(\beta_1+\beta_2)T}}{F(T)}, \tag{3}$$

$$f'(T) = -(\lambda_1 - \lambda_2)(\beta_1 + \beta_2) z_1 z_2 \frac{(\beta_1 P_1 - \beta_2 P_2)e^{-(\beta_1+\beta_2)T}}{F^2(T)},$$

$$F(T) = z_1 z_2 - P e^{-(\beta_1+\beta_2)T} > 0, \ 0 \leqslant T \leqslant \tau_m, \ \tau_m \geqslant 0,$$

where $\beta_1 = \alpha_1 + p\lambda_1$, $\beta_2 = \alpha_2 + q\lambda_2$, $P_1 = \lambda_1\alpha_2(1-p) + q\lambda_2(\lambda_1 + \alpha_1)$, $P_2 = \lambda_2\alpha_1(1-q) + p\lambda_1(\lambda_2 + \alpha_2)$, $P = \lambda_1\lambda_2(1-p-q)$, $z_1 z_2 = \lambda_1\lambda_2 + \lambda_1\alpha_2 + \lambda_2\alpha_1$; z_1, z_2 are defined in (1).

Lemma 1. *The derivative function $p_T'(\tau_m)$ is a positive function of the variable τ_m when $T = 0$ ($p_T'(\tau_m) > 0$).*

Proof. Because τ_m is any non-negative number ($\tau_m \geqslant 0$) then $p_T'(\tau_m)$ can be considered as a function of the variable τ_m. Inserting $T = 0$ into (3) we can obtain:

$$p_0'(\tau_m) = \frac{C}{(z_2 - z_1)A^2} \left\{ z_1 e^{-z_1\tau_m} (z_2 A - C) \right.$$

$$\left. - z_2 e^{-z_2\tau_m} (z_1 A - C) \right\}, \tau_m \geqslant 0, \tag{4}$$

where $A = P_1 + P_2$, $C = \lambda_1 P_1 + \lambda_2 P_2$; P_1, P_2 are defined in (3), $z_1 z_2$ are defined in (1).

The value of the derivative function (4) in point $\tau_m = 0$ is of the form:

$$p_0'(0) = p_0'(\tau_m = 0) = (C/A)^2 > 0, \tag{5}$$

where A, C are defined in (4).

The limit of the function $p_0'(\tau_m)$ when $\tau_m \to \infty$ is defined as follows:

$$p_0'(\infty) = \lim_{\tau_m \to \infty} p_0'(\tau_m) = \pm 0. \tag{6}$$

Let us treat the function $p_0'(\tau_m)$ $(\tau_m > 0)$ on the existence of zero. Let us consider the equation $p_0'(\tau_m) = 0$, which can be written in the form:

$$e^{-(z_2 - z_1)\tau_m} = \frac{z_1(z_2 A - C)}{z_2(z_1 A - C)}, \ \tau_m > 0. \tag{7}$$

Let us note $B = B_1/B_2$; $B_1 = z_1(z_2 A - C)$, $B_2 = z_2(z_1 A - C)$. We should emphasize that if $B_2 = 0$ then $z_2(z_1 A - C) = 0$, that is $z_1 A = C$. Then $z_1 A - C = (z_2 - z_1)A > 0$ and it follows from (4) that $p_0'(\tau_m) > 0$ $(\tau_m \geq 0)$.

Transforming B_1 we can obtain:

$$B_1 = \frac{1}{2} z_1^2 \Big\{ -2A + (\beta_1 + \beta_2)(z_1 + z_2)$$

$$+(\beta_1 + \beta_2)\sqrt{(\lambda_1 - \lambda_2 + \alpha_1 - \alpha_2)^2 + 4\alpha_1\alpha_2} \Big\} > \frac{1}{2} z_1^2 \Big\{ -2A + (\beta_1 + \beta_2)$$

$$\times (z_1 + z_2) + (\beta_1 + \beta_2)|\lambda_1 - \lambda_2 + \alpha_1 - \alpha_2| \Big\}.$$

It is not difficult to show that the first part of the last inequation always is strictly more than zero. Then $B_1 > 0$. Therefore two variants occur in Eq. (7): (1) $B_1 > 0$, $B_2 < 0$; (2) $B_1 > 0$, $B_2 > 0$. For the first variant $B < 0$ and Eq. (7) does not have a solution. For the second variant $B > 1$ because $B_1 - B_2 = (z_2 - z_1)C > 0$ and Eq. (7) does not have a solution either. It follows that the derivative function $p_0'(\tau_m)$ does not reach zero point for $\tau_m > 0$. Then (6) takes the form $p_0'(\infty) = +0$. Grouping (5), (6) and conclusion $p_0'(\tau_m) \neq 0$ $(\tau_m > 0)$ proves *Lemma 1*. *Lemma 1* is proved.

From *Lemma 1* it follows that the probability density function $p_T(\tau_m)$ is the increasing function of the variable T in the point $T = 0$ $(\tau_m \geq 0)$.

Lemma 2. *The derivative function* $p_T'(\tau_m)$ *is strictly more than zero when* $T = \tau_m$ $(p_{T=\tau_m}'(\tau_m) = p'(\tau_m) > 0, \tau_m \geq 0)$.

Proof. Inserting $T = \tau_m$ into (3) we obtain

$$p'(\tau_m) = \frac{1}{\beta_1 + \beta_2}\Big[C + (\lambda_1 - \lambda_2)(\beta_1 P_1 - \beta_2 P_2)$$

$$\times \frac{e^{-(\beta_1 + \beta_2)\tau_m}}{(z_1 z_2 - Pe^{-(\beta_1 + \beta_2)\tau_m})^2} X(\tau_m) \Big], \tag{8}$$

$$X(\tau_m) = C + P(z_1 + z_2)(1 - e^{-(\beta_1 + \beta_2)\tau_m}), \ \tau_m \geq 0,$$

where C is defined in (4), β_1, β_2, P_1, P_2, P are defined in (3), z_1, z_2 are defined in (1).

It is not difficult to show that $X(\tau_m) > 0$ for $\tau_m \geqslant 0$. The latter means that a sign of the derivative function $p'(\tau_m)$ is defined by the multiplier $(\beta_1 P_1 - \beta_2 P_2)$.

Let us consider $p'(\tau_m)$ as a function of the variable τ_m $(\tau \geqslant 0)$. Then we have:

$$p'(\tau_m = 0) = (C/A)^2 > 0, \quad p'(\tau_m = \infty) = C/(\beta_1 + \beta_2) > 0.$$

A sign of the derivative function $p'(\tau_m = 0) - p'(\tau_m = \infty) = C(\lambda_1 - \lambda_2)(\beta_1 P_1 - \beta_2 P_2)/(A^2(\beta_1 + \beta_2))$ is defined by a sign $(\beta_1 P_1 - \beta_2 P_2)$: if $(\beta_1 P_1 - \beta_2 P_2) > 0$ then $p'(\tau_m = 0) > p(\tau_m = \infty)$; if $(\beta_1 P_1 - \beta_2 P_2) < 0$ then $p'(\tau_m = 0) < p(\tau_m = \infty)$; if $(\beta_1 P_1 - \beta_2 P_2) = 0$ then $p'(\tau_m = 0) = p(\tau_m = \infty)$.

From the form of the derivative function $p'(\tau_m)$ it follows that if $(\beta_1 P_1 - \beta_2 P_2) \geqslant 0$ the function $p'(\tau_m) > 0$, therefore $p(\tau_m)$ is the increasing function of the variable T in the point $T = \tau_m(\tau_m \geqslant 0)$. The case when $(\beta_1 P_1 - \beta_2 P_2) < 0$ needs to be considered in more detail. For this case we should consider the second derivative function $p''(\tau_m)$ of $p(\tau_m)$. The derivative function $p''(\tau_m)$ with regard to (8) is of the form:

$$p''(\tau_m) = - (\lambda_1 - \lambda_2)(\beta_1 P_1 - \beta_2 P_2)z_1 z_2 \frac{e^{-(\beta_1 + \beta_2)\tau_m}}{F^3(\tau_m)} Y(\tau_m),$$

$$Y(\tau_m) = \left\{ C + (z_1 + z_2)P - (z_1 + z_2 + \beta_1 + \beta_2)Pe^{-(\beta_1 + \beta_2)\tau_m} \right\}, \quad \tau_m \geqslant 0,$$

(9)

where C is defined in (4), β_1, β_2, P_1, P_2, P, $F(T = \tau_m)$ are defined in (3), z_1, z_2 are defined in (1).

Because we consider the case when $(\beta_1 P_1 - \beta_2 P_2) < 0$ then the multiplier in (9): $-(\lambda_1 - \lambda_2)(\beta_1 P_1 - \beta_2 P_2)z_1 z_2 e^{-(\beta_1 + \beta_2)\tau_m}/F^3(\tau_m) > 0$; then a sign of the derivative function $p''(\tau_m)$ is defined by a sign of the expression for $Y(\tau_m)$. Let us treat the function $Y(\tau_m)$. We have: $Y(\infty) = C + (z_1 + z_2)P \geqslant 0$, $Y(0) = C - P(\beta_1 + \beta_2)$, in addition $Y(0) > 0$ or $Y(0) < 0$ or $Y(0) = 0$. The derivative function $Y'(\tau_m)$ is of the form:

$$Y'(\tau_m) = (\beta_1 + \beta_2)(z_1 + z_2 + \beta_1 + \beta_2)Pe^{-(\beta_1 + \beta_2)\tau_m}, \tau_m \geqslant 0.$$

A sign of the derivative function $Y'(\tau_m)$ depends on the multiplier P: (1) if $P < 0$ then $Y'(\tau_m) < 0$ $(\tau_m \geqslant 0)$, then $p''(\tau_m) > 0$ $(\tau_m \geqslant 0)$; (2) if $P = 0$ then $Y'(\tau_m) = C > 0$ $(\tau_m \geqslant 0)$, then $p''(\tau_m) > 0$ $(\tau_m \geqslant 0)$; (3) if $P > 0$ then $Y'(\tau_m) > 0$ $(\tau_m \geqslant 0)$, then $p''(\tau_m) > 0$ or $p''(\tau_m) < 0$ or $p''(\tau_m) = 0$ $(\tau_m \geqslant 0)$.

For the first two cases the derivative function $p'(\tau_m) > 0$ $(\tau_m \geqslant 0)$. If the third case is realized then there are some variants of the function $p'(\tau_m)$ $(\tau_m \geqslant 0)$ behaviour: (a) if $Y(0) = C - P(\beta_1 + \beta_2) > 0$ then $p''(\tau_m) > 0$ $(\tau_m \geqslant 0)$; then $p'(\tau_m) > 0$ $(\tau_m \geqslant 0)$; (b) if $Y(0) = C - P(\beta_1 + \beta_2) = 0$ then $p''(\tau_m) = 0$ $(\tau_m = 0)$, $p''(\tau_m) > 0$ $(\tau_m > 0)$; then $p'(\tau_m) > 0$ $(\tau_m \geqslant 0)$; (c) if $Y(0) = C - P(\beta_1 + \beta_2) < 0$ then behaviour of the derivative function $p''(\tau_m)$ is defined by the next situations: (1) $0 \leqslant \tau_m < \tau_m^*$, then $p''(\tau_m) < 0$; (2) $\tau_m = \tau_m^*$, then $p''(\tau_m) = 0$; (3) $\tau_m^* < \tau_m$, then $p''(\tau_m) > 0$.

Therefore when $Y(0) = C - P(\beta_1 + \beta_2) < 0$ is realized then the derivative function $p'(\tau_m)$ $(\tau_m > 0)$ reaches its global minimum at the point $\tau_m = \tau_m^*$. In addition the point τ_m^* is defined by the expression:

$$\tau_m^* = -\frac{1}{(\beta_1 + \beta_2)} \ln \frac{z_1 z_2 (z_1 + z_2 - \beta_1 - \beta_2)}{(z_1 + z_2 + \beta_1 + \beta_2)P}, \quad P > 0. \tag{10}$$

Inserting (10) into (8) we obtain the expression for the derivative function $p'(\tau_m)$ in the minimum point $\tau_m = \tau_m^*$:

$$p'(\tau_m^*) = \frac{1}{(\beta_1 + \beta_2)}$$
$$\times \left\{ C + (\lambda_1 - \lambda_2)(\beta_1 P_1 - \beta_2 P_2)\frac{(z_1 + z_2 - \beta_1 - \beta_2)^2}{4P(\beta_1 + \beta_2)} \right\}, \quad P > 0. \tag{11}$$

It can be shown that the derivative function defined in (11) is strictly more than zero ($p'(\tau_m^*) > 0, P > 0$). From the latter it follows that when the situation $Y(0) = C - P(\beta_1 + \beta_2) < 0$ is realized then the inequation $p'(\tau_m) > 0$, $(\tau_m \geqslant 0)$ takes place. *Lemma* 2 is proved.

Then we should consider the behaviour of the derivative function $p'_T(\tau_m)$ as a function of the variable T on the interval $(0, \tau_m)$.

Let us consider the equation $p'_T(\tau_m) = 0$ on root existence which can be transformed according to (3) into the form:

$$\Psi(T) = e^{-(z_2 - z_1)(\tau_m - T)}, \quad \Psi(T) = \frac{F_1(T)}{F_2(T)}, \quad 0 \leqslant T \leqslant \tau_m, \tag{12}$$

where $F_1(T)$, $F_2(T)$ are defined in (3). Because τ_m is as big as can be then let us consider the behaviour of the function $\Psi(T)$ when $T \geqslant 0$.

Let us emphasize that $e^{-(z_2 - z_1)\tau_m} \leqslant e^{-(z_2 - z_1)(\tau_m - T)} \leqslant 1, 0 \leqslant T \leqslant \tau_m$, $\tau_m \geqslant 0$.

Let us denote $h = 1 - p - q$, then $P = \lambda_1 \lambda_2 h$. The variable $(\lambda_1 - \lambda_2)(\beta_1 P_1 - \beta_2 P_2)$ can be written as follows:

$$(\lambda_1 - \lambda_2)(\beta_1 P_1 - \beta_2 P_2)$$
$$= -(\lambda_1 \lambda_2)^2 h^2 + \lambda_1 \lambda_2 [2z_1 z_2 - (\beta_1 + \beta_2)(z_1 + z_2)] h$$
$$- z_1 z_2 (z_1 - \beta_1 - \beta_2)(z_2 - \beta_1 - \beta_2) = x(h), \quad -1 \leqslant h \leqslant 1. \tag{13}$$

The function (13) reaches its null in the points $h = h_1$ and $h = h_2$; $h_1 = z_2(z_1 - \beta_1 - \beta_2)/\lambda_1 \lambda_2$, $h_2 = z_1(z_2 - \beta_1 - \beta_2)/\lambda_1 \lambda_2$, $h_1 < h_2$.

Let us investigate the function $F_1(T)$, $T \geqslant 0$ from the (12) which was defined in (3).

Lemma 3. *The function* $F_1(T) < 0, T \geqslant 0$.

Proof. We have
$$F_1(0) = z_1(\beta_1 + \beta_2)(z_2 A - C)C/A^2, \quad F_1(\infty) = \lim_{T \to \infty} F_1(T) = z_1(z_2 A - C) > 0$$
because $(z_2 A - C) > 0$ (*Lemma* 1).

Then $(z_2 A - C) = \lambda_1 \lambda_2 z_1 (h - h_1) > 0$. From this it follows that $h > h_1$ always.

For the next investigation of the function $F_1(T)$ for a sign we should present it as a function depending on the variable h $(F_1(T) = F_1(T, h))$:

$$F_1(T, h) = z_1 \left[(\beta_1 + \beta_2) z_1 z_2 - z_1 f(T, h) - f'(T, h) \right], \quad T \geqslant 0, \ -1 \leqslant h \leqslant 1,$$

$$f(T, h) = A + x(h) \frac{e^{-(\beta_1 + \beta_2)T}}{F(T, h)}, \tag{14}$$

$$f'(T, h) = -z_1 z_2 (\beta_1 + \beta_2) x(h) \frac{e^{-(\beta_1 + \beta_2)T}}{F^2(T, h)},$$

where $x(h)$ is defined in (13), A – in (4); β_1, β_2, $F(T, h)$ – in (3); z_1, z_2 – in (1).

A sign of the function $F_1(T, h)$ depends on a sign of the function $x(h)$ defined in (13). Let us consider all possible variants.

Let $x(h) = 0$. It is possible as follows from (13) if $h = h_1$ or $h = h_2$. Because always $h > h_1$ then the situation $h = h_1$ is excluded. Then for the situation $h = h_2$ we have: $F_1(T, h = h_2) = z_1^2 (z_2 - z_1)(\beta_1 + \beta_2) > 0, \ T \geqslant 0$.

Let $x(h) > 0$. The latter is possible if one of the following variants is realized: $0 < h_1 < h < h_2$; $0 = h_1 < h < h_2$; $h_1 < 0 < h < h_2$; $h_1 < h = 0 < h_2$; $h_1 < h < 0 < h_2$; $h_1 < h < h_2 = 0$; $h_1 < h < h_2 < 0$.

Let us consider the variant $0 < h_1 < h < h_2$. We have:

$$\frac{\partial F_1(T, h)}{\partial T} = \frac{z_1^2 z_2 (\beta_1 + \beta_2) x(h) e^{-(\beta_1 + \beta_2)T}}{F^3(T, h)}$$

$$\times \left\{ z_1 z_2 (z_1 - \beta_1 - \beta_2) - \lambda_1 \lambda_2 (z_1 + \beta_1 + \beta_2) h e^{-(\beta_1 + \beta_2)T} \right\},$$

$$T \geqslant 0, \ 0 < h_1 < h < h_2. \tag{15}$$

A sign of the derivative function (15) is defined by a sign of the function:

$$y_1(T, h) = z_1 z_2 (z_1 - \beta_1 - \beta_2)$$

$$- \lambda_1 \lambda_2 (z_1 + \beta_1 + \beta_2) h e^{-(\beta_1 + \beta_2)T}, \quad T \geqslant 0, \ 0 < h_1 < h < h_2. \tag{16}$$

Then because $h > 0$ the function (15) is the increasing function of the variable T (it increases from $y_1(T = 0, h) = z_1 z_2 (z_1 - \beta_1 - \beta_2) - \lambda_1 \lambda_2 (z_1 + \beta_1 + \beta_2) h$ to $y_1(T = \infty, h) = z_1 z_2 (z_1 - \beta_1 - \beta_2) > 0$). Then $y_1(T = 0, h) < 0$ for $0 < h_1 < h < h_2$. From this it follows that the function (16) crosses zero into the point:

$$T_1(h) = -\frac{1}{(\beta_1 + \beta_2)} \ln \frac{z_1 z_2 (z_1 - \beta_1 - \beta_2)}{\lambda_1 \lambda_2 (z_1 + \beta_1 + \beta_2) h}, \quad 0 < h_1 < h < h_2.$$

According to this the behaviour of the derivative function $F_1'(T, h)$ (defined in (15)) depending on the variable T will look as follows: $F_1'(T, h) < 0$, $0 \leqslant T < T_1(h)$; $F_1'(T, h) = 0, T = T_1(h)$; $F_1'(T, h) > 0, T > T_1(h)$. Therefore the function $F_1(T, h)$ reaches its global minimum at the point $T = T_1(h)$.

In addition $F_1(T = 0, h) > F_1(T = \infty, h), 0 < h_1 < h < h_2$. Let us show that the function $F_1(T, h)$ is strictly more than zero at the point $T = T_1(h)$ $(F_1(T = T_1(h), h) > 0, 0 < h_1 < h < h_2)$. We have:

$$F_1(T = T_1(h), h) = z_1\{(\lambda_1\lambda_2)^2 (z_1 + \beta_1 + \beta_2)^2 h^2 - \lambda_1\lambda_2(z_1 - \beta_1 - \beta_2)$$
$$\times \left[z_1(z_1 - \beta_1 - \beta_2)(z_2 - \beta_1 - \beta_2) + z_2(z_1 + \beta_1 + \beta_2)^2\right]h$$
$$+ z_1 z_2(z_1 - \beta_1 - \beta_2)^3 (z_2 - \beta_1 - \beta_2)\}, \quad 0 < h_1 < h < h_2. \quad (17)$$

The function (17) reaches its null at the point $h = h_1$. When $h > h_1$ the inequation $F_1(T = T_1(h), h) > 0$ takes place. Using the formulas (15), it can be shown that $F_1(T, h) > 0, T \geqslant 0$ for other variants which are realized for the situation $x(h) > 0$.

Let $x(h) < 0$. It is possible if one of the following variants is realized: $0 < h_1 < h_2 < h \leqslant 1; 0 = h_1 < h_2 < h \leqslant 1; h_1 < 0 < h_2 < h \leqslant 1; h_1 < h_2 = 0 < h \leqslant 1; h_1 < h_2 < 0 < h \leqslant 1; h_1 < h_2 < h = 0; h_1 < h_2 < h < 0$.

Using the formulas (15)–(17) it can be shown that for all variants described above the function $F_1(T, h) > 0, T \geqslant 0, h > h_2$. *Lemma 3* is proved.

Let us consider the function $F_2(T), T \geqslant 0$ from (12) which is defined in (3).

Lemma 4. *(1) The function $F_2(T) = 0, T \geqslant 0$ if $x(h) = 0$; (2) the function $F_2(T) < 0, T \geqslant 0$ if $x(h) > 0$.*

Proof. We have

$$F_2(0) = \lambda_1\lambda_2 z_2^2(\beta_1 + \beta_2)(h - h_2)C/A^2, \quad F_2(\infty) = \lim_{T \to \infty} F_2(T) = \lambda_1\lambda_2 z_2^2(h - h_2).$$

For the next investigation of the function $F_2(T)$ on a sign we should present it as a function depending on the variable h $(F_2(T) = F_2(T, h))$:

$$F_2(T, h) = z_2 \left[(\beta_1 + \beta_2)z_1 z_2 - z_2 f(T, h) - f^{'}(T, h)\right],$$
$$T \geqslant 0, \quad -1 \leqslant h \leqslant 1, \quad (18)$$

where $f(T, h), f^{'}(T, h)$ are defined in (14).

A sign of the function $F_2(T, h)$ depends on a sign of the function $x(h)$ defined in (13). Let us consider all possible variants.

Let $x(h) = 0$. It is possible as follows from (13) if $h = h_1$ or $h = h_2$. Because always $h > h_1$ (*Lemma 3*) then the situation $h = h_1$ is excluded. The for the situation $h = h_2$ we have: $F_2(T, h = h_2) = 0, T \geqslant 0$.

Let $x(h) > 0$. The latter is possible if one of the following variants is realized (*Lemma 3*): $0 < h_1 < h < h_2; 0 = h_1 < h < h_2; h_1 < 0 < h < h_2; h_1 < h = 0 < h_2; h_1 < h < 0 < h_2; h_1 < h < h_2 = 0; h_1 < h < h_2 < 0$.

Let us consider the variant $0 < h_1 < h < h_2$. We have:

$$\frac{\partial F_2(T, h)}{\partial T} = \frac{z_1 z_2^2(\beta_1 + \beta_2)x(h)e^{-(\beta_1+\beta_2)T}}{F^3(T, h)}$$
$$\times \left\{z_1 z_2(z_2 - \beta_1 - \beta_2) - \lambda_1\lambda_2(z_2 + \beta_1 + \beta_2)he^{-(\beta_1+\beta_2)T}\right\},$$
$$T \geqslant 0, \quad 0 < h_1 < h < h_2. \quad (19)$$

A sign of the derivative function (19) is defined by a sign of the function:

$$y_2(T, h) = z_1 z_2 (z_2 - \beta_1 - \beta_2) - \lambda_1 \lambda_2 (z_2 + \beta_1 + \beta_2) h e^{-(\beta_1+\beta_2)T},$$
$$T \geqslant 0, \ 0 < h_1 < h < h_2. \quad (20)$$

Then because $h > 0$ the function (20) is the increasing function of the variable T (it increases from $y_2(T = 0, h) = z_1 z_2 (z_2 - \beta_1 - \beta_2) - \lambda_1 \lambda_2 (z_2 + \beta_1 + \beta_2) h$ to $y_2(T = \infty, h) = z_1 z_2 (z_2 - \beta_1 - \beta_2) > 0$). There are two possible situations here: (a) $y_2(T = 0, h) < 0$ for $0 < h_1 < h < h_2$; (b) $y_2(T = 0, h) > 0$ for $0 < h_1 < h < h_2^*$; $y_2(T = 0, h) = 0$ for $h = h_2^*$; $y_2(T = 0, h) < 0$ for $h_2^* < h < h_2$, where $h_2^* = z_1 z_2 (z_2 - \beta_1 - \beta_2) [\lambda_1 \lambda_2 (z_2 + \beta_1 + \beta_2)]^{-1}$.

Let us consider the situation (a). Then the function (20) crosses zero at the point:

$$T_2(h) = -\frac{1}{(\beta_1 + \beta_2)} \ln \frac{z_1 z_2 (z_2 - \beta_1 - \beta_2)}{\lambda_1 \lambda_2 (z_2 + \beta_1 + \beta_2) h}, \ 0 < h_1 < h < h_2.$$

According to this the behaviour of the derivative function $F_2'(T, h)$ (defined in (19)) depending on the variable T will look as follows: $F_2'(T, h) < 0$, $0 \leqslant T < T_2(h)$; $F_2'(T, h) = 0, T = T_2(h)$; $F_2'(T, h) > 0, T > T_2(h)$. Therefore the function $F_2(T, h)$ reaches its global minimum at the point $T = T_2(h)$. In addition $F_2(T = 0, h) < 0$, $F_1(T = \infty, h) < 0$, $F_2(T = 0, h) < F(T = \infty, h)$, $0 < h_1 < h < h_2$. This all proves that $F_2(T, h) < 0$, $T \geqslant 0$ for the situation (a). It can be proved that $F_2(T, h) < 0$, $T \geqslant 0$ for the situation (b).

Using the formulas (19), (20) it can be shown that $F_2(T, h) < 0$, $T \geqslant 0$ for other variants which are realized for the situation $x(h) > 0$. *Lemma 4* is proved.

For the situation $x(h) < 0$ it takes place

Lemma 5. *The function $F_2(T) > 0$, $T \geqslant 0$ if $x(h) < 0$.*

Proof. The lemma can be proved by applying the formulas (19), (20) (similar to applying these formulas for *Lemma 4*) for the variants of parameter h changing which are described in *Lemma 3* for the situation when $x(h) < 0$. *Lemma 5* is proved.

Let us introduce into consideration the function

$$\Phi(T, h) = F_1(T, h) - F_2(T, h) = (z_2 - z_1)$$
$$\times \left[-z_1 z_2 (\beta_1 + \beta_2) + (z_1 + z_2) f(T, h) + f'(T, h) \right], \ T \geqslant 0, \ -1 \leqslant h \leqslant 1, \quad (21)$$

where $F_1(T, h)$, $f(T, h)$, $f'(T, h)$ are defined in (14), $F_2(T, h)$ – in (18).

Lemma 6. *The function $\Phi(T, h) > 0$, $T \geqslant 0$ if $x(h) \geqslant 0$.*

Proof. We have $\Phi(T = 0, h) = (z_2 - z_1)(\beta_1 + \beta_2)(C/A)^2 > 0$, $\Phi(T = \infty, h) = \lim\limits_{T \to \infty} \Phi(T, h) = (z_2 - z_1)C > 0$.

The sign of the function $\Phi(T, h)$ depends on the sign of the function $x(h)$ defined in (13).

Let $x(h) = 0$. It is possible as follows from (13) if $h = h_1$ or $h = h_2$. Because always $h > h_1$ (*Lemma 3*) then the situation $h = h_1$ is excluded. Then for the situation $h = h_2$ inserting h_2 into (21) we have: $\Phi(T, h = h_2) = (z_2 - z_1)(\beta_1 + \beta_2)z_1^2 > 0$, $T \geqslant 0$.

Let $x(h) > 0$. The latter is possible if one of the following variants is realized (*Lemma 3*): $0 < h_1 < h < h_2$; $0 = h_1 < h < h_2$; $h_1 < 0 < h < h_2$; $h_1 < h = 0 < h_2$; $h_1 < h < 0 < h_2$; $h_1 < h < h_2 = 0$; $h_1 < h < h_2 < 0$.

Let us consider the variant $0 < h_1 < h < h_2$. We have:

$$\frac{\partial \Phi(T, h)}{\partial T} = -\frac{z_1 z_2 (z_2 - z_1)(\beta_1 + \beta_2) x(h) e^{-(\beta_1+\beta_2)T}}{F^3(T, h)}$$

$$\times \left\{ z_1 z_2 (z_1 + z_2 - \beta_1 - \beta_2) - \lambda_1 \lambda_2 (z_1 + z_2 + \beta_1 + \beta_2) h e^{-(\beta_1+\beta_2)T} \right\},$$

$$T \geqslant 0, \ 0 < h_1 < h < h_2. \tag{22}$$

A sign of the derivative function (22) is defined by a sign of the function:

$$y(T, h) = z_1 z_2 (z_1 + z_2 - \beta_1 - \beta_2)$$
$$- \lambda_1 \lambda_2 (z_1 + z_2 + \beta_1 + \beta_2) h e^{-(\beta_1+\beta_2)T}, \ T \geqslant 0, \ 0 < h_1 < h < h_2. \tag{23}$$

Then because $h > 0$ the function (23) is the increasing function of the variable T (it increases from $y(T = 0, h) = z_1 z_2 (z_1 + z_2 - \beta_1 - \beta_2) - \lambda_1 \lambda_2 (z_1 + z_2 + \beta_1 + \beta_2) h$ to $y(T = \infty, h) = z_1 z_2 (z_1 + z_2 - \beta_1 - \beta_2) > 0$). There are three possible situations here: (a) $y(T = 0, h) > 0$ for $0 < h_1 < h < h_2$; (b) $y(T = 0, h = h_2) = 0$; (c) $y(T = 0, h) > 0$ for $0 < h_1 < h^*$; $y(T = 0, h) = 0$ for $h = h^*$; $y(T = 0, h) < 0$ for $h^* < h < h_2$, where $h^* = z_1 z_2 (z_1 + z_2 - \beta_1 - \beta_2) [\lambda_1 \lambda_2 (z_1 + z_2 + \beta_1 + \beta_2)]^{-1}$.

Let us consider the situation (a). Then the function (23) is a positive function of the variable T ($y(T = 0, h) > 0$, $T \geqslant 0$). According to this the derivative function (22) is a negative function of the variable T ($T \geqslant 0$). All these mean that the function (21) is decreasing from $\Phi(T = 0, h) > 0$ to $\Phi(T = \infty, h) > 0$. It follows from this that the function $\Phi(T, h) > 0$, $T \geqslant 0$ for the situation (a). Similarly it can be proved that the function $\Phi(T, h) > 0$, $T \geqslant 0$ for the situation (c).

Using the formulas (22), (23) it can be shown that for all the other variants described above (*Lemma 3*) for the situation $x(h) > 0$ the function $\Phi(T, h) > 0$, $T \geqslant 0$. *Lemma 6* is proved.

For the situation $x(h) < 0$ we have

Lemma 7. *The function* $\Phi(T, h) > 0$, $T \geqslant 0$ *if* $x(h) < 0$.

Proof. The lemma can be proved by applying the formulas (22), (23) (similar to applying these formulas for the *Lemma 6*) for the variants of parameter h changing which described in the *Lemma 3* for the situation when $x(h) < 0$. *Lemma 7* is proved.

Lemma 8. *The functions* $F_1(T, h)$, $F_2(T, h)$ *satisfy the inequation* $F_1(T, h) > F_2(T, h)$, $T \geqslant 0$, $-1 \leqslant h \leqslant 1$.

Proof. The lemma can be proved by applying the results of the *Lemmas 6, 7*. *Lemma 8* is proved.

Lemma 9. *The Eq. (12) does not have a solution.*

Proof. The lemma can be proved by applying the results of the *Lemmas 3, 4, 5* and *8*. *Lemma 9* is proved.

Lemma 10. *The derivative function* $p_T'(\tau_m)$ *defined in the formula (3) is a positive function of the variable* T $(p_T'(\tau_m) > 0,\ 0 \leqslant T \leqslant \tau_m,\ \tau_m \geqslant 0)$.

Proof. The lemma can be proved by applying the results of the *Lemmas 1, 2* and *9*. *Lemma 10* is proved.

Theorem 1. *The probability density function* $p_T(\tau_m)$ *is an increasing function of the variable* T $(0 \leqslant T \leqslant \tau_m,\ \tau_m \geqslant 0)$.

Proof. The theorem can be proved by applying the result of the *Lemma 10*. *Theorem 1* is proved.

Theorem 2. *The probability density function* $p_T(\tau_m)$ *reaches its global maximum at the point* $T = \tau_m$ $(0 \leqslant T \leqslant \tau_m,\ \tau_m \geqslant 0)$.

Proof. The theorem can be proved by applying the result of the *Theorem 1*. *Theorem 2* is proved.

Corollary 1. *The functions* $p_T(\tau^{(j)}) = 0$, $j = \overline{1, k}$ *from (2) are increasing functions of the variable* T $(0 \leqslant T \leqslant \tau_m,\ \tau_m \geqslant 0)$.

Proof. It follows from *Theorem 1*.

Corollary 2. *The likelihood function* $L(T | \tau^{(1)}, \tau^{(2)}, \ldots, \tau^{(k)})$ *defined in (2) reaches its global maximum at the point* $T = \tau_m$ *that the solution of the optimization problem (2) is the estimation of a dead time period duration* $\hat{T} = \tau_m$.

Proof. It follows from *Theorem 2*.

5   Conclusion and Future Research

The results obtained make it possible to solve the problem of a dead time period estimation without using a numerical computing: when observing (during the interval (t_0, t)) the flow of events the variables τ_k, $k = \overline{1, n}$ are being computed. After that it can be obtained $\tau_m = \min \tau_k$ $(k = \overline{1, n})$ and assumed $\hat{T} = \tau_m$.

Acknowledgments. The work is supported by Tomsk State University Competitiveness Improvement Program.

References

1. Gortsev, A., Sirotina, M.: Joint probability density function of modulated synchronous flow interval duration. In: Dudin, A., Nazarov, A., Yakupov, R. (eds.) ITMM 2014. CCIS, vol. 487, pp. 145–152. Springer, Heidelberg (2014)
2. Gortsev, A., Sirotina, M.: Joint probability density function of modulated synchronous flow interval duration under conditions of fixed dead time. In: Dudin, A., Nazarov, A., Yakupov, R. (eds.) ITMM 2015. CCIS, vol. 564, pp. 41–52. Springer, Heidelberg (2015). doi:10.1007/978-3-319-25861-4_4
3. Dudin, A.N., Klimenok, V.N.: Queue Systems With Correlated Flows, p. 175. Belorussian State University, Minsk (2000)
4. Basharin, G.P., Gajdamaka, U.V., Samujlov, K.E.: Mathematical theory of teletraffic and its applications to analysis of multiservice networks of the next ages. Autom. Comput. **2**, 11–21 (2013)
5. Kingman, J.F.C.: On doubly stochastic poisson process. In: Proceedings of Cambridge Phylosophical Society, vol. 60, no. 4, pp. 923–930. Cambridge University Press, Cambridge (1964)
6. Basharin, G.P., Kokotushkin, V.A., Naumov, V.A.: About the method of renewals of subnetwork computation: AN USSR, Techn. kibernetics, vol. 6, pp. 92–99 (1979)
7. Basharin, G.P., Kokotushkin, V.A., Naumov, V.A.: About the method of renewals of subnetwork computation: AN USSR, Techn. Kibernetics, vol. 1, pp. 55–61 (1980)
8. Lucantoni, D.M.: New results on the single server queue with a batch Markovian arrival process. Commun. Stat. Stochast. Models **7**, 1–46 (1991)
9. Card, H.C.: Doubly stochastic Poisson processes in artifical neural learning. IEEE Trans. Neural Netw. **9**(1), 229–231 (1998)
10. Bushlanov, I.V., Gortsev, A.M., Nezhel'skaya, L.A.: Estimating parameters of the synchronous twofold-stochastic flow of events. Autom. Remote Control **69**(9), 1517–1533 (2008)
11. Gortsev, A.M., Golofastova, M.N.: Optimal state estimation of modulated synchronous doubly stochastic flow of events: control, computation and informstics. Tomsk State Univ. J. **2**(23), 42–53 (2013)
12. Sirotina, M.N.: Optimal state estimation of modulated synchronous doubly stochastic flow of events in conditions of fixed dead time: control, computation and informstics. Tomsk State Univ. J. **1**(26), 63–72 (2014)
13. Bushlanov, I.V., Gortsev, A.M.: Optimal estimation of the states of a synchronous double stochastic flow of events. Avtomatika i Telemekhanika **9**, 40–51 (2004)
14. Bushlanov, I.V., Gortsev, A.M.: Optimal estimation of the states of a synchronous double stochastic flow of events. Autom. Remote Control **65**(9), 1389–1399 (2004)
15. Gortsev, A.M., Shmyrin, I.S.: Optimal estimation of the states of a synchronous double stochastic flow of events in the presence of a measurement errors of time instants. Autom. Remote Control **60**(1), 41–51 (1999)
16. Gortsev, A.M., Shmyrin, I.S.: Optimal state estimation of doubly stochastic flow under imprecise measurements of time instants. Dianzi Keji Daxue J. of the Univ. of Electron. Sci. and Technol. of China **27**(7), 52–66 (1998)
17. Bakholdina, M.A., Gortsev, A.M.: Optimal estimation of the states of modulated semi-synchronous integrated flow of events in condition of its incomplete observability. Appl. Math. Sci. **9**(29–32), 1433–1451 (2015)
18. Gortsev, A.M., Nezhel'skaya, L.A., Shevchenko, T.I.: Estimation of the states of an MC-stream of events in the presence of measurement errors. Russ. Phys. J. **36**(12), 1153–1167 (1993)

19. Bakholdina, M., Gortsev, A.: Joint probability density of the intervals length of the modulated semi-synchronous integrated flow of events and its recurrence conditions. In: Dudin, A., Nazarov, A., Yakupov, R., Gortsev, A. (eds.) ITMM 2014. CCIS, vol. 487, pp. 18–25. Springer, Heidelberg (2014)
20. Gortsev, A.M., Nezhel'skaya, L.A.: An asynchronous double stochastic flow with initiation of superflows events. Discrete Math. Appl. **21**(3), 283–290 (2011)
21. Vasil'eva, L.A., Gortsev, A.M.: Esimation of parameters of a double-stochastic flow of events under conditions of its incomplete observability. Autom. Remote Control **63**(3), 511–515 (2002)
22. Vasil'eva, L.A., Gortsev, A.M.: Parameter esimation of a doubly stochastic flow of events under incomplete observability. Avtomatika i Telemekhanika **3**, 179–184 (2002)
23. Gortsev, A.M., Shmyrin, I.S.: Optimal estimate of the parameters of a twice stochastic Poisson stream of events with errors in measuring times the events occur. Russ. Phys. J. **42**(4), 385–393 (1999)
24. Gortsev, A.M., Nezhel'skaya, L.A.: Estimate of parameters of synchronously alternating Poisson stream of events by the moment method. Telecommun. Radio Eng. (English translation of Elektrosvyaz and Radiotekhnika) **50**(1), 56–63 (1996)
25. Gortsev, A.M., Nezhel'skaya, L.A.: Estimation of the parameters of a synchro-alternating Poisson event flow by the method of moments. Radioteknika **40**(7–8), 6–10 (1995)
26. Gortsev, A.M., Klimov, I.S.: Estimation of the parameters of an alternating Poisson stream of events. Telecommun. Radio Eng. (English translation of Elektrosvyaz and Radiotekhnika) **48**(10), 40–45 (1993)
27. Gortsev, A.M., Klimov, I.S.: Estimation of intensity of Poisson stream of evnets for conditions under which it is partially unobservable. Telecommunations and Radio Engineering (English translation of Elektrosvyaz and Radiotekhnika) **47**(1), 33–38 (1992)
28. Gortsev, A.M., Solov'ev, A.A.: Joint probability density of interarrival interval of a flow of physical events with unextendable dead time period. Russ. Phys. J. **57**(7), 973–983 (2014)
29. Gortsev, A.M., Nezhel'skaya, L.A., Solov'ev, A.A.: Optimal state estimation in map event flows with unextendable dead time. Autom. Control **73**(8), 1316–1326 (2012)
30. Gortsev, A.M., Nissenbaum, O.V.: Estimation of the dead time period and parameters of an asynchronous alternative flow of events with unextendable dead time period. Russ. Phys. J. **48**(10), 1039–1054 (2005)
31. Gortsev, A.M., Nezhel'skaya, L.A.: Estimation of the dead time period and intensities of the synchronous double stochastic event flow. Radiotekhnika **10**, 8–16 (2004)
32. Vasil'eva, L.A., Gortsev, A.M.: Dead-time interval estimation of incompletely observable asynchronous bistochastic flow of events. Avtomatika i Telemekhanika **12**, 69–79 (2003)
33. Vasil'eva, L.A., Gortsev, A.M.: Estimation of the dead time of an asynchronous double stochastic flow of events under incomplete observability. Autom. Remote Control **64**(12), 1890–1898 (2003)
34. Gortsev, A.M., Nezhel'skaya, L.A.: Estimation of the dead-time period and parameters of a semi-synchronous double-stochastic stream of events. Meas. Tech. **46**(6), 536–545 (2003)
35. Gortsev, A.M., Parshina, M.E.: Estimation of parameters of an alternate stream of events in "dead" time conditions. Russ. Phys. J. **42**(4), 373–378 (1999)

36. Gortsev, A.M., Klimov, I.S.: Estimation of the non-observability period and intensity of Poisson event flow. Radiotekhnika **2**, 8–11 (1996)
37. Gortsev, A.M., Klimov, I.S.: An estimate for intensity of Poisson flow of events under the condition of its partial missing. Radiotekhnika **12**, 3–7 (1991)

TCP-NewReno Protocol Modification for MANET Networks

Denis Iovlev[1(✉)] and Anatoliy Korikov[1,2]

[1] Tomsk State University of Control Systems and Radioelectronics,
40 Lenina Prospect, Tomsk 634050, Russia
prodenx@t-sk.ru
[2] National Research Tomsk Polytechnic University,
30 Lenina Prospect, Tomsk 634050, Russia

Abstract. In recent years more and more attention has been paid to mobile ad hoc special purpose networks. The application of such networks expands the specter of opportunities in any sphere of human activity. The main objective of MANET networks, as well as any others, is data transmission. The TCP protocol provides for its execution. This article describes the features of MANET networks and contains results of model experiments with the TCP protocol under the influence of a high bit errors rate and changeable topology. Additionally, modifications for increasing the performance of the TCP-NewReno protocol are presented.

Keywords: MANET · TCP · Ad hoc · Protocol · Networks · Performance of TCP · Modification of TCP

1 Introduction

MANET is an autonomous wireless ad hoc special purpose network system consisting of independent nodes, which can move freely in any direction. Unlike wire and cellular networks, MANET has no fixed infrastructure and central management. The nodes can communicate directly if they are located in each others radio area, or through intermediate nodes, and each of them can be a transmission endpoint and router simultaneously.

The presence of mobility and the single-level principle of construction make MANET networks difficult to realize in real life. The difficulties lie in the developing of routing systems, resource management, etc.

At the moment there is much research in the field of routing [1–4] and transport protocols [5–8].

MANET networks [9,10] are suitable for use in combat, rescue operations and other situations, where it is not possible to use an expanded network infrastructure.

In such situations the possibility of data transmitting with delivery guarantee is necessary. For modern networks such a service is provided by the TCP protocol – the protocol of stream data transmitting with packet switching. It was intended

A. Dudin et al. (Eds.): ITMM 2016, CCIS 638, pp. 120–131, 2016.
DOI: 10.1007/978-3-319-44615-8_10

initially for networks with a low level of losses and fixed or weakly varying topology, which is natural for wired networks.

In research on the mechanisms of TCP congestion control in MANET [11] it was found that this protocol deals with its problems with insufficient efficiency in such networks. In the same research it was found that the TCP-NewReno protocol operates better in MANET.

The following research is dedicated to analysis of algorithms and parameters of TCP-NewReno protocol and its modifications for MANET networks.

Our research is based on imitating modeling on the basis of an event-discrete approach, which is carried out by MANET networks simulation. Such an approach allows us to research and develop original network protocols. NS-3 is recognized as one of the best event-discrete simulators for MANET modeling, researching and development of new protocols [12]. The NS-3.22 is used in our research. The main model parameters are given in [11]. Other parameters (the number of nodes, their initial location, rate, the number of UDP applications (User Datagram Protocol – the protocol of data delivery without logical connection establishment) and their transmission rate, duration of the simulated period) are specified by the conditions of the experiment.

2   Research

2.1   Problem Definition

There is an imitation model of the network consisting of 5 mobile nodes moving at any time. The physical and canal levels are realized according to the standard 802.11a (Wi-Fi). The communication method is half-duplex (the device transmits or receives simultaneously) with a flow capacity of 54 Mbps/sec. At the network level the IP4v protocol is used. The AODV protocol is responsible for routing. The simulation period lasts 250 s.

It is necessary to conduct an analysis of the parameters and TCP-NewReno protocols algorithms and to perform a modification of the protocol in order to increase its efficiency. Efficiency is understood as the amount of successfully transmitted data during the whole simulation period.

2.2   Theory

With an understanding of the TCP protocols work and MANET features, it is possible to make preliminary conclusions on their compatibility [10].

Bit Errors. In the ad hoc networks bit errors have a quite low level of BER, if the network is multilevel and communicating nodes are closer than the detection limit and nonmovable. In the cellular networks BER increases, because with the change in distance comes a change in the received power and translation parameters. An even greater increase in BER is caused by use of the single-rank principle of network organization, It adds medium access conflict problems, problems of hidden terminal, etc.

The high segment losses level makes the congestion control spend much time in states of congestion elimination and slow start. Because of this, frequent delays and pauses in transmission, and excessive retransmissions occur.

Route Errors. The high level of bit errors in the network leads to losses in the routing protocols packets. Because of this there may be routes breaks and changes. The packet with a TCP segment will be dropped at the intermediate node. As a result we have the same consequences as in the previous section.

Changeable Topology. The high mobility of the nodes in MANET networks leads to frequent shot-term breaks and changes of routes. This feature negatively influences the work of the TCP protocol because the connection parameters are selected principally for the certain route. But the change of the route leads to changes in its metric. And then the calculated parameters may not fit the new route. For example, the high value of CWND (Congestion Window) may cause network congestion and segment losses after route metric degradation, and the RTO high value (Retransmission Time-out) may cause downtime in transmission after route recovery.

Because of the changeable topology, the transmissioned packets may be received in the wrong order since segments are sent via different routes with different metrics. As the result, gaps in the stream could appear, which would lead to a decrease in the transmission speed and excessive data retranslation.

Multipath Routing. It promotes the distribution of the load on separate nodes and improves the translation speed, but a typical TCP protocol is unable to work with it. The use of such a type of routing leads to disordered entry of segments at the receiving end, the consequences of which are stated above.

2.3   Analysis of the TCP-NewReno Protocol Parameters and Algorithms

Medium Access Conflict. In wireless ad hoc networks the channel throughput is divided between all the nodes and that is why the conflicts appear in simultaneous medium access, and part of the transmitted data gets lost.

Conflicts lead to route errors. In order to determine the influence of the conflicts on TCP efficiency, the experiment with a network model was conducted, where 8 nonmovable nodes were arranged in the lattice sites (Fig. 1).

In order to eliminate the influence of the mobility of the nodes their location was fixed.

The shortest way between the server and the client has a length of 3 hops, but it does not mean that the route between them has the same length and consists of the same nodes. Also, the data from the C node was transmitted to the D node according to the UDP protocol with the speed of 5 mbps in order to create multiple signals from the nearby nodes (Fig 2).

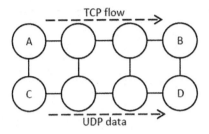

Fig. 1. Experiment illustration

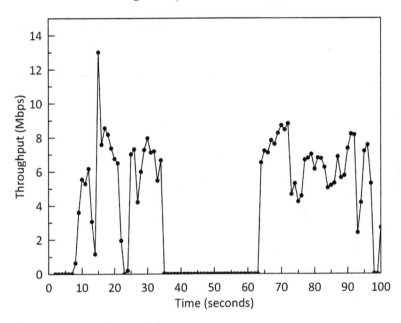

Fig. 2. TCP connection throughput as a function of time

The conditions of the experiment are presented in Fig. 3: (a) – route length change; (b) – sequence of numbers of segments sent from the client node; (c) – RTO change; (d) – instantaneous data stream speed.

Increasing the number of hops leads to performance degradation. The route of greater length passes through the nodes, which are used by the UDP application. It leads to additional data and confirmations delay, because of which TCP-NewReno enters the slow start mode.

In Fig. 3b one can see how test segments starting from the 32 s were sent in order to check the paths availability. Wherein, the RTO value increased every time. At the moment of 48.8 s the route went back to the previous state, but the TCP protocol was waiting for the confirmation of the last test segment. For this reason time was wasted, which affected the general performance.

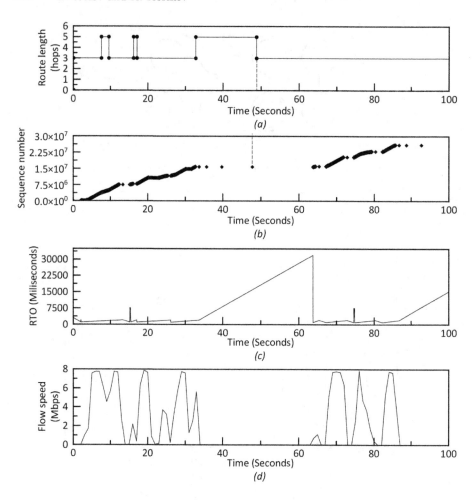

Fig. 3. TCP-NewReno protocol performance

Dynamic Topology. In the model experiment 15 nodes were moving in a random direction. The data was transmitting from the client to the server only. 4 simulation runs were conducted, in which the nodes velocity amounted to 2, 6, 10 and 20 m/s respectively.

In Fig. 4 the characteristic of the shortest way between the client and the server is presented. The marks on the graphic indicate the time moments of the path changes, and the broken line – its length. In every run the same pattern of behavior was used, so the duration of the simulated time period was decreasing inversely related to speed: 1500, 500, 300 and 150 s.

In the graphs of Fig. 4b, e, the data transmission speed depending on the time is shown. Let us note that the curves of Fig. 4b, e correlate with changes in the length of the route [11].

At a node speed equal to 2 m/s (Fig. 4b) the data stream was sent through the routes of 1, 2 and 3 hop length. Every change in the network leads to the data transmission speed dropping up to zero, after which recovery occurs with a delay.

Starting from the normalized time moment 0.34 (Fig. 4a), a distance reduction up to 1 hop and lost segments led to a long connection pause. The protocol did not have time to establish the necessary connection settings.

At 6 m/s (Fig. 4c) the path changes come faster, so the protocol does not have time to react to each of them. The attempts to eliminate nonexistent congestion and determine the availability of a route with the help of test segments lead to a weakening of data transmission in the period of normalized time from 0.16 to 0.37 and to its complete absence after 0.6, because the test segments are sent during the topology changes.

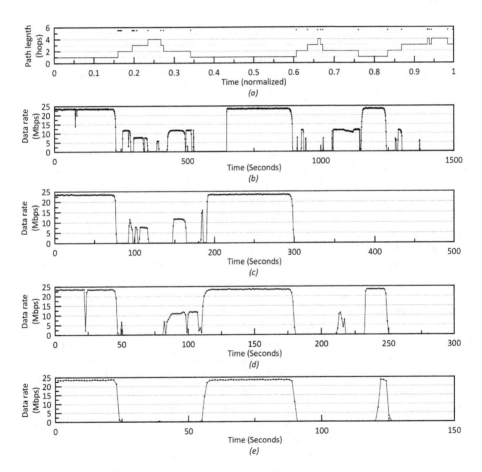

Fig. 4. TCP-NewReno protocol performance at different speeds of node movement

At 10 m/s (Fig. 4d) even greater weakening is observed in the first half of the runs time and, surprisingly, the appearance of the traffic in the second. The short-form route changes barely affected the connection settings. Unlike the previous run, it allowed the recommencement of transmission in the period of normalized time [0.76, 0.8]. The dissynchronization occurred between the test segments and topology changes.

At an even higher speed (Fig. 4e) data was not transmitted through the routes longer than one hop.

In each of the 4 cases the most problematic time for the TCP protocol starts from the moment 0.6 of normalized time. In order to check the data transmission possibility at this period of time all the 4 runs were repeated with the replacement of the TCP protocol by UDP. The obtained results are shown in Fig. 5. It follows from the figure that the network and link levels enabled the data transmission, which was not used by the TCP protocol.

The Complex Experiment. The interaction of the processes occurring in the network is manifested in the data transmission speed drops. Let us turn to the results of one of the experiments on choosing the congestion control mechanism in order to explain the causes of this phenomenon.

Let us take a look at CWND value dynamics (Fig. 6).

The instability in the changes of this parameter indicates the presence of congestion control switching states. The reason for this is the segments loss, the time of which is also shown in the graph. Moreover, the data loss did not occur for the reason of buffer overflow at any of the nodes, i.e. congestion did not appear in the network. The states of slow start and congestion elimination lead to data delay and unnecessary segments retransmission. The cause of the long pauses was discussed earlier. Let us take a look at the problem of unnecessary retransmissions.

During the elimination of the congestion, the segments with a sequence number equal to the last AckSeq are sent repeatedly. After that, with every received ACK (Acknowledgement) double, the value of CWND increases by one MSS (Maximum Segment Size) until the new AscSeq (Acknowledgement Sequence Number) is received. If it is equal to the maximum sequence number of the sent byte in the stream, than the congestion elimination is considered successful. Otherwise, the process continues. The examined algorithm leads to unnecessary retransmissions if the first sent segment in the state of congestion elimination is lost. In this case the transmitting side sends all the data from the buffer, then the RTO time-out expiration occurs and the mechanism goes to the slow start mode. This transition may cause the third attempt to send the existing data to the receiving end, since the ACK doubles received in the previous state do not indicate the sequence number of the received segments (Fig. 7).

The high CWND value during the route change leads to occurrence of congestion and, consequently, to the loss of large number of segments, which was revealed during the model experiment, in which a network consisting of 5 nodes was used.

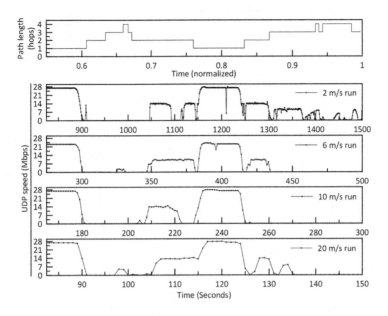

Fig. 5. UDP protocol performance at different speeds of node movement

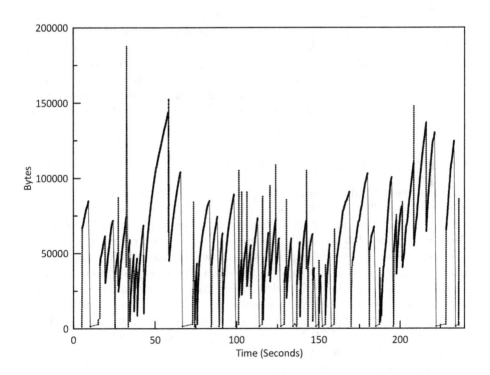

Fig. 6. Congestion window of the TCP-NewReno protocol in MANET

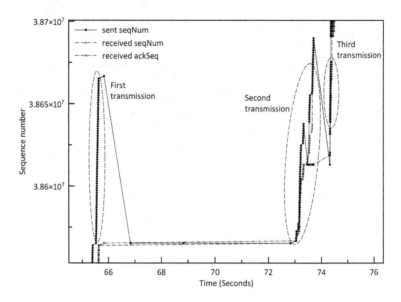

Fig. 7. Demonstration of the excess retransmissions

At the start moment the routes length between the client and the server was equal to one hop. After a while, the client stopped the data transmission in order to receive the remaining acknowledgements and to clear the entire congestion window. After that, the clients node shifted in such a way that the route to the server began to pass through the node, which was used to transmit another data stream, and the stream renews. The large number of segments quickly filled the remaining buffer space of the intermediate node and a part of the segments was lost (Fig. 8).

2.4   TCP-NewReno Protocol Modification

Some changes were added to congestion control in order to boost the performance of the protocol. The ssThresh (Slow Start Threshold) was reduced to MSS. RTO does not change after loss of the test segment. When receiving the ACK double the congestion window remains the same and the first unconfirmed segment is sent once again. After receiving a new acknowledgement CWND decreases or increases to such a value at which the free space of congestion window allows to send DelAckCount (Delay Acknowledgement Count) of the segments.

Modified congestion control was used in the re-runs of the experiments. In Fig. 9 a comparison of the results of the first experiment with the standard TCP-NewReno and modified TCP-NewReno is shown. It can be seen that the amount of successfully transmitted data increased.

Figure 10 presents the results of a 6 m/s run.

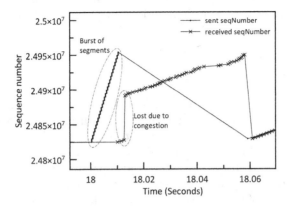

Fig. 8. Congestion appears after route change

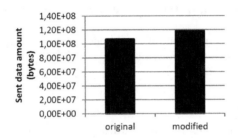

Fig. 9. Comparison of the original and the modified TCP-NewReno protocol

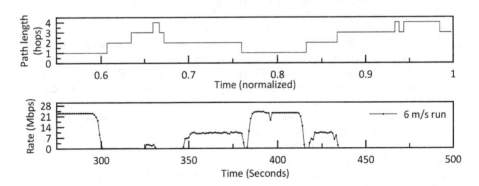

Fig. 10. The performance of the modified TCP-NewReno protocol at 6 m/s nodes movement speed

The fixed RTO value in the process of determining the presence of the path allowed the use of longer routes in a period of time, in which the original mechanism was not able to do so.

3   Conclusion

The experimental results described in Sect. 2, confirm the preliminary conclusions on the effectiveness of the TCP protocol application in MANET.

The protocol interprets the segments losses incorrectly because of the bit errors and errors of route that affect the TCP performance.

Frequent switching in the slow start and congestion elimination mode leads to the large number of unnecessary segment retransmissions.

At the high level of mobility the protocol is able to transmit the data through one-hop routes only, though the using of UDP in the same conditions showed the possibility of using longer paths.

The changes in the congestion control gave a boost to the protocol performance. The weakening of the onslaught on the network during the congestion elimination and reduction of the interval between the test segments improved the results.

In order to solve the problem of the unnecessary retransmissions it is necessary to make changes to the protocol itself.

References

1. Caro, G. D., Ducatelle, F., Gambardella, L. M.: AntHocNet: an ant-based hybridrouting algorithm for mobile ad hoc networks. Technical report No. IDSIA-25-04-2004, Dalle Molle Institute for Artificial Intelligence, Switzerland, August 2004. (Also appeared in the Proceedings of Parallel Problem Solving from Nature VIII, LNCS, vol. 3242, pp. 461–470. Springer (2004))
2. Mueller, S., Tsang, R.P., Ghosal, D.: Multipath routing in mobile ad hoc networks: issues and challenges. In: Calzarossa, M.C., Gelenbe, E. (eds.) MASCOTS 2003. LNCS, vol. 2965, pp. 209–234. Springer, Heidelberg (2004)
3. Xue, Y., Nahrstedt, K.: Providing fault-tolerant ad hoc routing service in adversarial environments. Wireless Pers. Commun. **29**, 367–388 (2004)
4. Elhadef, M., Boukerche, A., Elkadiki, H.: A distributed fault identication protocol for wireless and mobile ad hoc networks. J. parallel Distrib. Comput. **68**, 321–335 (2008)
5. Balakrishnan, H., Padmanabhan, V., Seshan, S., Katz, R.: A comparison of mechanisms for improving TCP performance over wireless links. ACM Transactions on Networking **5**(6), 756–769 (1997)
6. Gerla, M., Bagrodia, R., Zhang, L., et al.: TCP over wireless multihop protocols: simulation and experiments. In: IEEE ICC 1999, Vancouver, Canada, pp. 1089–1094, June 1999
7. Chandran, K., Raghunathan, S., Venkatesan, S., Prakash, R.: A feedback-based scheme for improving TCP performance in ad hoc wireless networks. IEEE Pers. Commun. **8**(1), 34–39 (2001)

8. Romanowicz, E.: TCP with Explicit Link Failure Notification. Department of Computer Science, York University, Toronto (2008)
9. Kumar, M., Mishra, R.: An overview of MANET: history, challenges and applications. Indian J. Comput. Sci. Eng. (IJCSE) **3**(1), 408–417 (2012)
10. Radhika, R.R.: Handbook of Mobile Ad Hoc Networks for Mobility Models, p. 1169. Springer, New York (2011)
11. Iovlev, D.I.: Vibor modeli TCP dlya setey MANET. Vliyanie odnourovnego principa organ-izacii setey na protocol TCP. Proc. Tomsk State Univ. Control Syst. Radioelectronics **3**, 123–127 (2015)
12. Romanov, S.V., Zholobov, A.N., Prozorov, D.E.: Simulyatory besprovodnykh MANET-setey. Infocommunication Technologies, vol. 3, pp. 28–33 (2012)

New Results for a Thinned Renewal Process

Viktor Ivnitskii[1] and Alexander Moiseev[2(✉)]

[1] Railway Research Institute, Moscow, Russia
ivnitsky.viktor@vniizht.ru
[2] Tomsk State University, Tomsk, Russia
moiseev.tsu@gmail.com

Abstract. In the paper, we present some analytical results obtained for probability characteristics of a thinned renewal process. The thinning is processed according to a given function which depends on evolution time and on the number of already thinned points. The characteristics are obtained in the form of Laplace–Stieltjes transforms which are defined by a system of recurrence integral equations. Some partial cases are also considered.

Keywords: Renewal process · Thinned point process · Time-depended function of thinning · Laplace–Stieltjes transform

1 Introduction

Thinning of random point processes plays an important role in the theory of random processes and in queueing theory. A thinned point process is a random process which is constructed on some other random point process by excluding some points according to a given probability, or given law (function), or by using additional random processes, etc. In queueing theory, such thinned processes can be used as independent models of processes forming different real fields, and also as the models of arrival processes for queueing systems.

In the Russian edition [1] of books [2,3], Yu.K. Belyaev remarked that the thinned Poisson process is still a Poisson process. A. Renyi [4] proved the first theorem about the thinning of a renewal process. The main idea is that we place a point from the source process into the thinned process with a constant probability q and do not place with probability $(1-q)$. A rate of the thinned process can be considered as constant if we make suitable changes in the time scale. If we make a thinning n times with different probabilities $q_1, \ldots, q_n$ where $n \to \infty$ and all $q_1, \ldots, q_n \to 0$, then the resulting thinned process converges to a Poisson process. Later, Yu.K. Belyaev made a generalization [5] of the theorem to the case of an arbitrary point process. In the book [6], this theorem was generalized to the case of a non-stationary asymptotically thinned process. A.D. Soloviev demonstrated [7] that the first epoch of a rare event in a regenerating process is asymptotically exponential with a parameter equal to 1 (if we use an appropriate scale).

© Springer International Publishing Switzerland 2016
A. Dudin et al. (Eds.): ITMM 2016, CCIS 638, pp. 132–139, 2016.
DOI: 10.1007/978-3-319-44615-8_11

All these papers are focused on obtaining asymptotic results under a condition of unlimited thinning. One main feature of the construction of these approaches is that a thinning process depends neither on time nor on the number of epochs before. In our paper, we use both parameters to involve them in the process of thinning of the source renewal point process.

Some other results regarding thinned point processes and their applications can be found in the books and papers [8–11]. The inverse (in a certain sense) problem of "thickening" of the point process is considered in the paper [12].

In Sect. 2, the mathematical model and the problem are formulated. A solution is given in Sect. 3. In Sect. 4, we consider a special case of thinning where the thinning probabilities depend on the number of thinned points and do not depend on the current time, but the arrival probabilities for source process depend both on the number of thinned points and on the last decision of the thinning process.

2   Problem Statement

Let us consider stationary renewal process (we call it as *source process*). Let $t_0 = 0$ be the initial time moment and the number of the points in the *thinned process* under construction is equal to 0. Let $F(x)$ be a cumulative distribution function (c.d.f.) for all epochs except the first one which has distribution function $F_0(x)$. The thinning of the considering renewal process is performed in the following way. Let the point arrive in the source process at the time moment $t > t_0$ and a number of the points in the thinned process before the moment t is equal to i. In this case we generate a point in the thinned process with given probability $P_i(t)$ where $i = 0, 1, 2, \ldots$ (and do not generate with probability $1 - P_i(t)$).

The aim of the study is a probability distribution of number of the points in the thinned process occurring before some arbitrary time moment t.

3   Solution

Let us denote by $\nu(t)$ a number of points in the thinned process occurring before the moment t and by $\xi(t)$ a residual time from the moment t until the next point in the source process. Let us consider the two-dimensional process $\zeta(t) = \{\nu(t), \xi(t)\}$. This process is Markovian because its evolution after the moment t depends only on values $\nu(t)$ and $\xi(t)$ but does not depend on its states before t. Let us denote $\varphi_i(t, x) = \mathrm{P}\{\nu(t) = i, \xi(t) < x\}$, $\varphi_i(t) = \varphi_i(t, \infty)$ for $i = 0, 1, 2, \ldots$. The problem is to find the probability distribution $\varphi_i(t)$ of the number of points in the thinned process occurring before the time moment t.

At the first stage, let us derive the distributions $\varphi_i(t, x)$. We can write the following system of differential equations for these functions:

$$\varphi_0(t + \Delta t, x - \Delta t) = \varphi_0(t, x) - \varphi_0(t, \Delta t) + \varphi_0(t, \Delta t)[1 - P_0(t)]F(x),$$

$$\varphi_i(t + \Delta t, x - \Delta t) = \varphi_i(t, x) - \varphi_i(t, \Delta t) + \varphi_{i-1}(t, \Delta t)P_{i-1}(t)F(x)$$
$$+ \varphi_i(t, \Delta t)[1 - P_i(t)]F(x) \text{ for } i > 0.$$

Then,, from this system, we can derive the following system of differential equations:

$$\frac{\partial \varphi_0(t, x)}{\partial t} - \frac{\partial \varphi_0(t, x)}{\partial x} = -\frac{\partial \varphi_0(t, 0)}{\partial x} + \frac{\partial \varphi_0(t, 0)}{\partial x}[1 - P_0(t)]F(x),$$

$$\frac{\partial \varphi_i(t, x)}{\partial t} - \frac{\partial \varphi_i(t, x)}{\partial x} = -\frac{\partial \varphi_i(t, 0)}{\partial x} + \frac{\partial \varphi_{i-1}(t, 0)}{\partial x}P_{i-1}(t)F(x) \tag{1}$$
$$+ \frac{\partial \varphi_i(t, 0)}{\partial x}[1 - P_i(t)]F(x) \text{ for } i > 0$$

with the initial condition

$$\varphi_0(0, x) = F_0(x). \tag{2}$$

Let us use the following notations:

$$\tilde{F}_0(s) = \int_0^\infty e^{-sx} dF_0(x), \quad \tilde{F}(s) = \int_0^\infty e^{-sx} dF(x), \tag{3}$$

$$\tilde{\varphi}_i(u, s) = \int_0^\infty \int_0^\infty e^{-ut} e^{-sx} d_x \varphi_i(t, x) dt, \quad \tilde{\varphi}_i'(u) = \int_0^\infty e^{-ut} \frac{\partial \varphi_i(t, 0)}{\partial x} dt, \tag{4}$$

$$\tilde{\psi}_i'(u) = \int_0^\infty e^{-ut} \frac{\partial \varphi_i(t, 0)}{\partial x} P_i(t) dt, \quad i = 0, 1, 2, \dots \tag{5}$$

We can prove the following theorem.

Theorem 1. *Laplace–Stieltjes transforms $\tilde{\varphi}_i(u, s)$ satisfy the following expressions*

$$\tilde{\varphi}_0(u, s) = \frac{1}{u - s} \left[-\tilde{\varphi}_0'(u) \left(1 - \tilde{F}(s)\right) - \tilde{\psi}_0'(u)\tilde{F}(s) + \tilde{F}_0(s) \right], \tag{6}$$

$$\tilde{\varphi}_i(u, s) = \frac{1}{u - s} \left[-\tilde{\varphi}_i'(u) \left(1 - \tilde{F}(s)\right) + \tilde{\psi}_{i-1}'(u)\tilde{F}(s) - \tilde{\psi}_i'(u)\tilde{F}(s) \right], i > 0, \tag{7}$$

where $\frac{\partial \varphi_i(t, 0)}{\partial x}$ in the expressions (4)–(5) for $\tilde{\varphi}_i'(u)$ and $\tilde{\psi}_i'(u)$ $(i = 0, 1, 2, \dots)$ are determined by the following system of recurrence integral equations

$$\left(1 - \tilde{F}(u)\right) \int_0^\infty e^{-ut} \frac{\partial \varphi_0(t, 0)}{\partial x} dt + \tilde{F}(u) \int_0^\infty e^{-ut} \frac{\partial \varphi_0(t, 0)}{\partial x} P_0(t) dt = \tilde{F}_0(u), \tag{8}$$

$$\left(1 - \tilde{F}(u)\right) \int_0^\infty e^{-ut} \frac{\partial \varphi_i(t,0)}{\partial x} dt + \tilde{F}(u) \int_0^\infty e^{-ut} \frac{\partial \varphi_i(t,0)}{\partial x} P_i(t) dt = \qquad (9)$$

$$\tilde{F}(u) \int_0^\infty e^{-ut} \frac{\partial \varphi_{i-1}(t,0)}{\partial x} P_{i-1}(t) dt, \quad i > 0.$$

Proof. Let us apply a Laplace–Stieltjes transform in the system (1)–(2), we obtain the system

$$\tilde{\varphi}_0(u,s)(u-s) = -\tilde{\varphi}_0'(u)\left(1 - \tilde{F}(s)\right) - \tilde{\psi}_0'(u)\tilde{F}(s) + \tilde{F}_0(s), \qquad (10)$$

$$\tilde{\varphi}_i(u,s)(u-s) = -\tilde{\varphi}_i'(u)\left(1 - \tilde{F}(s)\right) + \tilde{\psi}_{i-1}'(u)\tilde{F}(s) - \tilde{\psi}_i'(u)\tilde{F}(s), \quad i > 0, \ (11)$$

from which we can directly derive the expressions (6)–(7).

If we substitute $u = s$ into the equalities (10)–(11) we obtain the following expressions

$$-\tilde{\varphi}_0'(u)\left(1 - \tilde{F}(u)\right) - \tilde{\psi}_0'(u)\tilde{F}(u) + \tilde{F}_0(u) = 0,$$

$$-\tilde{\varphi}_i'(u)\left(1 - \tilde{F}(u)\right) + \tilde{\psi}_{i-1}'(u)\tilde{F}(u) - \tilde{\psi}_i'(u)\tilde{F}(u) = 0, \quad i > 0.$$

Substituting here expressions (3)–(5), we obtain the system (8)–(9). The theorem is proved.

The goal of the study is to find the probability distribution functions $\varphi_i(t)$. Using the notations

$$\tilde{\varphi}_i(u) = \int_0^\infty e^{-ut} \varphi_i(t); dt, \quad \tilde{P}_i(u) = \int_0^\infty e^{-ut} P_i(t) dt,$$

for $i = 0, 1, 2, \ldots$, we can make the following conclusion from Theorem 1.

Corollary 1. *Laplace transforms $\tilde{\varphi}_i(u)$ of the probability distribution functions $\varphi_i(t)$ for $i = 0, 1, 2, \ldots$ are determined by the following expressions:*

$$\tilde{\varphi}_0(u) = \frac{1}{u}\left[-\tilde{\psi}_0'(u) + \tilde{F}_0(u)\right], \qquad (12)$$

$$\tilde{\varphi}_i(u) = \frac{1}{u}\left[\tilde{\psi}_{i-1}'(u) - \tilde{\psi}_i'(u)\right], \quad i > 0. \qquad (13)$$

It is interesting to see what will happen if we consider partial cases of the model under study. The first one is when thinning probabilities do not depend on the number of the points occurring in the thinned process but depend on the current time t. In this case, we have $P_i(t) = P(t)$ for all $i = 0, 1, 2, \ldots$ Unfortunately, we obtain only a slightly simpler form for the functions $\tilde{\psi}_i'(u)$ in the expression (5) but we still need to use the expressions (12)–(13) to calculate probability distributions $\varphi_i(t)$.

In the contrary case, the thinning probabilities depend only on a number of points occurring in the thinned process before the current time moment t and do not depend on the value of t. We can obtain some results which are more simple than the results derived in this section. Such a case with some additional complicate condition is considered in the following section.

4   Special Case: Thinning Probabilities Depend only on Number of Points

Let us consider the case when the thinning probabilities $P_i(t)$ do not depend on the value of t. So, we can use them in the form P_i as a probability that any point of the source process generates a point in the thinned process if the number of points already occurring in the thinned process equals i. In this section we consider a more complex problem when arrivals of the points in the source process depend on the number of points in the thinned process too and, in addition, they depend on the last decision of the thinning: did the last arrival generate a thinned point or not. So, the arrivals and thinning process are described as follows.

The first point in the source process arrives at the random epoch with c.d.f. $F_0(x)$. Then, if some point occurs in the source process it may be thinned into the thinned process with the probability P_i where i is a number of points in the thinned process. If it does then the epoch of the next arrival in the source process has c.d.f. $F_i(x)$, $i \geq 1$. Otherwise (if the source point was not thinned into the thinned process), the epoch of the next arrival in the source process will have c.d.f. $G_i(x)$, $i \geq 0$. The goal of the study is the same as earlier – to find the probability distribution of the number of points occurring in the thinned process before the time moment $t > 0$ if at the initial moment this number is equal to zero.

Using the notations from Sect. 3, we can write the following system of differential equations for the functions $\varphi_i(s)$:

$$\varphi_0(t + \Delta t, x - \Delta t) = \varphi_0(t, x) - \varphi_0(t, \Delta t) + \varphi_0(t, \Delta t)[1 - P_0]G_0(x),$$

$$\varphi_i(t + \Delta t, x - \Delta t) = \varphi_i(t, x) - \varphi_i(t, \Delta t) + \varphi_{i-1}(t, \Delta t)P_{i-1}F_i(x) \qquad (14)$$
$$+ \varphi_i(t, \Delta t)[1 - P_i]G_i(x) \text{ for } i > 0.$$

We can derive the following system of differential equations from (14)

$$\frac{\partial \varphi_0(t, x)}{\partial t} - \frac{\partial \varphi_0(t, x)}{\partial x} = -\frac{\partial \varphi_0(t, 0)}{\partial x} + \frac{\partial \varphi_0(t, 0)}{\partial x}[1 - P_0]G_0(x),$$

$$\frac{\partial \varphi_i(t, x)}{\partial t} - \frac{\partial \varphi_i(t, x)}{\partial x} = -\frac{\partial \varphi_i(t, 0)}{\partial x} + \frac{\partial \varphi_{i-1}(t, 0)}{\partial x}P_{i-1}F_i(x) \qquad (15)$$
$$+ \frac{\partial \varphi_i(t, 0)}{\partial x}[1 - P_i]G_i(x) \text{ for } i > 0$$

with initial condition

$$\varphi_0(0, x) = F_0(x). \tag{16}$$

Let us use the following notations:

$$\tilde{F}_i(s) = \int_0^\infty e^{-sx} dF_i(x), \quad \tilde{G}_i(s) = \int_0^\infty e^{-sx} dG_i(x),$$

$$\tilde{\varphi}_i(u, s) = \int_0^\infty \int_0^\infty e^{-ut} e^{-sx} d_x \varphi_i(t, x) dt, \quad \tilde{\varphi}_i'(u) = \int_0^\infty e^{-ut} \frac{\partial \varphi_i(t, 0)}{\partial x} dt$$

for $i = 0, 1, 2, \ldots$ We can formulate the following theorem.

Theorem 2. *Laplace–Stieltjes transforms $\tilde{\varphi}_i(u, s)$ satisfy the following expressions*

$$\tilde{\varphi}_0(u, s) = \frac{1}{u - s} \left[-\tilde{\varphi}_0'(u) \left(1 - (1 - P_0)\tilde{G}_0(s) \right) + \tilde{F}_0(s) \right], \tag{17}$$

$$\tilde{\varphi}_i(u, s) = \frac{1}{u - s} \left[-\tilde{\varphi}_i'(u) \left(1 - (1 - P_i)\tilde{G}_i(s) \right) + P_{i-1}\tilde{\varphi}_{i-1}'(u)\tilde{F}_i(s) \right], \quad i > 0, \tag{18}$$

where $\tilde{\varphi}_i'(u)$ are determined by the following recurrence equations

$$\tilde{\varphi}_0'(u) = \frac{\tilde{F}_0(u)}{1 - (1 - P_0)\tilde{G}_0(u)}, \tag{19}$$

$$\tilde{\varphi}_i'(u) = \frac{\tilde{\varphi}_{i-1}'(u) P_{i-1} \tilde{F}_i(u)}{1 - (1 - P_i)\tilde{G}_i(u)}, \quad i > 0. \tag{20}$$

Proof. Let us apply a Laplace–Stieltjes transform in the system (15)–(16), we obtain the following equations

$$\tilde{\varphi}_0(u, s)(u - s) = -\tilde{\varphi}_0'(u) \left[1 - (1 - P_0)\tilde{G}_0(s) \right] + \tilde{F}_0(s), \tag{21}$$

$$\tilde{\varphi}_i(u, s)(u - s) = -\tilde{\varphi}_i'(u) \left[1 - (1 - P_i)\tilde{G}_i(s) \right] + P_{i-1}(u)\tilde{\varphi}_{i-1}'(u)\tilde{F}(s), \quad i > 0, \tag{22}$$

from which we directly derive the expressions (17)–(18).

If we substitute $u = s$ into the equalities (21)–(22) we obtain the following expressions

$$-\tilde{\varphi}_0'(u) \left[1 - (1 - P_0)\tilde{G}_0(u) \right] + \tilde{F}_0(u) = 0,$$

$$-\tilde{\varphi}_i'(u) \left[1 - (1 - P_i)\tilde{G}_i(u) \right] + P_{i-1}\tilde{\varphi}_{i-1}'(u)\tilde{F}_i(u) = 0, \quad i > 0.$$

These expressions give us the recurrence Eqs. (19)–(20). The theorem is proved.

The goal of the study is to find the probability distribution functions $\varphi_i(t)$. Using the notations $\tilde{\varphi}_i(u) = \int_0^\infty e^{-ut}\varphi_i(t)dt$ for $i = 0, 1, 2, \ldots$, we can make the following conclusion from Theorem 2.

Corollary 2. *Laplace transforms $\tilde{\varphi}_i(u)$ of the probability distribution functions $\varphi_i(t)$ are determined by the following expressions*

$$\tilde{\varphi}_0(u) = \frac{1}{u}\left[\tilde{\varphi}_0'(u)P_0 + 1\right],$$

$$\tilde{\varphi}_i(u) = \frac{1}{u}\left[-P_i\tilde{\varphi}_i'(u) + P_{i-1}\tilde{\varphi}_{i-1}'(u)\right], \quad i > 0,$$

where functions $\tilde{\varphi}_i'(u)$ are determined by the system (19)–(20).

5   Conclusions

In the paper, we have considered a thinning process which is performed on a source renewal point process according to given functions $P_i(t)$. These functions depend both on the time t of the process evolution and on the number i of the points already thinned in the resulting process. In Sect. 4, we have considered a special case of thinning when thinning probabilities do not depend on time and on the number of thinned points but at the same time the probability of the future arrival significantly depend on the last thinning decision. All results are presented in the form of Laplace–Stieltjes transforms.

The results of the paper may be useful both for solving practical problems with the appropriate models and for using as models of arrival processes for queueing systems and networks.

References

1. Cox, D.R., Smith, W.L.: Renewal Theory. Soviet Radio, Moscow (1967). (in Russian)
2. Cox, D.R.: Renewal Theory. Wiley, New York (1962)
3. Cox, D.R., Smith, W.L.: Queues. Wiley, New York (1961)
4. Rényi, A.: A characterization of poisson processes. Magyar Tud. Akad. Mat. Kutato Int. Kozl. **1**(4), 519–527 (1956). (in Hungarian)
5. Belyaev, Y.: Limit theorems for rearing flows. Probab. Theor. Appl. **8**(2), 175–184 (1963). (in Russian)
6. Gnedenko, B.V., Kovalenko, I.N.: Introduction in Queueing Theory. KomKniga, Moscow (2005). (in Russian)
7. Soloviev, A.D.: Asymptotic behaviour of the first epoch of rare event. Eng. Cybern. **6**, 79–89 (1971). (in Russian)
8. Streit, R.L.: Poisson Point Processes: Imaging, Tracking, and Sensing. Springer, New York (2010)
9. Serfozo, R.: A Course in Applied Stochastic Processes. Springer, New York (2009)

10. Assunção, R.M., Ferrari, P.A.: Independence of thinned processes characterizes the poisson process: an elementary proof and a statistical application. TEST **16**(2), 333–345 (2007)
11. Kushnir, A.O.: Asymptotic behavior of a renewal process thinned by an alternating process. Cybern. Syst. Anal. **29**(1), 20–25 (1993)
12. Gurel-Gurevich, O., Peled, R.: Poisson thickening. Israel J. of Math. **196**(1), 215–234 (2013)

A Retrial Queueing System with Renewal Input and Phase Type Service Time Distribution

Valentina Klimenok[(⊠)]

Department of Applied Mathematics and Computer Science,
Belarusian State University, 220030 Minsk, Belarus
vklimenok@yandex.ru

Abstract. In this paper, a single-server retrial queue with renewal input, phase type service time distribution and a constant retrial rate is analyzed. A constant retrial rate is typical for some real world systems where the intensity of individual retrials is inversely proportional to the number of customers in the orbit or only one customer from the orbit is allowed to make the retrials. A distinguishing feature of the system under consideration is an arbitrary distribution of inter-arrival times and phase type service time distribution while the vast majority of previous research is devoted to retrial systems with a stationary Poisson input or Markovian extensions and exponentially distributed service times. We derive the stationary distributions of the system states and the Laplace-Stieltjes transform of the sojourn time distribution.

Keywords: Renewal input · Phase-type distribution · Constant retrial rate · Stationary distribution · Sojourn time

1 Introduction

Retrial queues take into account the phenomenon that a customer who cannot get service immediately upon arrival goes to so-called orbit and retries for the service again after a random time. Most retrial queues assume a classical retrial policy, i.e. each customer in the orbit seeks service independently. An overview of the literature on such retrial queues can be found in [1–4].

However, there are situations in telecommunication networks which do not satisfy the assumption about classical retrial policy. Sometimes information about the state of the server is not available to the customer at the head of the queue and he/she repeats the attempts to reach the server at random times. In other cases the server takes time to check whether the transmission facility is available or the server requires time to search for a specified customer. In such cases the retrial rate is independent of the number of customers (if any) in the orbit. The constant retrial policy was introduced by Fayolle [5] who modeled a telephone exchange system. Later, retrial queueing systems with a constant retrial rate were studied by many researchers, see, e.g., [6–11].

© Springer International Publishing Switzerland 2016
A. Dudin et al. (Eds.): ITMM 2016, CCIS 638, pp. 140–150, 2016.
DOI: 10.1007/978-3-319-44615-8_12

However, practically all research in retrial queues with a constant retrial rate has been conducted for cases when the arrival flow is described by a stationary Poisson process or its Markovian extensions. To the best of our knowledge, only paper [12] has dealt with a retrial queue with arbitrary distribution of inter-arrival times. In that paper, the $GI/M/1$ queue with a constant retrial rate was investigated. However, real computer networks almost never satisfy the assumption about exponential distribution of service time. In the present paper we consider an extension of the model investigated in [12] to the case of phase type distribution (PH distribution) of service time. We construct an embedded Markov chain describing the process of the system states, derive the ergodicity condition and present an algorithm for calculation of the stationary distribution of the system states. The stationary distribution of the sojourn time is solved in terms of a Laplace-Stieltjes transform. Analysis is based on the matrix technique.

2   Model Description

We consider $GI/PH/1$ retrial queue with a constant retrial rate. The inter-arrival times in the input flow are independent random variables with general distribution $A(t)$, Laplace-Stieltjes transform $A^*(s) = \int\limits_0^\infty e^{-st} dA(t)$, $Re\ s \geq 0$, and finite first moment $a = \int\limits_0^\infty t \, dA(t)$.

The service time of a customer by the server has PH type distribution with irreducible representation (β, S). It means the following. The service time is interpreted as the time until the continuous time Markov chain $m_t, t \geq 0$, with the state space $\{1, \ldots, M+1\}$ reaches the single absorbing state $M+1$. Transitions of the chain $m_t, t \geq 0$, within the state space $\{1, \ldots, M\}$ are defined by the sub-generator S while the intensities of transitions into the absorbing state are defined by the vector $\mathbf{S}_0 = -S\mathbf{e}$. At the service beginning epoch, the state of the process $m_t, t \geq 0$, is chosen within the state space $\{1, \ldots, M\}$ according to the probabilistic row vector β. It is assumed that the matrix $S + \mathbf{S}_0\beta$ is an irreducible one. The service rate is calculated as $\mu = -(\beta S^{-1}\mathbf{e})^{-1}$. For more information about the PH type distribution, its properties, partial cases, and suitability for approximation of a variety of probability distributions arising in modeling real-life systems see, e.g., [13].

If an arriving (primary) customer meets an idle server, it starts the service immediately. In the contrary case, it goes to some virtual place called orbit and stands at the end of the queue of the so-called repeat customers. The first customer in the queue makes repeated attempts to get service in a random amount of time until it succeeds in entering the service. The intervals between two successive retrials are independent exponentially distributed random variables with parameter γ. The orbit capacity is assumed to be unlimited.

The goal of the paper is to obtain the stationary distribution of the system states and the stationary distribution of the sojourn time.

3   Embedded Markov Chain

We define the system state at time t as (0) if the system is empty at this time. In the contrary case, the state of the system is defined as a pair (i, m) where i is the number of customers in the system (in the orbit and in the service, if any), $m = 0$ if the server is idle and $m = \tilde{m}$ if the server is busy at time t and the service process is at the phase $\tilde{m}, \tilde{m} = 1, \ldots, M$.

Let t_n denote the instant of the nth arrival. We define an embedded Markov chain as $\xi_n = \{i_{t_n-0}, m_{t_n+0}\}$, $n \geq 1$. To be able to derive the transition probabilities of the chain ξ_n, $n \geq 1$, we need to know the distribution of the number of customers (primary and repeated) that complete their service during an inter-arrival time. To calculate such a distribution we introduce the notion of generalized service time of a customer from the orbit. The generalized service time consists of an exponentially distributed with parameter γ phase during which the customer reaches the idle server and a proper service time of this customer by the server.

Then the generalized service time distribution of the repeat customer can be described as phase-type (PH) distribution with the irreducible representation $(\tilde{\beta}, \tilde{S})$ where $\tilde{\beta} = (1, \mathbf{0}_M)$ and $\tilde{S} = \begin{pmatrix} -\gamma & \gamma\beta \\ 0 & S \end{pmatrix}$. Here $\mathbf{0}_M$ is a row vector of size M. For the detailed description of PH distribution see, e.g. [13]. In our case the matrix $\tilde{S}$ describes the transition rates of the phases of generalized service that do not lead to service completion. The vector $\tilde{S}_0 = -\tilde{S}\mathbf{e}$, where $\mathbf{e}$ is a column vector of units, describes the rates of transitions leading to service completion. The vector $\tilde{\beta}$ indicates that the service of the repeated customer always starts from the first phase 0. For further use we introduce the notation $D(z) = \tilde{S} + \tilde{S}_0\tilde{\beta}z$, $|z| \leq 1$. It is easy to see that the matrix $D(1)$ is a generator of the Markov chain that describes the transitions of the generalized service phases during the inter-arrival time intervals belonging to a busy period.

The service time of a primary customer who found an idle server can be considered as the phase of the generalized service time corresponding to the proper service time. It is evident that the service time of such a customer has PH distribution with irreducible representation $(\hat{\mathbf{e}}, \tilde{S})$ where $\hat{\mathbf{e}} = (0, \beta)$.

Let $P(k, t)$, $k \geq 0$, be the matrix probability that k renewals occur in the renewal process defined by the above PH distribution. It is well-known that the matrices $P(k, t)$, $k \geq 0$, are defined by the expansion

$$\sum_{k=0}^{\infty} P(k, t)z^k = e^{D(z)t}, \; |z| \leq 1. \tag{1}$$

Let us denote by $P_{i,j}$ transition probability matrix of the chain ξ_n, $n \geq 1$, from the states corresponding to the number i of customers in the system to the states with the number j of customers, $i, j \geq 0$.

Theorem 1. *The transition probability matrix of the chain* ξ_n, $n \geq 1$, *has the following block structure*

$$P = (P_{i,j})_{i,j \geq 0} = \begin{pmatrix} B_0 & A_0 & 0 & 0 & \cdots \\ B_1 & A_1 & A_0 & 0 & \cdots \\ B_2 & A_2 & A_1 & A_0 & \cdots \\ \vdots & \vdots & \vdots & \vdots & \ddots \end{pmatrix}$$

where the matrices A_k, B_k, $k \geq 0$, *are defined by*

$$A_k = C \int_0^\infty P(k,t)dA(t), \quad B_k = C \sum_{l=k+1}^\infty \int_0^\infty P(l,t)dA(t)\mathbf{e}\hat{\mathbf{e}}.$$

Here $C = \begin{pmatrix} 0 & \beta \\ \mathbf{0}^T & I_M \end{pmatrix}$, $\hat{\mathbf{e}} = (0, \beta)$, I_M *is an identity matrix of size* M, $\mathbf{0}$ *is a row vector of zeroes.*

Proof. To clarify the expressions for transition probability matrices A_k and B_k, we note the following probabilistic interpretation of the matrices appearing in these expressions.

The (m, m')th entry of the matrix $P(k, x)$ is the conditional probability that, in the interval $[0, x)$ belonging to the busy period of the system, k customers will be served and the state of the PH generalized service process at the moment x is m' conditional the state of this process at the moment 0 was m. So, the integral $\int_0^\infty P(k, t)dA(t)$ defines the matrix probability that k customers will be served during an inter-arrival time.

The matrix C defines the probabilities of jumps of the phase of a generalized service time at an arrival epoch. If the server is idle, with probability $(\beta)_m$ the phase is changed from 0 to m, $m = 1, \ldots, M$, and the service of the arriving customer starts. In the contrary case, the phase of a customer who is being served is not changed and the service of this customer is continued. The above explains why the integrals in the expressions for A_k, B_k are multiplied on the left on the matrix C. In the expression for B_k, the integrals are multiplied on the right by the vector $\mathbf{e}\hat{\mathbf{e}}$ to specify the phase of the generalized service time after the moment of arrival of a customer to the idle system.

Corollary 1. *The process* ξ_n, $n \geq 1$, *belongs to the class of* $GI/M/1$ *type Markov chains, see [13].*

Proof of the corollary follows from the structure of the transition probability matrix P.

4 Stationary Distribution

In this section we focus on the derivation of the ergodicity condition and calculation of the stationary distribution of the chain ξ_n.

Let us present the matrix $(D(1))^k, k \geq 1$, in the block form $(D(1))^k = \begin{pmatrix} h_k & \mathbf{h}_k \\ \mathbf{f}_k & F_k \end{pmatrix}$ where $\mathbf{f}_k$ is a column vector of size M and F_k is $M \times M$ matrix.

For further use we need to calculate the blocks F_k and $\mathbf{f}_k$ of the matrices $D^k(1), \ k \geq 1$.

Lemma 1

$$F_k = (-1)^{k-1} S(\gamma \mathcal{B} - S)^{k-1}, \quad \mathbf{f}_k = -F_k \mathbf{e}, \ k \geq 1, \tag{2}$$

where $\mathcal{B} = \mathbf{e}\beta$.

Proof. Formula (2) is derived by induction.

Theorem 2. *Stationary distribution of the Markov chain ξ_n, $n \geq 1$, exists if and only if the following inequality holds*

$$(x_1\beta + \mathbf{x}_2)(\gamma \mathcal{B} - S)^{-1}\Psi \, S_0 + \frac{a_1\mu\gamma}{\mu + \gamma} > 1 \tag{3}$$

where

$$\mathbf{x}_2 = \beta(I - \Psi[I - (I - \mathcal{B})(I - \Psi)]^{-1}, \ x_1 = 1 - \mathbf{x}_2\mathbf{e}, \tag{4}$$

$$\Psi = S(\gamma \mathcal{B} - S)^{-1}[A^*(\gamma \mathcal{B} - S) - I]. $$

Proof. Using the results of [13] for $GI/M/1$ type Markov chains, it is possible to formulate the ergodicity condition in terms of the generating function $A(z) = \sum\limits_{i=0}^{\infty} A_i z^i, \ |z| \leq 1$.

The chain is ergodic if and only if the inequality

$$\mathbf{x}A'(1)\mathbf{e} > 1 \tag{5}$$

is satisfied where the vector $\mathbf{x} = (x_1, \mathbf{x}_2)$ is the unique solution of the system

$$\mathbf{x}A(1) = \mathbf{x}, \tag{6}$$

$$\mathbf{x}\mathbf{e} = 1. \tag{7}$$

This condition has an abstract form and is not amenable to interpretation in terms of the system parameters. Taking into account these shortcomings of the general approach we will derive the ergodicity condition in form (3).

It follows from (1) and Theorem 1 that

$$A(z) = C \int\limits_0^{\infty} e^{D(z)t} dA(t). \tag{8}$$

Using expression (8) for $A(z)$, we can write system (6) in the form

$$(0, x_1\boldsymbol{\beta} + \mathbf{x}_2) \int_0^\infty (I + \sum_{k=1}^\infty \frac{D^k(1)t^k}{k!}) dA(t) = (x_1, \mathbf{x}_2). \qquad (9)$$

Using Lemma 1, we reduce (9) to the following system

$$(x_1\boldsymbol{\beta} + \mathbf{x}_2) \int_0^\infty \sum_{k=1}^\infty \mathbf{f}_k \frac{t^k}{k!} dA(t) = x_1,$$

$$(x_1\boldsymbol{\beta} + \mathbf{x}_2)(I + \int_0^\infty \sum_{k=1}^\infty F_k \frac{t^k}{k!} dA(t) = \mathbf{x}_2.$$

Substituting in this system expressions (2) for $\mathbf{f}_k$ and F_k, calculating sums of series and using the notion Ψ given in the statement of the theorem, we arrive to the system

$$(x_1\boldsymbol{\beta} + \mathbf{x}_2)\Psi\mathbf{e} = x_1, \qquad (10)$$

$$(x_1\boldsymbol{\beta} + \mathbf{x}_2)(I - \Psi) = \mathbf{x}_2. \qquad (11)$$

In such a way, system (6) is reduced to systems (10) and (11). Taking into account the normalizing Eq. (7), we see that Eq. (10) is a linear combination of equations from (11). Thus, system (6) is equivalent to system (11) and systems (6) and (7) is reduced to system (11), (7). From (11), (7) we obtain expressions (4) for x_1 and $\mathbf{x}_2$.

Now consider inequality (5). The left-hand side of this inequality is transformed as follows:

$$\mathbf{x}A'(1)\mathbf{e} = (0, x_1\boldsymbol{\beta} + \mathbf{x}_2) \int_0^\infty [e^{D(z)t}]'|_{z=1} dA(t)\mathbf{e} = (0, x_1\boldsymbol{\beta} + \mathbf{x}_2) \int_0^\infty [e^{D(z)t}]'|_{z=1} dA(t)\mathbf{e}$$

$$= (0, x_1\boldsymbol{\beta} + \mathbf{x}_2) \int_0^\infty [tI + \sum_{k=1}^\infty \frac{D^k(1)t^{k+1}}{(k+1)!}] D'(1) dA(t)\mathbf{e}$$

$$= ((x_1\boldsymbol{\beta}+\mathbf{x}_2) \int_0^\infty \sum_{k=1}^\infty \mathbf{f}_k \frac{t^{k+1}}{(k+1)!} dA(t), (x_1\boldsymbol{\beta}+\mathbf{x}_2) \int_0^\infty [tI + \sum_{k=1}^\infty F_k \frac{t^{k+1}}{(k+1)!}] dA(t))\tilde{\mathbf{S}}_0$$

$$\times (x_1\boldsymbol{\beta} + \mathbf{x}_2) \int_0^\infty [tI + \sum_{k=1}^\infty F_k \frac{t^{k+1}}{(k+1)!}] dA(t)\mathbf{S}_0.$$

Then inequality (5) can be write in the form

$$(x_1\beta + \mathbf{x}_2) \int_0^\infty [tI + \sum_{k=1}^\infty F_k \frac{t^{k+1}}{(k+1)!}] dA(t) \mathbf{S}_0 > 1. \tag{12}$$

Substituting in (12) expressions (2) for F_k after some algebraic transformation we obtain ergodicity condition (3).

Corollary 1. *In the case of exponential service time the stationary distribution of the Markov chain ξ_n, $n \geq 1$, exists if and only if the inequality*

$$\frac{\mu}{\mu+\gamma} \left\{ \frac{\mu}{\mu+\gamma} \left[1 - A^*(\mu+\gamma) \right] + a\gamma \right\} > 1$$

is fulfilled.

In the following we assume that the ergodicity condition (3) holds. Let us denote the stationary probabilities of the Markov chain ξ_n, $n \geq 1$, by

$$\pi(i,m) = \lim_{n\to\infty} P\{i_{t_n-0} = i, m_{t_n+0} = m\}, \ i \geq 0, m = \overline{0,M}.$$

Let us denote $\boldsymbol{\pi}_i = (\pi(i,0), \pi(i,1), \dots, \pi(i,M)), \ i \geq 0$.

Theorem. *The stationary probability vectors of the Markov chain ξ_n, $n \geq 1$, are calculated by*

$$\boldsymbol{\pi}_i = \boldsymbol{\pi}_0 R^i, \ i \geq 0,$$

where the matrix R is the minimal nonnegative solution of the matrix equation

$$R = \sum_{j=0}^\infty R^j A_j,$$

and the vector $\boldsymbol{\pi}_0$ is the unique solution of the following system of linear algebraic equation:

$$\boldsymbol{\pi}_0 \sum_{j=0}^\infty R^j B_j = \boldsymbol{\pi}_0, \quad \boldsymbol{\pi}_0 (I - R)^{-1} \mathbf{e} = 1.$$

Proof of the theorem follows from the results for $GI/M/1$ type Markov chains, see [13].

5   Laplace-Stieltjes Transform of the Stationary Distribution of the Sojourn Time

The sojourn time of an arriving customer is equal to the proper service time if the server is idle at the arriving epoch. Otherwise, the customer goes to the orbit and the structure of its sojourn time is much more complicated. In this case the derivation of the desired Laplace-Stieltjes transform (LST) is a difficult problem. So, in this section our main efforts will be directed to solving this problem.

Let $\boldsymbol{\varphi}_i(u)$, $Re\ u \geq 0$, be the column vector of size M whose mth entry, $(\boldsymbol{\varphi}_i(u))_m$, is the LST of the stationary distribution of the sojourn time of an arriving customer which finds the busy server in phase m and $i-1$ customers in the orbit, $i \geq 1, m = \overline{1, M}$.

Let us introduce the vector generating function $\boldsymbol{\Phi}(z, u) = \sum_{i=1}^{\infty} \boldsymbol{\varphi}_i(u)z^i$, $0 < |z| \leq 1$.

Lemma 2. *The vector generating function $\boldsymbol{\Phi}(z, u)$ satisfies the following equation:*

$$\boldsymbol{\Phi}(z, u)\left[I - A^*(uI - S) - H(u)[A_1^*(u) - A_2^*(u)](\mathbf{S}_0 \otimes I_{M+1})\left(\boldsymbol{\beta}\right)zI_M \right]$$

$$= zH(u)\left[[(uI - S)^{-1} \otimes I_{M+1}][I - A_1^*(u)] - [I_M \otimes (uI - D(z))^{-1}][I - A_2^*(u)] \right]$$

$$\times (\mathbf{S}_0 \otimes I_{M+1})\tilde{\mathbf{S}}_0 + [A_1^*(u) - A_2^*(u)](\mathbf{S}_0 \otimes I_{M+1})\left(\mathbf{0}\right)\mathbf{e}_M\ \boldsymbol{\beta}(uI - S)^{-1}\mathbf{S}_0 \quad (13)$$

where

$$H(u) = (I_M \otimes \tilde{\boldsymbol{\beta}})([S \oplus (-D(z))]^{-1},\ A_1^*(u) = A^*(uI - S) \otimes I_{M+1},$$

$$A_2^*(u) = I_M \otimes A^*(uI - D(z)).$$

Proof. First, we derive the following recursive formulas for the vector functions $\boldsymbol{\varphi}_i(u)$, $i \geq 1$:

$$\boldsymbol{\varphi}_i(u) = \sum_{k=0}^{\infty} \left[\int_0^{\infty} e^{(S-uI)y}dA(y) \right]^k \quad (14)$$

$$\times \left\{ \int_0^{\infty} e^{-ut} \int_0^t e^{Sx}\mathbf{S}_0dx\tilde{\boldsymbol{\beta}} \sum_{n=0}^{i-1} P(n, t-x)dA(t)\tilde{\boldsymbol{\beta}}^T \boldsymbol{\beta}\boldsymbol{\varphi}_{i-n}(u) \right.$$

$$+ \int_0^{\infty} e^{-ut} \int_0^t e^{Sx}\mathbf{S}_0dx\tilde{\boldsymbol{\beta}} \sum_{n=0}^{i-1} P(n, t-x)dA(t)\left(\begin{matrix}\mathbf{0}\\I_M\end{matrix}\right)\boldsymbol{\varphi}_{i-n-1}(u)$$

$$\left. + \int_0^{\infty} e^{-ut} \int_0^t e^{Sx}\mathbf{S}_0dx\tilde{\boldsymbol{\beta}}P(i-1, t-x)\tilde{\mathbf{S}}_0(1 - A(t))dt \right\}$$

where $\quad \boldsymbol{\varphi}_0(u) = \mathbf{e}_M\boldsymbol{\beta}(uI - S)^{-1}\mathbf{S}_0$.

In the derivation of (14) we used the probabilistic interpretation of the LST, see, e.g. [14]. We assume that, independently of the system operation, the stationary Poisson flow of so-called catastrophes arrives. Let u, $u > 0$ be the rate of this flow. Then $(\boldsymbol{\varphi}_i(u))_m$ can be interpreted as the probability of no catastrophe arrival during the sojourn time of an arriving customer who found the busy server in phase m and $i-1$ customers in the orbit, $i \geq 1, m = \overline{1, M}$.

This allows us to derive the expression for $\varphi_i(u)$ by means of probabilistic reasonings.

Let a tagged primary customer find the busy server with $i - 1$ customers waiting in the orbit. In this case, the customer is placed at the end of the queue in the orbit. Without loss of generality, we assume that the arrival epoch of the tagged customer is $t_0 = 0$. Let us denote $t_1, t_2, \ldots$ the moments of further arrivals.

Let us analyze the structure of the sojourn time of the tagged customer. First of all, this time includes a residual service time of a customer which occupies the server at time t_0. During the residual service time, k, $k \geq 0$, new primary customers can arrive into the system. This means that the service of a customer who occupied the server at time t_0 is completed in the interval (t_k, t_{k+1}). We assume that the length of this interval is equal to t and service completion occurs at the moment $t_k + x$ where $x < t$.

During the residual inter-arrival time $t - x$ the following events affecting the sojourn time of the tagged customer may occur:

(1) n, $n = \overline{0, i-1}$, customers from the orbit will be served and the $(n+1)th$ customer has made no attempt to reach the server. In this case the generalized service of the $(n+1)th$ customer is at the phase $m = 0$ at the next arrival epoch t_{k+1};

(2) n, $n = \overline{0, i-1}$, customers from the orbit will be served and the $(n+1)th$ customer has reached the server earlier than t_{k+1} but service has not ended by the epoch t_{k+1}. Let service at the phase m, $m = \overline{1, M}$, the epoch t_{k+1}. It means that the generalized service of the $(n+1)th$ customer is at the phase m at the epoch t_{k+1};

(3) n, $n = \overline{0, i-1}$, customers staying in the queue ahead of the tagged customer and this customer itself are served by the arrival epoch t_{k+1}.

It is evident that in scenario (1) the distribution of the remaining sojourn time of the tagged customer after the moment t_{k+1} coincides with the distribution of the sojourn time of a primary customer which finds, upon arrival, a busy server and $i - n$ customers in the orbit. So, the vector probability of no catastrophe arrival during the remaining sojourn time is equal to $\varphi_{i-n}(u)$. Analogously we come to the conclusion that, in scenario (2), such a vector probability is equal to $\varphi_{i-n-1}(u)$.

Using the above reasoning, we derive expression (14) for the vector probability of no catastrophe arrival during the sojourn time of the tagged customer. In this expression the rth summand in the brackets pre-multiplied by $\sum\limits_{k=0}^{\infty} \left[\int\limits_0^{\infty} e^{(S-uI)y} dA(y) \right]^k$ is the vector probability of no catastrophe arrival during the sojourn time of the tagged customer under scenario (r), $r = 1, 2, 3$.

By multiplying Eq. (14) by z^i and summing up, after tedious algebra we derive formula (13).

Let $\mathbf{v}(u)$, $Re\ u \geq 0$, be the column vector of size M whose mth entry is the LST of the stationary distribution of the sojourn time of a customer in the system.

Theorem 7. *The vector LST* $\mathbf{v}(u)$ *of the stationary distribution of the sojourn time in the system is calculated as*

$$\mathbf{v}(u) = \boldsymbol{\pi}_0 \boldsymbol{\varphi}_0(u) + \sum_{i=1}^{\infty} \boldsymbol{\pi}^i \tilde{\boldsymbol{\beta}}^T \tilde{\boldsymbol{\beta}} \boldsymbol{\varphi}_0(u) + \sum_{i=1}^{\infty} \boldsymbol{\pi}^i \begin{pmatrix} \mathbf{0} \\ I_M \end{pmatrix} \boldsymbol{\varphi}_i(u). \tag{15}$$

Proof. Formula (15) follows immediately from the formula of total probability.

6   Conclusion

In this paper, the $GI/PH/1$ retrial queue with a constant retrial rate has been studied. The necessary and sufficient condition for the existence of stationary distribution has been derived and the algorithm for calculating the steady state probabilities has been presented. The Laplace-Stieltjes transform of the stationary distribution of the sojourn time has been derived. Using formula for this transform, the moments of the sojourn time can be obtained in a direct way. All the presented results agree with the results for the $GI/M/1$ retrial queue with a constant retrial rate and for the $GI/PH/1$ queue with waiting room. To check the results for a queue with waiting room, one should to move to the limit $\gamma \to \infty$ in the corresponding formulas. The results can be extended to the case of more general semi-Markovian arrival process.

Acknowledgments. This research was supported by Basic Science Research Program through the National Research Foundation of Korea (NRF) funded by the Ministry of Education (Grant No. 2014K2A1B8048465) and by Belarusian Republican Foundation of Fundamental Research (Grant No. F15KOR-001).

References

1. Kulkarni, V.G., Liang, H.M.: Retrial queues revisited. In: Dshalalow, J.H. (ed.) Frontiers in Queueing: Models and Applications in Science and Engineering, pp. 19–34. CRC Press, Boca Raton (1997)
2. Falin, G., Templeton, J.: Retrial Queues. Chapman and Hall, London (1997)
3. Gomez-Corral, A.: A bibliographical guide to the analysis of retrial queues through matrix analytic techniques. Ann. Oper. Res. **141**, 163–191 (2006)
4. Artalejo, J.R., Gomez-Corral, A.: Retrial Queueing Systems: A Computational Approach. Springer, Heidelberg (2008)
5. Fayolle, G.: A simple telephone exchange with delayed feedback. In: Boxma, O.J., Cohen, I.W., Tijms, M.C. (eds.) Teletraffic Analysis and Computer Performance Evaluation, pp. 245–253. North-Holland, Amsterdam (1986)
6. Farahmand, K.: Single line queue with repeated demands. Queueing Syst. **6**, 223–228 (1990)
7. Choi, B.D., Shin, Y.W., Ahn, W.C.: Retrial queues with collision arising from unslotted CSMA/CD protocol. Queueing Syst. **11**, 335–356 (1992)
8. Dudin, A.N., Klimenok, V.I.: A retrial $BMAP/SM/1$ system with linear repeated requests. Queueing Syst. **34**, 47–66 (2000)

9. Artalejo, J., Gomez-Corral, A., Neuts, M.F.: Analysis of multiserver queues with constant retrial rate. Eur. J. Oper. Res. **135**, 569–581 (2001)
10. Li, H., Zhao, Y.Q.: A retrial queue with constatnt retrial rate, server downs and impatient customers. Stoch. Models **21**, 531–550 (2005)
11. Efrosinin, D., Winkler, A.: Queuing system with a constant retrial rate, non-reliable server and threshhold-based recovery. Eur. J. Oper. Res. **210**, 594–605 (2011)
12. Kim, C.S., Klimenok, V., Dudin, A.: A G/M/1 retrial queue with constant retrial rate. TOP **22**, 509–529 (2014)
13. Neuts, M.: Matrix-geometric Solutions in Stochastic Models - An Algorithmic Approach. Johns Hopkins University Press, Baltimore (1981)
14. van Dantzig, D.: Chaines de Markof dans les ensembles abstraits et applications aux processus avec regions absorbantes et au probleme des boucles. Ann. de l'Inst. H. Pioncare **14**(fasc. 3), 145–199 (1955)

Influence of the Pipeline Effect on the Delay of the Multipacket Message in Multilink Transport Connection

Vladimir Kokshenev, Pavel Mikheev$^{(\boxtimes)}$, Sergey Suschenko,
and Roman Tkachyov

National Research Tomsk State University, Lenina street, 36, 634050 Tomsk, Russia
doka.patrick@gmail.com, ssp.inf.tsu@gmail.com

Abstract. A stochastic model is proposed of the process of information transfer of multipacket messages in a multilink transport connection with distortion of the packets on separate areas of hops, differentiated by accounts of the pipeline effect. We investigate the influence of the level of distortions of protocol data units in a separate part of transport connections and the duration of the timeout of non-use of an end-to-end acknowledgment in probability-time characteristics of the transport protocol.

Keywords: Transport connection · Multipacket message · Pipeline effect · Delay · Duration of timeout · The probability of distortion

1 Introduction

The operational characteristics of modern computer networks are largely determined by the transport protocol and its parameters — window width and length of the timeou t waiting end-to-end Acknowledgements [1,2]. Modeling the transport connection and the analysis of its probability-time characteristics in various conditions were shown in [1–15]. Modern transport protocols contain a wide variety of mechanisms for congestion control [3]. There is a wide range of studies [3–9] in the field of control parameters of the transport protocol in order to prevent and circumvent congestion that focuses on constructing models of diagnostics over various indicators [3] and adaptation of protocol parameters to changing network load and connectivity, the level of loss, activity interactive subscribers and other data transmission conditions. In [10], an analysis of the impact of reliability of transmission of packets in individual links and lock of finite buffer memory in the transit nodes of the transport connection on its performance. The most important indicator of the quality of customer service is the delay of user messages within the transport connection, which is largely determined

This work is performed according to state order No. 1.511.2014/K of the Ministry of Education and Science of the Russian Federation.

© Springer International Publishing Switzerland 2016
A. Dudin et al. (Eds.): ITMM 2016, CCIS 638, pp. 151–161, 2016.
DOI: 10.1007/978-3-319-44615-8_13

by the pipeline effect, manifested in the transmission multipacket messages in multi data link path in the form of parallel transfer of different parts of a message by different parts of the way. Well-known approaches [11–15] to analyze the delay of subscriber messages allow studying of the influence of the time-out duration waiting for the Acknowledgment and packet size on operating characteristics process of data transfer in a determined [11–13], or stochastic [14,15] data transmission path. In these studies suggested methods of analysis of operating characteristics of the multi-packet messaging process in a deterministic and single-packet messaging in the multi data link path when moving end-to-end acknowledgments in the information and service packets of oncoming traffic. However, the results were obtained with significant constraints on the protocol parameters and conditions of transfer. In addition, according to end-to-end delay multi-packet messages in a stochastic multi-data link transport a connection is not obtained. A natural extension and generalization of the results [11–15] is the study of the mechanism of pipeline transfer multi-packet messages in the multi-data link paths with distortion in some parts of the hop. In this paper we suggest an approach to building a distribution of time-transfer multi-packet messages in a multi-data link virtual channel consisting of two sections and three hops with a distortion, which is the basis for an analysis of the influence of the length of the path, the data transfer size multi-packet communication, distortion and duration of end-to-end timeout for acknowledgment of non-use of probability-time characteristics of data transportation.

2   Transport Connections Model

When applying the factors of distortion of protocol data units and the crucial feedback mechanism [2] (retransmissions of corrupted blocks) on the pipeline effect, taking place for the process of transferring information flows on multi-link (multi-phase) paths, a virtual connection can be interpreted as a stochastic pipeline, whose processing time for individual phases is random. As the level of pass-through multi-packet messaging application systems urgent task is to determine the duration of end-to-end acknowledgment timeout, it becomes an important aspect of a probability description of the process of information transfer through the transport level. A comprehensive description of this pipeline specifies the time distribution of end-to-end delivery of application messages to the recipient, which allows the obtaining of a probabilisty-time characteristic of the protocol to control procedures of the transport level. Let us consider the process of transmission of a message consisting of $N > 1$ packets, in a transport compound consisting of $D > 1$ parts of a hop with the same performance. We assume that the information packets of the message and acknowledgment carried in the information packets counter flow, are the same size, are transmitted in each link according to the administering procedure for start-stop protocol [2], and the transmission cycle of the packet in each link from the beginning of its conclusion in the line of communication till the moment of acknowledgment at the data link level is T. The likelihood of distortion of the n-th packet of the message on the d-th plot of hops will be denoted by R_{nd}, $n = \overline{1, N}$, $d = \overline{1, D}$. Then the time error-free transmission of the n-th packet on the d-th cross-site connection is a random quantity that is a multiple of the cycle duration T and distributed according to geometric law with parameter $1 - R_{nd}$. It is also assumed that uploading the packet at each site hop of the virtual channel begins only after it has been transferred without distortion in the previous area of the connecting path. We believe that end-to-end transport of data is organized as follows. To send a message to a remote destination and receiving from it a return acknowledgment is allocated a time-out duration of S intervals of size T. If an acknowledgment is not received at the time of the timeout, the sender re-organizes end-to-end transmission. The number of end-to-end retransmissions is believed to be unlimited.

3   Probabilistic-Time Characteristics of the Process of Message Delivery to the Addressee

To find the probability of delivery of a multipacket message to the addressee by a multilink virtual channel precisely for $k \geq N + D - 1$ intervals of duration T. We define a probability function $p(k, N, D)$ using probability distortion R_{nd} for the set of parameters $N = 2$, $D = 2$. Since the process of transmission (including retransmission) of the first and second packets of the message on the second and first sections of hops is combined, there are two possible scenarios when the first packet on the second site is transferred either before or after the correct completion of the transmission of the second packet in the first link:

$$p(k, 2, 2) = \prod_{n,d=1}^{2,2} (1 - R_{nd}) \sum_{i=0}^{k-3} R_{11}^i \sum_{j=0}^{k-3-i} R_{21}^j$$
$$\times \left\{ \sum_{l=0}^{j-1} R_{12}^l R_{22}^{k-3-i-j} + \sum_{m=0}^{k-3-i-j} R_{12}^{j+m} R_{22}^{k-3-i-j-m} \right\}.$$

For a statistically flat uniform data path and uniform information flow $R_{nd} = R$, $n = 1, 2$, $d = 1, 2$ this dependence of the probability of delivery time of the message to the remote subscriber using the expressions finite sums revealing-power functions [16] is converted into the following form:

$$p(k, 2, 2) = (1 - R)^3 R^{k-3} \left[2 \binom{k-1}{2} - (k-2) \frac{1+R}{1-R} + \frac{R(1+R)(1 - R^{k-2})}{(1-R)^2} \right].$$

For sets of parameters $N = 3$, $D = 2$ $N = 2$, $D = 3$ the variety of events is even higher, since this increases the pipeline effect of combining transmissions of different packets of a message at different parts of the way. When $N = 3$, $D = 2$ the first packet can be transmitted in the second link until the transmission is completed in the first phase path or the second or the third packet, either after stopping at a transit node of the third package. In addition, the second message packet can reach the destination either before or after the transfer of the third package in the first link. Taking into account the combinatorics of the probabilistic trajectories of events receiving different packets of the message by the addressee, the function of probability delivery of a three packet massage on a two-link path within a given time $k \geq N + D - 1$ is defined as:

$$
p(k,3,2) = \prod_{n,d=1}^{3,2} (1 - R_{nd}) \sum_{i=0}^{k-4} R_{11}^{i} \sum_{j=0}^{k-4-i} R_{21}^{j} \sum_{l=0}^{k-4-i-j} R_{31}^{l} \left\{ \sum_{m=0}^{j-1} R_{12}^{m} \left(\sum_{g=0}^{l-1} R_{22}^{g} \right. \right.
$$

$$
\times R_{32}^{k-4-i-j-l} + \sum_{g=0}^{k-4-i-j-l} R_{22}^{g+l} R_{32}^{k-4-i-j-l-g} \bigg) + \sum_{m=0}^{l-1} R_{12}^{m+j} \left(\sum_{g=0}^{l-1-m} R_{22}^{g} R_{32}^{k-4-i-j-l} \right.
$$

$$
\left. + \sum_{g=0}^{k-4-i-j-l} R_{22}^{g+l-m} R_{32}^{k-4-i-j-l-g} \right) + \sum_{m=0}^{k-4-i-j-l} R_{12}^{m+j+l} \sum_{g=0}^{k-4-i-j-l-m} R_{22}^{g} R_{32}^{k-4-i-j-l-m-g} \bigg\}.
$$

For the set of parameters $N = 2$, $D = 3$ a probability function is determined similar to the transmission of the first packet to the second and third sections of the hops either before or after the transmission of the second packet or the first or in the second link of the transport connection:

$$
p(k,2,3) = \prod_{n,d=1}^{2,3} (1 - R_{nd}) \sum_{i=0}^{k-4} R_{11}^{i} \sum_{j=0}^{k-4-i} R_{21}^{j} \left\{ \sum_{m=0}^{j-1} R_{12}^{m} \sum_{l=0}^{k-4-i-j} R_{22}^{l} \left(\sum_{g=0}^{j-m+l-1} R_{13}^{g} \right. \right.
$$

$$
\times R_{23}^{k-4-i-j-l} + \sum_{g=0}^{k-4-i-j-l} R_{13}^{l+g+j-m} R_{23}^{k-4-i-j-l-g} \bigg) + \sum_{m=0}^{k-4-i-j} R_{12}^{j+m} \sum_{l=0}^{k-4-i-j-m} R_{22}^{l}
$$

$$
\times \left(\sum_{g=0}^{l-1} R_{13}^{g} R_{23}^{k-4-i-j-m-l} + \sum_{g=0}^{k-4-i-j-m-l} R_{13}^{l+g} R_{23}^{k-4-i-j-m-l-g} \right) \bigg\}.
$$

When $R_{nd} = R$, using the correlation [16], it is easy to verify that these equations coincide:

$$
p(k,3,2) = p(k,2,3) = (1-R)^4 R^{k-4} \left\{ 3 \binom{k-1}{3} - 2 \binom{k-2}{2} \frac{1+R}{1-R} \right.
$$

$$
\left. \times \left(1 + \frac{R^{k-2}}{2} \right) + (k-2) \frac{R(1+2R)(1-R^{k-3})}{(1-R)^2} \right\}.
$$

Thus, there is a space-time symmetry of the process of information transfer of a uniform stream of packets in a statistically uniform data path, which consists of the invariance of probability-time characteristics of the delivery process multipacket messages in multilink virtual channels to mutually symmetrical values of N and D.

We get analytical dependence for the distribution function of the time of communicating the messages to the recipient $P(S, N, D)$. According to the meaning the probability of message delivery to the addressee within a specified time time-out S is:

$$P(S, N, D) = \sum_{k=N+D-1}^{S} p(N, D, k).$$

For these values of N and D are true:

$$P(S, 2, 2) = (1 - R^S)(1 - R^{S-2}) - S(S - 2)(1 - R)^2 R^{S-2};$$

$$P(S, 3, 2) = P(S, 2, 3) = (1 - R^{S-3})(1 - 3R^{S-1} + 2R^S) - (S-3)(1-R)^3 R^{S-3}$$

$$\times \left\{ \frac{S}{2}\left(S - 1 - \frac{1}{1-R}\right) + \frac{1+R^2}{(1-R)^2} - \frac{R^{S-1}}{1-R}\left(\frac{S}{2} - \frac{1+R}{1-R}\right) \right\}.$$

Hence it is easy to verify that if the minimum duration of the timeout with $S = N + D - 1$ the probability of message delivery respectively is $P(S, 2, 2) = (1 - R)^4$ and $P(S, 3, 2) = (1 - R)^6$, and in the case of a timeout of unlimited duration ($S = \infty$) the receipt of the message by the subscriber is a determine event.

We find the average time $\bar{N}(S, N, D)$ of message delivery to the addressee, expressed as a duration T, at time S, subject to be recieved by the remote subscriber:

$$\bar{N}(S, N, D) = \frac{\bar{n}(S, N, D)}{P(S, N, D)}, \quad \bar{n}(S, N, D) = \sum_{k=N+D-1}^{S} kp(k, N, D).$$

For sets of values for the message size and the data length of the transmission path $N = D = 2$; $N = 3$, $D = 2$ and $N = 2$, $D = 3$, we obtain:

$$\bar{n}(S, 2, 2) = \frac{1 - R^{S-2}}{1 - R^2}\left[3 + 4R - (3 - 2R^2)R^S\right] - (S - 2)R^{S-2}\left[S(1 - R)(1\right.$$

$$\left. + S(1 - R)) + (1 + R)^2 - R^S\right];$$

$$\bar{n}(S, 3, 2) = \bar{n}(S, 2, 3) = \frac{1 - R^{S-3}}{1 - R}\left[4 + \frac{2R}{1 + R} - \frac{R^2}{(1 + R)^3} - R^{S-1}\right.$$

$$\times \frac{12 + 16R - 10R^2 - 18R^3 + 3R^4 + 6R^5}{(1 + R)^3}\right] - (S - 3)(1 - R)^3 R^{S-3}$$

$$\times \left[\frac{(S - 3)^3}{2} + 4(S - 3)^2 + (S - 3)\frac{21 - 38R + 23R^2}{2(1 - R)^2}\right.$$

$$+ \frac{11 - 20R + 29R^2 - 14R^3}{(1 - R)^3}\right] + (S - 3)(1 - R)^2 R^{2S-4}\left[\frac{(S - 3)^2}{2} + (S - 3)\right.$$

$$\times \frac{9 + 4R - 4R^2}{2(1 - R^2)} + \frac{26 + 24R - 19R^2 - 16R^3 + 3R^4}{2(1 - R^2)^2}\right].$$

For timeout minimum duration $S = D + N - 1$, the average conventional time of message delivery to the addressee coincides with the duration of the timeout. When $S = \infty$ the average delivery time of the message takes the form:

$$\bar{N}(\infty, 2, 2) = \frac{3}{1 - R} + \frac{R}{1 - R^2};$$

$$\bar{N}(\infty, 3, 2) = \bar{N}(\infty, 2, 3) = \frac{4}{1 - R} + \frac{R}{1 - R^2}\left(2 - \frac{R}{(1 + R)^2}\right).$$

Thus, in the case of an unlimited duration of timeout the average time of message delivery to the recipient is determined by the sum of average delays of packets in a determine pipeline [11] at the transmission time on each phase equal to the average delay of the packet in a separate link $1/(1 - R)$, and a value proportional to the intensity of distortion: $R/(1 - R^2)$.

4   Analysis of the End-To-End Operating Performance of Message Transmission

We get the dependency for the probability of delivery of a message of length N to the subscriber from a remote sender on to the D hops, and getting end-to-end aknowledgement, packed in the information package, exactly for $k \geq N + 2D - 1$ intervals of duration T. This characteristic is determined by the set of all variants of the probability of transmitting the message to the recipient and the probability that the source of notification notifies about the successful transmission for a specified time:

$$f(k, N, D) = \sum_{i=N+D-1}^{k-D} p(i, N, D)p(k - i, 1, D).$$

For these values of the set of parameters N, D and type of dependence for $p(k, 1, D)$ [14] the probability function of time end-to-end has the form:

$$f(k, 2, 2) = (1 - R)^5 R^{k-5}\left\{2\binom{k-1}{4} - \binom{k-2}{3}\frac{R(1+R)}{(1-R)^2}\right.$$

$$\left. - (k-3)\frac{R^2(1+R)}{(1-R)^3} + \frac{R^2(1+R)}{(1-R)^4}(1 - R^{k-3})\right\};$$

$$f(k, 3, 2) = (1-R)^6 R^{k-6}\left\{3\binom{k-1}{5} - 2\binom{k-2}{4}\frac{1+R}{1-R} + \frac{k}{3}\binom{k-4}{2}\right.$$

$$\times \frac{R(1+2R)}{(1-R)^2} + \frac{R^2(1 - R^{k-5})}{(1-R)^4}\left(\frac{3+7R+2R^2}{1-R} - (k-4)\right.$$

$$\times (3+4R-2R^2)\bigg) - \frac{(k-5)R^{k-3}}{(1-R)^3}\left(\frac{3+7R+2R^2}{1-R} + (k-4)\frac{R(1-R)}{2}\right)\bigg\};$$

$$f(k,2,3) = (1-R)^7 R^{k-7} \left\{ 3\binom{k-1}{6} - 2\binom{k-2}{5}\frac{1+R}{1-R} + \frac{k+1}{4} \right.$$

$$\times \binom{k-4}{3}\frac{R(1+2R)}{(1-R)^2} - \frac{R^2(1-R^{k-6})}{(1-R)^4}\left(\frac{R(6+12R+3R^2)}{(1-R)^2}\right.$$

$$\left. - (k-5)\left(\frac{3+7R+2R^2}{1-R} - (k-4)\frac{3+4R-2R^2}{2}\right)\right)$$

$$\left. + \frac{(k-6)R^{k-4}}{(1-R)^4}\left(\frac{R(6+12R+3R^2)}{1-R} - (k-5)\frac{3+4R-3R^2}{2}\right)\right\}.$$

Now we find the probability of end-to-end transfer multipacket messages for a given time $S \geq N + 2D - 1$.

$$F(S,N,D) = \sum_{k=N+2D-1}^{S} f(k,N,D).$$

Finally, the function of distribution of time end-to-end transmission with different N and D takes the following form:

$$F(S,2,2) = (1-R^{S-4})(1-R^S) - \frac{S(S-4)}{12}(1-R)^4 R^{S-4}\left((S-2)^2 + 3\frac{(1+R)^2}{(1-R)^2}\right);$$

$$F(S,3,2) = (1-R^{S-5})(1-10R^{S-1}+2R^S) + (S-5)(1-R)^2 R^{2S-6}\left(\frac{S-4}{2}\right.$$

$$+\frac{4}{1-R}\right) - (S-5)(1-R)^5 R^{S-5}\left(\frac{(S-5)^4}{40} + \frac{(S-5)^3(7-8R)}{24(1-R)}\right.$$

$$+\frac{(S-5)^2(31-64R+41R^2)}{24(1-R)^2} + \frac{(S-5)(65-158R+217R^2-100R^3)}{24(1-R)^3}$$

$$\left. +\frac{161+309R+541R^2-859R^3+286R^4}{60(1-R)^4}\right);$$

$$F(S,2,3) = (1-R^{S-6})(1+15R^{S-1}-2R^S) - (S-6)(1-R)^2 R^{2S-7}\left(\frac{S-5}{2}\right.$$

$$\cdot +\frac{5}{1-R}\right) - (S-6)(1-R)^6 R^{S-6}\left(\frac{(S-6)^5}{240} + \frac{(S-6)^4(17-19R)}{240(1-R)}\right.$$

$$+\frac{(S-6)^3(23-48R+29R^2)}{48(1-R)^2} + \frac{(S-6)^2(279-215R+265R^2-113R^3)}{48(1-R)^3}$$

$$+\frac{S-6}{120(1-R)^4}\left(362-1023R+1727R^2-1833R^3+587R^4\right)$$

$$\left. +\frac{167-352R+668R^2-982R^3+1343R^4-304R^5}{60(1-R)^5}\right).$$

We find the average waiting time of end-to-end acknowledgement at the time of the timeout $S \geq N + 2D - 1$ upon the receipt of the message by the sender:

$$\bar{S}(S, N, D) = \frac{\bar{s}(S, N, D)}{F(S, N, D)}, \qquad \bar{s}(S, N, D) = \sum_{k=N+2D-1}^{S} k f(k, N, D).$$

Hence, we obtain the dependence of the average conditional waiting time of the end-to-end receipt:

$$\bar{s}(S, 2, 2) = \frac{1 - R^{S-4}}{1 - R} \left(5 + \frac{R}{1+R} - R^S \frac{5 - 4R^2}{1+R} \right) + (S - 4)R^{2S-4} - (S - 4)$$

$$\times \frac{(1-R)^4 R^{S-4}}{12} \left((S-4)^4 + (S-4)^3 \frac{13 - 12R}{1 - R} + (S-4)^2 \frac{63 - 102R + 55R^2}{(1-R)^2} \right.$$

$$+ \frac{S-4}{(1-R)^3} \left(143 - 268R + 269R^2 - 120R^3 \right) + \frac{152 - 218R + 272R^2 - 258R^3 + 124R^4}{(1-R)^4} \right);$$

$$\bar{s}(S, 3, 2) = \frac{1 - R^{S-5}}{1 - R} \left(6 + \frac{2R}{1+R} - \frac{R^2}{(1+R)^3} - \frac{R^{S-1}}{(1+R)^3} \left(60 + 108R - 9R^2 - 98R^3 \right. \right.$$

$$\left. - 30R^4 + 10R^5 \right) \right) + \frac{(S-5)(1-R)^2}{2} R^{2S-6} \left((S-5)^2 + (S-5)\frac{15 + 8R - 6R^2}{1 - R^2} \right.$$

$$+ \frac{74 + 84R - 37R^2 - 44R^3 + 5R^4}{(1 - R^2)^2} \right) - (S-5)(1-R)^5 R^{S-5} \left(\frac{(S-5)^5}{40} + (S-5)^4 \right.$$

$$\times \frac{53 - 55R}{120(1-R)} + (S-5)^3 \frac{73 - 144R + 81R^2}{24(1-R)^2} + (S-5)^2 \frac{251 - 663R + 757R^2 - 305R^3}{24(1-R)^3}$$

$$+ (S-5)\frac{1136 - 3139R + 5336R^2 - 4869R^3 + 1536R^4}{60(1-R)^4}$$

$$+ \frac{513 - 1012R + 2018R^2 - 3557R^3 + 3173R^4 - 775R^5}{30(1-R)^5} \right);$$

$$\bar{s}(S, 2, 3) = \frac{1 - R^{S-6}}{1 - R} \left(7 + \frac{2R}{1+R} - \frac{R^2}{(1+R)^3} + \frac{R^{S-1}}{(1+R)^3} \left(105 + 196R - 7R^2 \right. \right.$$

$$\left. - 177R^3 - 66R^4 + 12R^5 \right) \right) - \frac{(S-6)(1-R)^2}{2} R^{2S-7} \left((S-6)^2 + (S-6)\frac{18 + 10R - 7R^2}{1 - R^2} \right.$$

$$+ \frac{1}{(1 - R^2)^2} \left(107 + 126R - 49R^2 - 64R^3 + 6R^4 \right) \right) - (S - 6)(1 - R)^6 R^{S-6}$$

$$\times \left(\frac{(S-6)^6}{240} + \frac{(S-6)^5(24 - 25R)}{240(1 - R)} + \frac{(S-6)^4}{240(1 - R)^2} \left(234 - 469R + 259R^2 \right) \right.$$

$$+ (S-6)^3 \frac{240 - 669R + 740R^2 - 287R^3}{48(1 - R)^3} + (S - 6)^2$$

$$\times \frac{(3489 - 11291R + 18279R^2 - 15161R^3 + 456R^4)}{240(1 - R)^4} + \frac{S - 6}{120(1-R)^5} \left(2868 - 9007R \right.$$

$$\left. + 17903R^2 - 23667R^3 + 17293R^4 - 4130R^5 \right) + \frac{1}{120(1 - R)^6} \left(2458 - 5340R \right.$$

$$\left. \left. + 11448R^2 - 19192R^3 + 28337R^4 - 21600R^5 + 3888R^6 \right) \right).$$

Let us analyze the average end-to-end delay of multipacket messages. The time of protocol of the transmission cycle end-to-end transmission consists of the time taken to pass messages to a remote subscriber and time of the acknowledgement receipt. Because of the failure to obtain acknowledgement during the timeout, the sender retransmits the message, and the number of retransmissions is not limited, the average end-to-end delay, expressed in number of intervals of duration T is:

$$\bar{T}(N, D, S) = \sum_{i=1}^{\infty} \left\{ (i-1)S + \bar{S}(S, N, D) \right\} \left\{ 1 - F(S, N, D) \right\}^{i-1} F(S, N, D)$$

$$= \frac{S(1 - F(S, N, D))}{F(S, N, D)} + \bar{S}(S, N, D).$$

At a minimum timeout duration $S_m = N + 2D - 1$ average end-to-end delay is given by:

$$\bar{T}(N, D, S_m) = \frac{N + 2D - 1}{(1 - R)^{DN + D}}.$$

With unlimited growth of S delay coincides with the average conditional expectation of the acknowledgement, and leads to the relation:

$$\bar{T}(2, 2, \infty) = \frac{5}{1 - R} + \frac{R}{1 - R^2};$$

$$\bar{T}(3, 2, \infty) = \frac{6}{1 - R} + \frac{2R}{1 - R^2} - \frac{R^2}{(1 - R^2)(1 - R)^2};$$

$$\bar{T}(2, 3, \infty) = \frac{7}{1 - R} + \frac{2R}{1 - R^2} - \frac{R^2}{(1 - R^2)(1 - R)^2}.$$

The first term in these expressions determines the delay in the determine pipeline multipacket messages and end-to-end return acknowledgement receipt when the transmission time in a separate phase is $1/(1 - R)$ and the corresponding average transmission time of a packet is a separate inter-node connection.

The contribution is proportional to the remaining terms of R and R^2, and the real level of distortion in the high-quality channels of communication can be neglected. From Table 1 it is easy to see that for the three-fold excess of S over a minimum duration of the timeout S_m and $R \leq 0.5$ the value of end-to-end delay practically coincides with $\bar{T}(N, D, \infty)$. Hence, for practical calculations when $S \geq 3S_m$ and low distortion R as the average transmission time multipacket messages in the multi-link path, you can use an analytical expression for the delay in the determine pipeline with the time of transmission in a separate phase, equal to the average delay of a packet in a separate link:

$$\bar{T}(N, D, S) = \frac{N + 2D - 1}{1 - R}.$$

Table 1. The distribution of mean end-to-end message delay from the probability of distortion of the packet for the size of the timeout, multiples of the minimum duration S_m

$\bar{T}(N, D, S)$	R				
	0.3	0.4	0.5	0.6	0.7
$\bar{T}(2, 2, S_m)$	42.50	107.17	320.00	1220.70	6858.71
$\bar{T}(2, 2, 2S_m)$	7.87	10.58	16.83	34.51	105.87
$\bar{T}(2, 2, 3S_m)$	7.87	8.90	11.38	17.09	35.49
$\bar{T}(2, 2, 4S_m)$	7.87	8.81	10.74	14.20	23.46
$\bar{T}(2, 2, \infty)$	7.87	8.81	10.67	13.44	18.04
$\bar{T}(3, 2, S_m)$	104.08	357.22	1536.00	9155	91449.47
$\bar{T}(3, 2, 2S_m)$	9.68	13.48	23.50	57.76	245.55
$\bar{T}(3, 2, 3S_m)$	9.18	10.95	14.10	22.20	53.56
$\bar{T}(3, 2, 4S_m)$	9.17	10.86	13.25	17.62	30.71
$\bar{T}(3, 2, \infty)$	9.17	10.86	13.19	16.66	22.41
$\bar{T}(2, 3, S_m)$	173.47	694.60	3584.00	26702.88	355636.84
$\bar{T}(2, 3, 2S_m)$	11.08	15.36	27.35	72.00	353.52
$\bar{T}(2, 3, 3S_m)$	10.60	12.59	16.04	25.18	63.46
$\bar{T}(2, 3, 4S_m)$	10.60	12.52	15.23	20.03	34.77
$\bar{T}(2, 3, \infty)$	10.60	12.52	15.19	19.16	25.75

5   Conclusion

In this article, we construct a stochastic model of the transfer process of a multi-packet subscriber message in a multi-link transport connection, differentiated by the accounts of pipeline effect in the tract with distortion in some parts of the hops. The proposed model allows us to analyze the effect of the duration of the timeout on the probability-time characteristics of the transport protocol. We detected the property of space-time symmetry of the stochastic process of information transfer a uniform stream of packets in a statistically uniform data path, which manifests itself in the invariance of probability-time performance of message delivery to a remote subscriber to mutually symmetrical values of the size of the message N and the length of transport connection D. It is established that the main contribution to the limiting value of the average message delay in a transport connection with the distortion in some parts of hops corresponding to the unlimited duration of the timeout, making the transmission of multipacket messages and upon receiving return acknowledgement in the determine pipeline and transmission time in a separate phase, equal to the average delay of the packet. The contribution of other components in end-to-end delay is proportional to the intensity of the packet distortion of R, which for high-quality communication channels can be neglected. It is shown that at three times the size the timeout of idling of end-to-end acknowledgements over a minimum duration of S_m and low distortion in communication channels of a transport connection for practical calculations, in most cases, you can use the ratio of marginal delay in the stochastic pipeline.

References

1. Fall, K., Stevens, R.: TCP/IP Illustrated, Volume 1: The Protocols, 2nd edn, p. 1017. Addison-Wesley Professional Computing Series (2012)
2. Boguslavskii, L.B.: Upravlenie potokami dannykh v setyakh EVM (Controlling Data Flows in Computer Networks), p. 168. Energoatomizdat, Moscow (1984)
3. Calleari, C., Giordano, S., Pagano, M., Pepe, T.: A survey of congestion control mechanisms in Linux TCP Communications in Computer and Information Science. In: Vishnevsky, V.; Kozyrev, D.; Larionov. A. (eds.) DCCN: Distributed Computer and Communication Networks 17th International Conference, vol. 2014, pp. 28–24, Moscow, Russia, 7–10 October 2013. Revised Selected Papers
4. Dunaitsev, R.A., Kucheryavaya, E.A.: Improved and supplemented PFTK-model of TCP Reno. Telecommunications, no. 3. pp. 27–31 (2005)
5. Bogoiavlenskaia, O.: Discrete model of TCP congestion control algorithm with round dependent loss rate. In: Balandin, S., Andreev, S., Koucheryavy, Y. (eds.) NEW2AN/ruSMART 2015. LNCS, vol. 9247, pp. 190–197. Springer, Heidelberg (2015)
6. Giordano, S., Pagano, M., Russo, F., Secchi, R.: Modeling TCP startup performance. J. Math. Sci. 200(4), 424–431 (2014)
7. Kravets, O.Y.: Mathematical modeling of parameterized TCP protocol. Autom. Remote Control. 74(7), 1218–1224 (2013)
8. Wang, J., Wen, J., Han, Y., Zhang, J., Li, C., Xiong, Z.: Achieving high throughput and TCP Reno fairness in delay-based TCP over large networks. Front. Comput. Sci. 8(3), 426–439 (2014)
9. Nikitinskiy, M.A., Chalyy, D.J.: Performance analysis of trickles and TCP transport protocols under high-load network conditions. Autom. Control Comput. Sci. 47(7), 359–365 (2013)
10. Sushchenko, S.P.: Lockout of buffer storage: its effect on the performance of a ladder-type data-transfer track. Autom. Remote Control. 60(7), 958–969 (1999)
11. Sushchenko, S.P.: Method of rational choice of the packet length of packet switching network. Avtom. Vychisl. Tekh., no. 3. pp. 24–28 (1984)
12. Sushchenko, S.P.: Parametric optimization of packet switching networks. Avtom. Vychisl. Tekh., no. 2., pp. 43–49 (1985)
13. Sushchenko, S.P.: Analysis of end-to-end message delay in multi-tier virtual channel. Avtom. Vychisl. Tekh., no. 3. pp. 52–64 (1989)
14. Sushchenko, S.P.: Effect of duration end-to-end timeout on the data latency in a virtual channel. Avtom. Vychisl. Tekh., no. 6., pp. 36–40 (1991)
15. Sushchenko, S.P.: Analysis of the influence of duration of end-to-end timeout on the operational characteristics of a virtual channel. Avtom. Vychisl. Tekh., no. 4., pp. 43–66 (1995)
16. Sushchenko, S.P.: Analytical computability of sums of compositions of exponential and power functions. In: kN.: Mathematical Modeling and Probability Theory, Collected Scientific Works of TSU, pp. 253–256. TSU publishing house, Tomsk (1998)

Analysis of LRU Cache Trees with a Power Law Reference Distribution

Udo R. Krieger[1(✉)] and Natalia M. Markovich[2]

[1] Otto-Friedrich-Universität, 96045 Bamberg, Germany
udo.krieger@ieee.org
[2] Russian Academy of Sciences, 117997 Moscow, Russia
markovic@ipu.rssi.ru

Abstract. We investigate the performance of a LRU cache replacement policy. Regarding the hit and miss ratios of Zipf-distributed frequencies of object requests in a cascade of LRU caches, new explicit, computationally tractable formulae are derived.

Keywords: LRU cache performance · Che's approximation · Cache trees

1 Modeling LRU Cache Trees

Following Che et al. [3] and Fricker et al. [4], we consider a hierarchical cascade of LRU caches in an access and backbone infrastructure of a packet-switched next generation network (see also [1]).

We assume that we have four layers in a tree-like hierarchy:

- *Layer 3*: The consumers consist of clients with a very small browser cache. They are not modelled and just work as requestors of M different Poisson object streams with the rates $\Lambda_{*i} = \Lambda^{(1)} \cdot p_i$ and individual selection probabilities $p_i = K/i^\alpha$, $K^{-1} = \sum_{i=1}^{M} 1/i^\alpha$, $i \in \mathcal{C} = \{1, \ldots, M\}$, governed by a Zipf law [2] with tail index $\alpha > 0$. They constitute an overall Poisson stream with the rate $\Lambda^{(1)} = \sum_{i=1}^{M} \Lambda_{*i}$.
- *Layer 2*: This layer of leaf caches at level 1 consists of a network of N LRU caches with the capacities $C(L_1) = (C_1, \ldots, C_N)$ in the access network of an ISP. Each cache m serves M Poisson flows f_{mj} of object requests with rates λ_{mj}. They are arising from all M object types $j \in \mathcal{C}$ stemming from the clients at layer 3. The superimposed Poisson stream of cache m has the rate $\Lambda_m = \sum_{j=1}^{M} \lambda_{mj}$. It arises from the splitting $\Lambda_m = \Lambda^{(1)} \cdot l(1)_m$ of the overall Poisson traffic of the clients with the rate $\Lambda^{(1)}$ into the offered load Λ_m of each cache m with probabilities $l(1)_m$.

  The individual miss and hit ratios of each stream i are given by $\eta_{mi}^{(1)}, H_{mi}^{(1)} = 1 - \eta_{mi}^{(1)}$ and the corresponding overall variants of cache m at level 1 by $\eta_m^{(1)}, H_m^{(1)} = 1 - \eta_m^{(1)}$, respectively.

© Springer International Publishing Switzerland 2016
A. Dudin et al. (Eds.): ITMM 2016, CCIS 638, pp. 162–167, 2016.
DOI: 10.1007/978-3-319-44615-8_14

– *Layer 1*: This layer comprises one level 0 backbone cache of capacity C_0 with an overall miss ratio $\overline{\eta}^{(0)}$ and hit ratio $\overline{H}^{(0)} = 1 - \overline{\eta}^{(0)}$.
– *Layer 0*: It comprises a fully interconnected network of servers offering all requested objects, e.g. provided by a CDN-like set of data centers.

2   Performance Analysis of LRU Cache Cascades

In the following we formulate a computationally tractable, matrix-oriented framework that can be used to determine all relevant quantities of our cache tree model by appropriate measurement and estimation procedures. First we define the matrix of the rates of the Poisson arrival streams f_{ij} at cache $i \in L_1 = \{1, \ldots, N\}$ of level 1 associated with object references of type $j \in \mathcal{C} = \{1, \ldots, M\}$ by $\lambda = (\lambda_{ij})_{\{i=1,\ldots,N; j=1,\ldots,M\}} \in (\mathbb{R}^+)^{N \times M}$. Its i-th row $\lambda_{i\cdot} = e_i^t \cdot \lambda = (\lambda_{i1}, \ldots, \lambda_{iM})$, where e_i is the ith unit vector, determines the vector of all object-specific arrival rates to cache i. Then the overall rate of the Poisson process arriving at cache i is given by $\Lambda_i = \lambda_{i\cdot} \cdot e = \sum_{j=1}^{M} \lambda_{ij} = e_i^t \cdot \lambda \cdot e$ where e is the vector of all ones. As the arrival streams to each cache i are derived from a heavy-tailed popularity distribution with selection probabilities $p_i = K \cdot 1/i^\alpha$, $K^{-1} = \sum_{i=1}^{M} 1/i^\alpha$, of Zipf type, we get the representation $\lambda_{i\cdot} = (\Lambda_i p_1, \ldots, \Lambda_i p_M) = \Lambda_i \cdot p^t$, $i = 1, \ldots, N$ of the rows of the rate matrix λ in terms of the overall arrival rate Λ_i and the type selection vector $p^t = (p_1, \ldots, p_M) \in (\mathbb{R}^+)^M$. We denote the corresponding rate vector by $\Lambda^t = (\Lambda_1, \ldots, \Lambda_N)$. The superposition of all request processes for objects of type $j \in \mathcal{C}$ at all level 1 caches i constitutes a Poisson process with the rate $\Lambda^{(1)} = \sum_{i=1}^{N} \sum_{j=1}^{M} \lambda_{ij} = e^t \cdot \Lambda \cdot p^t \cdot e = e^t \cdot \lambda \cdot e$. Then we get the representation $\lambda_{ij} = \Lambda_i \cdot p_j$, hence, $\lambda = \Lambda \cdot p^t$ of the arrival rate matrix λ in terms of the rate vector Λ at level 1 caches and the type selection probability vector p. The splitting of the overall Poisson stream at caches of level 1 into the Poisson arrival streams with the rates Λ_i can be described by a vector $l(1)^t = (l(1)_1, \ldots, l(1)_N)$ of splitting probabilities $l(1)_i = \Lambda_i/\Lambda^{(1)} = e_i^t \cdot \lambda \cdot e/e^t \cdot \lambda \cdot e$ for $i = 1, \ldots, N$, hence, $l(1)^t = \Lambda^t/\Lambda^{(1)}$.

Considering the object flows of type $j \in \mathcal{C}$ at a cache $i \in L_1 = \{1, \ldots, N\}$ of level 1, we describe the hit ratios $H_{ij}^{(1)}$ and corresponding miss ratios $\eta_{ij}^{(1)} = 1 - H_{ij}^{(1)}$ by a blocking matrix $B^{(1)} = \left(\eta_{ij}^{(1)}\right)_{\{i=1,\ldots,N; j=1,\ldots,M\}}$ with Che's approximation $\eta_{ij}^{(1)} \in (0,1)$ of the miss ratio at cache i (cf. [3,4]) represented in terms of $\theta_{ij} = 1 - \left(\eta_{ij}^{(1)}\right)^{1/C_i}$ by $(B^{(1)})_{ij} = \eta_{ij}^{(1)} = (1 - \theta_{ij})^{C_i}$. According to Little's law the miss ratio is determined by the ratio of the rejected proportion $\lambda_{ij} \cdot (B^{(1)})_{ij}$ of the requests with the total rate $\Lambda_i = \sum_{j=1}^{M} \lambda_{ij}$ and its proportion $\lambda_{ij} = \Lambda_i \cdot p_j$ of offered type $j \in \mathcal{C}$ traffic

$$\eta_{ij}^{(1)} = 1 - H_{ij}^{(1)} = \frac{\lambda_{ij} \cdot (B^{(1)})_{ij}}{\lambda_{ij}} = \frac{\Lambda_i \cdot p_j \cdot (1 - \theta_{ij})^{C_i}}{\Lambda_i \cdot p_j} = (1 - \theta_{ij})^{C_i}. \quad (1)$$

Then the miss ratio $\eta_{.j}^{(1)}$ arising from the superimposed Poisson traffic of type $j \in \mathcal{C}$ at all caches of level 1 with the total rate $\Lambda_{*j} = \sum_{i=1}^{N} \lambda_{ij} = \sum_{i=1}^{N} \Lambda_i \cdot p_j = \Lambda^{(1)} \cdot p_j$ is given by $\left(\eta_{.1}^{(1)}, \ldots, \eta_{.M}^{(1)}\right) = l(1)^t \cdot B^{(1)} = \Lambda^t \cdot B^{(1)} / \Lambda^{(1)}$.

Due to Little's law the overall miss ratio of cache i is determined by the relative blocking rates $(B^{(1)})_{ij} = \eta_{ij}^{(1)}$ of all individual object streams of types $j \in \mathcal{C}$ with miss traffic rates $\lambda_{ij} \cdot (1 - \theta_{ij})^{C_i}$ compared to the overall arrival rate $\Lambda_i = \sum_{j=1}^{M} \lambda_{ij}$ at cache i, i.e. by the ratio of the rejected proportion of the requests to the total rate at cache i:

$$\eta_i^{(1)} = 1 - H_i^{(1)} = \frac{\sum_{j=1}^{M} \lambda_{ij}(B^{(1)})_{ij}}{\Lambda_i} = \left(\left(B^{(1)} \odot (e \cdot p^t)\right) e\right)_i = \left(B^{(1)} \cdot p\right)_i (2)$$

Here we use the Hadamard matrix product $(A \odot B)_{i,j} = A_{ij} \cdot B_{ij}$ for the entry-wise multiplication of two matrices $A = (A_{ij}), B = (B_{ij})$ of equal dimensions. Then the vector of the miss ratios of all caches at level 1 is determined by

$$\eta^{(1)} = \begin{pmatrix} \eta_1^{(1)} \\ \vdots \\ \eta_M^{(1)} \end{pmatrix} = \left(B^{(1)} \odot (e \cdot p^t)\right) \cdot e = B^{(1)} \cdot p.$$

We realize that the miss ratio $\eta_i^{(1)}$ of a level 1 cache i can be associated with a superposition of truncated geometric distributions $\widehat{G}_{ij}(k) = g_{ij} \cdot \theta_{ij} \cdot (1 - \theta_{ij})^k = s_{ij} \cdot G_{ij}(k)$, $G_{ij}(k) = p_j \cdot (1 - \theta_{ij})^k$, $k = 0, \ldots, C_i$, with normalization constants $(g_{ij})^{-1} = \sum_{k=0}^{C_i} G_{ij}(k) = 1 - (1 - \theta_{ij})^{C_i+1}$, $(s_{ij})^{-1} = (g_{ij})^{-1} \cdot p_j / \theta_{ij}$.

The overall miss ratio $\overline{\eta}^{(1)}$ of level 1 caches $L_1 = \{1, \ldots, N\}$ is determined by

$$\overline{\eta}^{(1)} = \frac{\Lambda^t}{\Lambda^{(1)}} \cdot \eta^{(1)} = \frac{\Lambda^t \cdot B^{(1)} \cdot p}{\Lambda^{(1)}} = l(1)^t \cdot B^{(1)} \cdot p \tag{3}$$

The missing proportions $\lambda(1)_{ij} = \lambda_{ij} \cdot (B^{(1)})_{ij}$ of the original traffic $\lambda(2) = \lambda = (\lambda_{ij})_{\{i,j\}}$ offered to cache i of object type j at level 1 is approximated by a new Poisson process with the rate $\lambda(1) = (\lambda(1)_{ij})_{\{i,j\}}$ and routed as a load to cache 0 at level 0 (cf. [3]). We use again the associative, commutative and distributive Hadamard matrix product $(A \odot B)_{i,j} = A_{ij} \cdot B_{ij}$ for the entrywise multiplication of two real matrices to define this new load matrix

$$\lambda(1) = \left(\lambda_{ij} \cdot \eta_{ij}^{(1)}\right)_{\{i=1,\ldots,N;j=1,\ldots,M\}} = \lambda \odot B^{(1)} = B^{(1)} \odot \lambda. \tag{4}$$

This representation enables an efficient matrix-vector computation, e.g., in Matlab by the '.*' operator.

We define a routing matrix $R = \left(R_{(i,k),(j,l)}\right)$ that models the routing of type l traffic of cache j at level 1 to type k of cache i at level 0. As the type classes are not modified and all level 1 caches route to cache 0, we get $R = \left(R_{\{0\} \times \{1,\ldots,M\}}, \ldots, R_{\{0\} \times \{1,\ldots,M\}}\right) = (I_M, \ldots, I_M) = (e(N)^t \otimes I_M)$ where $e(N) \in \mathbb{R}^N$ denotes the vector of all ones, $I_M \in \mathbb{R}^{M \times M}$ the identity matrix, and $\otimes$ the Kronecker product $A \otimes B = (A_{ij}B)$. The input-output relation of the routing chains is given by $y = R \cdot x$ with an input load vector $x = \begin{pmatrix} (\lambda(1)_{1\cdot})^t \\ \vdots \\ (\lambda(1)_{N\cdot})^t \end{pmatrix}$.

It is defined by the rows $\lambda(1)_{i\cdot}$ corresponding to cache i of the missing traffic matrix $\lambda(1) = \begin{pmatrix} \lambda(1)_{1\cdot} \\ \vdots \\ \lambda(1)_{N\cdot} \end{pmatrix}$, now sorted as column vectors. Then the output

$y = R \cdot x = \sum_{i=1}^{N} (\lambda(1)_{i\cdot})^t = \lambda(1)^t \cdot e(N) = \lambda_0^t = (\lambda_{01}, \ldots, \lambda_{0M})^t$ is the type-based superposition of the missing traffic rates and determines the arrival rates λ_{0j} of type j to cache 0.

If we interpret the vector $e(N)$ of all ones as submatrix $\widehat{R}_{\{1,\ldots,M\} \times \{0\}}$ of the adjacency matrix of the directed graph Γ describing the two-level tree of caches $L_1 = \{1, \ldots, N\}$, $L_0 = \{0\}$ at levels 1 and 0, respectively, then we get

$$\lambda_0 = (\lambda_{01}, \ldots, \lambda_{0M}) = y^t = e(N)^t \cdot \lambda(1) = \widehat{R}^t \cdot \lambda(1) \tag{5}$$

as a simple representation in terms of the missing traffic rates $\lambda(1)$ of level 1. This scheme can be easily extended to arbitrary cache hierarchies and routing schemes.

We conclude that the matrix representation

$$\lambda(1)_{ij} = (\lambda \odot B^{(1)})_{ij} = \left((\Lambda p^t) \odot B^{(1)}\right)_{ij} = \Lambda_i p_j (1 - \theta_{ij})^{C_i} = \Lambda_i \cdot G_{ij}(C_i) \tag{6}$$

holds with the matrix $G = (G_{ij}(C_i))_{\{i,j\}} = \left(p_j \cdot (1 - \theta_{ij})^{C_i}\right)_{\{i,j\}} = (e \cdot p^t) \odot B^{(1)} = B^{(1)} \odot (e \cdot p^t)$. Then the proportion of the offered load $\lambda(0) = \lambda_0$ to cache 0 of type $j \in \{1, \ldots, M\}$ $\lambda_{0j} = \sum_{i=1}^{N} \Lambda_i \cdot p_j \cdot (1 - \theta_{ij})^{C_i} = \Lambda^t \cdot (G \cdot e_j)$ yields the simple matrix representation

$$\lambda_0 = \widehat{R}^t \cdot \lambda(1) = e(N) \left((\Lambda \cdot p^t) \odot B^{(1)}\right) = \Lambda^t \cdot \left(B^{(1)} \odot (e \cdot p^t)\right) = \Lambda^t \cdot G \tag{7}$$

related to the type-based arrival rates λ_{0j} of the superimposed Poisson process routed to cache 0.

Then the aggregated arrival rate $\Lambda^{(0)}$ of the Poisson process of missing cache requests reaching level 0 is given by

$$\Lambda^{(0)} = \sum_{j=1}^{M} \lambda_{0j} = \lambda_0 \cdot e = \Lambda^t \cdot G \cdot e = \Lambda^t \left(B^{(1)} \odot (e \cdot p^t)\right) e = \Lambda^t \cdot B^{(1)} \cdot p. \tag{8}$$

The selection probability of type j at the cache of level 0 is determined by the ratio of the arrival rate of type j to the overall arrival rate at the single cache of level 0 $p_j^{(0)} = \lambda_{0j}/\Lambda^{(0)}$. It yields the selection vector

$$\left(p^{(0)}\right)^t = \frac{\Lambda^t \cdot G}{\Lambda^t \cdot G \cdot e} = \frac{e^t \cdot \left((\Lambda \cdot p^t) \odot B^{(1)}\right)}{\Lambda^t \cdot B^{(1)} \cdot p} = \frac{(\Lambda^t \cdot B^{(1)}) \odot p^t}{\Lambda^t \cdot B^{(1)} \cdot p}. \tag{9}$$

Following the single LRU cache analysis (1), the miss ratio $\eta_{0j}^{(0)}$ of object flows of type j at cache 0 is determined as blocking $(B^{(0)})_{0j}$ by the ratio of the missing traffic to the offered traffic of type j, approximated by Che's approximation [3], and represented as

$$\eta_{0j}^{(0)} = \frac{\lambda_{0j} \cdot (B^{(0)})_{0j}}{\lambda_{0j}} = (1 - \theta_{0j})^{C_0}. \tag{10}$$

This representation (7) to (10) allows a simultaneous computation of all the miss ratios $\eta_{0j}^{(0)}$, e.g., by a simple Matlab routine.

The total miss rate $\overline{\eta}^{(0)}$ of cache 0 at the level 0 is determined by

$$\overline{\eta}^{(0)} = \sum_{j=1}^{M} \frac{\lambda_{0j}}{\lambda_0 \cdot e} \cdot (B^{(0)})_{0j} = \frac{1}{\Lambda^t \cdot B^{(1)} \cdot p} \sum_{j=1}^{M} \sum_{i=1}^{N} \Lambda_i \cdot (1 - \theta_{0j})^{C_0} \cdot (1 - \theta_{ij})^{C_i} \cdot p_j$$

and yields the following matrix representation:

$$\overline{\eta}^{(0)} = \frac{1}{\Lambda^{(0)}} \cdot \left(\lambda_0 \odot B^{(0)} \right) \cdot e = \frac{\left(\left(\Lambda^t \cdot B^{(1)} \right) \odot B^{(0)} \right) \cdot p}{\Lambda^t \cdot B^{(1)} \cdot p} = B^{(0)} \cdot p^{(0)} \tag{11}$$

We realize the structural equivalence to relation (2) describing the miss ratios $\eta^{(1)}$ of level 1.

Due to Little's law the total miss ratio η_j of type j traffic handled by the complete two-level cache hierarchy is determined by the ratio of the rates of the overall miss traffic of type j to the arriving Poisson traffic of type j, i.e.

$$\eta_j = \frac{\lambda_{0j}}{\sum_{i=1}^{N} \lambda_{ij}} \cdot (B^{(0)})_{0j} = \frac{\left(\left(\Lambda^t \cdot B^{(1)} \right) \odot B^{(0)} \right) \cdot e_j}{\Lambda^{(1)}}. \tag{12}$$

Thus, we get the matrix representation

$$(\eta_1, \ldots, \eta_M) = \frac{\left(\Lambda^t \cdot B^{(1)} \right) \odot B^{(0)}}{\Lambda^{(1)}} = \left(l(1)^t \cdot B^{(1)} \right) \odot B^{(0)}. \tag{13}$$

The total miss ratio $\overline{\eta}$ of the complete two-level cache hierarchy is determined by the ratio of the rates of the overall miss traffic to the arriving Poisson traffic:

$$\overline{\eta} = \sum_{j=1}^{M} \frac{\lambda_{0j}}{\Lambda^{(1)}} \cdot (B^{(0)})_{0j} = \left(\left(l(1)^t \cdot B^{(1)} \right) \odot B^{(0)} \right) \cdot p \tag{14}$$

This representation of the overall miss ratio $\overline{\eta}$ of the cache cascade clearly reveals the distribution of the load among the caches at level 1 by the cache selection vector $l(1)^t$, the blocking $B^{(1)}, B^{(0)}$ by cache misses at level 1 and 0, respectively, and the superposition of the different object types by the selection vector p.

References

1. Blefari-Melazzi, N., et al.: A general, tractable and accurate model for a cascade of LRU caches. IEEE Commun. Lett. **18**(5), 877–880 (2014)
2. Breslau, L., Cao, P., Fan, L., Phillips, G., Shenker, S.: Web caching and Zipf-like distributions: Evidence and implications. In: Proceedings of the INFOCOM 1999, pp. 126–134, IEEE Press, Piscataway (1999)
3. Che, H., Tung, Y., Wang, Z.: Hierarchical web caching systems: modeling, design and experimental results. IEEE JSAC **20**(7), 1305–1314 (2002)
4. Fricker, C., Robert, P., Roberts, J.: A versatile and accurate approximation for LRU cache performance. In: Proceedings 24th International Teletraffic Congress, ITC 2012, pp. 8: 1–8: 8 (2012)

Stationary Distribution Insensitivity of a Closed Queueing Network with Non-active Customers and Multi-regime Service

Julia Kruk[1,2](✉) and Yuliya Dudovskaya[1,2]

[1] Belarusian National Technical University, Minsk, Belarus
juls1982@list.ru, dudovskaya@gmail.com
[2] Francisk Skorina Gomel State University, Gomel, Belarus

Abstract. Stationary functioning of an exponential closed queueing network with temporarily non-active customers and multi-regime service is considered. Non-active customers are located in queues of network systems, not being serviced. For a customer, the opportunity of passing from its ordinary state to the temporarily non-active state (and backwards) is provided. Each system can operate in several regimes corresponding to different degrees of its efficiency. Service times are independent exponentially distributed random values. Times of functioning in regimes are independent random distributed values. Stationary distribution insensitivity is established.

Keywords: Closed queueing network · Non-active customers · Multi-regime service · Stationary distribution insensitivity

1 Introduction

Presently attention to queueing theory is mainly stimulated by the need to apply the results of this theory to important practical problems. Nowadays important research efforts are devoted to the problem of queueing systems reliability. In practical terms, it is important to consider several different approaches: a queueing system can break down totally or partially. Yu.V. Malinkovsky introduced into consideration open and closed queuing networks with multi-regime service strategies. Queueing systems in multi-regime networks can operate in several regimes. Each regime corresponds to a certain degree of service efficiency. Herewith, the problem of customer reliability becomes relevant too. Indeed not only the queueing system can break down. Customers may also lose some of their quality indicators. A queueing network with temporarily non-active customers is a model with customers which are partly unreliable. The necessity of their research was caused by practical considerations, because such networks allow us to consider models with partially unreliable customers. Non-active customers are located in queues of network systems, not being serviced. For a customer, the opportunity of passing from its ordinary state to the temporarily non-active

© Springer International Publishing Switzerland 2016
A. Dudin et al. (Eds.): ITMM 2016, CCIS 638, pp. 168–176, 2016.
DOI: 10.1007/978-3-319-44615-8_15

state (and backwards) is provided. Non-active customers can be interpreted as customers with a defect that makes them unfit for service. G. Tsitsiashvili and M. Osipova [1,2] have researched an open exponential queueing network with temporarily non-active customers and have obtained the form of stationary distribution.

The standard assumption in analysis of classical Jackson and Gordon-Newell queueing networks [3,4] is that service times are independent exponentially distributed random values. But real numerous statistical data prove the opposite. Therefore there is an actual problem of researching open and closed queueing networks with random distributed service times. Currently, this problem attracts the increasing attention of researchers. The first result about stationary distribution insensitivity belongs to B.A. Sevastyanov, who has observed queueing system M/G/m/0 and has proved stationary distribution insensitivity [5]. BCMP-theorem (Baskett, Chandy, Muntz, Palacios) [6] is the first result about stationary distribution insensitivity for queueing networks. We have generalized the result [1,2] in the case of random distributed service times [7–9]. We have established stationary distribution insensitivity with respect to functional form of service time distribution.

V.A. Ivnitsky [10] has considered a rather interesting class of queueing networks with not "temporal" but so-called "energetic" interpretation of customer service. For this type of queueing networks every service operation is characterized by the random variable of work for customer service. Stationary distribution insensitivity with respect to a functional form of distribution of work quantity for customer service has been established for different classes of open and closed queueing networks [10] and for open and closed queueing networks with temporarily non-active customers [11,12].

Closed multi-regime queueing networks with non-active customers were researched in [13,14]. Stationary distribution has been obtained in [14]. [13] has generalized the result of [11] in the case of multi-regime service. Times of functioning at regimes in the model from [13] were exponentially distributed values. Stationary distribution insensitivity with respect to functional form of distribution of work quantity for customer service has been established [13].

This paper considers stationary functioning of a closed queueing network with non-active customers and multi-regime service strategies. Each system can operate at several regimes corresponding to different degrees of its efficiency. In contrast to [11,13] this paper considers closed queueing network with exponential distributed times of service, herewith it is assumed that times of functioning at regimes are random distributed values. Stationary distribution insensitivity with respect to functional form of distribution of time of functioning at regimes is established.

2   Queueing Network Description

A closed queueing network with the set of systems $J = \{1, 2, \ldots, N\}$ is considered. M customers are circulating in the network. Non-active customers are located in queues of network systems, not being serviced. There are input Poisson flows of signals with rates ν_i and φ_i, $i \in J$. When arriving at the system

$i \in J$ the signal with rate ν_i induces an ordinary customer, if any, to become non-active. When arriving at the system $i \in J$ the signal with rate φ_i induces an non-active customer, if any, to become ordinary. Signals do not need service.

Let $n_i(t), n_i'(t)$ be the numbers of ordinary and non-active customers at the system $i \in J$ at time t accordingly and $n_i''(t)$ – the number of service regime.

A stochastic process $z(t) = ((n_i(t), n_i'(t), n_i''(t)), \ i \in J)$ is considered. The space of states for process $z(t)$ is $Z = \{(z = (n_1, n_1', n_1''), \ldots, (n_N, n_N', n_N''))|$ $n_i, n_i' \geq 0, \sum_{i \in J}(n_i + n_i') = M, n_i'' = \overline{0, r_i}, i \in J\}$.

Numbering of ordinary customers in the system queue is made from the "tail" of the queue to the device. Non-active customers in the queue of the system $i \in J$ are numbered as follows: a customer which has become non-active in the last turn, has number n_i'. When arriving at the system $i \in J$ the signal with rate ν_i induces an ordinary customer with number 1 to become a non-active customer with number $n_i' + 1$. When arriving at the system $i \in J$ the signal with rate φ_i induces a non-active customer with number n_i' to become an ordinary customer with number 1. So, the set of customers' numbers in the system $i \in J$ is $(1, \ldots, n_i', 1, \ldots, n_i)$.

The discipline of service is $FCFS$. Service times are independent exponentially distributed random values with rates $\mu_i, i \in J$.

Each system can operate in several regimes corresponding to different degrees of efficiency. The times of functioning in regimes are independent random distributed values with functions of distribution $\Phi_i(n_i'', t)$, $(\Phi_i(n_i'', 0) = 0)$ and expected values $\eta_i < \infty, i \in J$. Switching is possible only to neighboring regimes. The speed of regime switching is $\sigma_i + \rho_i, (\sigma_i, \rho_i > 0, i \in J)$.

System i has a single device, which can operate at $r_i + 1$ regimes. Let us denote 0 as a basic service regime, then the device is switched to the regime 1. For states (n_i, n_i', n_i''), where $0 \leq n_i'' \leq r_i$, the time of functioning in the regime n_i'' is a random distributed value, then the device is switched to the regime $n_i'' + 1$ with the probability $\frac{\sigma_i I_{n_i'' < r_i}}{\sigma_i I_{n_i'' < r_i} + \rho_i I_{n_i'' > 0}}$ or to the regime $n_i'' - 1$ with the probability $\frac{\rho_i I_{n_i'' > 0}}{\sigma_i I_{n_i'' < r_i} + \rho_i I_{n_i'' > 0}}$. The time of functioning in the regime r_i is a random distributed value, then the device is switched to the regime $r_i - 1$. During regime switching the number of customers does not change. Switching from the regime $n_i'' - 1$ to n_i'' can be interpreted as a partial working capacity decline. Transition from regime n_i'' to regime $n_i'' - 1$ means the recovery of working capacity, which was lost after switching from regime $n_i'' - 1$ to n_i''.

After completing the of service process in the system $i \in J$ the customer passes to the system $j \in J$ with the probability $p_{i,j}$ $(\sum_{j=1}^{N} p_{i,j} = 1)$. Let $p_{i,i} = 0, i \in J$.

A traffic equations system is:

$$\varepsilon_i = \sum_{j=1}^{N} \varepsilon_j p_{j,i}, \ \ i \in J. \tag{1}$$

It has been proved [4] that a traffic equations system has a unique non-trivial solution up to a constant.

3   Stationary Distribution Insensitivity

We consider a closed queueing network with multi-regime service strategies. The times of functioning in regimes are independent random distributed values. In this instance $z(t)$ is not a Markov process.

Let us denote by $\xi_{i,n_i''}(t)$ the rest time of functioning in the regime n_i'' from time t to the time of regime switching, $\xi(t) = (\xi_{1,n_1''}(t), \ldots, \xi_{N,n_N''}(t))$.

So we introduce into consideration a Markov process $\zeta(t) = (z(t), \xi(t))$, where $\xi(t) = (\xi_{1,n_1''}(t), \ldots, \xi_{N,n_N''}(t))$.

Let us denote by

$$F(z, x) = F(z, x_{1,n_1''}, \ldots, x_{N,n_N''})$$

$$= \lim_{t \to \infty} P\{z(t) = z, \xi_{1,n_1''}(t) < x_{1,n_1''}, \ldots, \xi_{N,n_N''}(t) < x_{N,n_N''}\}, \ z \in Z,$$

$$x_{i,n_i''} \in \mathbb{R} \ \forall \, i \in J, \ n_i'' = \overline{0, r_i}.$$

Functions $F(z, x)$ are called stationary functions of $\zeta(t)$ distribution.

A model of closed multi-regime queueing network with temporarily non-active customers has beenwas considered in [14]. The times of functioning at regimes were independent exponentially distributed random values. The following theorem has been proved.

Theorem 1. *The Markov process $z(t)$ is ergodic. The stationary distribution of the process is:*

$$p((n_1, n_1', n_1''), \ldots, (n_N, n_N', n_N'')) = G^{-1}(M, N)p_1(n_1, n_1', n_1'')p_2(n_2, n_2', n_2'') \times \cdots$$
$$\times p_N(n_N, n_N', n_N''), \tag{2}$$

where

$$p_i(n_i, n_i', n_i'') = \left(\frac{\varepsilon_i}{\mu_i}\right)^{n_i} \left(\frac{\varepsilon_i \nu_i}{\mu_i \varphi_i}\right)^{n_i'} \left(\frac{\sigma_i}{\rho_i}\right)^{n_i''},$$

ε_i *is the traffic equations system solution. $G(M, N)$ is a normalizing constant.*

In our case for the closed queueing network with non-active customers, multi-regime service strategies and random distributed times of functioning in regimes the following theorem is true.

Theorem 2. *The Markov process $\zeta(t)$ is ergodic. The stationary functions of the process $\zeta(t)$ distribution are:*

$$F(z, x) = G^{-1}(M, N)p_1(n_1, n_1', n_1'')p_2(n_2, n_2', n_2'') \ldots p_N(n_N, n_N', n_N'') \tag{3}$$

$$\times \prod_{i=1}^{N} \frac{1}{\eta_i} \int_0^{x_{i,n_i''}} (1 - \Phi_i(n_i'', u))du, \ z \in Z,$$

where

$$p_i(n_i, n_i', n_i'') = \left(\frac{\varepsilon_i}{\mu_i}\right)^{n_i} \left(\frac{\varepsilon_i \nu_i}{\mu_i \varphi_i}\right)^{n_i'} \left(\frac{\sigma_i}{\rho_i}\right)^{n_i''}, \tag{4}$$

ε_i is the traffic equations system solution (1). $G(M, N)$ is a normalizing constant, which can be found from the following condition

$$\sum_{((n_1, n_1', n_1''), \ldots, (n_N, n_N', n_N'')) \in Z} p((n_1, n_1', n_1'), \ldots, (n_N, n_N', n_N'')) = 1. \qquad (5)$$

$p((n_1, n_1', n_1'), \ldots, (n_N, n_N', n_N''))$ is found by means of (2).

Proof. Let us denote by $e_i \in Z$ – the vector which coordinates equal 0 with the exception of $(n_i, n_i', n_i'') = (1, 0, 0)$, let us denote by $e_i' \in Z$ – the vector which coordinates equal 0 with the exception of $(n_i, n_i', n_i'') = (0, 1, 0)$, analogically let us denote by $e_i'' \in Z$ – the vector which coordinates equal 0 with the exception of $(n_i, n_i', n_i'') = (0, 0, 1)$, $i \in J$.

We consider the process $\zeta(t)$. In the case of exponentially distributed times of functioning in regimes the process $z(t)$ is ergodic by ergodic Markov theorem. The process $\zeta(t)$ is also ergodic because $\zeta(t)$ is obtained from $z(t)$ by adding continuous components.

The process $\zeta(t)$ can change its states due to incoming signals or regime switching. Such changes we call spontaneous changes.

Let us suppose that h is a small time interval and consider the probability

$$P\{z(t+h) = z, \xi_{1,n_1''}(t+h) < x_{1,n_1''}, \ldots, \xi_{N,n_N''}(t+h) < x_{N,n_N''}\}.$$

This event may occur in the following ways:

1. From the moment t during time h there were no spontaneous changes and the service in any system was not over. The probability of this event is

$$P\{z(t) = z, h \le \xi_{1,n_1''}(t) < x_{1,n_1''} + h, \ldots, h \le \xi_{N,n_N''}(t) < x_{N,n_N''} + h\}$$
$$\times (1 - \sum_{i=1}^{N}(\nu_i I_{n_i>0} + \varphi_i I_{n_i'>0} + \mu_i I_{n_i>0})h + o(h)).$$

2. During time h a customer was serviced in the system $j \in J$ and was routed to the system $i \in J$. There were no spontaneous changes.

$$P\{z(t) = z - e_i + e_j, h \le \xi_{1,n_1''}(t) < x_{1,n_1''} + h,$$
$$\ldots, h \le \xi_{N,n_N''}(t) < x_{N,n_N''} + h\} \times (\mu_j p_{j,i} h + o(h)) I_{n_i>0}.$$

3. During time h an informational signal with rate ν_i arrived at the system $i \in J$. There were no other spontaneous changes. No customer was serviced.

$$P\{z(t) = z + e_i - e_i', h \le \xi_{1,n_1''}(t) < x_{1,n_1''} + h,$$
$$\ldots, h \le \xi_{N,n_N''}(t) < x_{N,n_N''} + h\} \times (\nu_i h + o(h)) I_{n_i'>0}.$$

4. During time h an informational signal with rate φ_i arrived at the system $i \in J$. There were no other spontaneous changes. No customer was serviced.

$$P\{z(t) = z - e_i + e_i', h \le \xi_{1,n_1''}(t) < x_{1,n_1''} + h,$$
$$\ldots, h \le \xi_{N,n_N''}(t) < x_{N,n_N''} + h\}(\varphi_i h + o(h)) I_{n_i>0}.$$

5. During time h the service regime of the system i was increased by 1. There were no other spontaneous changes. No customer was serviced.

$$P\{z(t) =z - e_i'', h \le \xi_{1,n_1''}(t) < x_{1,n_1''} + h,$$
$$\ldots, \xi_{i,n_i''-1}(t) < h, \ldots, h \le \xi_{N,n_N''}(t) < x_{N,n_N''} + h\}$$
$$\times (\Phi_i(n_i'', x_{i,n_i''} + \theta h) \frac{\sigma_i I_{n_i'' < r_i}}{\sigma_i I_{n_i'' < r_i} + \rho_i I_{n_i'' > 0}} + o(h)) I_{n_i'' > 0}.$$

6. During time h the service regime of the system i was decreased by 1. There were no other spontaneous changes. No customer was serviced.

$$P\{z(t) =z + e_i'', h \le \xi_{1,n_1'}(t) < x_{1,n_1'} + h,$$
$$\ldots, \xi_{i,n_i''+1}(t) < h, \ldots, h \le \xi_{N,n_N''}(t) < x_{N,n_N''} + h\}$$
$$\times (\Phi_i(n_i'', x_{i,n_i''} + \theta h) \frac{\rho_i I_{n_i'' > 0}}{\sigma_i I_{n_i'' < r_i} + \rho_i I_{n_i'' > 0}} + o(h)) I_{n_i'' < r_i}.$$

Hereinbefore $0 < \theta < 1$.

7. During time h there were more than two changes of queueing network condition. This probability is $o(h)$.

Therefore

$$P\{z(t + h) = z, \xi_{1,n_1''}(t + h) < x_{1,n_1''}, \ldots, \xi_{N,n_N''}(t + h) < x_{N,n_N''}\}$$
$$= P\{z(t) = z, h \le \xi_{1,n_1''}(t) < x_{1,n_1''} + h, \ldots, h \le \xi_{N,n_N''}(t) < x_{N,n_N''} + h\}$$
$$\times (1 - \sum_{i=1}^{N} (\nu_i I_{n_i > 0} + \varphi_i I_{n_i' > 0} + \mu_i I_{n_i > 0}) h + o(h))$$
$$+ P\{z(t) = z - e_i + e_j, h \le \xi_{1,n_1''}(t) < x_{1,n_1''} + h,$$
$$\ldots, h \le \xi_{N,n_N''}(t) < x_{N,n_N''} + h\} \times (\mu_j p_{j,i} h + o(h)) I_{n_i > 0}$$
$$+ P\{z(t) = z + e_i - e_i', h \le \xi_{1,n_1''}(t) < x_{1,n_1''} + h,$$
$$\ldots, h \le \xi_{N,n_N''}(t) < x_{N,n_N''} + h\} \times (\nu_i h + o(h)) I_{n_i' > 0}$$
$$+ P\{z(t) = z - e_i + e_i', h \le \xi_{1,n_1''}(t) < x_{1,n_1''} + h,$$
$$\ldots, h \le \xi_{N,n_N''}(t) < x_{N,n_N''} + h\} \times (\varphi_i h + o(h)) I_{n_i > 0}$$
$$+ P\{z(t) = z - e_i'', h \le \xi_{1,n_1'}(t) < x_{1,n_1'} + h,$$ \quad (6)
$$\ldots, \xi_{i,n_i''-1}(t) < h, \ldots, h \le \xi_{N,n_N''}(t) < x_{N,n_N''} + h\}$$
$$\times (\Phi_i(n_i'', x_{i,n_i''} + \theta h) \frac{\sigma_i I_{n_i'' < r_i}}{\sigma_i I_{n_i'' < r_i} + \rho_i I_{n_i'' > 0}} + o(h)) I_{n_i'' > 0}$$
$$+ P\{z(t) = z + e_i'', h \le \xi_{1,n_1''}(t) < x_{1,n_1''} + h,$$
$$\ldots, \xi_{i,n_i''+1}(t) < h, \ldots, h \le \xi_{N,n_N''}(t) < x_{N,n_N''} + h\}$$
$$\times (\Phi_i(n_i'', x_{i,n_i''} + \theta h) \frac{\rho_i I_{n_i'' > 0}}{\sigma_i I_{n_i'' < r_i} + \rho_i I_{n_i'' > 0}} + o(h)) I_{n_i'' < r_i} + o(h).$$

Each probability from (6) may be expressed in terms of functions

$$F_t(z,x) = P\{z(t) = z, \xi_{1,n_1''}(t) < x_{1,n_1''}, \ldots, \xi_{N,n_N''}(t) < x_{N,n_N''}\}.$$

Consider the decomposition of $F_t(z,x)$ in Taylor series, taking into consideration that

$$P\{z(t) = z, h \le \xi_{i,n_i''}(t) < x_{i,n_i''} + h, \, i \in J\}$$

$$= F_t(z, x_{i,n_i''} + h, \, i \in J) - \sum_{k=1}^{N} F_t(z, x_{i,n_i''} + h, \, i \in J, i \ne k; x_{k,n_k'} \times 0 + h) + \ldots$$

$$+ F_t(z, x_{i,n_i''} \times 0 + h, \, i \in J).$$

Therefore

$$P\{z(t) = z, h \le \xi_{i,n_i''}(t) < x_{i,n_i''} + h, \, i \in J\}$$

$$= F_t(z, x_{i,n_i''}, \, i \in J) + \sum_{i=1}^{N} \frac{\partial F_t(z, x_{i,n_i''}, \, i \in J)}{\partial x_{i,n_i''}} \times h$$

$$- \sum_{i=1}^{N} \left(\frac{\partial F_t(z,x)}{\partial x_{i,n_i''}}\right)_{x_{i,n_i''}=0} \times h + o(h).$$

We consider $\Phi_i(n_i'', x_{i,n_i''} + \theta h)$ as a function of the variable θ, use its decomposition in a Taylor series and let t tend to infinity. So we obtain the following differential equations system:

$$F(z,x) = F(z,x) + h \sum_{i=1}^{N} \left(\frac{\partial F(z,x)}{\partial x_{i,n_i''}} - \left(\frac{\partial F(z,x)}{\partial x_{i,n_i''}}\right)_{x_{i,n_i''}=0}\right)$$

$$- \left(\sum_{i=1}^{N} (\nu_i I_{n_i>0} + \varphi_i I_{n_i'>0} + \mu_i I_{n_i>0}) h + o(h)\right) F(z,x)$$

$$+ h \sum_{j=1}^{N} \sum_{i=1, i \ne j}^{N} p_{j,i} F(z + e_j - e_i, x)(\mu_j h + o(h)) I_{n_i>0} \qquad (7)$$

$$+ \sum_{i=1}^{N} F(z + e_i - e_i', x)(\nu_i h + o(h)) I_{n_i'>0}$$

$$+ \sum_{i=1}^{N} F(z - e_i + e_i', x)(\varphi_i h + o(h)) I_{n_i>0}$$

$$+ h \sum_{i=1}^{N} \Phi_i(n_i'', x_{i,n_i''}) \left(\frac{\partial F(z - e_i'', x)}{\partial x_{i,n_i''-1}}\right)_{x_{i,n_i''-1}=0} \times \frac{\sigma_i I_{n_i''<r_i}}{\sigma_i I_{n_i''<r_i} + \rho_i I_{n_i''>0}} I_{n_i''>0}$$

$$+ h \sum_{i=1}^{N} \Phi_i(n_i'', x_{n_i''}) \left(\frac{\partial F(z + e_i'', x)}{\partial x_{i,n_i''+1}}\right)_{x_{i,n_i''+1}=0} \times \frac{\rho_i I_{n_i''>0}}{\sigma_i I_{n_i''<r_i} + \rho_i I_{n_i''>0}} I_{n_i''<r_i} + o(h).$$

Subtracting $F(z, x)$ from both sides of (7), dividing both sides of (7) by h and letting h tend to zero, we obtain the following differential equations system:

$$
F(z, x) \sum_{i=1}^{N} \left(\nu_i I_{n_i > 0} + \varphi_i I_{n_i' > 0} + \mu_i I_{n_i > 0} \right)
$$

$$
= \sum_{i=1}^{N} \left(\frac{\partial F(z, x)}{\partial x_{i, n_i''}} - \left(\frac{\partial F(z, x)}{\partial x_{i, n_i''}} \right)_{x_{i, n_i''} = 0} \right)
$$

$$
+ \sum_{j=1}^{N} \sum_{i=1, i \neq j}^{N} \mu_j p_{j,i} F(z + e_j - e_i, x) I_{n_i > 0}
$$

$$
+ \sum_{i=1}^{N} F(z + e_i - e_i', x) \nu_i I_{n_i' > 0} + \sum_{i=1}^{N} F(z - e_i + e_i', x) \varphi_i I_{n_i > 0}
$$

$$
+ \sum_{i=1}^{N} \Phi_i(n_i'', x_{n_i''}) \left(\frac{\partial F(z - e_i'', x)}{\partial x_{i, n_i'' - 1}} \right)_{x_{i, n_i'' - 1} = 0} \times \frac{\sigma_i I_{n_i'' < r_i}}{\sigma_i I_{n_i'' < r_i} + \rho_i I_{n_i'' > 0}} I_{n_i'' > 0}
$$

$$
+ \sum_{i=1}^{N} \Phi_i(n_i'', x_{n_i''}) \left(\frac{\partial F(z + e_i'', x)}{\partial x_{i, n_i'' + 1}} \right)_{x_{i, n_i'' + 1} = 0} \times \frac{\rho_i I_{n_i'' > 0}}{\sigma_i I_{n_i'' < r_i} + \rho_i I_{n_i'' > 0}} I_{n_i'' < r_i}. \tag{8}
$$

Let us divide (8) into local balance equations:

$$
F(z, x) \left(\nu_i I_{n_i > 0} + \varphi_i I_{n_i' > 0} \right) = F(z + e_i - e_i', x) \nu_i I_{n_i' > 0} + F(z - e_i + e_i', x) \varphi_i I_{n_i > 0}, \tag{9}
$$

$$
F(z, x) \mu_i I_{n_i > 0} = \sum_{j=1, j \neq i}^{N} \mu_j p_{j,i} F(z + e_j - e_i, x) I_{n_i > 0}, \tag{10}
$$

$$
\left(\frac{\partial F(z, x)}{\partial x_{i, n_i''}} \right)_{x_{i, n_i''} = 0} - \frac{\partial F(z, x)}{\partial x_{i, n_i''}}
$$

$$
= \Phi_i(n_i'', x_{n_i''}) \left(\left(\frac{\partial F(z - e_i'', x)}{\partial x_{i, n_i'' - 1}} \right)_{x_{i, n_i'' - 1} = 0} \times \frac{\sigma_i I_{n_i'' < r_i}}{\sigma_i I_{n_i'' < r_i} + \rho_i I_{n_i'' > 0}} I_{n_i'' > 0} \right.
$$

$$
\left. + \left(\frac{\partial F(z + e_i'', x)}{\partial x_{i, n_i'' + 1}} \right)_{x_{i, n_i'' + 1} = 0} \times \frac{\rho_i I_{n_i'' > 0}}{\sigma_i I_{n_i'' < r_i} + \rho_i I_{n_i'' > 0}} I_{n_i'' < r_i} \right). \tag{11}
$$

Substituting functions $F(z, x)$, determined by means of (3)–(5) into local balance Eqs. (9)–(11) and considering traffic equation system (1), we obtain the identity. $\square$

Let us denote by $\{p(z), z \in Z\}$ the stationary distribution of the process $z(t)$. From the foregoing theorem, considering equality $p(z) = F(z, +\infty)$, we obtain the following corollary.

Corollary 1. *Process $z(t)$ is ergodic and has stationary distribution*

$$
p(z) = G^{-1}(M, N) p_1(n_1, n_1', n_1'') p_2(n_2, n_2', n_2'') \ldots p_N(n_N, n_N', n_N''), \ z \in Z,
$$

which does not depend on the functional form of $\Phi_i(s, x)$, $i \in J$. Probabilities $p_i(n_i, n_i', n_i'')$ may be found by means of (4).

4 Conclusion

We have considered the stationary functioning of a closed queueing network with temporarily non-active customers and multi-regime service strategies. An expression for stationary distribution has been obtained. Stationary distribution insensitivity has been established. The research results have practical importance and may be used for investigations of real networks.

References

1. Tsitsiashvili, G.Sh., Osipova, M.: Distributions in Stochastic Network Models. Nova Publishers Inc., New York (2008)
2. Tsitsiashvili, G.Sh., Osipova, M.: Queueing models with different schemes of customers transformations. In: 19th International Conference "Mathematical Methods for Increasing Efficiency of Information Telecommunication Networks", pp. 128–133. BSU, Minsk (2007)
3. Jackson, J.R.: Network of waiting lines. Oper. Res. **4**, 518–521 (1957)
4. Gordon, W.J., Newell, G.F.: Closed Queueing Networks with Exponential Servers. Oper. Res. **15**, 252–267 (1967)
5. Sevastyanov, B.A.: An ergodic theorem for markov processes and its application to telephone systems with refusals. Theor. Probab. Appl. **2**, 104–112 (1957)
6. Baskett, F.: Open, closed and mixed networks of queues with different classes of customers. J. Assoc. Comput. Mach. **22**, 248–260 (1975)
7. Boyarovich, Y.S.: The stationary distribution invariance of states in a closed queueing network with temporarily non-active customers. Autom. Remote Control. **73**, 1616–1623 (2012)
8. Bojarovich, J., Malinkovsky, Y.V.: Stationary distribution invariance of an open queueing network with temporarily non-active customers. J. Control Comput. Sci. **20**, 62–70 (2012). Tomsk State University
9. Bojarovich, J., Malinkovsky, Y.V.: Stationary distribution invariance of an open queueing network with temporarily non-active customers. In: Dudin, A., Klimenok, V., Tsarenkov, G., Dudin, S. (eds.) BWWQT 2013. CCIS, vol. 356, pp. 26–32. Springer, Heidelberg (2013)
10. Ivnitsky, V.A.: Theory of Queueing Networks. Fizmatlit, Moscow (2004)
11. Bojarovich, J., Dudovskaya, Y.: Stationary distribution insensitivity of a closed queueing network with non-active customers. In: Dudin, A., Nazarov, A., Yakupov, R., Gortsev, A. (eds.) ITMM 2014. CCIS, vol. 487, pp. 50–58. Springer, Heidelberg (2014)
12. Kruk, Y.S., Dudovskaya, Y.E.: Insensitivity of the stationary distribution of state probabilities in an open network with non-active customers. Autom. Remote Control. **76**, 2168–2178 (2015)
13. Kruk, J., Dudovskaya, Y.: Stationary distribution insensitivity of a closed multi-regime queueing network with non-active customers. In: Dudin, A., et al. (eds.) ITMM 2015, CCIS. LNCS, vol. 564, pp. 373–383. Springer, Heidelberg (2015)
14. Kruk, J.S., Dudovskaya, Y.: Stationary distribution of a closed queueing network with non-active customers and multimode service strategies. In: Proceedings of the National Academy of Sciences of Belarus. Physic and Mathematics Series. vol. 1, pp. 54–57 (2015)

A Unified Framework for Analyzing Closed Queueing Networks in Bike Sharing Systems

Quan-Lin Li$^{(\boxtimes)}$, Rui-Na Fan, and Jing-Yu Ma

School of Economics and Management Sciences, Yanshan University,
Qinhuangdao 066004, People's Republic of China
liquanlin@tsinghua.edu.cn

Abstract. During the last decade bike sharing systems have emerged as a public transport mode in urban short trips in more than 500 major cities around the world. For the mobility service mode, many challenges from its operations are not well addressed yet, for example, how to develop bike sharing systems to be able to effectively satisfy the fluctuating demands both for bikes and for vacant lockers. To this end, it is key to conduct performance analysis of bike sharing systems. This paper first describes a large-scale bike sharing system. Then the bike sharing system is abstracted as a closed queueing network with multi-class customers, where the virtual customers and the virtual nodes are set up, and the service rates as well as the relative arrival rates are established. Finally, this paper gives a product-form solution to the steady state joint probabilities of queue lengths, and provides a performance analysis of the bike sharing system. Therefore, this paper provides a unified framework for analyzing closed queueing networks in the study of bike sharing systems. We hope that the methodology and results of this paper can be applicable in the study of more general bike sharing systems.

Keywords: Bike sharing system · Closed queueing network · Product-form solution · Problematic station

1 Introduction

During the last decade bike sharing systems are fast increasing as a public transport mode in urban short trips, and have been launched in more than 500 major cities around the world. Also, bike sharing systems offer a low cost and environmental protection mobility service through sharing one-way use. Nowadays, bike sharing systems are regarded as an effective way to jointly solve traffic congestion, parking difficulties, traffic noise, air pollution and so forth. DeMaio [3] reviewed the history, impacts, models of provision and future of bike sharing systems. Larsen [12] reported that over 500 major cities host advanced bike sharing systems with a combined fleet of more than half a million bikes up to April 2013. A synthesis of the literature for bike sharing systems was given by Fishman et al. [5] and Labadi et al. [11]. At the same time, for some countries or cities developing bike sharing systems, readers may refer to, such as, Europe, the Americas and Asia by Shaheen

© Springer International Publishing Switzerland 2016
A. Dudin et al. (Eds.): ITMM 2016, CCIS 638, pp. 177–191, 2016.
DOI: 10.1007/978-3-319-44615-8_16

et al. [22], the European OBIS Project by Janett and Hendrik [10], France by Faye [4], China by Tang et al. [24], London by Lathia et al. [13], Montreal by Morency et al. [16], and a number of well-known cities by Shu et al. [23].

In operations of bike sharing systems, a crucial question is the ability not only to meet fluctuating demand for renting bikes at each station but also to provide enough vacant lockers to allow the renters to return the bikes at their destinations. Since the number of bikes packed in each station is always randomly dynamically changing, this causes an unpredictable imbalance, such as, some stations contain more bikes but others are seriously short of available bikes. Such a randomly dynamic imbalance of bikes distributed among stations often leads to occurrence of problematic stations (i.e., full or empty stations). Note that problematic stations reflect a common challenge faced by bike sharing systems in practice due to the stochastic and time-inhomogeneous nature of both customer arrivals and bike returns, thus the probability of problematic stations has been regarded as a main factor to measure the satisfaction of customers and even to estimate the quality of service. Obviously, how to effectively reduce the probability of problematic stations becomes a key way to improve the satisfaction of customers and further to promote the quality of service. Therefore, it is a major task to develop effective methods for computing the probability of problematic stations in the study of bike sharing systems.

Queueing theory and Markov processes are very useful for computing the probability of problematic stations, and more generally, analyzing performance measures of the bike sharing systems. However, available works on such a research line are still few up to now. We would like to refer readers to four classes of recent literature as follows. **(a) Simple queues:** Leurent [14] used the $M/M/1/C$ queue to study a vehicle-sharing system in which each station contains an additional waiting room which helps those customers arriving at a problematic station, and analyzed performance measures of this system in terms of geometric distribution. Schuijbroek et al. [20] evaluated the service level by means of the transient distribution of the $M/M/1/C$ queue, and the service level is used to establish some optimal models to discuss the inventory rebalancing and vehicle routing. Raviv et al. [18] and Raviv and Kolka [17] employed the transient distribution of a time-inhomogeneous $M(t)/M(t)/1/C$ queue to compute the expected number of bike shortages at each station. **(b) The mean-field theory:** Fricker et al. [7] considered a space inhomogeneous bike sharing system with different clusters, and expressed the minimal proportion of problematic stations within each cluster. For a space homogeneous bike sharing system, Fricker and Gast [6] used the $M/M/1/K$ queue to provide a more detailed analysis for some simple mean-field models (including the *power of two choices*), derived a closed-form solution to the minimal proportion of problematic stations, and compared the incentives and redistribution mechanisms. Fricker and Tibi [8] studied the central limit and local limit theorems for independent (perhaps non identically distributed) random variables which effectively support analysis of a generalized Jackson network with a product-form solution; and used these obtained results to evaluate performance measures of space inhomogeneous bike sharing systems, where its asymptotics give a complete picture for equilibrium state analysis of the locally space homogeneous bike sharing systems. Li et al. [15] provided a mean-field queueing method to study a large-scale bike sharing system through using a combination of, for example, the virtual time-inhomogeneous queue, the mean-field equations, the martingale limit, the nonlinear birth-death process, numerical computation of the fixed point, and numerical analysis for the steady state probability of problematic stations. **(c) Queueing networks:** Savin et al. [19] used a loss network as well as admission control to discuss capacity allocation of a rental model with two classes of customers, and studied the revenue management and fleet sizing decision in the rental system. Adelman [1] applied a closed queueing network to set up an internal pricing mechanism for managing a fleet of service units, and also used a nonlinear flow model to discuss the price-based policy for the vehicle redistribution. George and Xia [9] provided a queueing network method in the study of vehicle rental systems, and determined the optimal number of parking spaces for each rental location. **(d) Markov decision processes:** Stochastic optimization and Markov decision processes are applied to analysis of bike sharing systems. From a dynamic price mechanism, Waserhole and Jost [25] used closed queueing networks to propose a Markov decision model of a bike sharing system. To overcome the curse of dimensionality in the Markov decision process with a high dimension, they established a fluid approximation that computes a static policy and gave an upper bound on the potential optimization. Such a fluid approximation for the Markov decision processes of the bike sharing systems was further developed in Waserhole and Jost [26, 27] and Waserhole et al. [28].

The main purposes of this paper are to provide a unified framework for analyzing closed queueing networks in the study of bike sharing systems. This framework of closed queueing networks is interesting, difficult and challenging from three crucial features: (a) Stations and roads have very different physical attributes, but all of them are abstracted as indistinguishable nodes in closed queueing networks; (b) the service discipline of the stations is First Come First Served (abbreviated as FCFS), while the service discipline of the roads is Processor Sharing (abbreviated as PS); and (c) the virtual customers (i.e., bikes) in the stations are of a single class, while the virtual customers (i.e., bikes) on the roads are of two classes, and their classes may change on the roads according to the first bike-return or at least two successive bike-returns due to the full stations, respectively. For such a closed queueing network, this paper provides a detailed analysis both for establishing a product-form solution to the steady state joint probabilities of queue lengths, and for computing the steady state probability of problematic stations, more generally, for analyzing performance measures of the bike sharing system. The main contributions of this paper are twofold. The first contribution is to describe a large-scale bike sharing system and to provide a unified framework for analyzing closed queueing networks through establishing some basic factors: The service rates from stations or roads; and the routing matrix as well as the relative arrival rates to stations or roads. Notice that the basic factors play a key role in the study of closed queueing networks. The second contribution of this paper is to provide a product-form solution to the steady state joint probabilities of queue lengths in the closed queueing network, and give performance analysis of the bike sharing system in terms of the steady state joint probabilities.

The remainder of this paper is organized as follows. In Sect. 2, we describe a large-scale bike sharing system with N different stations and with at most $N(N-1)$ different roads. In Sect. 3, we provide a unified framework for analyzing closed queueing networks in the study of bike sharing systems, and also compute the service rates, the routing matrix, and the relative arrival rates. In Sect. 4, we give a product-form solution to the steady state joint probabilities of queue lengths in the closed queueing network, and analyze performance measures of the bike sharing system by means of the steady state joint probabilities. Some concluding remarks are given in Sect. 5.

2   Model Description

In this section, we describe a large-scale bike sharing system with N different stations and with at most $N(N-1)$ different roads due to the riding-bike directed connection between any two stations. To analyze such a bike sharing system, we provide a unified framework for analyzing closed queueing networks in the study of bike sharing systems.

In a large-scale bike sharing system, a customer arrives at a station, rents a bike, and uses it for a while; then she returns the bike to a destination station, and immediately leaves this system. Obviously, for any customer renting and using a bike, her first return-bike time is different from those return-bike times that she has successively returned the bike for at least two times due to arriving at full stations. At the same time, it is easy to understand that for any customer, her first road selection as well as her first riding-bike speed are different from those of having successively returned her bike for at least two times. Also, it is noted that the customer must return her bike to a station, then she can immediately leave the bike sharing system.

Now, we describe the bike sharing system, including operation mechanism, system parameters and mathematical notation, as follows:

(1) Stations and roads: There are N different stations and at most $N(N-1)$ different roads, where the $N(N-1)$ roads are observed from the fact that there must exist a direct road from a station to another station. In addition, we assume that at the initial time $t = 0$, every station has C bikes and K parking places, where $1 \leq C < K < \infty$; and $NC \geq K$, which means that some of the NC bikes can result in at least a full station.

(2) Customer arrival process: The arrivals of outside customers at the ith station are a Poisson process with arrival rate $\lambda_i > 0$ for $1 \leq i \leq N$.

(3) The first riding-bike time: Once an outside customer arrives at the ith station, she immediately goes to rent a bike. If there is no bike in the ith station (i.e., the ith station is empty), then the customer directly leaves this bike sharing system. If there is at least one bike in the ith station, then the customer rents a bike, and then goes to Road $i \rightarrow j$. We assume that for $j \neq i$ with $1 \leq i, j \leq N$, the customer at the ith station rides the bike into Road $i \rightarrow j$

with probability $p_{i,j}$ for $\sum_{j\neq i}^{N} p_{i,j} = 1$; and her riding-bike time from the ith station to the jth station (i.e., riding on Road $i \to j$) is an exponential random variable with riding-bike rate $\mu_{i,j} > 0$, where the expected riding-bike time is $1/\mu_{i,j}$.

(4) The bike return times:

The first return – When the customer completes her short trip on the above Road $i \to j$ (see Assumption (3)), she needs to return her bike to the jth station. If there is at least one available parking position (i.e., a vacant docker), then the customer directly returns her bike to the jth station, and immediately leaves this bike sharing system.

The second return – If no parking position is available at the jth station, then she has to ride the bike to another station l_1 with probability α_{j,l_1} for $l_1 \neq j$ for $\sum_{l_1 \neq j}^{N} \alpha_{j,l_1} = 1$; and her riding-bike time from the jth station to the l_1th station (i.e., riding on Road $j \to l_1$) is an exponential random variable with riding-bike rate $\xi_{j,l_1} > 0$. If there is at least one available parking position, then the customer directly returns her bike to the l_1th station, and immediately leaves this bike sharing system.

The third return – If no parking position is available at the l_1th station, then she has to ride the bike to another station l_2 with probability α_{l_1,l_2} for $l_2 \neq l_1$ for $\sum_{l_2 \neq l_1}^{N} \alpha_{l_1,l_2} = 1$; and her riding-bike time from the l_1th station to the l_2th station (i.e., riding on Road $l_1 \to l_2$) is an exponential random variable with riding-bike rate $\xi_{l_1,l_2} > 0$. If there is at least one available parking position, then the customer directly returns her bike to the l_2th station, and immediately leaves this bike sharing system.

The $(k+1)$st return for $k \geq 3$ – We assume that this bike has not been returned at any station yet through k consecutive return processes. In this case, the customer has to try her $(k+1)$st lucky return. Notice that the customer goes to the l_kth station from the l_{k-1}th full station with probability α_{l_{k-1},l_k} for $l_k \neq l_{k-1}$ for $\sum_{l_k \neq l_{k-1}}^{N} \alpha_{l_{k-1},l_k} = 1$; and her riding-bike time from the l_{k-1}th station to the l_kth station (i.e., riding on Road $l_{k-1} \to l_k$) is an exponential random variable with riding-bike rate $\xi_{l_{k-1},l_k} > 0$. If there is at least one available parking position, then the customer directly returns her bike to the l_kth station, and immediately leaves this bike sharing system; otherwise she has to continuously try another station again.

We further assume that the returning-bike process is persistent in the sense that the customer must find a station with an empty position to return her bike, because the bike is public property so no one can make it her own.

It is seen from the above description that the parameters: $p_{i,j}$ and $\mu_{i,j}$ for $j \neq i$ and $1 \leq i, j \leq N$, of the first return, may be different from the parameters: $\alpha_{i,j}$ and $\xi_{i,j}$ for $j \neq i$ and $1 \leq i, j \leq N$, of the kth return for $k \geq 2$. Note that such an assumption with respect to these different parameters is actually reasonable because the customer possibly has more things (for example, tourism, shopping, visiting friends and so on) in the first return process, but she has only one return task during the k successive return processes for $k \geq 2$.

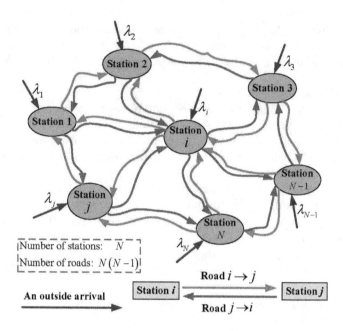

Fig. 1. The physical structure of the bike sharing system

(5) The departure discipline: The customer departure has two different cases: (a) An outside customer directly leaves the bike sharing system if she arrives at an empty station; or (b) if one customer rents and uses a bike, and she finally returns the bike to a station, then the customer completes her trip, and immediately leaves the bike sharing system.

We assume that the customer arrival and riding-bike processes are independent, and also all the above random variables are independent of each other. For such a bike sharing system, Fig. 1 provides some physical interpretation.

3   A Closed Queueing Network

In this section, we first provide a closed queueing network to express the bike sharing system, as seen in Fig. 1. Then we determine the service rates, the routing matrix, and the relative arrival rates of the closed queueing network. Note that there are two classes of customers on the $N(N-1)$ roads.

In the bike sharing system described in the above section, there are NC bikes, N stations and $N(N-1)$ roads. Now, we abstract the bike sharing system as a closed queueing network as follows:

(1) Virtual nodes: Although the stations and roads have different physical attributes such as functions, and geographical topologies, the stations and roads are all regarded as the same nodes in the closed queueing network.

(2) Virtual customers: The bikes at the stations or roads are described as follows:

Abstract: The virtual customers are abstracted by the bikes, which are either parked in the stations or ridden on the roads. Note that the total number of bikes in the bike sharing system is fixed as NC due to the fact that bikes can neither enter nor leave this system, thus the bike sharing system can be regarded as a closed queueing network.

Multiple classes: From Assumption (2) in Sect. 2, it is seen that there is only one class of customers in the nodes abstracted from the stations. From Assumptions (3) and (4) in Sect. 2, we understand that there are two different classes of customers in the nodes abstracted from the roads, where the first class of customers is the bikes ridden on the roads for the first time; while the second class of customers is the bikes which are successively ridden on at least two different roads due to the full station.

(3) Service disciplines: The First Come First Served (or FCFS) is used in the nodes abstracted from the stations; while new processor sharing (or PS) is used in the nodes abstracted from the roads.

In the above closed queueing network, let $Q_i(t)$ be the number of bikes parked in ith station at time $t \geq 0$ for $1 \leq i \leq N$, and $R_{k,l}^{(r)}(t)$ the number of bikes of class r ridden on Road $k \to l$ at time t for $r = 1, 2$, and $k \neq l$ with $1 \leq k, l \leq N$. We write

$$\mathbf{X}(t) = (\mathbf{L}_1(t), \mathbf{L}_2(t), \ldots, \mathbf{L}_{N-1}(t), \mathbf{L}_N(t)),$$

where for $1 \leq i \leq N$

$$\mathbf{L}_i(t) = \left(Q_i(t) \, ; R_{i,1}^{(1)}(t) \, , R_{i,1}^{(2)}(t) \, ; R_{i,2}^{(1)}(t) \, , R_{i,2}^{(2)}(t) \, ; \ldots ; R_{i,i-1}^{(1)}(t) \, , R_{i,i-1}^{(2)}(t) \, ; \right.$$
$$\left. R_{i,i+1}^{(1)}(t) \, , R_{i,i+1}^{(2)}(t) \, ; R_{i,i+2}^{(1)}(t) \, , R_{i,i+2}^{(2)}(t) \, ; \ldots ; R_{i,N}^{(1)}(t) \, , R_{i,N}^{(2)}(t) \right).$$

Obviously, $\{\mathbf{X}(t) : t \geq 0\}$ is a Markov process of size $N(2N-1)$ due to the exponential and Poisson assumptions of this bike sharing system.

Now, we describe the state space of the Markov process $\{\mathbf{X}(t) : t \geq 0\}$. It is seen from Sect. 2 that

$$0 \leq Q_i(t) \leq K, 1 \leq i \leq N, \tag{1}$$

$$0 \leq R_{k,l}^{(r)}(t) \leq NC, \quad r = 1, 2, \ k \neq l, \ 1 \leq k, \ l \leq N, \tag{2}$$

and

$$\sum_{i=1}^{N} Q_i(t) + \sum_{k=1}^{N} \sum_{l \neq k}^{N} R_{k,l}^{(1)}(t) + \sum_{k=1}^{N} \sum_{l \neq k}^{N} R_{k,l}^{(2)}(t) = NC. \tag{3}$$

From (1) to (3), it is seen that the state space of the Markov process $\{\mathbf{X}(t) : t \geq 0\}$ of size $N(2N - 1)$ is given by

$$\Omega = \left\{ \overrightarrow{n} : 0 \leq n_i \leq K, 0 \leq m_{k,l}^{(1)}, m_{k,l}^{(2)} \leq NC, \right.$$

$$\left. \sum_{i=1}^{N} n_i + \sum_{k=1}^{N} \sum_{l \neq k}^{N} m_{k,l}^{(1)} + \sum_{k=1}^{N} \sum_{l \neq k}^{N} m_{k,l}^{(2)} = NC \right\},$$

where

$$\overrightarrow{n} = (\mathbf{n}_1, \mathbf{n}_2, \ldots, \mathbf{n}_{N-1}, \mathbf{n}_N),$$

and for $1 \leq i \leq N$

$$\mathbf{n}_i = \left(n_i; m_{i,1}^{(1)}, m_{i,1}^{(2)}; m_{i,2}^{(1)}, m_{i,2}^{(2)}; \ldots; m_{i,i-1}^{(1)}, m_{i,i-1}^{(2)}; \right.$$
$$\left. m_{i,i+1}^{(1)}, m_{i,i+1}^{(2)}; m_{i,i+2}^{(1)}, m_{i,i+2}^{(2)}; \ldots; m_{i,N}^{(1)}, m_{i,N}^{(2)} \right).$$

Note that $m_{k,l} = m_{k,l}^{(1)} + m_{k,l}^{(2)}$ is the total number of bikes being ridden on Road $k \rightarrow l$ for $k \neq l$ with $1 \leq k, l \leq N$, and also the state space Ω contains $(K + 1)^N (NC + 1)^{2N(N-1)}$ elements.

To compute the steady state joint probabilities of $N(2N - 1)$ queue lengths in the bike sharing system, it is seen from Chap. 7 in Bolch et al. [2] that we need to determine the service rate, the routing matrix and the relative arrival rate for each node in the closed queueing network.

(a) The service rates

From Fig. 2, it is seen that the service rates of the closed queueing network are given from two different cases as follows:

Case one: The node is one of the N stations

The departure process of bikes from the ith station, renting at the ith station and immediately ridden on one of the $N - 1$ roads (such as, Road $i \rightarrow l$ for $l \neq i$ with $1 \leq l \leq N$), is Poisson with service rate

$$b_i = \lambda_i \mathbf{1}_{\{1 \leq n_i \leq K\}} \sum_{l \neq i}^{N} p_{i,l} = \lambda_i \mathbf{1}_{\{1 \leq n_i \leq K\}} \tag{4}$$

by means of the condition: $\sum_{l \neq i}^{N} p_{i,l} = 1$.

Case two: The node is one of the $N(N - 1)$ roads

In this case, two different processor sharing queueing processes of Road $i \rightarrow l$ (with two classes of different customers) are explained in Fig. 2. Now we describe the service rates with respect to the two classes of different customers as follows:

The first class of customers: The departure process of bikes from Road $i \rightarrow l$, rented from Station i and being ridden on Road $i \rightarrow l$ for the first time, is Poisson with service rate

$$b_{i,l}^{(1)} = m_{i,l}^{(1)} \mu_{i,l}. \tag{5}$$

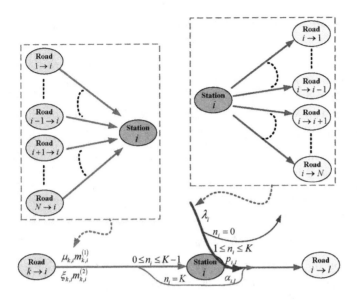

Fig. 2. The queueing processes in the closed queueing network

<u>The second class of customers:</u> The departure process of bikes from Road $i \rightarrow l$, having successively been ridden on the roads for at least two times but now on Road $i \rightarrow l$, is Poisson with service rate

$$b_{i,l}^{(2)} = m_{i,l}^{(2)} \xi_{i,l}. \tag{6}$$

(b) The routing matrix and the relative arrival rates

Now we compute the relative arrival rate of each node in the closed queueing network. As opposed to the service rates analyzed above, it is more complicated to determine the relative arrival rates by means of the routing matrix.

Based on Chap. 7 in Bolch et al. [2], we denote by $e_i(n_i)$ and $e_{i,l}^{(r)}(m_{i,l}^{(r)})$ the relative arrival rates of the ith station with n_i parking bikes, and of Road $i \rightarrow l$ with $m_{i,l}^{(r)}$ riding bikes of class r, respectively. We write

$$\mathbb{E} = \{\overrightarrow{e}(\overrightarrow{n}) : \overrightarrow{n} \in \Omega\},$$

where

$$\overrightarrow{e}(\overrightarrow{n}) = (\mathbf{e}_1(\overrightarrow{n}), \mathbf{e}_2(\overrightarrow{n}), \ldots, \mathbf{e}_{N-1}(\overrightarrow{n}), \mathbf{e}_N(\overrightarrow{n})),$$

and for $1 \leq i \leq N$

$$\mathbf{e}_i(\overrightarrow{n}) = (e_i(n_i); e_{i,1}^{(1)}(m_{i,1}^{(1)}), e_{i,1}^{(2)}(m_{i,1}^{(2)}); \ldots; e_{i,i-1}^{(1)}(m_{i,i-1}^{(1)}), e_{i,i-1}^{(2)}(m_{i,i-1}^{(2)});$$
$$e_{i,i+1}^{(1)}(m_{i,i+1}^{(1)}), e_{i,i+1}^{(2)}(m_{i,i+1}^{(2)}); \ldots; e_{i,N}^{(1)}(m_{i,N}^{(1)}), e_{i,N}^{(2)}(m_{i,N}^{(2)})).$$

Now we introduce two useful notations: $\overrightarrow{g}_i$ and $\overrightarrow{g}_{i,l}^{(r)}$ as follows:

$\overrightarrow{g}_i$: A unit row vector of size $N(2N-1)$, which is given by the method of replacing elements from $\overrightarrow{n}$ to $\overrightarrow{g}_i$, that is, corresponding to the row vector $\overrightarrow{n}$, the element n_i is replaced by one, while all other elements of the vector $\overrightarrow{n}$ are replaced by zeros.

$\overrightarrow{g}_{i,l}^{(r)}$: A unit row vector of size $N(2N-1)$, which is given by the method of replacing elements from $\overrightarrow{n}$ to $\overrightarrow{g}_{i,l}^{(r)}$, that is, corresponding to the row vector $\overrightarrow{n}$, the element $m_{i,l}^{(r)}$ is replaced by one, while all other elements of the vector $\overrightarrow{n}$ are replaced by zeros.

To compute the vector $\overrightarrow{e}(\overrightarrow{n})$, we first need to give the routing matrix $\mathbf{P}$ of the closed queueing network as follows:

$$\mathbf{P} = (P_{\overrightarrow{n},\overrightarrow{n}'})_{\overrightarrow{n},\overrightarrow{n}' \in \Omega},$$

where the routing matrix $\mathbf{P}$ is of order $(K+1)^N (NC+1)^{2N(N-1)}$, and the element $P_{\overrightarrow{n},\overrightarrow{n}'}$ is computed from the following three cases:

Case one: From a station to a road

For $1 \leq i, l \leq N$ with $l \neq i$, we observe a transition route from the ith station to Road $i \to l$. If a rented bike leaves the ith station and enters Road $i \to l$, then $1 \leq n_i \leq K$, and there is a two-element change: $(n_i, m_{i,l}^{(1)}) \to (n_i - 1, m_{i,l}^{(1)} + 1)$. Thus we obtain that for $1 \leq n_i \leq K$

$$P_{\overrightarrow{n},\overrightarrow{n}'} = P_{\overrightarrow{n},\overrightarrow{n} - \overrightarrow{g}_i + \overrightarrow{g}_{i,l}^{(1)}} = p_{i,l}$$

by means of Assumption (3) of Sect. 2. There are $NK(N-1)$ such elements with $P_{\overrightarrow{n},\overrightarrow{n}'} = P_{\overrightarrow{n},\overrightarrow{n} - \overrightarrow{g}_i + \overrightarrow{g}_{i,l}^{(1)}} = p_{i,l}$ in the closed queueing network.

Case two: From a road to a station

For $r = 1, 2$ and $1 \leq k, i, l \leq N$ with $i \neq k$ and $l \neq i$, we observe a transition route from Road $k \to i$ to the ith station. If a riding bike of class r leaves Road $k \to i$, then either it enters the ith station if $0 \leq n_i \leq K - 1$; or it goes to Road $i \to l$ if $n_i = K$.

In the former case (the riding bike of class r enters the ith station if $0 \leq n_i \leq K - 1$), we obtain for $0 \leq n_i \leq K - 1$, there is a two-element change: $(m_{k,i}^{(r)}, n_i) \to (m_{k,i}^{(r)} - 1, n_i + 1)$, hence this results in for $0 \leq n_i \leq K - 1$

$$P_{\overrightarrow{n},\overrightarrow{n}'} = P_{\overrightarrow{n},\overrightarrow{n} - \overrightarrow{g}_{k,i}^{(r)} + \overrightarrow{g}_i} = 1,$$

since the end of Road $k \to i$ is only the ith station. There are $2N^2(N-1)CK$ such elements with $P_{\overrightarrow{n},\overrightarrow{n}'} = P_{\overrightarrow{n},\overrightarrow{n} - \overrightarrow{g}_{k,i}^{(r)} + \overrightarrow{g}_i} = 1$ in the closed queueing network.

Case three: From a road to another road

In the latter case (the riding bike of class r goes to Road $i \to l$ if $n_i = K$), we see that there is a two-element change: $(m_{k,i}^{(r)}, m_{i,l}^{(2)}) \to (m_{k,i}^{(r)} - 1, m_{i,l}^{(2)} + 1)$. Thus we obtain that for $n_i = K$

$$P_{\overrightarrow{n},\overrightarrow{n}'} = P_{\overrightarrow{n},\overrightarrow{n} - \overrightarrow{g}_{k,i}^{(r)} + \overrightarrow{g}_{i,l}^{(2)}} = \alpha_{i,l}$$

by means of Assumption (4) of Sect. 2. There are $2N^3(N-1)^2C^2$ such elements with $P_{\overrightarrow{n},\overrightarrow{n}'} = P_{\overrightarrow{n},\overrightarrow{n}-\overrightarrow{g}_{k,i}^{(r)}+\overrightarrow{g}_{i,l}^{(2)}} = \alpha_{i,l}$ in the closed queueing network.

In summary, the above analysis gives

$$P_{\overrightarrow{n},\overrightarrow{n}'} = \begin{cases} P_{\overrightarrow{n},\overrightarrow{n}-\overrightarrow{g}_i+\overrightarrow{g}_{i,l}^{(1)}} = p_{i,l}, & \text{if } 1 \le n_i \le K, & (\text{station} \rightarrow \text{road}) \\ P_{\overrightarrow{n},\overrightarrow{n}-\overrightarrow{g}_{k,i}^{(r)}+\overrightarrow{g}_i} = 1, & \text{if } 0 \le n_i \le K-1, & (\text{road} \rightarrow \text{station}) \\ P_{\overrightarrow{n},\overrightarrow{n}-\overrightarrow{g}_{k,i}^{(r)}+\overrightarrow{g}_{i,l}^{(2)}} = \alpha_{i,l}, & \text{if } n_i = K, & (\text{road} \rightarrow \text{road, a full station}) \\ 0, & \text{otherwise.} \end{cases}$$

At the same time, the minimal number of zero elements in the routing matrix $\mathbf{P}$ is given by

$$\left[(K+1)^N (NC+1)^{2N(N-1)}\right]^2 - NK(N-1) - 2N^2(N-1)CK - 2N^3(N-1)^2C^2.$$

This also shows that there exist more zero elements in the routing matrix $\mathbf{P}$.

We write a row vector

$$\overrightarrow{\mathfrak{R}} = \left(\overrightarrow{e}(\overrightarrow{n}) : \overrightarrow{e}(\overrightarrow{n}) \in \mathbb{E}\right),$$

where

$$\mathbb{E} = \left\{\overrightarrow{e}(\overrightarrow{n}) : \overrightarrow{n} \in \Omega\right\}.$$

Theorem 1. *The routing matrix* $\mathbf{P}$ *is irreducible and stochastic (i.e.,* $\mathbf{P1} = \mathbf{1}$*, where* $\mathbf{1}$ *is a column vector of ones), and there exists a unique positive solution to the following system of linear equations*

$$\begin{cases} \overrightarrow{\mathfrak{R}} = \overrightarrow{\mathfrak{R}} \, \mathbf{P}, \\ \left(\overrightarrow{\mathfrak{R}}\right)_1 = 1, \end{cases}$$

where $\left(\overrightarrow{\mathfrak{R}}\right)_1$ *is the first element of the row vector* $\overrightarrow{\mathfrak{R}}$.

Proof: The outline of this proof is described as follows. It is well-known that the routing structure of the closed queueing network indicates that the routing matrix $\mathbf{P}$ is stochastic, and the accessibility of each station or road of the bike sharing system shows that the routing matrix $\mathbf{P}$ is irreducible. Thus the routing matrix $\mathbf{P}$ is not only irreducible but also stochastic. Note that the size of the routing matrix $\mathbf{P}$ is $(K+1)^N (NC+1)^{2N(N-1)}$, it follows from Theorem 1.1 (a) and (b) of Chap. 1 in Seneta [21] that the left eigenvector $\overrightarrow{\mathfrak{R}}$ of the irreducible stochastic matrix $\mathbf{P}$ corresponding to the maximal eigenvalue 1 is more than 0, that is, $\overrightarrow{\mathfrak{R}} > 0$, and $\overrightarrow{\mathfrak{R}}$ is unique for $\left(\overrightarrow{\mathfrak{R}}\right)_1 = 1$. This completes this proof. ∎

4   A Product-Form Solution and Performance Analysis

In this section, we first provide a product-form solution to the steady state joint probabilities of $N(2N-1)$ queue lengths in the closed queueing network. Then we analyze performance measures of the bike sharing system by means of steady state joint probabilities.

Note that

$$\mathbf{X}(t) = (\mathbf{L}_1(t), \mathbf{L}_2(t), \ldots, \mathbf{L}_{N-1}(t), \mathbf{L}_N(t)),$$

where for $1 \leq i \leq N$

$$\mathbf{L}_i(t) = \left(Q_i(t); R_{i,1}^{(1)}(t), R_{i,1}^{(2)}(t); R_{i,2}^{(1)}(t), R_{i,2}^{(2)}(t); \ldots; R_{i,i-1}^{(1)}(t), R_{i,i-1}^{(2)}(t); \right.$$
$$\left. R_{i,i+1}^{(1)}(t), R_{i,i+1}^{(2)}(t); R_{i,i+2}^{(1)}(t), R_{i,i+2}^{(2)}(t); \ldots; R_{i,N}^{(1)}(t), R_{i,N}^{(2)}(t) \right).$$

At the same time, $\{\mathbf{X}(t) : t \geq 0\}$ is an irreducible continuous-time Markov process on state space Ω which contains $(K+1)^N (NC+1)^{2N(N-1)}$ states. Therefore, the Markov process $\{\mathbf{X}(t) : t \geq 0\}$ is irreducible and positive recurrent. In this case, we set

$$\pi(\overrightarrow{n}) = \lim_{t \to +\infty} P\left\{ Q_i(t) = n_i, 1 \leq i \leq N; R_{k,l}^{(1)}(t) = m_{k,l}^{(1)}, R_{k,l}^{(2)}(t) = m_{k,l}^{(2)}, \right.$$
$$\left. 1 \leq k, l \leq N \text{ with } k \neq l, \sum_{i=1}^{N} n_i + \sum_{r=1,2} \sum_{k=1}^{N} \sum_{l \neq k}^{N} m_{k,l}^{(r)} = NC \right\}.$$

(a) A product-form solution to the steady state joint probabilities

The following theorem provides a product-form solution to the steady state joint probability $\pi(\overrightarrow{n})$ for $\overrightarrow{n} \in \Omega$; while its proof is easy by means of Chap. 7 in Bolch et al. [2] and is omitted here.

Theorem 2. *For the closed queueing network of the bike sharing system, the steady state joint probability $\pi(\overrightarrow{n})$ is given by*

$$\pi(\overrightarrow{n}) = \frac{1}{G} \prod_{i=1}^{N} F(n_i) \prod_{k=1}^{N} \prod_{l \neq k}^{N} m_{k,l}! H^{(1)}\left(m_{k,l}^{(1)}\right) H^{(2)}\left(m_{k,l}^{(2)}\right),$$

where $\overrightarrow{n} \in \Omega$, $m_{k,l} = m_{k,l}^{(1)} + m_{k,l}^{(2)}$,

$$F(n_i) = \begin{cases} \left[\frac{e_i(n_i)}{\lambda_i} \right]^{n_i}, & 1 \leq n_i \leq K, \\ 1, & n_i = 0, \end{cases}$$

$$H^{(1)}\left(m_{k,l}^{(1)}\right) = \begin{cases} \frac{1}{m_{k,l}^{(1)}!} \left[\frac{e_{k,l}^{(1)}\left(m_{k,l}^{(1)}\right)}{m_{k,l}^{(1)} \mu_{k,l}} \right]^{m_{k,l}^{(1)}}, & 1 \leq m_{k,l}^{(1)} \leq NC, \\ 1, & m_{k,l}^{(1)} = 0, \end{cases}$$

$$H^{(2)}\left(m_{k,l}^{(2)}\right) = \begin{cases} \dfrac{1}{m_{k,l}^{(2)}!}\left[\dfrac{e_{k,l}^{(2)}\left(m_{k,l}^{(2)}\right)}{m_{k,l}^{(2)}\xi_{k,l}}\right]^{m_{k,l}^{(2)}}, & 1 \le m_{k,l}^{(2)} \le NC, \\ 1, & m_{k,l}^{(2)} = 0, \end{cases}$$

and **G** *is a normalization constant, given by*

$$\mathbf{G} = \sum_{\overrightarrow{n}\in\Omega}\prod_{i=1}^{N}F\left(n_i\right)\prod_{k=1}^{N}\prod_{l\neq k}^{N}m_{k,l}!H^{(1)}\left(m_{k,l}^{(1)}\right)H^{(2)}\left(m_{k,l}^{(2)}\right).$$

(b) Performance analysis

Now we consider three key performance measures of the bike sharing system in terms of the steady state joint probability $\pi\left(\overrightarrow{n}\right)$ for $\overrightarrow{n}\in\Omega$.

(1) The steady state probability of problematic stations

In the study of bike sharing systems, it is a key task to compute the steady state probability of problematic stations. To this end, our aim is to care for the ith station with respect to its full or empty cases. Thus the steady state probability $\Im$ of problematic stations is given by

$$\Im = \sum_{i=1}^{N}P\{n_i = 0 \text{ or } n_i = K\} = \sum_{i=1}^{N}P\{n_i = 0\} + \sum_{i=1}^{N}P\{n_i = K\}$$

$$= \sum_{i=1}^{N}\sum_{\overrightarrow{n}\in\Omega \ \& \ n_i=0}\pi\left(\overrightarrow{n}\right) + \sum_{i=1}^{N}\sum_{\overrightarrow{n}\in\Omega \ \& \ n_i=K}\pi\left(\overrightarrow{n}\right).$$

(2) The means of steady state queue lengths

The steady state mean of the number of bikes parked *at* the ith station is given by

$$\mathbf{Q}_i = \sum_{\overrightarrow{n}\in\Omega \ \& \ 1\le n_i\le K}n_i\pi\left(\overrightarrow{n}\right), \ 1 \le i \le N,$$

and the steady state mean of the number of bikes ridden on the $N\left(N-1\right)$ roads is given by

$$\mathbf{Q}_0 = NC - \sum_{i=1}^{N}\left[\sum_{\overrightarrow{n}\in\Omega \ \& \ 1\le n_i\le K}n_i\pi\left(\overrightarrow{n}\right)\right],$$

or

$$\mathbf{Q}_0 = \sum_{r=1,2}\sum_{k=1}^{N}\sum_{l\neq k}^{N}\sum_{\overrightarrow{n}\in\Omega \ \& \ 1\le m_{k,l}^{(r)}\le NC}m_{k,l}^{(r)}\pi\left(\overrightarrow{n}\right).$$

5 Concluding Remarks

In this paper, we provide a unified framework for analyzing closed queueing networks in the study of bike sharing systems, and show that this framework of closed queueing networks is interesting, difficult and challenging. We describe and analyze a closed queueing network corresponding to a large-scale bike sharing system, and specifically, we provide a product-form solution to the steady state joint probabilities of $N(2N-1)$ queue lengths, which leads to being able to calculate the steady state probability of problematic stations, and more generally, to analyze performance measures of this bike sharing system. We hope that the methodology and results of this paper can be applicable in the study of more general bike sharing systems by means of closed queueing networks. Along these lines, there are a number of interesting areas for potential future research, for example:

- Developing effective algorithms for computing the routing matrix, the relative arrival rates, and the steady state joint probabilities of queue lengths;
- analyzing bike sharing systems with Markovian arrival processes (MAPs) of customers to rent bikes, and phase type (PH) riding-bike times on the roads;
- considering the heterogeneity of bike sharing systems with an irreducible graph with stations, roads and their connections;
- discussing repositioning bikes by trucks in bike sharing systems with information technologies; and
- applying periodic MAPs, periodic PH distributions, or periodic Markov processes to studying time-inhomogeneous bike sharing systems.

Acknowledgements. Q.L. Li was supported by the National Natural Science Foundation of China under grant No. 71271187 and No. 71471160, and the Fostering Plan of Innovation Team and Leading Talent in Hebei Universities under grant No. LJRC027.

References

1. Adelman, D.: Price-directed control of a closed logistics queueing network. Oper. Res. **55**(6), 1022–1038 (2007)
2. Bolch, G., Greiner, S., de Meer, H., Trivedi, K.S.: Queueing Networks and Markov Chains: Modeling and Performance Evaluation with Computer Science Applications. Wiley, Hoboken (2006)
3. DeMaio, P.: Bike-sharing: history, impacts, models of provision, and future. J. Public Transp. **12**(4), 41–56 (2009)
4. Faye, V.: French Network of Bike: Cities and Bikesharing Systems in France. le Club des Villes Cyclables, Paris (2008)
5. Fishman, E., Washington, S., Haworth, N.: Bike share: a synthesis of the literature. Transport Rev. **33**(2), 148–165 (2013)
6. Fricker, C., Gast, N.: Incentives and redistribution in homogeneous bike-sharing systems with stations of finite capacity. EURO J. Transp. Logistics, 1–31 (2014)
7. Fricker, C., Gast, N., Mohamed, A.: Mean field analysis for inhomogeneous bike-sharing systems. DMTCS Proc. **1**, 365–376 (2012)
8. Fricker, C., Tibi, D.: Equivalence of ensembles for large vehicle-sharing models, pp. 1–28 (2015). arXiv Preprint: arXiv:1507.07792
9. George, D.K., Xia, C.H.: Fleet-sizing and service availability for a vehicle rental system via closed queueing networks. Eur. J. Oper. Res. **211**(1), 198–207 (2011)
10. Janett, B., Hendrik, M.: Optimising bike-sharing in European cities: A Handbook. OBIS Project (2011)

11. Labadi, K., Benarbia, T., Barbot, J.P., Hamaci, S., Omari, A.: Stochastic Petri net modeling, simulation and analysis of public bicycle sharing systems. IEEE Trans. Autom. Sci. Eng. **12**(4), 1380–1395 (2015)
12. Larsen, J.: Bike-sharing programs hit the streets in over 500 cities worldwide. Earth Policy Institute, vol. 25 (2013)
13. Lathia, N., Ahmed, S., Capra, L.: Measuring the impact of opening the London shared bicycle scheme to casual users. Transp. Res. Part C **22**(1), 88–102 (2012)
14. Leurent, F.: Modelling a vehicle-sharing station as a dual waiting system: stochastic framework and stationary analysis. HAL Id: hal-00757228, pp. 1–19 (2012)
15. Li, Q.L., Chen, C., Fan, R.N., Xu, L., Ma, J.Y.: Queueing analysis of a large-scale bike sharing system through mean-field theory, pp. 1–50 (2016). arXiv Preprint: arXiv:1603.09560
16. Morency, C., Trépanier, M., Godefroy, F.: Insight into the Montreal bikesharing system. In: TRB-Transportation Research Board Annual Meeting, Washington, USA (2011)
17. Raviv, T., Kolka, O.: Optimal inventory management of a bikesharing station. IIE Trans. **45**(10), 1077–1093 (2013)
18. Raviv, T., Tzur, M., Forma, I.A.: Static repositioning in a bike-sharing system: models and solution approaches. EURO J. Trans. Logistics **2**(3), 187–229 (2013)
19. Savin, S., Cohen, M., Gans, N., Katala, Z.: Capacity management in rental businesses with two customer bases. Oper. Res. **53**(4), 617–631 (2005)
20. Schuijbroek, J., Hampshire, R., van Hoeve, W.J.: Inventory rebalancing and vehicle routing in bike-sharing systems. Technical report 2013–2, Tepper School of Business, Carnegie Mellon University, pp. 1–27 (2013)
21. Seneta, E.: Non-negative Matrices and Markov Chains. Springer, New York (1981)
22. Shaheen, S., Guzman, S., Zhang, H.: Bikesharing in Europe, the Americas, and Asia: past, present, and future. Transp. Res. Rec. J. Transp. Res. Board **2143**, 159–167 (2010)
23. Shu, J., Chou, M.C., Liu, Q., Teo, C.P., Wang, I.L.: Models for effective deployment and redistribution of bicycles within public bicycle-sharing systems. Oper. Res. **61**(6), 1346–1359 (2013)
24. Tang, Y., Pan, H., Shen, Q.: Bike-sharing systems in Beijing, Shanghai, and Hangzhou and their impact on travel behavior. In: The 90th Annual Meeting of the Transportation Research Board, Washington, D.C. (2011)
25. Waserhole, A., Jost, V.: Vehicle sharing system pricing regulation: transit optimization of intractable queuing network. HAL Id: hal-00751744, pp. 1–20 (2012)
26. Waserhole, A., Jost, V.: Vehicle sharing system pricing regulation: a fluid approximation. HAL Id: hal-00727041, pp. 1–35 (2013)
27. Waserhole, A., Jost, V.: Pricing in vehicle sharing systems: optimization in queuing networks with product forms. EURO J. Transp. Logistics, 1–28 (2014)
28. Waserhole, A., Jost, V., Brauner, N.: Pricing techniques for self regulation in vehicle sharing systems. Electron. Notes Discrete Math. **41**, 149–156 (2013)

Switch-Hysteresis Control of the Production Process in a Model with Perishable Goods

Klimentii Livshits$^{(\boxtimes)}$ and Ekaterina Ulyanova

Tomsk State University, Tomsk, Russia
kim47@mail.ru, katerina_tomsk@sibmail.com

Abstract. In this paper we obtained the probability density function of perishable goods stock under switch-hysteresis control and random sales. The optimal parameters of algorithm control for the case of switch control of production and exponentially distributed purchase amounts were found.

Keywords: Perishable goods · Hysteresis control · Probability density function

1   Introduction

Models and methods of queueing theory [1] are widely used in various areas and, in particular, can be used to analyze inventory management problems with a limited shelf life, which have been intensively studied in recent years. Let us cite, for example, review articles on this subject – S.K. Goyal, B.C. Giri [2], M. Bakker, J. Riezebos, R.H. Teunter [3]. Also worth noting are papers by V.K. Mishra [4], R. Begum, S.K. Sahu, R. [5], where the authors consider models of inventory management of continuously deteriorating goods under the assumption of a known demand function. In V. Sharma's and R. R. Chaudhary's paper [6] a model is considered where demand is a known function of time, while the deterioration process is random and follows a Weibull distribution. In K. Tripathy's and U. Mishra's paper [7] a model is considered where demand is a known function of price. The relationship between the problems of queueing theory and inventory control theory is clearly seen in the works of M. Schwarz and H. Daduna [8].

2   Mathematical Model of the Problem

We consider a single-line queueing system (Fig. 1) with arrival rate c of perishable goods which are input to the system. We assume that the arrival process can be approximated in such a way that c units arrive per unit time.

The goods continuously deteriorate as they are stored. Let $S(t)$ be the amount of goods at time t. We assume that in a small time interval Δt loss is equal to $kS(t)\Delta t$. The service, which in the work will be called sales, is provided by the parties with random size x, where the values of purchases x are an independent

© Springer International Publishing Switzerland 2016
A. Dudin et al. (Eds.): ITMM 2016, CCIS 638, pp. 192–206, 2016.
DOI: 10.1007/978-3-319-44615-8_17

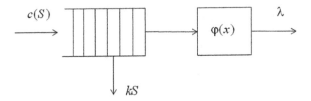

Fig. 1. Mathematical model

random variable with a probability density function $\varphi(x)$, mean $M\{x\} = a$ and the second moment $M\{x^2\} = a_2$. The selling times follow the Poisson process with intensity λ that depends on a selling price b. For the given price b and, hence, sales process intensity λ the average amount of goods $\bar{S}(t)$ is defined as

$$\bar{S}(t) = S(0)e^{-kt} + \frac{c - \lambda a}{k}\left(1 - e^{-kt}\right).$$

Thus if $c - \lambda a > 0$ and $t \gg 1$ we have a constant stock of unsold goods which is undesirable. If $c - \lambda a \le 0$ we have an unsatisfied demand. Therefore, we need to control either selling price b (this problem was considered in [9,10]), or the rate of goods arrival c depending on the current stock.

In this paper we assume that production is controlled in the following way. The two boundary values for the stock of goods are set, S_1 and S_2, besides $S_2 > S_1$. For $S < S_1$ a production rate c_0 is established, for $S > S_2$ a production rate $c_1 < c_0$ is established. For $S_1 \le S \le S_2$ the production rate will be either $c = c_0$ or $c = c_1$ depending on the trajectory which the process $S(t)$ followed when it entered this domain. If it crossed the lower bound S_1 upwards then $c = c_0$, while if it crossed the upper bound S_2 downwards, then $c = c_1$. Thus the production rate $c = c_1$ is set as soon as $S(t)$ reaches S_2 and lasts until the stock falls to S_1. The domain $S_1 \le S \le S_2$ is in fact what we call the domain of hysteresis stock control. In accordance with this, the intensity of the production rate at any given moment is given by

$$c(S) = \left\{ \begin{array}{ll} c_0, & S < S_1, \\ c_0 \ or \ c_1, & S_1 \le S \le S_2, \\ c_1, & S > S_2. \end{array} \right\} \tag{1}$$

It is natural to assume that $c_0 - \lambda a > 0$. Finally, there may be a situation when the current demand cannot be fully satisfied by the current stock of goods. In such case we assume that $S(t) < 0$. The orders are satisfied in order of arrival.

The main goal of this paper is to determine the probability density function of the stock of goods in this model under several additional assumptions.

Let us denote

$$P_0(S,t)dS = \Pr\{S \le S(t) < S + dS, \ c(t) = c_0\}, \quad S < S_1,$$
$$P_2(S,t)dS = \Pr\{S \le S(t) < S + dS, \ c(t) = c_1\}, \quad S > S_2,$$
$$P_{i1}(S,t)dS = \Pr\{S \le S(t) < S + dS, \ c(t) = c_i\}, \quad i = 0, 1, \ S_1 \le S \le S_2.$$

Obviously, the probability density function of the stock of goods $P(S,t)$ is determined by the relation

$$P(S,t) = \left\{ \begin{array}{ll} P_0(S,t), & S < S_1, \\ P_{01}(S,t) + P_{11}(S,t), & S_1 \le S \le S_2, \\ P_2(S,t), & S > S_2. \end{array} \right\} \tag{2}$$

Theorem 1. *If $P_i(S,t), P_{i1}(S,t)$ are differentiable in t, $SP_i(S,t), SP_{i1}(S,t)$ are differentiable in S, then functions $P_i(S,t), P_{i1}(S,t)$ satisfy the following system of Kolmogorov equations*

$$\frac{\partial P_2(S,t)}{\partial t} = -\lambda P_2(S,t) - \frac{\partial}{\partial S}((c_1 - kS)P_2(S,t))$$
$$+\lambda \int_0^\infty P_2(S+x,t)\varphi(x)dx, \quad S > S_2, \tag{3}$$

$$\frac{\partial P_{11}(S,t)}{\partial t} = -\lambda P_{11}(S,t) - \frac{\partial}{\partial S}((c_1 - kS)P_{11}(S,t))$$
$$+\lambda \int_0^{S_2-S} P_{11}(S+x,t)\varphi(x)dx + \lambda \int_{S_2-S}^\infty P_2(S+x,t)\varphi(x)dx, \quad S_1 \le S \le S_2, \tag{4}$$

$$\frac{\partial P_{01}(S,t)}{\partial t} = -\lambda P_{01}(S,t) - \frac{\partial}{\partial S}((c_0 - kS)P_{01}(S,t))$$
$$+\lambda \int_0^{S_2-S} P_{01}(S+x,t)\varphi(x)dx, \quad S_1 \le S \le S_2, \tag{5}$$

$$\frac{\partial P_0(S,t)}{\partial t} = -\lambda P_0(S,t) - \frac{\partial}{\partial S}((c_0 - kI(S)S)P_0(S,t))$$
$$+\lambda \int_0^{S_1-S} P_0(S+x,t)\varphi(x)dx + \lambda \int_{S_1-S}^{S_2-S} P_{01}(S+x,t)\varphi(x)dx$$
$$+\lambda \int_{S_1-S}^{S_2-S} P_{11}(S+x,t)\varphi(x)dx + \lambda \int_{S_2-S}^\infty P_2(S+x,t)\varphi(x)dx, \quad S < S_1, \tag{6}$$

where $I(x)$ is a step unit function.

Proof. Let us start with domain $S > S_2$. We shall consider two close moments of time t and $t + \Delta t$, where $\Delta t \ll 1$. Under the given assumptions the conditional probability

$$P\left\{S_2 \le S(t+\Delta t) < z, c(t+\Delta t) = c_1 \,|\, S(t) = S,\, c(t) = c_1\right\}$$
$$= (1 - \lambda \Delta t) I(z - S - (c_1 - kS)\Delta t)$$
$$+ \lambda \Delta t \int_0^{S-S_2} I(z - S + x)\varphi(x)dx + o(\Delta t). \tag{7}$$

Thus for $z \ge S_2$ probability

$$P\left\{S_2 \le S(t+\Delta t) < z, c(t+\Delta t) = c_1\right\}$$
$$= (1 - \lambda \Delta t) \int_{S_2}^{\infty} I(z - S - (c_1 - kS)\Delta t) P_2(S,t)dS \tag{8}$$
$$+ \lambda \Delta t \int_{S_2}^{\infty} \int_0^{S-S_2} I(z - S + x)\varphi(x)dx\, P_2(S,t)dS + o(\Delta t).$$

For $z \ge S_2$ and a small Δt the integral

$$\int_{S_2}^{\infty} I(z - S - (c_1 - kS)\Delta t) P_2(S,t)dS = \int_{S_2}^{z-(c_1-kz)\Delta t + o(\Delta t)} P_2(S,t)dS$$
$$= \int_{S_2}^{z} P_2(S,t)dS - P_2(z,t)(c_1 - kz)\Delta t + o(\Delta t)$$

and the integral

$$\int_{S_2}^{\infty} \int_0^{S-S_2} I(z - S + x)\varphi(x)dx\, P_2(S,t)dS = \int_0^{\infty} \varphi(x) \int_{S_2+x}^{z+x} P_2(S,t)dS dx.$$

Substituting the expressions above into (8), differentiating with respect to z and taking the limit $\Delta t \to 0$ we derive Eq. (3).

Consider now the domain $S_1 \le S \le S_2$. In this domain two options are possible, $c(S) = c_1$ and $c(S) = c_0$. First, we will consider the case $c(S) = c_1$. For $z < S_2$ conditional probabilities

$$P\left\{S_1 \le S(t+\Delta t) < z, c(t+\Delta t) = c_1 \,|\, S(t) = S < S_2,\, c(t) = c_1\right\}$$
$$= (1 - \lambda \Delta t) I(z - S - (c_1 - kS)\Delta t) + \lambda \Delta t \int_0^{S-S_1} I(z - S + x)\varphi(x)dx + o(\Delta t),$$

$$P\left\{S_1 \le S(t+\Delta t) < z, c(t+\Delta t) = c_1 \,|\, S(t) = S > S_2,\, c(t) = c_1\right\}$$
$$= \lambda \Delta t \int_0^{S-S_1} I(z - S + x)\varphi(x)dx + o(\Delta t).$$

Therefore, the probability

$$P\{S_1 \le S(t + \Delta t) < z, c(t + \Delta t) = c_1\}$$

$$= (1 - \lambda \Delta t) \int_{S_1}^{S_2} I(z - S - (c_1 - kS)\Delta t) P_{11}(S, t) dS$$

$$+ \lambda \Delta t \int_{S_1}^{S_2} \int_{0}^{S-S_1} I(z - S + x)\varphi(x) dx P_{11}(S, t) dS \qquad (9)$$

$$+ \lambda \Delta t \int_{S_2}^{\infty} \int_{0}^{S-S_1} I(z - S + x)\varphi(x) dx P_2(S, t) dS + o(\Delta t).$$

For $z \ge S_1$ and a small Δt the integral

$$\int_{S_1}^{S_2} I(z - S - (c_1 - kS)\Delta t) P_{11}(S, t) dS = \int_{S_1}^{z - (c_1 - kz) + o(\Delta t)} P_{11}(S, t) dS$$

$$= \int_{S_1}^{z} P_{11}(S, t) dS - P_{11}(z, t)(c_1 - kz)\Delta t + o(\Delta t),$$

the integrals

$$\int_{S_2}^{\infty} \int_{0}^{S-S_1} I(z - S + x)\varphi(x) dx P_2(S, t) dS = \int_{S_2}^{\infty} \int_{S-z}^{S-S_1} \varphi(x) dx P_2(S, t) dS,$$

$$\int_{S_1}^{S_2} \int_{0}^{S-S_1} I(z - S + x)\varphi(x) dx P_{11}(S, t) dS$$

$$= \int_{S_1}^{S_2} \int_{0}^{S-S_1} \varphi(x) dx P_{11}(S, t) dS - \int_{z}^{S_2} \int_{0}^{S-z} \varphi(x) dx P_{11}(S, t) dS.$$

Substituting the expressions above into (9), differentiating with respect to z and taking the limit $\Delta t \to 0$ we derive Eq. (4).

Let us consider now the case, when $c(S) = c_0$ In this case, the transitions from the domain $S > S_2$ are impossible, so they set to $c(S) = c_1$. Therefore we have a single conditional probability

$$P\{S_1 \le S(t + \Delta t) < z, c(t + \Delta t) = c_0 \mid S(t) = S < S_2, c(t) = c_0\}$$

$$= (1 - \lambda \Delta t) I(z - S - (c_0 - kS)\Delta t) + \lambda \Delta t \int_{0}^{S-S_1} I(z - S + x)\varphi(x) dx + o(\Delta t),$$

Consequently, the probability

$$P\left\{S_1 \leq S(t + \Delta t) < z, c(t + \Delta t) = c_0\right\}$$

$$= (1 - \lambda \Delta t) \int_{S_1}^{S_2} I(z - S - (c_0 - kS)\Delta t) P_{01}(S, t) dS$$

$$+ \lambda \Delta t \int_{S_1}^{S_2} \int_0^{S - S_1} I(z - S + x)\varphi(x) dx P_{01}(S, t) dS + o(\Delta t).$$

(10)

Transforming the integrals in the expression (10), differentiating and taking the limit $\Delta t \to 0$, we derive Eq. (5).

Let us consider finally the domain $z \leq S_1$. The transitions in this domain are possible from both the area $S > S_2$ and $S_1 \leq S \leq S_2$. In this case the probability

$$P\left\{S(t + \Delta t) < z, c(t + \Delta t) = c_0\right\}$$

$$= (1 - \lambda \Delta t) \int_{-\infty}^{S_1} I(z - S - (c_0 - kI(S)S)\Delta t) P_0(S, t) dS$$

$$+ \lambda \Delta t \int_{-\infty}^{S_1} \int_0^{\infty} I(z - S + x)\varphi(x) dx P_0(S, t) dS$$

(11)

$$+ \lambda \Delta t \int_{S_1}^{S_2} \int_{S - S_1}^{\infty} I(z - S + x)\varphi(x) dx \left[P_{01}(S, t) + P_{11}(S, t)\right] dS$$

$$+ \lambda \Delta t \int_{S_2}^{\infty} \int_{S - S_1}^{\infty} I(z - S + x)\varphi(x) dx P_2(S, t) dS + o(\Delta t),$$

because if $S < 0$ there is no products.

Given that

$$\int_{-\infty}^{S_1} \int_0^{\infty} I(z - S + x)\varphi(x) dx P(S, t) dS = \int_{-\infty}^{z} P(S, t) dS + \int_{z}^{S_1} P(S, t) \int_{S - z}^{\infty} \varphi(x) dx dS,$$

transforming the integrals in the expression (11), differentiating and taking the limit $\Delta t \to 0$, we derive Eq. (6).

The solution of the system (3)–(6) must, apparently, satisfy the following normalising condition

$$\int_{-\infty}^{S_1} P_0(S, t) dS + \int_{S_1}^{S_2} (P_{01}(S, t) + P_{11}(S, t)) dS + \int_{S_2}^{\infty} P_2(S, t) dS = 1.$$

(12)

and must satisfy additional conditions

$$P_0(S_1) = P_{01}(S_1), \quad P_{11}(S_1) = 0,$$

(13)

which result from the following. On the trajectory with $c(S) = c_1$ the point $S = S_1$ is achievable only from above. Therefore

$$P\{S_1 \leq S(t + \Delta t) < S_1 + dz, c(t + \Delta t) = c_1\}$$

$$= \lambda \Delta t \int_{S_1}^{S_2} \int_{0}^{S - S_1} I(S_1 + dz - S + x)\varphi(x)dx P_{11}(S, t)dS$$

$$+ \lambda \Delta t \int_{S_2}^{\infty} \int_{0}^{S - S_1} I(S_1 + dz - S + x)\varphi(x)dx P_2(S, t)dS + o(\Delta t).$$

whence under $\Delta t \to 0$ follows, that $P_{11}(S_1) = 0$. The second relation (13) is a consequence of

$$P_{01}(S_1 + 0, t + \Delta t) = (1 - \lambda \Delta t)P_0(S_1 - 0, t) + O(\Delta t).$$

3   Asymptotic Expression for Probability Density Function of the Stock of Goods

To obtain the exact solution for the system (3)–(6) is possible only in exceptional cases. It is, therefore, of interest to construct an approximate solution of the system (3)–(6) in the steady state under some additional asymptotic assumptions. Let us introduce parameter $\varepsilon \ll 1$ and assume that the parameters c_0, c_1 and k satisfy the following conditions:

$$c_0 = (1 + \alpha \varepsilon)\lambda a, \quad c_1 = (1 + \beta \varepsilon)\lambda a, \quad k = k_0 \varepsilon^2, \tag{14}$$

where $\beta < \alpha$. The first two relations in (14) mean that the amount of products produced per unit time almost equals the mean of sales per unit time for any S. The last relation in (14) means that the ratio $(c_0 - \lambda a)/k \gg 1$, i.e. products in storage deteriorate rather slowly. It is natural to assume that threshold values S_1 and S_2, which define the domain of hysteresis control of production, depend on ε. More precisely we assume that under $\varepsilon \to 0$ the threshold values $S_1(\varepsilon)$ and $S_2(\varepsilon) \to \infty$, but there exist finite limits

$$\lim_{\varepsilon \to 0} S_1(\varepsilon) = z_1, \quad \lim_{\varepsilon \to 0} S_2(\varepsilon) = z_2. \tag{15}$$

Let us consider first the domain $S > S_2$. Let us denote $P_2(S) = \lim_{t \to \infty} P_2(S, t)$. To find the solution of (3) we introduce the function $f_2(S, \varepsilon)$ by relation

$$P_2(S) = \varepsilon f_2(\varepsilon S, \varepsilon) \tag{16}$$

and assume that the function $f_2(S, \varepsilon)$ is twice differentiable in S and is uniformly continuous in ε. Substituting (16) into (3) and making the change of variables $\varepsilon S = z$, we see that function $f_2(z, \varepsilon)$ satisfies the equation

$$\lambda f_2(z, \varepsilon) = -\frac{d}{dz}((\varepsilon c_1 - kz)f_2(z, \varepsilon)) + \lambda \int_{0}^{\infty} f_2(z + \varepsilon x, \varepsilon)\varphi(x)dx = 0. \tag{17}$$

Taking Taylor expansion of $f_2(z+\varepsilon x, \varepsilon)$ with respect to the first argument, considering the first three terms of the sum and taking into consideration relation (14), we get

$$-\frac{d}{dz}((\beta\lambda a - k_0 z)f_2(z,\varepsilon)) + \frac{\lambda a_2}{2}\frac{d^2}{dz^2}f_2(z,\varepsilon) + \frac{o(\varepsilon^2)}{\varepsilon^2} = 0. \tag{18}$$

Let us denote

$$f_2(z) = \lim_{\varepsilon \to \infty} f_2(z,\varepsilon). \tag{19}$$

Passing to the limit when $\varepsilon \to 0$ in (18) we see that the function $f_2(z)$ satisfies the equation

$$\frac{d^2}{dz^2}f(z) - \frac{2}{\lambda a_2}\frac{d}{dz}((\beta\lambda a - k_0 z)f_2(z)) = 0. \tag{20}$$

Boundary condition $P_2(+\infty) = 0$ yields $f_2(+\infty) = 0$, taking into account which, we obtain

$$f_2(z) = Ae^{-\frac{(\beta\lambda a - k_0 z)^2}{\lambda a_2 k_0}}. \tag{21}$$

Wherefrom

$$f_2(z,\varepsilon) = Ae^{\frac{(\beta\lambda a - k_0 z)^2}{\lambda a_2 k_0}} + O(\varepsilon). \tag{22}$$

The constant of integration A is to be determined from the conditions of crosslinking solutions.

Let us consider the domain $S_1 \leq S \leq S_2$. Let us denote $P_{i1}(S) = \lim_{t \to \infty} P_{i1}(t,S)$. To find the solution of Eqs. (4) and (5) we introduce the functions $f_{i1}(S,\varepsilon)$ by relation

$$P_{i1} = \varepsilon f_{i1}(\varepsilon S, \varepsilon), \quad i = 0,1, \tag{23}$$

and assume that the functions $f_{i1}(S,\varepsilon)$ are twice differentiable in S and are uniformly continuous in ε. Substituting (23) into (4) and making the change of variables $\varepsilon S = z$, we see that the function $f_{11}(z,\varepsilon)$ satisfies the equation

$$\lambda f_{11}(z,\varepsilon) = -\frac{d}{dz}((\varepsilon c_1 - kz)f_{11}(z,\varepsilon)) + \lambda \int_0^\infty f_{11}(z + \varepsilon x, \varepsilon)\varphi(x)dx$$
$$+ R(z,\varepsilon) = 0. \tag{24}$$

where

$$R(z,\varepsilon) = \int_{\frac{z_2-z}{\varepsilon}}^\infty f_2(z + \varepsilon x, \varepsilon)\varphi(x)dx - \lambda \int_{\frac{z_2-z}{\varepsilon}}^\infty f_{11}(z + \varepsilon x, \varepsilon)\varphi(x)dx. \tag{25}$$

The functions $f_2(z,\varepsilon)$ and $f_{11}(z,\varepsilon)$ are differentiable and hence bounded. Therefore, as $z \neq z_2$

$$R(z,\varepsilon) \leq const \int_{\frac{z_2-z}{\varepsilon}}^\infty \varphi(x)dx \leq const\frac{\varepsilon^2}{(z_2-z)^2}\int_{\frac{z_2-z}{\varepsilon}}^\infty x^2\varphi(x)dx = O(\varepsilon^2),$$

since the second moment $M\{x^2\} = a_2$ exists by the assumptions of the model. Hence we are not taking into account the member $R(z, \varepsilon)$ in (24). Let us denote

$$f_{11}(z) = \lim_{\varepsilon \to 0} f_{11}(z, \varepsilon). \tag{26}$$

Similarly the derivation of relation (20) we obtain, that function $f_{11}(z)$ satisfies the equation

$$\frac{d^2}{dz^2} f_{11}(z) - \frac{2}{\lambda a^2} \frac{d}{dz}((\beta\lambda a - k_0 z) f_{11}(z)) = 0. \tag{27}$$

The solution of Eq. (27) has the following form

$$f_{11}(z) = \left[B_1 + B_2 \int_{z_1}^{z} e^{\frac{(\beta\lambda a - k_0 x)^2}{\lambda a_2 k_0}} dx \right] e^{\frac{(\beta\lambda a - k_0 z)^2}{\lambda a_2 k_0}}. \tag{28}$$

Constants B_1 and B_2 can be derived from the following consideration. There is, firstly, $P_{11}(S_1) = \varepsilon f_{11}(\varepsilon S_1, \varepsilon) = 0$. Passing to the limit when $\varepsilon \to 0$, we obtain $f_{11}(z_1) = 0$. Therefore $B_1 = 0$. Secondly, function $f_{11}(z, \varepsilon)$ must satisfy the initial equation under $z = z_2$. From Eqs. (3) and (4) under $S = S_2$ we get

$$\lambda P_2(S_2) + \frac{d}{dS}((c_1 - kS)P_2(S))\big|_{S=S_2} \\ = \lambda P_{11}(S_2) + \frac{d}{dS}((c_1 - kS)P_{11}(S))\big|_{S=S_2}. \tag{29}$$

By taking the limit as $\varepsilon \to 0$ in (29) we have $f_2(z_2) = f_{11}(z_2)$ or

$$A = B_2 \int_{z_1}^{z_2} e^{\frac{(\beta\lambda a - k_0 x)^2}{\lambda a_2 k_0}} dx. \tag{30}$$

Let us consider now to the definition of the function $f_{01}(z, \varepsilon)$, which follows by (5) satisfies the equation

$$\lambda f_{01}(z, \varepsilon) = -\frac{d}{dz}((\varepsilon c_0 - kz)f_{01}(z, \varepsilon) + \lambda \int_0^{\infty} f_{01}(z + \varepsilon x, \varepsilon)\varphi(x)dx \\ + R(z, \varepsilon) = 0, \tag{31}$$

where for $z \neq z_2$ $R(z, \varepsilon) = o(\varepsilon^2)$. Therefore this term in the Eq. (31) will not be taken into account. Similarly to the above, we can show that

$$f_{01}(z) = \lim_{\varepsilon \to 0} f_{01}(z, \varepsilon) \tag{32}$$

satisfies a second-order differential equation

$$\frac{d^2}{dz^2} f_{01}(z) - \frac{2}{\lambda a_2} \frac{d}{dz}((\alpha\lambda a - k_0 z) f_{01}(z)) = 0, \tag{33}$$

the general solution of which has the form

$$f_{01}(z) = \left[D_1 + D_2 \int_z^{z_2} e^{\frac{(\alpha\lambda a - k_0 x)^2}{\lambda a_2 k_0}} dx \right] e^{-\frac{(\alpha\lambda a - k_0 z)^2}{\lambda a_2 k_0}}. \tag{34}$$

For $S = S_2$ the Eq. (5) gives

$$\lambda P_{01}(S_2) + \frac{d}{dS}((c_0 - kS)P_{01}(s))|_{s=S_2} = 0. \tag{35}$$

By taking the limit as $\varepsilon \to 0$ in (35), we obtain $f_{01}(z_2) = 0$. Wherefrom $D_1 = 0$. Thus,

$$f_{01}(z, \varepsilon) = D_2 \int_z^{z_2} e^{\frac{(\alpha\lambda a - k_0 x)^2}{\lambda a_2 k_0}} dx e^{-\frac{(\alpha\lambda a - k_0 z)^2}{\lambda a_2 k_0}} + O(\varepsilon).$$

Finally, let us consider the domain $S < S_1$. We introduce the function $f_0(S, \varepsilon)$ by relation

$$P_0(S) = \varepsilon f_0(\varepsilon S, \varepsilon). \tag{36}$$

Substituting (36) into (6) we obtain function $f_0(z, \varepsilon)$ in domain $z < z_1$ defined by the equation

$$\lambda f_0(z, \varepsilon) = -\frac{d}{dz}((\varepsilon c_0 - kI(z)z)f_0(z, \varepsilon) + \lambda \int_0^\infty f_0(z + \varepsilon x, \varepsilon)\varphi(x)dx \tag{37}$$

$$+ R(z, \varepsilon) = 0,$$

where for $z < z_1$ $R(z, \varepsilon) = o(\varepsilon^2)$ and therefore the last term in (37) will not be taken into account. Similarly to the above, we can show that

$$f_0(z) = \lim_{\varepsilon \to 0} f_0(z, \varepsilon) \tag{38}$$

satisfies the following equation

$$\frac{d^2}{dz^2} f_0(z) - \frac{2}{\lambda a_2} \frac{d}{dz}((\alpha\lambda a - k_0 I(z)z)f_0(z)) = 0. \tag{39}$$

Boundary condition $P_0(-\infty) = 0$ yields $f_0(-\infty) = 0$. Therefore from Eq. (39) we obtain

$$\frac{d}{dz} f_0(z) - \frac{2}{\lambda a_2}((\alpha\lambda a - k_0 I(z)z)f_0(z)) = 0. \tag{40}$$

Wherefrom

$$f_0(z) = \begin{cases} Fe^{-\frac{(\alpha\lambda a - k_0 z)^2}{\lambda a_2 k_0}}, & 0 < z < z_1, \\ Fe^{-\frac{(\alpha\lambda a)^2}{\lambda a_2 k_0}} e^{\frac{2\alpha a}{a_2} z}, & z < 0. \end{cases} \tag{41}$$

The condition $P_{01}(S_1) = P_0(S_1)$ yields $f_{01}(z_1) = F_0(z_1)$. Wherefrom

$$F = D_2 \int_{z_1}^{z_2} e^{\frac{(\alpha\lambda a - k_0 x)^2}{\lambda a_2 k_0}} dx. \tag{42}$$

Finally, for $S = S_1$ from the system of Eqs. (4)–(6), taking into consideration that $P_0(S_1) = P_{01}(S_1)$ and $P_{11}(S_1) = 0$, we have

$$\frac{d}{dz}((\varepsilon c_0 - kz)f_0(z, \varepsilon))\big|_{z=z_1} = \frac{d}{dz}(\varepsilon c_0 - kz)f_{01}(z, \varepsilon))\big|_{z=z_1}$$
$$+ \frac{d}{dz}((\varepsilon c_1 - kz)f_{11}(z, \varepsilon))\big|_{z=z_1}. \tag{43}$$

Computing the expressions included in (43) and taking the limit as $\varepsilon \to 0$, we obtain that $D_2 = B_2$. Thus, we finally get, that functions $f_i(z)$ and $f_{ij}(z)$ are defined by system of relations

$$f_2(z) = B \int_{z_1}^{z_2} e^{\frac{(\beta\lambda a - k_0 x)^2}{\lambda a_2 k_0}} dx e^{-\frac{(\beta\lambda a - k_0 z)^2}{\lambda a_2 k_0}}, \quad z > z_2, \tag{44}$$

$$f_{11}(z) = B \int_{z_1}^{z} e^{\frac{(\beta\lambda a - k_0 x)^2}{\lambda a_2 k_0}} dx e^{-\frac{(\beta\lambda a - k_0 z)^2}{\lambda a_2 k_0}}, \quad z_1 \le z \le z_2, \tag{45}$$

$$f_{01}(z) = B \int_{z}^{z_2} e^{\frac{(\alpha\lambda a - k_0 x)^2}{\lambda a_2 k_0}} dx e^{-\frac{(\alpha\lambda a - k_0 z)^2}{\lambda a_2 k_0}}, \quad z_1 \le z \le z_2, \tag{46}$$

$$f_0(z) = \begin{cases} B \int_{z_1}^{z_2} e^{\frac{(\alpha\lambda a - k_0 x)^2}{\lambda a_2 k_0}} dx e^{-\frac{(\alpha\lambda a - k_0 z)^2}{\lambda a_2 k_0}}, & 0 \le z \le z_1, \\ B \int_{z_1}^{z_2} e^{\frac{(\alpha\lambda a - k_0 x)^2}{\lambda a_2 k_0}} dx e^{-\frac{(\alpha\lambda a)^2}{\lambda a_2 k_0}} e^{\frac{2\alpha a}{a_2} z}, & z < 0. \end{cases} \tag{47}$$

Constant B is defined from normalizing condition, which under $\varepsilon \to 0$ takes the form

$$\int_{-\infty}^{z_1} f_0(z)dz + \int_{z_1}^{z_2} (f_{01}(z) + f_{11}(z)) \, dz + \int_{z_2}^{\infty} f_2(z)dz = 1. \tag{48}$$

Finally we obtain since $\varepsilon \gg 1$ the probability density function of the stock of goods has the form

$$P(S) = \begin{cases} \varepsilon B \int\limits_{z_1}^{z_2} e^{\frac{(\beta\lambda a - k_0 x)^2}{\lambda a_2 k_0}} dx e^{-\frac{(\beta\lambda a - k_0 \varepsilon S)^2}{\lambda a_2 k_0}} + O(\varepsilon), & S > S_2, \\[4mm] \varepsilon B \left[\int\limits_{\varepsilon S}^{z_2} e^{\frac{(\alpha\lambda a - k_0 x)^2}{\lambda a_2 k_0}} dx e^{-\frac{(\alpha\lambda a - k_0 \varepsilon S)^2}{\lambda a_2 k_0}} \right. \\[4mm] \left. + \int\limits_{z_1}^{\varepsilon S} e^{\frac{(\beta\lambda a - k_0 x)^2}{\lambda a_2 k_0}} dx e^{-\frac{(\beta\lambda a - k_0 \varepsilon S)^2}{\lambda a_2 k_0}} \right] + O(\varepsilon), & S_1 \le S \le S_2, \quad (49) \\[4mm] \varepsilon B \int\limits_{z_1}^{z_2} e^{\frac{(\alpha\lambda a - k_0 x)^2}{\lambda a_2 k_0}} dx e^{-\frac{(\alpha\lambda a - k_0 \varepsilon S)^2}{\lambda a_2 k_0}} + O(\varepsilon), & 0 \le S \le S_1, \\[4mm] \varepsilon B \int\limits_{z_1}^{z_2} e^{\frac{(\alpha\lambda a - k_0 x)^2}{\lambda a_2 k_0}} dx e^{-\frac{(\alpha\lambda a)^2}{\lambda a_2 k_0}} e^{\frac{2\alpha a}{a_2}\varepsilon S} + O(\varepsilon), & S < 0. \end{cases}$$

4   Switch Control of Production for the Case of Exponentially Distributed Purchase Amounts

Let us consider the simplest case of switch control of production (threshold $S_1 = S_2$) when sales are distributed exponentially

$$\varphi(x) = \frac{1}{a} e^{-\frac{x}{a}} . \qquad (50)$$

In the steady state Eqs. (3)–(6), which define probability density function of the stock of goods, take the form

$$\lambda P_2(S) + \frac{d}{dS}((c_1 - kS)P_2(S)) - \frac{\lambda}{a} e^{\frac{S}{a}} \int\limits_S^\infty P_2(x) e^{-\frac{x}{a}} dx = 0, \quad S \ge S_1, \qquad (51)$$

$$\lambda P_0(S) + \frac{d}{dS}((c_0 - kI(S)S)P_0(S)) - \frac{\lambda}{a} e^{\frac{S}{a}} \int\limits_S^{S_1} P_0(x) e^{-\frac{x}{a}} dx$$

$$- \frac{\lambda}{a} e^{\frac{S}{a}} \int\limits_{S_1}^\infty P_2(x) e^{-\frac{x}{a}} dx = 0, \quad S < S_1. \qquad (52)$$

Equation (51) can be differentiated and represented as the following differential equation

$$\frac{d^2}{dS^2}((c_1 - kS)P_2(S)) - \frac{d}{dS}\left(\frac{c_1 - kS - \lambda a}{a} P_2(S)\right) = 0. \qquad (53)$$

The general solution of the Eq. (56) has the form

$$P_2(S) = \left[A + A_1 \int\limits_{S_1}^S e^{-\frac{x}{a}}(c_1 - kx)^{-\frac{\lambda}{k}} dx \right] e^{\frac{S}{a}}(c_1 - kS)^{\frac{\lambda}{k}-1}. \qquad (54)$$

Constants A and A_1 must now be selected so that expression (54) satisfies the initial Eq. (51). Substituting (54) into (51) and taking into account that by the assumptions of the model the condition $c_1 - kS > 0$, should be fulfilled, we are convinced that $A_1 = 0$.

Equation (54) can be differentiated and represented as the following differential equation

$$\frac{d^2}{dS^2}((c_0 - kI(S)S)P_0(S)) - \frac{d}{dS}\left(\frac{c_0 - kI(S)S - \lambda a}{a}P_0(S)\right) = 0. \tag{55}$$

Wherefrom, the subject to boundary condition $P_0(-\infty) = 0$, we obtain

$$P_0(S) = \begin{cases} Be^{\frac{S}{a}}(c_0 - kS)^{\frac{\lambda}{k} - 1} & , 0 \leq S \leq S_1, \\[2mm] Bc_0^{\frac{\lambda}{k} - 1}\,e^{\frac{c_0 - \lambda a}{c_0 a}S} & , \quad S < 0. \end{cases} \tag{56}$$

To obtain the relationship between the constants A and B we may proceed as follows. Equations (51) and (52) can be integrated in their domain and we sum up the results. We obtain

$$(c_1 - kS_1)^{\frac{\lambda}{k}}P_2(S_1) = (c_0 - kS_1)P_0(S_1). \tag{57}$$

Substituting the expressions (54) and (56) into (57), we obtain

$$A = B\gamma, \qquad \gamma = \frac{(c_0 - kS_1)^{\frac{\lambda}{k}}}{(c_1 - kS_1)^{\frac{\lambda}{k}}}. \tag{58}$$

The constant B now can be obtained from the normalizing condition.

Consider now the case, then $c_1 = 0$. Obviously, in this case $P_2(S) = 0$, and probability density function $P_0(S)$ is determined by the equation

$$\lambda P_0(S) + \frac{d}{dS}((c_0 - kI(S)S)P_0(S)) - \frac{\lambda}{a}e^{\frac{S}{a}}\int_S^{S_1} P_0(x)e^{-\frac{x}{a}}\,dx = 0, \quad S < S_1, \tag{59}$$

the solution of which has the form (56).

Let $c_0 = (1 + \theta)\lambda a$, where $\theta > 0$. Let us denote $\mu = \lambda/k$. Parameter μ shows how the product sales rate exceeds the rate of its loss. Then

$$P_0(S) = \begin{cases} De^{\frac{S}{a}}\left(1 - \frac{S}{\mu(1 + \theta)a}\right)^{\mu - 1} & , 0 \leq S \leq S_1, \\[2mm] De^{\frac{\theta}{(1+\theta)a}S} & , \quad S < 0, \end{cases} \tag{60}$$

where

$$D = \left[\frac{(1 + \theta)a}{\theta} + \int_0^{S_1} e^{\frac{S}{a}}\left(1 - \frac{S}{\mu(1 + \theta)a}\right)^{\mu - 1}\,dS\right]^{-1}. \tag{61}$$

The dependence of probability density function $P_0(S)$ of S is given on the Fig. 2. The parameters $\mu = 20$, $\theta = 0.5$, $a = 1.5$.

In the model considered mean profit R per unit time on the condition that the product unit cost per unit time equals 1, will be equal to

$$R = \lambda a b \int_0^\infty P_0(S)dS - c_0,$$

where b is the selling price of the product unit. Or

$$R = \lambda a b - \lambda a \left(1 + \theta + \frac{b}{1 + \frac{\theta}{(1+\theta)a} \int_0^{S_1} e^{\frac{S}{a}} \left(1 - \frac{S}{\mu(1+\theta)a}\right)^{\mu-1} dS} \right). \qquad (62)$$

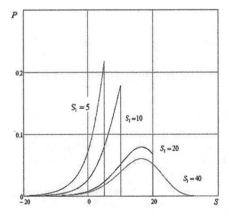

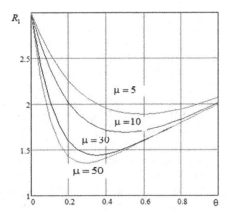

Fig. 2. Relationship between the probability density $P_0(S)$ and the stock size S.

Fig. 3. Relationship between R_1 and parameter θ.

From (62) it follows, that for all values of other parameters, the optimal value $S_{1 opt.} = \mu(1+\theta)a$ and for maximizing R we should minimize the following expression by θ

$$R_1 = 1 + \theta + b \left(1 + \mu\theta \int_0^1 e^{\mu(1+\theta)z}(1 - z)^{\mu-1}dz \right)^{-1}. \qquad (63)$$

The relationship between R_1 and θ is given in Fig. 3, the selling price $b = 2$.

Thus, the values of the considered parameters, the production rate c_0 must be about 30 % higher than the mean value of the sales λa.

5 Conclusion

In this paper we have obtained equations for the probability density function of perishable goods stock under switch-hysteresis control of the production rate. Under arbitrary distribution values of sales we have obtained the approximate solutions of these equations in the case when the rate of production "is almost the same" as the rate of its sales and the products deteriorate rather slowly under storage. In the case of the exponential distribution of sales and under the switch control of the production rate the exact solution of equations was found and the problem of selecting the optimal production rate was solved, thus maximizing the average profit per unit time. A similar approach can be used to investigate other models of the production and sales control of perishable goods, in particular, a model with continuous dependence of the rate of production from the current stock of products.

References

1. Nazarov, A.A., Moisseeva, S.P.: Method of Asymptotic Analysis in the Queuing Theory. NTL, Tomsk (2006). (in Russian)
2. Goyal, S.K., Giri, B.C.: Recent trends in modeling of deteriorating inventory. Eur. J. Oper. Res. **134**(1), 1–16 (2001)
3. Bakker, M., Riezebos, J., Teunter, R.H.: Review of inventory systems with deterioration since 2001. Eur. J. Oper. Res. **221**, 275–284 (2012)
4. Mishra, V.K.: An inventory model of instantaneous deteriorating items with controllable deterioration rate for time dependent demand and holding cost. J. Ind. Eng. Manage. **6**(2), 495–506 (2013)
5. Begum, R., Sahu, S.K., Sahoo, R.R.: An inventory model for deteriorating items with quadratic demand and partial backlogging. Br. J. Appl. Sci. Technol. **2**(2), 112–131 (2012)
6. Sharma, V., Chaudhary, R.: An inventory model for deteriorating items with Weibull deterioration with time dependent demand and shortages. Res. J. Manage. Sci. **2**(3), 28–30 (2013)
7. Tripathy, C.K., Mishra, U.: An inventory model for Weibull deteriorating items with price dependent demand and time-varying holding cost. Appl. Math. Sci. **4**(44), 2171–2179 (2010)
8. Schwarz, M., Daduna, H.: Queuing systems with inventory managements with random lead times and with backordering. Math. Methods Oper. Res. **64**, 383–414 (2006)
9. Livshits, K., Ulyanova, E.: Switch-hysteresis control of the selling times flow in a model with perishable goods. In: Dudin, A., Nazarov, A., Yakupov, R. (eds.) ITMM 2015. CCIS, vol. 564, pp. 263–274. Springer, Heidelberg (2015)
10. Livshits, K., Ulyanova, E.: Diffusion approximation of the production and selling of perishable products. Russ. Phys. J. **58**(11/2), 281–285 (2015). (in Russian)

On the Use of a Bridge Process in a Conditional Monte Carlo Simulation of Gaussian Queues

Oleg Lukashenko[1,2(✉)], Evsey Morozov[1,2], and Michele Pagano[3]

[1] Institute of Applied Mathematical Research, Karelian Research Centre RAS,
Petrozavodsk, Russia
emorozov@karelia.ru
[2] Petrozavodsk State University, Petrozavodsk, Russia
lukashenko-oleg@mail.ru
[3] University of Pisa, Pisa, Italy
m.pagano@iet.unipi.it

Abstract. In spite of their low frequency, rare events often play a major role in determining systems performance. In most cases they can be analysed only through simulation with *ad-hoc* techniques since traditional Monte Carlo approaches are quite inefficient in terms of simulation length and/or estimation accuracy. Among rare event simulation techniques, conditional Monte Carlo is an interesting approach as it always leads to variance reduction. Unfortunately, it is often impossible, or at least very difficult, to find a suitable conditioning strategy. To tackle this issue, the applicability of a bridge process is proposed in the case of queueing systems with Gaussian inputs. In more detail, overflow probability and busy-period length are investigated and the analytical expressions of the corresponding estimators are derived. Finally, the effectiveness of the proposed approach is investigated through simulations.

Keywords: Gaussian processes · Conditional Monte Carlo · Bridge process · Rare events

1 Introduction

By definition, rare events are events that occur with a low frequency and are associated with *large deviations* from normal system behaviour. Such events typically determine severe consequences in several natural and societal frameworks, such as major earthquakes, floods, solar flares, industrial accidents and financial market crashes to name just a few. In teletraffic engineering, rare events are related to packet losses in Quality of Service (QoS) supporting network architectures (see, for instance [8], where loss rates below 10^{-9} are considered for

E. Morozov—This work is partially supported by the Russian Foundation for Basic research, projects 15–07–02341 A, 15–07–02354 A,15–07–02360 A and by the Program of strategic development of Petrozavodsk State University.
M. Pagano—This work is partially supported by the PRA 2016 research project 5GIOTTO funded by the University of Pisa.

A. Dudin et al. (Eds.): ITMM 2016, CCIS 638, pp. 207–220, 2016.
DOI: 10.1007/978-3-319-44615-8_18

high–quality video) or to long lasting busy periods, that might lead to burst of losses (unless suitable congestion control techniques are employed).

Even for elementary single server queueing systems, analytical results have been derived only when a Markovian structure can be identified in the system. Unfortunately, Joseph (long range dependence) and Noah (infinite variance syndrome) effects are well known features of packet traffic [21], for which Markovian models cannot account (at least in a parsimonious and physically-understandable way). Indeed, these effects can be better described by self-similar models, as originally highlighted by Mandelbrot (who proposed such terminology) in the framework of hydrology [14], and have a deep impact in terms of network dimensioning and QoS issues [2].

The first works on fractal queueing theory date back to the mid 1990s [16], but in most cases only logarithmic asymptotics are available, typically derived in the framework of Large Deviation Theory (LDT) [3], while exact asymptotics have been found only in special cases and often depend on parameters (for instance, Pickands' constants) that can be estimated only numerically (see [15] and references therein). Hence, discrete-event simulation often becomes the only possible way for performance evaluation.

Simulation is a powerful tool to study the performance of complex systems with an arbitrary level of detail, but the traditional approach, known in the literature as crude Monte-Carlo (MC) and implemented by most of the network simulators, becomes highly inefficient when the event of interest gets rarer and rarer. Indeed, for a fixed accuracy of the estimates the length of the simulation is inversely proportional to the target probability, that can assume values of the order of 10^{-9} as mentioned above. Moreover, in practical situations one sample may include generation of a huge amount (of the order of millions or more, depending on the time horizon and the complexity of the system) of random numbers.

Hence, in the literature special techniques have been proposed for rare event simulation, including parallel/distributed simulation (that might become very expensive to get significant speed-ups) and Variance Reduction techniques, aiming at achieving the desired accuracy with a lower number of samples [18]. It is well-known that, unlike crude MC simulation, the latter generally require some additional information about the behaviour of the system, such as the one provided (although in an asymptotic and eventually approximate form) by LDT. For instance, the optimal choice of the thresholds in RESTART (REpetitive Simulation Trials After Reaching Threshold) is based on the knowledge of the target probability [20] and in Importance Sampling (IS) under an improper choice of the *change of measure* [7] the variance may even grow infinitely [6], although, when applied properly, enormous variance reduction (several orders of magnitude with respect to crude MC) can be obtained. Moreover, *optimal* changes of measure are known only for very simple queueing systems and heuristic approaches (such as the exchange of arrival and service rates in Markovian tandem queues or the use of the *most-likely path* in Gaussian queues) often work poorly for arbitrary choices of the system parameters.

In this paper we focus on an alternative approach, known as Conditional Monte Carlo. In a nutshell, the target probability is expressed as a conditional expectation and this always leads to variance reduction. Unfortunately, it is often impossible, or at least very difficult, to find a suitable conditioning quantity. In the case of Gaussian processes, it is easy to overcome this issue by expressing the target probability as a function of the corresponding *bridge* process, as originally proposed in [4] for the estimation of the overflow probability. Note that Gaussianity of the input process is not a severe limitation, at least in the teletraffic framework. Indeed, Gaussian processes can be seen as the resulting aggregate process of many superimposed independent sources and are able to take into account the long memory properties of real traffic, while keeping a relatively simple and elegant description.

The contribution of the paper is two–fold: on the one hand, the properties of the Bridge MC approach are presented as a direct consequence of general results for Conditional MC, on the other hand the applicability and the effectiveness of the method are investigated in two different scenarios of practical relevance: packet overflow and duration of busy periods.

The rest of the paper is organized as follows. In Sect. 2 we introduce the general theory beyond rare event simulation and recall the basic features of the Conditional MC method, while in the following Section we present the Bridge MC approach, deriving the analytical expressions for the estimators of overflow probability and long duration of high activity periods. Then, simulation results for both case studies are discussed in Sect. 4 and the main contributions of the paper are summarized in the Conclusions.

2   Crude Monte Carlo vs. Conditional Monte Carlo

Given a random process X, let us consider estimating the probability

$$\pi = \mathbb{P}(X \in A) \; := \; \mathbb{E}\, I(X \in A) \tag{1}$$

for some Borel set A of the paths of the process X, where I denotes the indicator function. To estimate π by crude MC simulation, we should generate N independent sample paths $X_1, ..., X_N$ of the process X and calculate the sample mean

$$\widetilde{\pi}_N \; := \; \frac{1}{N} \sum_{n=1}^{N} I_n \, , \tag{2}$$

where $I_n := I\,(X_n \in A)$. The estimator (2) is unbiased and its variance is given by

$$\mathbb{V}ar\,[\widetilde{\pi}_N] \; = \; \frac{\pi(1-\pi)}{N} \, .$$

Although the estimator variance decreases with the target probability π, this does not imply an accurate estimate in the case of rare events. Indeed, a typical measure of the goodness of the estimator is represented by the relative error (RE):

$$\mathrm{RE}\left[\tilde{\pi}_N\right)] := \frac{\sqrt{\mathbb{V}ar\left[\tilde{\pi}_N\right]}}{\mathbb{E}\tilde{\pi}_N} \sim \frac{1}{\sqrt{\pi N}} \quad \text{as } \pi \to 0 \tag{3}$$

that diverges for small values of the target probability. In other words, the RE of the standard MC estimator is unbounded when the event becomes rare and the number of samples required to get a bounded RE is inversely proportional to π.

To overcome this problem, different rare event simulation techniques have been proposed in the literature [18]. In particular, as already discussed in Sect. 1, variance reduction techniques aim at modifying the estimator (2) in order to reduce its variance, hence requiring fewer samples for the desired accuracy. Bounded RE represents the ideal case, in which the accuracy of the estimation does not depend on the rarity of the event. However, such a criterion is rarely verified and in many practical situations the RE grows, but not as fast as in the crude MC case.

An approach that always leads to variance reduction is the method of simulation by conditioning, known in the literature as conditional MC [18]. Let us denote by Z the indicator function of the target event, i.e.

$$Z = I(X \in A),$$

and assume that there is an auxiliary random variable Y correlated with Z such that $\mathbb{E}[Z|Y]$ is available in explicit form. Then, π can be rewritten as

$$\pi = \mathbb{P}(X \in A) := \mathbb{E}\left[\mathbb{E}[Z|Y]\right], \tag{4}$$

where the outer expectation is taken with respect to Y, while $\mathbb{E}[Z|Y]$ is known by hypothesis. Hence, the corresponding unbiased estimator of π becomes

$$\hat{\pi}_N = \frac{1}{N}\sum_{n=1}^{N} \mathbb{E}[Z|Y_n], \tag{5}$$

where $Y_1, ..., Y_N$ are N samples of Y.

It is easy to verify that the variance of (5) is always less than the variance of (2) since

$$\mathbb{V}ar[Z] = \mathbb{E}[\mathbb{V}ar[Z|Y]] + \mathbb{V}ar[\mathbb{E}[Z|Y]]. \tag{6}$$

The latter result is a really attractive feature of conditional MC, not valid, in general, for variance reduction techniques (in spite of their name). For instance, in case of IS a well-known heuristic is that changes of measure suggested by LDT can reduce variance by many orders of magnitude (with speed-up factors higher than conditional MC), but in some cases such estimators can have variance that decreases at a slower rate than the crude MC estimator, variance that increases with the rarity of the event, and even infinite variance [6].

The main drawback of conditional MC is that, in general, it is not easy to find a random variable Y for which $\mathbb{E}[Z|Y]$ is available in explicit form. However, in the case of Gaussian processes, widely used in teletraffic modelling, such a problem can be easily solved through the introduction of the corresponding *bridge* process, as discussed in the next section.

In the rest of the paper, we will assume that $\{X_t, t \in \mathbb{R}_+\}$ is a centered Gaussian process with stationary increments and that its variance $v_t := \mathbb{V}arX_t$ is an increasing function. Such assumptions are quite mild and do not limit the practical applicability of our method: indeed, stationarity of the increments is a standard prerequisite in traffic modelling and the condition on v_t holds for a wide range of processes, including the ones typically used in teletraffic theory: Fractional Brownian Motion (FBM) [19], superposition of independent FBMs [19] and Integrated Ornstein-Uhlenbeck process (IOU) [9]. It is also worth mentioning that in the definition of the bridge process a key role is played by the covariance function, which is unambiguously defined by the variance function v_t:

$$\Gamma_{s,t} := \mathbb{E}[X_t X_s] = \frac{1}{2}\left(v_t + v_s - v_{|t-s|}\right). \tag{7}$$

3 Bridge Monte Carlo Estimator

Bridge Monte Carlo (BMC) is a special case of the conditional MC method, in which conditioning is expressed in terms of the *bridge* $Y := \{Y_t\}$ of the Gaussian input process X_t. The bridge process is obtained by conditioning X to reach a certain level at some fixed (deterministic) time τ:

$$Y_t = X_t - \psi_t X_\tau, \tag{8}$$

where ψ_t is expressed via the covariance function as

$$\psi_t := \frac{\Gamma_{t,\tau}}{\Gamma_{\tau,\tau}}.$$

It is easy to see that $\psi_t > 0$ for all t due to the assumption on v_t; moreover, we note that the process Y is independent of X_τ since they are jointly Gaussian and uncorrelated:

$$\mathbb{E}[X_\tau Y_t] = \Gamma_{\tau,t} - \frac{\Gamma_{t,\tau}}{\Gamma_{\tau,\tau}}\Gamma_{\tau,\tau} = 0.$$

In the following subsections, it will be shown that the estimators of overflow probability and the busy period length have the same structure, in accordance with the general structure given by Eq. (4).

3.1   Estimation of the Overflow Probability

BMC has been originally proposed by some of the authors in [4], where the stationary overflow probability in a single server queue with constant service rate C was written as the expectation of a function of the bridge. Indeed, by the continuous–time version of the Lindley recursion, the overflow probability (i.e., the probability that the stationary workload Q exceeds some threshold level B) can be written as

$$\mathbb{P}_{\text{overflow}} := \mathbb{P}(Q > B) = \mathbb{P}\left(\sup_{t \in \mathbb{R}_+} (X_t - \varphi_t) > 0\right), \tag{9}$$

where $\varphi_t := B + rt$ and $r := C - m > 0$, being $m > 0$ the average input rate. Indeed, for sake of simplicity, we assumed that X_t is a centered process, which describes the traffic fluctuation with respect to the average arrival rate and hence the cumulative arrival process during the interval $[0, t]$ can be written as

$$A_t = mt + X_t.$$

Taking advantage of the definition of the bridge process, we have (see [12] for a detailed analysis)

$$\mathbb{P}_{\text{overflow}} = \mathbb{P}\left(\sup_{t \in \mathbb{R}_+} (Y_t + \psi_t X_\tau - \varphi_t) > 0\right) = \mathbb{P}\left(\inf_{t \in \mathbb{R}_+} (\varphi_t - Y_t - \psi_t X_\tau) < 0\right)$$

which can be rewritten as

$$\mathbb{P}_{\text{overflow}} = \mathbb{P}\left(X_\tau > \overline{Y}\right), \tag{10}$$

where

$$\overline{Y} := \inf_{t \in \mathbb{R}_+} \frac{\varphi_t - Y_t}{\psi_t}. \tag{11}$$

3.2   Estimation of the Duration of High Activity Periods via BMC

As a second application of the BMC approach, we consider the estimation of the following probability

$$\pi(\mathbb{T}) := \mathbb{P}\left(\forall t \in \mathbb{T}: X_t > t\right), \tag{12}$$

where $\mathbb{T} = [0, T] \subseteq \mathbb{R}_+$ for $T \to \infty$. Persistence phenomena typical of self-similar processes (Joseph effect) imply that the arrival rate can remain at relatively high values for a considerable amount of time, making the estimation of such probability, closely related to the duration of busy periods [17], particularly relevant in the teletraffic framework.

Note that our target probability (12) is equivalent to the following probability

$$\pi(\mathbb{T}) = \mathbb{P}\left(\inf_{t \in \mathbb{T}} (X_t - t) > 0\right),$$

which has the same structure as (9); hence, (10) still holds under a proper definition of $\overline{Y}$ (see [13] for the detailed derivation of the result):

$$\overline{Y} := \sup_{t \in T} \frac{t - Y_t}{\psi_t}. \tag{13}$$

3.3 General Expression of the BMC Estimator

In this subsection we derive the common expression of the BMC estimator, taking advantage of the Gaussianity of the input process. Since

$$X_t =_d \mathcal{N}(0, v_t) =_d \sqrt{v_t}\,\mathcal{N}(0, 1),$$

where $=_d$ stands for stochastic equivalence, it is easy to show that the considered probabilities, denoted as π, can be rewritten as follows:

$$\pi = \mathbb{P}\left(X_\tau \geq \overline{Y}\right) = \mathbb{E}\left[I(X_\tau > \overline{Y})|\overline{Y}\right] \tag{14}$$

$$= \mathbb{E}\left[\Phi\left(\frac{\overline{Y}}{\sqrt{v_\tau}}\right)\right], \tag{15}$$

where independence between $\overline{Y}$ and X_τ is used and Φ denotes the tail distribution of a standard normal variable.

Hence, given an i.i.d. sequence $\{\overline{Y}^{(i)},\ i = 1, ..., N\}$ distributed as $\overline{Y}$ (for a proper choice of $\overline{Y}$, depending on the considered target probability), the estimator of π is defined as follows:

$$\widehat{\pi}_N^{\mathrm{BMC}} = \frac{1}{N}\sum_{i=1}^{N}\Phi\left(\frac{\overline{Y}^{(i)}}{\sqrt{v_\tau}}\right). \tag{16}$$

Note that, in accordance with (15), the BMC approach is actually a special case of the conditional MC method. Hence,

$$\mathbb{V}ar[Z] \geq \mathbb{V}ar\left[\mathbb{E}[Z|\overline{Y}]\right]$$

and we can expect that the BMC estimator implies variance reduction (comparing to the Crude MC simulation) for any value of the target probability (and not only in case of rare events!).

Moreover, as proved in [5] for the overflow probability, the BMC approach leads to a variance reduction also with respect to *single-twist IS* (in which the change of measure corresponds to a shift of the input process X_t by a deterministic path η_t) even when the twist is chosen as the *most-likely path to overflow*, highlighting that the heuristic application of LDT-based approaches may not lead to optimal solutions as already mentioned in Sect. 2.

4   Simulation Results

BMC can be applied to any Gaussian process, whose variance v_t is an increasing function. In the following we present the simulation results for FBM, the most widely used self-similar traffic model [16]. In this case

$$v_t = t^{2H} \qquad H \in (0,1)$$

where H denotes the Hurst parameter and, for $H \in (0.5, 1)$, the increments process is LRD. FBM is widely used in traffic modelling for its simplicity and for the theoretical motivations, originally presented in [19]: indeed, FBM arises as the scaled limit process when the cumulative workload is a superposition of on-off sources with mutually independent heavy-tailed on and/or off periods.

Note that in all simulations we have to consider realizations of the input process on a discrete grid; moreover, the overflow probability cannot be estimated over an infinite horizon and the set $\mathbb{R}_+$ is actually replaced by a discrete lattice $\{1, ..., T\}$ for some finite T. The proper choice of discretization step and simulation horizon goes beyond the goal of this paper; for a detailed analysis of such issues see Chap. 8 of [15] and [11].

In the following subsections, simulation results are presented to highlight the effectiveness and accuracy of the BMC estimator in both scenarios. In all cases we considered $H = 0.8$ (a *typical* value for traffic traces) and, unless otherwise stated, the estimation is carried over $N = 10^4$ sample paths.

4.1   Overflow Probability

It is well–known that no explicit expressions are available for the overflow probability in the case of general (correlated) Gaussian input. The only analytical results deal with asymptotic conditions, known in the literature as "large buffer regime" (when $B \to \infty$) and "many source regime" (when a large number n of i.i.d. flows are merged and queueing resources – buffer size and service rate – are scaled accordingly). Such results are typically derived in the LDT framework and provide just logarithmic asymptotics (see [12] and references therein), while the calculation of exact asymptotics is a much more complicated problem and some results have been drawn only for special processes, such as FBM (see [15] and references therein).

Figures 1 and 2 refer to the "many sources regime" with buffer size $B = nb$ (where $b = 0.3$) and $r = 0.1$. In this case $N = 10^6$ sample paths have been used for the estimations and the results are compared with the exact asymptotic (in discrete time) derived in [10]:

$$\mathbb{P}_n \sim \Phi\left(V(t^*)\sqrt{n}\right) \qquad n \to \infty, \tag{17}$$

where

$$V(t) = \frac{b + rt}{\sqrt{v_t}} \tag{18}$$

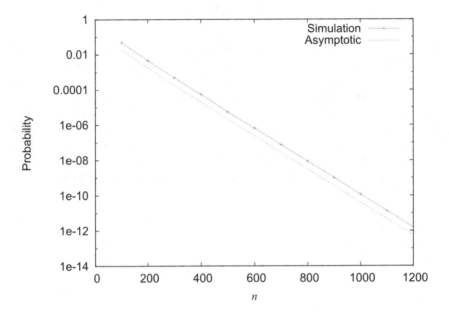

Fig. 1. Many sources regime: simulation vs. asymptotic

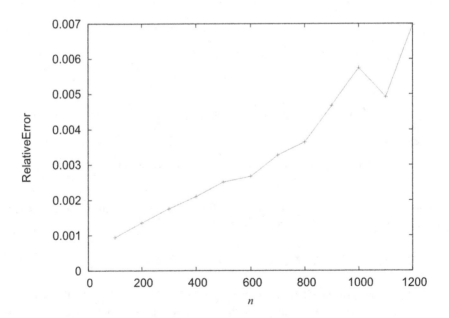

Fig. 2. Many sources regime: relative Error

and

$$t^* = \frac{H}{1-H} \cdot \frac{b}{r}$$

denotes the *most-likely time to overflow*. Heuristic arguments suggest setting the conditioning point τ in the BMC algorithm equal to t^*; indeed, it can be seen that the values of τ in a neighbourhood of t^* minimize the variance of the estimator [12], and so in the following simulations $\tau = t^*$ is used.

In more detail, Fig. 1 highlights the good consistency between theoretical values and simulation results over a wide range of the overflow probability values, while Fig. 2 shows the behavior of the RE of the BCM estimator: although the latter is not bounded, it grows slowly, and for probabilities of the order of 10^{-12} is still less than 1 % (compare the values in the above mentioned Figures).

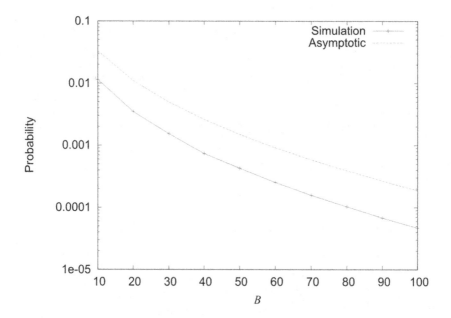

Fig. 3. Large buffer regime: simulation vs. asymptotic

Figure 3 refers to the large buffer regime (with $r = 1$) and compares the simulation results with the classical LDT bound [15]

$$\log \mathbb{P}_B \sim -\frac{V^2(\tau)}{2} \qquad B \to \infty, \tag{19}$$

where the expression of $V(\tau)$ is given by (18) with B instead on b. The slightly larger distance between the curves (with respect to Fig. 1) is not surprising since LDT provides only logarithmic asymptotics, which may differ from the *real* values by some subexponential (at ∞) function.

4.2   Duration of High Activity Periods

The distribution of high activity periods duration is less investigated than the overflow probability and it is not surprising that only a few asymptotic results based on LDT are available. For instance in the case of FBM input

$$\lim_{T \to \infty} \frac{1}{T^{2-2H}} \, \log \mathbb{P} \left(\forall t \in (0, T] : \, X_t > t \right)$$

$$= \lim_{n \to \infty} \frac{1}{n} \log \mathbb{P} \left(\forall t \in (0, 1] : \, \frac{X_t}{\sqrt{n}} > t \right) = - \inf_{f \in \mathscr{B}} I(f), \quad (20)$$

where

$$\mathscr{B} := \{ f \in \mathcal{R} : f(r) > r, \, \forall r \in (0, 1] \},$$

I denotes the rate function and $\mathcal{R}$ is the reproducing kernel Hilbert space associated with the distribution of FBM (see [1] for more details).

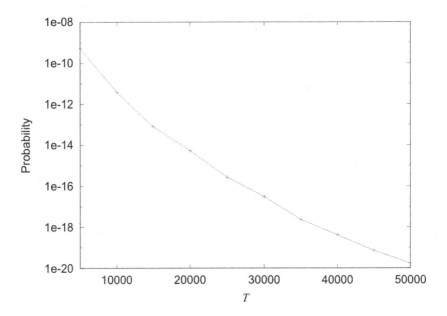

Fig. 4. Effect of T on $\pi(\mathbb{T})$

Figure 4 shows the dependence of $\pi(T)$ on the time duration T: in accordance with (20), the target probability exhibits an exponential decay. Also in this case (see Fig. 5), the relative error grows very slowly, and even for target probabilities of the order of 10^{-20} is still less than 30 %.

Finally, Fig. 6 highlights that the relative error decreases when the conditioning point τ approaches the interval length T (in our case $T = 5000$), as can be easily expected since our estimations cover long intervals and the event becomes rare as $T \to \infty$.

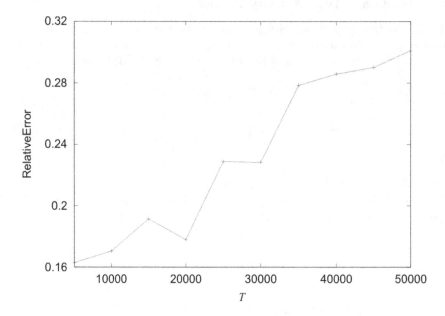

Fig. 5. Effect of the interval length on the relative error

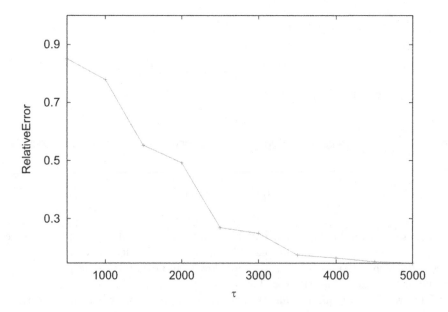

Fig. 6. Effect of the conditioning point on the relative error

5 Conclusions

Rare event simulation is an important research framework that lies between pure simulation and analytical techniques. Indeed, some additional knowledge of the underlying processes is needed in order to define efficient approaches that lead to variance reduction with respect to crude MC. In this paper, we analysed a variant of the conditional MC method based on the use of the bridge process (hence named Bridge MC) and showed how the BMC method, originally proposed for the estimation of the overflow probability, can be applied to the distribution of the length of high-activity periods.

The BMC simulation approach exploits the Gaussian nature of the input process (independence is equivalent to uncorrelatedness) and relies on the properties of bridges. Indeed, the bridge is independent of the random value of the original process at the conditioning time; hence, the target probability is obtained through a classical MC estimator, applied to an *adequate* conditional expectation (whose expression depends on the specific problem under investigation) of the bridge, that can be calculated via the tail function.

Several simulation experiments have been carried out to highlight the effectiveness of the BMC estimator in the various scenarios involving overflow events (in different asymptotic regimes) as well as long periods of high activity. As further research steps, it is possible to study the influence of the discretization step on simulation results (it is known in the literature that discrete and continuous time asymptotics may differ) and the asymptotic efficiency of the estimator.

References

1. Deuschel, J.D., Stroock, D.W.: Large Deviations. Academic Press, San Diego (1989)
2. Erramilli, A., Narayan, O., Willinger, W.: Experimental queueing analysis with long-range dependent packet traffic. IEEE/ACM Trans. Netw. 4(2), 209–233 (1996)
3. Ganesh, A., O'Connell, N., Wischik, D.: Big Queues. Lecture Notes in Mathematics. Springer, Heidelberg (2004)
4. Giordano, S., Gubinelli, M., Pagano, M.: Bridge Monte-Carlo: a novel approach to rare events of Gaussian processes. In: Proceedings of the 5th St. Petersburg Workshop on Simulation, St. Petersburg, Russia, pp. 281–286 (2005)
5. Giordano, S., Gubinelli, M., Pagano, M.: Rare events of Gaussian processes: a performance comparison between bridge Monte-Carlo and importance sampling. In: Koucheryavy, Y., Harju, J., Sayenko, A. (eds.) NEW2AN 2007. LNCS, vol. 4712, pp. 269–280. Springer, Heidelberg (2007)
6. Glasserman, P., Wang, Y.: Counterexamples in importance sampling for large deviations probabilities. Ann. Appl. Probab. 7(3), 731–746 (1997)
7. Heidelberger, P.: Fast simulation of rare events in queueing and reliability models. ACM Trans. Model. Comput. Simul. 5(1), 43–85 (1995)
8. Kouvatsos, D.D.: Performance Evaluation and Applications of ATM Networks. The Springer International Series in Engineering and Computer Science. Springer, New York (2006)

9. Kulkarni, V., Rolski, T.: Fluid model driven by an Ornstein-Uhlenbeck process. Probab. Eng. Inf. Sci. **8**, 403–417 (1994)
10. Likhanov, N., Mazumdar, R.: Cell loss asymptotics in buffers fed with a large number of independent stationary sources. J. Appl. Probab. **36**, 86–96 (1999)
11. Lukashenko, O., Morozov, E.: On convergence rate to stationarity of queues with general gaussian input. In: Remke, A., Manini, D., Gribaudo, M. (eds.) ASMTA 2015. LNCS, vol. 9081, pp. 130–142. Springer, Heidelberg (2015)
12. Lukashenko, O.V., Morozov, E.V., Pagano, M.: Performance analysis of Bridge Monte-Carlo estimator. Trans. KarRC RAS **3**, 54–60 (2012)
13. Lukashenko, O., Morozov, E., Pagano, M.: On conditional Monte Carlo estimation of busy period probabilities in Gaussian queues. In: Vishnevsky, V., et al. (eds.) DCCN 2015. CCIS, vol. 601, pp. 280–288. Springer, Heidelberg (2016). doi:10.1007/978-3-319-30843-2_29
14. Mandelbrot, B.B., Wallis, J.R.: Noah, Joseph, and operational hydrology. Water Resour. Res. **4**(5), 909–918 (1968)
15. Mandjes, M.: Large Deviations of Gaussian Queues. Wiley, Chichester (2007)
16. Norros, I.: On the use of fractional Brownian motion in the theory of connectionless networks. IEEE J. Sel. Areas Commun. **13**(6), 953–962 (1995)
17. Norros, I.: Busy periods for fractional Brownian storage: a large deviation approach. Adv. Perform. Anal. **2**, 1–19 (1999)
18. Ross, S.M.: Simulation. Elsevier, San Diego (2006)
19. Taqqu, M.S., Willinger, W., Sherman, R.: Proof of a fundamental result in self-similar traffic modeling. Comput. Commun. Rev. **27**, 5–23 (1997)
20. Villén-Altamirano, M., Villén-Altamirano, J.: The rare event simulation method RESTART: efficiency analysis and guidelines for its application. In: Kouvatsos, D.D. (ed.) Next Generation Internet: Performance Evaluation and Applications. LNCS, vol. 5233, pp. 509–547. Springer, Heidelberg (2011)
21. Willinger, W., Taqqu, M.S., Leland, W.E., Wilson, D.V.: Self-Similarity in high-speed packet traffic: analysis and modeling of ethernet traffic measurements. Stat. Sci. **10**(1), 67–85 (1995)

Stationary Distribution of Queueing Networks with Countable Set of Types of Batch Negative Customer Arrivals

Yury Malinkovsky[1,2(✉)]

[1] Francisk Skorina Gomel State University, Gomel, Belarus
malinkovsky@gsu.by, yury.malinkovsky@gmail.com
[2] National Research Tomsk State University, Tomsk, Russia

Abstract. Stationary distribution of a queueing network with countable set of types of batch negative customer arrivals is analyzed. The sufficient condition for ergodicity of the isolated node is established. Stationary product-form distribution of network states is found. The given network model is a generalization of a classic G-network model on the case of countable types of negative customers.

Keywords: Queueing network · Positive customers · Negative customers · Ergodicity · Stationary distribution · Product-form

1 Isolated Node

The queueing system (isolated node) and the queueing network with finite quantity of the negative customer types were studied in [1]. The aim of this paper is to transfer the results of paper [1] on the case of countable quantity of the negative customer types. We consider a queueing system with an exponential single server and countable mutually independent arriving Poisson flows: positive customers with intensity λ^+ and negative customers of countable types with intensities $\lambda_1^-, \lambda_2^-, \ldots$ respectively. An arriving negative customer flow with number l instantly deletes (kills) exactly l positive customers if there is such a quantity in the system and deletes all positive customers if there are fewer than l customers in the system ($l = 1, 2, \ldots$). Negative customers and deleted positive customers instantly leave the system and do not exert influence on the system's behavior. We suppose that

$$\sum_{t=1}^{\infty} t\lambda_t^- < \infty.$$

System state $n(t)$ at moment t is the quantity of the positive customers in the system. Obviously $n(t)$ is a Markov chain with continuous time and state space Z_+. If its stationary distribution $\{p(n), \ n = 0, 1, \ldots\}$ exists then it satisfies the system of equilibrium equations

$$\lambda^+ p(0) = (\mu + \lambda_1^- + \lambda_2^- + \ldots)p(1) + (\lambda_2^- + \lambda_3^- + \ldots)p(2) + (\lambda_3^- + \lambda_4^- + \ldots)p(3) + \ldots,$$
$$(\lambda^+ + \mu + \lambda_1^- + \lambda_2^- + \ldots)p(n) = \lambda^+ p(n-1) + (\mu + \lambda_1^-)p(n+1)$$
$$+\lambda_2^- p(n+2) + \lambda_3^- p(n+3) + \ldots, \qquad n = 1, 2, \ldots.$$

© Springer International Publishing Switzerland 2016
A. Dudin et al. (Eds.): ITMM 2016, CCIS 638, pp. 221–227, 2016.
DOI: 10.1007/978-3-319-44615-8_19

It is easy to check that these equilibrium equations are equivalent to equilibrium equations for vertical sections in a transition graph:

$$\lambda^+ p(n) = (\mu + \lambda_1^- + \lambda_2^- + \ldots)p(n+1) + (\lambda_2^- + \lambda_3^- + \ldots)p(n+2)$$
$$+(\lambda_3^- + \lambda_4^- + \ldots)p(n+3) + \ldots, \qquad n = 0, 1, \ldots. \tag{1}$$

This is a homogenous linear difference equation of infinite order. The partial solution to (1) we are looking for in the form $p(n) = z^n$. Denoting $\Lambda^-(z) = \sum_{t=1}^{\infty} \lambda_t^- z^t$ and substituting $p(n) = z^n$ in Eq. (1) we obtain the characteristic equation

$$g(z) = \sum_{l=1}^{\infty} z^l \sum_{s=l}^{\infty} \lambda_s^- + \mu z - \lambda^+ = \sum_{s=1}^{\infty} \lambda_s^- \sum_{l=1}^{s} z^l + \mu z - \lambda^+$$

$$= z \frac{\Lambda^-(z) - \Lambda^-(1)}{z - 1} + \mu z - \lambda^+ = 0, \tag{2}$$

We will prove the sufficiency of condition

$$\rho = \frac{\lambda^+}{\mu + \sum_{t=1}^{\infty} t\lambda_t^-} < 1 \tag{3}$$

for ergodicity of the process $n(t)$. By this $g(0) = -\lambda^+ < 0$ and $g(1) = \left[\Lambda^-(z)\right]'_{z=1} + \mu - \lambda > 0$ on the strength (3). The function $g(z)$ is continuous in segment $[0,1]$. So the root of Eq. (2) $z_0 \in (0,1)$. Hence equilibrium equation (1) has the solution $p(n) = Cz_0^n$. From normalization condition one coincides with geometric distribution:

$$p(n) = (1 - z_0)z_0^n, \qquad n = 0, 1, \ldots. \tag{4}$$

We will use Foster ergodic theorem [2]. For an irreducible conservative regular Markov chain with continuous time to be ergodic it is necessary and sufficient that the system of equilibrium equations has a nonzero solution $\sum_{n=0}^{\infty} |p(n)| < \infty$. Equation (2) as we have seen has root $z_0 \in (0,1)$ when condition (3) is satisfied and it is being that (4) is the partial solution to equilibrium equations (1). The series $\sum_{n=0}^{\infty} |p(n)|$ converges as a sum of the geometric progression members with a ratio less than one. Obviously the chain is irreducible and conservative. Regularity follows from leaving rate $q(n)$ of process $n(t)$ from state n bounded by $\lambda^+ + \mu + \sum_{t=1}^{\infty} t\lambda_t^-$ [3]. Hence the condition (3) is sufficient for ergodicity $n(t)$ and when (3) holds then ergodic distribution has the form (4).

The condition (3) is also necessary for ergodicity of process $n(t)$. To show this we can use reasonings analogous to reasonings in paper [1]. But to establish that we must base a point of great nicety of passage to the limit as $T \to \infty$. An alternative approach is research of the zeros of generating function $P(z) = \sum_{n=1}^{\infty} p(n)z^n$.

denominator in circle $|z| < 1$, using Rouche's theorem and its Klimenok modification [4] and the Bernshtein theorem conserning absolute monotonicity of the generating function. But we will not prove the necessity of condition (3) for the ergodicity of process $n(t)$ because our general aim concerns the queueing network process.

We resume proved result.

Theorem 1. *The Markov chain $n(t)$ is regular. It is ergodic if inequality (3) holds. The stationary distribution of chain has a geometric distribution form (4) in this case.* $\Box$

2   Queueing Network

We consider a queueing network consisting of N single-line exponential nodes with service rate μ_i for the server of node i $(i = \overline{1, N})$. An infinite (countable) quantity of mutually independent Poisson flows arrive from outside to the network. More specifically a positive (usual) customer flow with parameter Λ_i and a countable number of negative customer flows with parameters λ_{il} arrive in node i $(i = \overline{1, N}, l = 1, 2, \ldots)$. The negative customers do not demand service. The arriving negative customer flow l instantly deletes (kills) exactly l positive customers if there is such a quantity in the node i and deletes all the positive customers if there are fewer than l customers in node i $(i = \overline{1, N}, l = 1, 2, \ldots)$. Negative customers and deleted positive customers instantly leave the network and do not exert influence on the network behavior. The positive customer served in node i instantly and independently of other customers moves to node j as positive with probability p_{ij}^{+}, as negative customer flow with number l with probability p_{ijl}^{-}, or arrives to the network with probability p_{i0} $(i, j = \overline{1, N}, l = 1, 2, \ldots)$ and $\sum_{j=1}^{N} \left(p_{ij}^{+} + \sum_{l=1}^{\infty} p_{ijl}^{-} \right) + p_{i0} = 1$ $(i = \overline{1, N})$. The quantity of places for the waiting of positive customers is unbounded. For distinctness we suppose that the positive customers are served in order of their arrival moments. We assume

$$\sum_{t=1}^{\infty} t\lambda_{it}^{-} < \infty, \qquad i = \overline{1, N}.$$

We will describe the state of network by random vector

$$\mathbf{n}(t) = (n_1(t), n_2(t), \ldots, n_N(t)),$$

where $n_i(t)$ is a quantity of positive customers in node i at time t. Because primitive assumptions about entering flows and service times distributions $\mathbf{n}(t)$ is a multidimensional Markov chain with continuous time and state space $X = Z_+^N$ where $Z_+ = \{0, 1, \ldots\}$. Assume $\mathbf{n}(t)$ is irreducible. For example we can assume all $\Lambda_i > 0$ and for every i exists l such that $\lambda_{il} > 0$. Our purpose is to establish the ergodic condition and to determine the stationary distribution.

We consider an isolated node believing customer flows arrive with rates like those rates of corresponding flows in the network (which is not Poisson). We add index i as first index corresponding to the node number to all notations for an isolated node of Sect. 1. The characteristic equation (2) with substituting of root z_{i0} of (2) become identity

$$\sum_{l=1}^{\infty}\sum_{s=l}^{\infty}\lambda_{is}^{-}z_{0i}^{l} + \mu_i z_{i0} - \lambda_i^{+} = 0. \tag{9}$$

If ergodic condition

$$\rho_i = \frac{\lambda_i^{+}}{\mu_i + \sum_{t=1}^{\infty} t\lambda_{it}^{-}} < 1, \quad i = \overline{1, N}, \tag{10}$$

holds it follows according to the results of Sect. 1 that the stationary distribution of isolated node has the form

$$p_i(n_i) = (1 - z_{i0})z_{i0}^{n_i}, \qquad n_i = 0, 1, \dots. \tag{11}$$

Hence the probability of full server employment in steady-state is z_{i0}. So flow intensities of positive and negative customers in the network satisfy the next traffic equations system:

$$\lambda_i^{+} = \Lambda_i + \sum_{j=1}^{N} \mu_j z_{j0} p_{ji}^{+}, \qquad i = \overline{1, N}, \tag{12}$$

$$\lambda_{il}^{-} = \lambda_{il} + \sum_{j=1}^{N} \mu_j z_{j0} p_{jil}^{-}, \qquad l = 1, 2, \dots, \quad i = \overline{1, N}. \tag{13}$$

By the continuity theorem of implicit function and Shauder generalization of Brauer fixed point theorem on the infinite dimensional case we can prove a positive solution to the traffic equations system (12) and (13) exists.

If the steady-state distribution $\{p(\mathbf{n}\}$ of Markov chain $\mathbf{n}(t)$ exists then one satisfies the global balance equations

$$p(\mathbf{n})\sum_{i=1}^{N}[\Lambda_i + (\mu_i + \lambda_{i1} + \lambda_{i2} + \dots)I_{\{n_i \neq 0\}}]$$

$$= \sum_{i=1}^{N}\Big\{p(\mathbf{n} - \mathbf{e}_i)\Lambda_i I_{\{n_i \neq 0\}} + p(\mathbf{n} + \mathbf{e}_i)[\mu_i p_{i0} + \lambda_{i1} + (\lambda_{i2} + \lambda_{i3} + \dots +)I_{\{n_i=0\}}]$$

$$+ \sum_{l=2}^{\infty}p(\mathbf{n} + l\mathbf{e}_i)[\lambda_{il} + (\lambda_{il+1} + \lambda_{il+2} + \dots)I_{\{n_i=0\}}] + \sum_{j=1}^{N}\Big[p(\mathbf{n} + \mathbf{e}_j - \mathbf{e}_i)\mu_j p_{ji}^{+}I_{\{n_i \neq 0\}}$$

$$+ \sum_{l=1}^{\infty}p(\mathbf{n} + \mathbf{e}_j + l\mathbf{e}_i)\mu_j(p_{jil}^{-} + (p_{jil+1}^{-} + p_{jil+2}^{-} + \dots)I_{\{n_i=0\}})$$

$$+ p(\mathbf{n} + \mathbf{e}_j)\mu_j(p_{ji1}^{-} + p_{ji2}^{-} + \dots)I_{\{n_i=0\}}\Big]\Big\}, \qquad \mathbf{n} \in Z_{+}^{N}. \tag{14}$$

Here $\mathbf{e}_i$ is a unit vector of direction i and I_A is an indicator of event A equal to 1 if event A occurs and to 0 if event A does not occur.

The main result has the next form.

Theorem 2. *Markov process* $\mathbf{n}(t)$ *is regular and if inequalities (10) hold then it is ergodic. Its stationary distribution* $\{p(\mathbf{n}\}$ *is defined by*

$$p(\mathbf{n}) = \prod_{i=1}^{n} p_i(n_i), \qquad \mathbf{n} \in Z_+^N,$$

where $p_i(n_i)$ *and* ρ_i *are defined by equalities (11) and (10) respectively,* z_{i0}, $i = \overline{1, N}$, *are the roots of Eq. (9) belonging to interval* $(0, 1)$.

Proof. We have $I_{\{n_i=0\}} = 1 - I_{\{n_i \neq 0\}}$, so (14) shapes

$$p(\mathbf{n}) \sum_{i=1}^{N} \left[\Lambda_i + (\mu_i + \lambda_{i1} + \lambda_{i2} + \ldots) I_{\{n_i \neq 0\}} \right]$$

$$= \sum_{i=1}^{N} \Big\{ p(\mathbf{n} - \mathbf{e}_i) \Lambda_i I_{\{n_i \neq 0\}} + p(\mathbf{n} + \mathbf{e}_i)(\mu_i p_{i0} + \lambda_{i1} + \lambda_{i2} + \ldots)$$

$$- p(\mathbf{n} + \mathbf{e}_i)(\lambda_{i2} + \lambda_{i3} + \ldots) I_{\{n_i \neq 0\}} + \sum_{l=2}^{\infty} p(\mathbf{n} + l\mathbf{e}_i)(\lambda_{il} + \lambda_{il+1} + \ldots)$$

$$- \sum_{l=2}^{\infty} p(\mathbf{n} + l\mathbf{e}_i)(\lambda_{il+1} + \lambda_{il+2} + \ldots) I_{\{n_i \neq 0\}}$$

$$+ \sum_{j=1}^{N} \Big[p(\mathbf{n} + \mathbf{e}_j - \mathbf{e}_i) \mu_j p_{ji}^+ I_{\{n_i \neq 0\}} + \sum_{l=1}^{\infty} p(\mathbf{n} + \mathbf{e}_j + l\mathbf{e}_i) \mu_j (p_{jil}^- + p_{ji2}^- + \ldots)$$

$$- \sum_{l=1}^{\infty} p(\mathbf{n} + \mathbf{e}_j + l\mathbf{e}_i) \mu_j (p_{jil+1}^- + p_{jil+2}^- + \ldots) I_{\{n_i \neq 0\}} + p(\mathbf{n} + \mathbf{e}_j) \mu_j (p_{ji1}^- + p_{ji2}^- + \ldots)$$

$$- p(\mathbf{n} + \mathbf{e}_j) \mu_j (p_{ji1}^- + p_{ji2}^- + \ldots) I_{\{n_i \neq 0\}} \Big] \Big\}, \qquad \mathbf{n} \in Z_+^N. \tag{15}$$

We partition this equation into local balance equations. The sum of terms in the left side of (15) including factor $I_{\{n_i \neq 0\}}$ equates to the same sum on the right side of (15). After the sum of terms on the left side of (15) not containing factor $I_{\{n_i \neq 0\}}$ equates to the same sum on the right side of (15):

$$p(\mathbf{n}) \sum_{i=1}^{N} \Lambda_i = \sum_{i=1}^{N} \Big\{ p(\mathbf{n} + \mathbf{e}_i)(\mu_i p_{i0} + \lambda_{i1} + \lambda_{i2} + \ldots) + \sum_{l=2}^{\infty} p(\mathbf{n} + l\mathbf{e}_i)(\lambda_{il} + \lambda_{i2} + \ldots)$$

$$+ \sum_{j=1}^{N} \Big[\sum_{l=1}^{\infty} p(\mathbf{n} + \mathbf{e}_j + l\mathbf{e}_i) \mu_j (p_{jil}^- + p_{ji2}^- + \ldots) + p(\mathbf{n} + \mathbf{e}_j) \mu_j (p_{ji1}^- + p_{ji2}^- + \ldots) \Big] \Big\}. \tag{16}$$

$$p(\mathbf{n}) \sum_{i=1}^{N} (\mu_i + \lambda_{i1} + \lambda_{i2} + \ldots) = \sum_{i=1}^{N} \Big\{ p(\mathbf{n} - \mathbf{e}_i) \Lambda_i - p(\mathbf{n} + \mathbf{e}_i)(\lambda_{i2} + \lambda_{i3} + \ldots)$$

$$- \sum_{l=2}^{\infty} p(\mathbf{n} + l\mathbf{e}_i)(\lambda_{il+1} + \lambda_{il+2} + \ldots) + \sum_{j=1}^{N} \Big[p(\mathbf{n} + \mathbf{e}_j - \mathbf{e}_i) \mu_j p_{ji}^+$$

$$- \sum_{l=1}^{\infty} p(\mathbf{n} + \mathbf{e}_j + l\mathbf{e}_i) \mu_j (p_{jil+1}^- + p_{jil+2}^- + \ldots) - p(\mathbf{n} + \mathbf{e}_j) \mu_j (p_{ji1}^- + p_{ji2}^- + \ldots) \Big] \Big\}.$$

We partition the previous equations into more detailed balance equations:

$$p(\mathbf{n})(\mu_i + \lambda_{i1} + \lambda_{i2} + \ldots) = p(\mathbf{n} - \mathbf{e}_i)\Lambda_i - p(\mathbf{n} + \mathbf{e}_i)(\lambda_{i2} + \lambda_{i3} + \ldots)$$

$$-\sum_{l=2}^{\infty} p(\mathbf{n} + l\mathbf{e}_i)(\lambda_{il+1} + \ldots + \lambda_{iT}) + \sum_{j=1}^{N}\Big[p(\mathbf{n} + \mathbf{e}_j - \mathbf{e}_i)\mu_j p_{ji}^{+}$$

$$-\sum_{l=1}^{\infty} p(\mathbf{n} + \mathbf{e}_j + l\mathbf{e}_i)\mu_j(p_{jil+1}^{-} + p_{jil+2}^{-} + \ldots) - p(\mathbf{n} + \mathbf{e}_j)\mu_j(p_{ji1}^{-} + p_{ji2}^{-} + \ldots)\Big]. \quad (17)$$

Let probabilities $p_i(n_i)$ be defined by equalities (13). We will prove that

$$p(\mathbf{n}) = p_1(n_1)p_2(n_2)\ldots p_N(n_N), \qquad \mathbf{n} \in Z_{+}^{N}, \tag{18}$$

is solution of local balance equations (16) and (17), that is global balance equations (14). We will divide both sides of (16) on $p(\mathbf{n})$ and use (10), (11), (13)–(15):

$$\sum_{i=1}^{N}\Lambda_i = \sum_{i=1}^{N}\Big\{z_{i0}(\mu_i p_{i0} + \lambda_{i1} + \lambda_{i2} + \ldots) + \sum_{l=2}^{\infty}z_{i0}^{l}(\lambda_{il} + \lambda_{il+1} + \ldots)$$

$$+\sum_{j=1}^{N}\Big[\sum_{l=1}^{\infty}z_{j0}z_{i0}^{l}\mu_j(p_{jil}^{-} + p_{jil+1}^{-} + \ldots) + z_{j0}\mu_j(p_{ji1}^{-} + p_{ji2}^{-} + \ldots)\Big]\Big\}$$

$$= \sum_{i=1}^{N}\Big[z_{i0}\mu_i p_{i0} + \sum_{l=1}^{\infty}z_{i0}^{l}(\lambda_{il} + \lambda_{il+1} + \ldots)$$

$$+\sum_{l=1}^{\infty}z_{i0}^{l}(\lambda_{il}^{-} - \lambda_{il} + \lambda_{il+1}^{-} - \lambda_{il+1} + \ldots +) + \lambda_{i1}^{-} - \lambda_{i1} + \lambda_{i2}^{-} - \lambda_{i2} + \ldots\Big]$$

$$= \sum_{i=1}^{N}\Big[z_{i0}\mu_i p_{i0} + \sum_{l=1}^{\infty}z_{i0}^{l}(\lambda_{il} + \lambda_{il+1} + \ldots) + \lambda_{i1}^{-} - \lambda_{i1} + \lambda_{i2}^{-} - \lambda_{i2} + \ldots\Big]$$

$$= \sum_{i=1}^{N}\Big[z_{i0}\mu_i p_{i0} + \lambda_{i}^{+} - \mu_i z_{i0} + \lambda_{i1}^{-} - \lambda_{i1} + \lambda_{i2}^{-} - \lambda_{i2} + \ldots\Big]$$

$$= \sum_{i=1}^{N}\Big[z_{i0}\mu_i p_{i0} + \Lambda_i + \sum_{j=1}^{N}z_{j0}\mu_j p_{ji}^{+} - \mu_i z_{i0} + \lambda_{i1}^{-} - \lambda_{i1} + \lambda_{i2}^{-} - \lambda_{i2} + \ldots\Big]$$

$$= \sum_{i=1}^{N}\Big[z_{i0}\mu_i p_{i0} + \Lambda_i + \sum_{j=1}^{N}z_{j0}\mu_j\Big(1 - \sum_{l=1}^{\infty}p_{jil}^{-} - p_{j0}\Big) - \mu_i z_{i0} + \lambda_{i1}^{-} - \lambda_{i1} + \lambda_{i1+1}^{-} - \lambda_{i1+1} + \ldots\Big]$$

$$= \sum_{i=1}^{N}\Big[\Lambda_i + \lambda_{i1}^{-} - \lambda_{i1} + \lambda_{i2}^{-} - \lambda_{i2} + \ldots - \sum_{l=1}^{\infty}\sum_{j=1}^{N}z_{j0}\mu_j p_{jil}^{-}\Big]$$

$$= \sum_{i=1}^{N}\Big[\Lambda_i + \lambda_{i1}^{-} - \lambda_{i1} + \lambda_{i2}^{-} - \lambda_{i2} + \ldots - \sum_{l=1}^{\infty}(\lambda_{il}^{-} - \lambda_{il})\Big] = \sum_{i=1}^{N}\Lambda_i,$$

that is (17) becomes identity. In much the same way we check the implementation of local balance equation (18):

$$\mu_i + \lambda_{i1} + \lambda_{i1+1} + \ldots = \frac{\Lambda_i}{z_{i0}} - z_{i0}(\lambda_{i2} + \lambda_{i3} + \ldots) - \sum_{l=2}^{\infty} z_{i0}^l(\lambda_{il+1} + \lambda_{il+2} + \ldots)$$

$$+\sum_{j=1}^{N}\left[\frac{z_{j0}}{z_{i0}}\mu_j p_{ji}^+ - \sum_{l=1}^{\infty} z_{j0} z_{i0}^l \mu_j(p_{jil+1}^- + p_{jil+2}^- + \ldots) - z_{j0}\mu_j(p_{ji1}^- + p_{ji1+1}^- + \ldots)\right]$$

$$= \frac{\Lambda_i}{z_{i0}} - \sum_{l=1}^{\infty} z_{i0}^l(\lambda_{il+1} + \lambda_{il+2} + \ldots) + \frac{\lambda_i^+ - \Lambda_i}{z_{i0}}$$

$$-\sum_{l=1}^{\infty} z_{i0}^l(\lambda_{il+1}^- - \lambda_{il+1} + \lambda_{il+2}^- - \lambda_{il+2} + \ldots) - (\lambda_{i1}^- - \lambda_{i1} + \lambda_{i1+1}^- - \lambda_{i1+1} + \ldots)$$

$$= \frac{\lambda_i^+}{z_{i0}} - \sum_{l=0}^{\infty} z_{i0}^l(\lambda_{il+1} + \lambda_{il+2} + \ldots) + \lambda_{i1} + \lambda_{i2} + \ldots$$

$$= \mu_i + \frac{1}{z_{i0}}\sum_{l=1}^{\infty}\sum_{s=l}^{\infty}\lambda_{is}^- z_{i0}^l - \sum_{l=0}^{\infty}\sum_{s=l+1}^{\infty}\lambda_{is}^- z_{i0}^l + \lambda_{i1} + \lambda_{i2} + \ldots = \mu_i + \lambda_{i1} + \lambda_{i2} + \ldots.$$

Using Foster ergodic theorem [2] completes the proof. □

3 Conclusion

We have considered the stationary functioning of an open queueing network with batch arrivals of negative customers. The expression for stationary distribution has been derived in product form. The given network model is a generalization of a classic G-network model [5] on the case of infinite (countable) types of negative customers. The research results have practical importance and may be used for real networks investigation.

References

1. Malinkovsky, Y.: Stationary distribution of the queueing networks with batch negative customer arrivals. In: Dudin, A., Nazarov, A., Yakupov, R. (eds.) ITMM 2015. CCIS, vol. 564, pp. 53–63. Springer, Switzerland (2015)
2. Bocharov, P.P., Pechinkin, A.V.: Queueing Theory, 529 p. RUDN, Moskow (1995). (in Russian)
3. Gikhman, I.I., Skorohod, A.V.: Introduction in Theory of Stochastic Processes, 568 p. Nauka, Moskow (1977). (in Russian)
4. Klimenok, V.: On the modification of Rouche's theorem for the queueing theory problems. Queueing Syst. **38**, 431–434 (2001)
5. Gelenbe, E.: Product-form queueing networks with negative and positive customers. J. Appl. Prob. **28**, 656–663 (1991)

Probability Properties of Interest Rate Models

Gennady Medvedev[(✉)]

Department of Applied Mathematics and Computer Science,
Belarusian State University, 220030 Minsk, Belarus
medvedevga@bsu.by

Abstract. The processes of short-term interest rates generate changes in most market indices, as well as form the basis of determining the value of marketable assets and commercial contracts. They play a special role in calculating the term structure of the yield. Therefore, the development of mathematical models of these processes is extremely interesting for financial analysts and researchers of market issues. There are many versions of change of short-term risk-free interest rates in the framework of the theory of diffusion processes. However, there is still no such model, which would be the basis for building a term structure of yields close to that existing in a real financial market. It is interesting to analyze the existing models in order to clarify features of models in a probabilistic sense in more detail than has been done by their creators and users. Such an analysis will be made here for the family of models used by the authors in three well-known papers [1–3], where they were applied for the fitting of the real time series of yield.

Keywords: Yield · Short-term risk-free interest rates · Term structure

1 Introduction

All the models considered belong to the class of diffusion models, that generate processes $X(t)$, described by the equation

$$dX(t) = \mu(X(t))dt + \sigma(X(t))dW(t), \ t > t_0, \ X(t_0) = X_0, \tag{1}$$

where a specific determination of drift $\mu(x)$ and volatility $\sigma(x)$ defines one or another particular model. Some models, such as models: Vasicek, Cox-Ingersoll-Ross, geometric Brownian motion, Ahn-Gao, are well documented in the literature, but nevertheless their properties are listed here for convenience of comparison with other, less well-known or not investigated models. The analysis is the first part of the work devoted to the explanation of the most suitable short-term rate models to determine the term structure of a zero-coupon yield that is reproducing the actually observed yield, as far as possible, the best way.

2 The Vasicek Model [4]

For $\mu(x) = k(\theta - x)$, $\sigma^2(x) = \sigma^2$ the Eq. (1) generates the Ornstein-Uhlenbeck process that is known in finance literature as the Vasicek model. Probability

© Springer International Publishing Switzerland 2016
A. Dudin et al. (Eds.): ITMM 2016, CCIS 638, pp. 228–237, 2016.
DOI: 10.1007/978-3-319-44615-8_20

density for this process is normal with the expectation $E[X] = \theta$ and the variance $Var[X] = \dfrac{\sigma^2}{2k}$:

$$f(x) = \sqrt{\frac{k}{\pi}\frac{1}{\sigma}}e^{-k\frac{(x-\theta)^2}{\sigma^2}}. \tag{2}$$

3   The CIR Model [5]

When the functions of drift and volatility are $\mu(x) = k(\theta - x)$ and $\sigma^2(x) = \sigma^2 x$ from (1) for the short-term interest rate $r(t)$ a nonnegative process is obtained. In financial literature such a process is named the Cox-Ingersoll-Ross model (the CIR model).

$$dr(t) = k(\theta - r(t))dt + \sqrt{2kD\frac{r(t)}{\theta}}dW(t),$$

where θ and D are the stationary expectation and variance respectively.

The CIR process has a gamma distribution with the scale parameter $c = \dfrac{2k}{\sigma^2}$ and the form parameter $q = \dfrac{2k\theta}{\sigma^2}$. So

$$f(x) = \frac{c^q x^{q-1}}{\Gamma(q)}e^{-cx}, \quad q > 0, \ x > 0. \tag{3}$$

The moments of this distribution are calculated by the formula

$$E[X^m] = \frac{\Gamma(m+q)}{c^m \Gamma(q)},$$

and important numerical characteristics are the expectation $E[X]$, the variance $Var[X]$, the skewness S and the kurtosis K:

$$E[X] = \frac{q}{c} = \theta,$$

$$Var[X] \equiv D = \frac{q}{c^2} = \frac{\sigma^2 \theta}{2k},$$

$$S \equiv \frac{E\left[(X - E[X])^3\right]}{Var[X]^{\frac{3}{2}}} = 2\sqrt{q},$$

$$K \equiv \frac{E\left[(X - E[X])^4\right]}{Var[X]^2} = 3 + \frac{6}{q}.$$

4   The Duffie-Kan Model [6]

In the Duffie-Kan model the rate $r(t)$ is generated by Eq. (1) with functions $\mu(x) = k(\theta - x)$ and $\sigma(x) = \sqrt{\gamma x + \delta} \equiv \sqrt{2kD\dfrac{x - r_0}{\theta - r_0}}$:

$$dr(t) = (\alpha r(t) + \beta)dt + \sqrt{\gamma r(t) + \delta}dW(t), \quad \gamma r(0) + \delta > 0,$$

where $k = -\alpha > 0$, $\theta = -\dfrac{\beta}{\alpha} > 0$, $D = \dfrac{\beta\gamma - \alpha\delta}{2\alpha^2} > 0$, $r_0 = -\dfrac{\delta}{\gamma} < \theta$.

The process $r(t)$ has the stationary probability density $f(x)$ which is a shifted gamma density with the shift parameter r_0, the scale parameter c and the form parameter q, i.e.

$$f(x) = \frac{c^q(x - r_0)^{q-1}}{\Gamma(q)}e^{-c(x-r_0)}, \quad r_0 < x < \infty, \qquad (4)$$

where $q = \dfrac{(\theta - r_0)^2}{D}$, $c = \dfrac{(\theta - r_0)}{D} > 0$, r_0 is the limit bottom value of interest rate $r(t)$.

The important numerical characteristics of the stationary density

$$E[X] = \frac{q}{c} = \theta,$$

$$Var\,[X] \equiv D = \frac{q}{c^2},$$

$$S = 2\sqrt{q},$$

$$K = 3 + \frac{6}{q}.$$

5   The Ahn-Gao Model [2]

In the Ahn-Gao model it is assumed that drift and volatility are nonlinear functions $\mu(x) = k(\theta - x)x$ and $\sigma^2(x) = \sigma^2 x^3$. Such a process has the stationary probability density $f(x)$ of form

$$f(x) = \frac{c^q}{\Gamma(q)x^{1+q}}e^{-\frac{c}{x}}, \quad x > 0, \qquad (5)$$

where the scale parameter $c = \dfrac{2k\theta}{\sigma^2}$ and the form parameter $q = 2 + \dfrac{2k}{\sigma^2}$. The process of the Ahn-Gao model can be obtained from the CIR process by transformation $X_{AG} = \dfrac{1}{X_{CIR}}$. The important numerical characteristics of the stationary density of process are determined by formulae

$$E[X] = \frac{c}{q - 1} = \frac{2k\theta}{2k + \sigma^2},$$

$$Var\,[X] = \frac{c^2}{(q - 1)^2(q - 2)} = \frac{2k\sigma^2\theta^2}{(2k + \sigma^2)^2},$$

$$S = 4\frac{\sqrt{q - 2}}{q - 3},$$

$$K = 3\frac{(q - 2)(q + 5)}{(q - 3)(q - 4)}.$$

6   The BDT Model [7]

The Black-Derman-Toy (BDT) model

$$dr(t) = [\alpha_1 r(t) - \alpha_2 r(t) \ln r(t)]dt + \beta r(t)dW(t), \ \alpha_2 > 0,$$

by transformation $Y(t) = \ln r(t)$ reduces to linear form

$$dY(t) = \left(\alpha_1 - \frac{\beta^2}{2} - \alpha_2 Y(t) \right) dt + \beta dW(t).$$

This equation allows a stationary solution and process $Y(t)$ that is found in explicit form

$$Y(t) = \frac{1}{\alpha_2} \left(\alpha_1 - \frac{\beta^2}{2} \right) + \xi(t), \ \xi(t) = \beta \int\limits_{-\infty}^{t} e^{-\alpha_2 s} dW(s),$$

where $\xi(t)$ is a stochastic Gaussian process with zero expectation, variance $Var\left[\xi(t)\right] = \frac{\beta^2}{2\alpha_2}$ and covariance $Cov\left[t_1, t_2\right] = \frac{\beta^2}{2\alpha_2} e^{-\alpha_2 |t_2 - t_1|}$. Thus the BDT model generates a log-normal process and allows a stationary regime. The leading stationary moments of the interest rate are calculated by formulae

$$E[r] = e^{\frac{1}{\alpha_2} \left(\alpha_1 - \frac{\beta^2}{4} \right)},$$

$$Var\,[r] = (\lambda - 1)e^{\frac{2}{\alpha_2} \left(\alpha_1 - \frac{\beta^2}{4} \right)}, \ \lambda = e^{\frac{\beta^2}{2\alpha_2}},$$

$$S = (\lambda + 2)\sqrt{\lambda - 1},$$

$$K = \lambda^4 + 2\lambda^3 + 3\lambda^2 - 3.$$

7   The Ait-Sahalia Model [8]

Ait-Sahalia has tested the based models of short interest rates (including those described here) by fitting them to the actually time series of rates. It was found that an acceptable level of goodness-of-fit of all these rates was rejected because of the drift and volatility properties. As a result he proposed the following functions of drift and diffusion

$$\mu(r) = \alpha_0 + \alpha_1 r + \alpha_2 r^2 + \alpha_{-1}\frac{1}{r}, \ \sigma^2(r) = \beta_0 + \beta_1 r + \beta_2 r^2.$$

In this model, the non-linear functions of drift and diffusion allow a wide variety of forms. To $\sigma^2(r) > 0$ for any r, it is necessary that the diffusion function parameters ensure the fulfilment of inequalities

$$\beta_0 > 0, \ \beta_2 > 0, \ \gamma^2 \equiv 4\beta_0\beta_2 - \beta_1^2 \geqslant 0.$$

Relevant to this function a probability density is given by the expression

$$f(x) = Nx^B(\beta_0 + \beta_1 x + \beta_2 x^2)^{C-1} e^{Ax + \text{Garctg}\,(E+Fx)}, \ x > 0,$$

where N is the normalization constant,

$$A = \frac{2\alpha_2}{\beta_2} < 0, \ B = \frac{2\alpha_{-1}}{\beta_0} > 0, \ C = \frac{\alpha_1}{\beta_2} - \frac{\alpha_2\beta_1}{\beta_2^2} - \frac{\alpha_{-1}}{\beta_0},$$

$$G = \frac{2\left(2\alpha_0 + \dfrac{\alpha_2\beta_1^2}{\beta_2^2} - \dfrac{\alpha_1\beta_1}{\beta_2} - \dfrac{2\alpha_2\beta_0}{\beta_2} - \dfrac{\alpha_{-1}\beta_1}{\beta_0}\right)}{\gamma}, \ E = \frac{\beta_1}{\gamma}, \ F = \frac{\beta_2}{\gamma}.$$

Since the density $f(x)$ at $x \to 0$ has order $O(x^B)$, $B > 0$, and at $x \to \infty$ its order is $O(x^{B+C}e^{Ax})$, $A < 0$, then for every finite m the moments $E[X^m]$ exist, but their analytical expressions cannot be obtained, and they can be calculated only numerically.

8   The CKLS Model [1]

In the Chan-Karolyi-Longstaff-Sanders (CKLS) model it is assumed that $\mu(x) = k(\theta - x)$, $\sigma^2(x) = \sigma^2 x^3$. It turns out that a random process corresponding to this model has a stationary density

$$f(x) = \frac{n}{x^3} e^{-c\left(\left(\frac{\theta}{x}\right)^2 - 2\frac{\theta}{x}\right)}, \ x > 0, \tag{6}$$

where $c = \dfrac{k}{\theta\sigma^2}$, n is normalization constant. Note that such random process has only the first stationary moment $E[X] = \theta$.

9   The Unrestricted Model I [2]

In "unrestricted model I"

$$dr = (\alpha_1 + \alpha_2 r + \alpha_3 r^2)dt + \sqrt{\alpha_4 + \alpha_5 r + \alpha_6 r^3}dW \tag{7}$$

all the preceding models are embedded, that is, at a certain setting parameters $\{\alpha\}$ can get any of the previous models. The table in this case has the form

Restrictions of parameters	Model	Equation of processes
$\alpha_3 = \alpha_5 = \alpha_6 = 0$	Vasicek	$dr = k(\theta - r)dt + \sigma dW$
$\alpha_3 = \alpha_4 = \alpha_6 = 0$	CIR	$dr = k(\theta - r)dt + \sigma\sqrt{r}dW$
$\alpha_3 = \alpha_6 = 0$	Duffie-Kan	$dr = k(\theta - r)dt + \sqrt{\alpha + \beta r}dW$
$\alpha_1 = \alpha_4 = \alpha_5 = 0$	Ahn-Gao	$dr = k(\theta - r)rdt + \sigma r^{1.5}dW$
$\alpha_3 = \alpha_4 = \alpha_5 = 0$	CKLS	$dr = k(\theta - r)dt + \sigma r^{1.5}dW$

Stationary probability density "unrestricted I" process has the form

$$f(x) = \frac{c(w)}{\sigma^2(x)} e^{w \int\limits^{x} \frac{2\mu(u)}{\sigma^2(u)} du} = \frac{c(w)}{\alpha_4 + \alpha_5 x + \alpha_6 x^3} e^{w \int\limits^{x} \frac{2(\alpha_1 + \alpha_2 u + \alpha_3 u^2)}{\alpha_4 + \alpha_5 u + \alpha_6 u^3} du},$$

where $c(w)$ is the normalization constant, w is a fixed number from the set of possible values of a random process, the specific value of which does not play some role.

Getting the explicit form of expression for $f(x)$ is possible, but it will be quite cumbersome in a general case, and we restrict ourselves to the case when the values of the parameters $\{\alpha\}$ provide the performance properties of the probability density $f(x)$. First, we note that the volatility of the real process needs to be a real function, so $\sigma^2(r) = \alpha_4 + \alpha_5 r + \alpha_6 r^3 \geqslant 0$ for all values of r. At the same time analytic properties of the probability density depend on the type of the roots of equation $\alpha_4 + \alpha_5 r + \alpha_6 r^3 = 0$, $\alpha_6 > 0$. The sign of the discriminant $\Delta = \left(\dfrac{\alpha_5}{3\alpha_6}\right)^3 + \left(\dfrac{\alpha_4}{2\alpha_6}\right)^2$ specifies the number of real and complex roots of the equation. When $\Delta > 0$, there is one real and two complex conjugate roots. When $\Delta < 0$, there are three different real roots. When $\Delta = 0$, real roots are multiples.

Let $\Delta > 0$ and the real root is $r = r_0$, then we can write

$$\alpha_4 + \alpha_5 r + \alpha_6 r^3 = \alpha_6 (r - r_0)\left(r^2 + pr + q\right),$$

where r_0, p and q are a relatively sophisticated analytical expression and because of that are not listed here. However, if $\alpha_4 = 0$, then $r_0 = 0$, $p = 0$, $q = \dfrac{\alpha_5}{\alpha_6}$. In this case, the probability density is given by

$$f(x) = \frac{c(w)}{\alpha_6 x \left(x^2 + \dfrac{\alpha_5}{\alpha_6}\right)} e^{w \int\limits^{x} \frac{2(\alpha_1 + \alpha_2 u + \alpha_3 u^2)}{\alpha_6 u \left(u^2 + \frac{\alpha_5}{\alpha_6}\right)} du}$$

$$= n x^{\frac{2\alpha_1}{\alpha_5} - 1} \left(\alpha_6 x^2 + \alpha_5\right)^{\frac{\alpha_3}{\alpha_6} - \frac{\alpha_1}{\alpha_5} - 1} e^{\frac{2\alpha_2}{\sqrt{\alpha_5 \alpha_6}} \operatorname{arctg}\left[x\sqrt{\frac{\alpha_6}{\alpha_5}}\right]}, \quad (8)$$

where n is the normalization constant. For the existence of the probability density its parameters must satisfy the inequalities: $\dfrac{\alpha_1}{\alpha_5} > 1$, $\dfrac{\alpha_3}{\alpha_6} < 1$. In order to at the same time there exist stationary moments it is necessary for the expectation $\dfrac{\alpha_3}{\alpha_6} < 0.5$, for variance $\dfrac{\alpha_3}{\alpha_6} < 0$, for the third moment $\dfrac{\alpha_3}{\alpha_6} < -0.5$ and for the fourth moment $\dfrac{\alpha_3}{\alpha_6} < -1$.

If $\Delta < 0$, denote the roots of the equation $r_0 > r_1 > r_2$ so

$$\alpha_4 + \alpha_5 r + \alpha_6 r^3 = \alpha_6 (r - r_0)(r - r_1)(r - r_2).$$

Then the probability density is expressed in the form

$$f(x) = n(x - r_0)^{\frac{2(\alpha_1 + \alpha_2 r_0 + \alpha_3 r_0^2)}{\alpha_6(r_0 - r_1)(r_0 - r_2)} - 1}$$

$$\times (x - r_1)^{-\frac{2(\alpha_1 + \alpha_2 r_1 + \alpha_3 r_1^2)}{\alpha_6(r_0 - r_1)(r_1 - r_2)} - 1} (x - r_2)^{\frac{2(\alpha_1 + \alpha_2 r_2 + \alpha_3 r_2^2)}{\alpha_6(r_0 - r_2)(r_1 - r_2)} - 1}. \quad (9)$$

In this case the inequalities must be performed

$$2(\alpha_1 + \alpha_2 r_0 + \alpha_3 r_0^2) > \alpha_6(r_0 - r_1)(r_0 - r_2), \quad \frac{\alpha_3}{\alpha_6} < 1.$$

For the existence of the m-th moment other than that necessary to perform the conditions $\dfrac{m}{2} + \dfrac{\alpha_3}{\alpha_6} < 1$. Unfortunately, the analytical expression of the normalization constant n and moments $E[r^m]$ is very cumbersome and they includes hypergeometric functions. Under these assumptions the process with such a density has a bottom line equal to the largest root, i.e. $r(t) \geqslant r_0$.

10   The Unrestricted Model II [1]

In the "unrestricted model II" process of short rate follows the equation

$$dr = k(\theta - r)dt + \sigma r^\gamma dW, \ \gamma > 0. \quad (10)$$

Therefore $\mu(x) = k(\theta - x)$, $\sigma^2(x) = \sigma^2 x^{2\gamma}$ and the stationary density $f(x)$ has form

$$f(x) = \frac{n}{x^{2\gamma}} e^{\frac{1}{x^{2\gamma}}\left(\frac{qx}{1-2\gamma} - \frac{cx^2}{2-2\gamma}\right)}, \ x > 0, \quad (11)$$

where $q = \dfrac{2k\theta}{\sigma^2}$, $c = \dfrac{2k}{\sigma^2}$, n is the normalization constant. The values of parameter γ, allowing the convergence of the integral of $f(x)$ on the interval $(0, \infty)$, determined by the inequality $\gamma > 0.5$. At the same time, there are two critical points: $\gamma = 0.5$ (in this case, the model is transformed into a short-term rate model CIR) and $\gamma = 1$, when the probability density is reduced to a form that corresponds to process of the Brennan-Schwartz model [9]

$$f(x) = \frac{q^{1+c}}{x^{2+c}\Gamma(1+c)} e^{-\frac{q}{x}}, \ x > 0. \quad (12)$$

When $\gamma = 1.5$, the "unrestricted model II" is known as the CKLS model. The Vasicek model is also a model embedded in the "unrestricted model II" at $\gamma = 0$. For existence of moments of order m, it is necessary the fulfilment of inequality $2\gamma > m + 1$. Unfortunately, the expression for the probability density in the general case does not allow the calculation of moments in analytical form, although for referred particular cases they are simply calculated. For the model CIR

$$E[X^m] = \frac{\Gamma(m+q)}{c^m \Gamma(q)},$$

for Brennan-Schwartz model

$$E[X^m] = q^m \frac{\Gamma(1+c-m)}{\Gamma(1+c)},$$

the moments of order m exist if the inequality $m < 1 + c$ is fulfilled. So that

Model	γ	$E[X]$	$Var[X]$	Skewness	Kurtosis
Vasicek	0	θ	$\dfrac{\sigma^2}{2k}$	0	3
CIR	0.5	$\dfrac{q}{c} = \theta$	$\dfrac{q}{c^2} = \dfrac{\sigma^2\theta}{2k}$	$2\sqrt{q}$	$3 + \dfrac{6}{q}$
Brennan-Schwartz	1.0	$\dfrac{q}{c} = \theta$	$\dfrac{\theta^2}{c-1}$	$\dfrac{4\sqrt{c-1}}{c-2}$	$\dfrac{3(c-1)(c+6)}{(c-2)(c-3)}$
CKLS	1.5	$\dfrac{q}{c} = \theta$	not exist	not exist	not exist

Even before the appearance of the "unrestricted model II" models were used, which then turned out to be special cases of this model. This is the model of the CIR (1980) [10], which is obtained from the Eq. (10), if we assume that $\gamma = 1.5$ and $k = 0$. Another particular version is the CEV model, i.e. the model of constant elasticity of variance that was proposed J. Cox and S. Ross (1976) [11], as in Eq. (10) made $\theta = 0$. The properties of the processes generated by these models can be understood by considering the limiting transition $k \to 0$ in the first model or $\theta \to 0$ in the second. When k and θ are still finite the stationary regimes in the models exist and the probability density of processes for these models is expressed in the form (11). However, in the limiting case $k = 0$ or $\theta = 0$ stationary regimes of processes no longer exist, and the probability density cannot be expressed in the form (11), and can be obtained as solutions of partial differential equations

$$\frac{\partial f(x,t|y,s)}{\partial t} - \frac{1}{2}\frac{\partial^2[\sigma^2 x^3 f(x,t|y,s)]}{\partial x^2} = 0$$

for the CIR model (1980) and

$$\frac{\partial f(x,t|y,s)}{\partial t} + \beta \frac{\partial[xf(x,t|y,s)]}{\partial x} - \frac{\sigma^2}{2}\frac{\partial^2[x^{2\gamma} f(x,t|y,s)]}{\partial x^2} = 0$$

for the CEV model at the boundary condition for both equations

$$\lim_{t \to s} f(x,t|y,s) = \delta(x-y).$$

Unfortunately, these equations cannot be solved analytically, but we can say that for $k = 0$ or $\theta = 0$ the process generated by the Eq. (10) becomes unsteady for the CIR model (1980) with the constant expectation and increasing with time variance, and for the CEV model changing with time both the expectation and the variance.

The other non-stationary models are as following.

11   The Merton Model [12]

$$dr(t) = \alpha dt + \sigma dW(t)$$

generates a nonstationary Gaussian process

$$r(t) = r(0) + \alpha t + \sigma W(t)$$

with a linearly varying expectation and linearly increasing variance

$$E\left[r|r(0)\right] = r(0) + \alpha t, \ Var\left[r\right] = \sigma^2 t.$$

12   The Dothan Model [13]

The equation of the Dothan model

$$dr = \sigma r dW$$

is solved in explicit form:

$$r(t) = r(0)e^{-0.5\sigma^2 t + \sigma W(t)},$$

which implies that a random process generated by the model has a log-normal distribution and is non-stationary. The expectation is steady, but the variance increases exponentially with time

$$E[r|r(0)] = r(0), \ Var\left[r|r(0)\right] = r(0)^2 \left(e^{\sigma^2 t} - 1\right).$$

13   The GBM Model [14]

The GBM model is a model of process geometric Brownian motion

$$dr = \beta r dt + \sigma r dW$$

was introduced into the modern financial analysis by P. Samuelson (1965). It generates a non-stationary process of geometric Brownian motion

$$r(t) = r(0)e^{(\beta - 0.5\sigma^2)t + \sigma W(t)}.$$

In this case, the probability density of the interest rate is log-normal. Unlike BDT model, which also generates a log-normal process, moments of $r(t)$ in the GBM model is not constant but increases exponentially with time, in particular,

$$E[r|r(0)] = r(0)e^{\beta t},$$
$$Var\left[r|r(0)\right] = r(0)^2(\lambda - 1)e^{2\beta t}, \ \lambda = e^{\sigma^2 t},$$
$$S = (\lambda + 2)\sqrt{\lambda - 1},$$
$$K = \lambda^4 + 2\lambda^3 + 3\lambda^2 - 3.$$

Expressions for skewness and kurtosis formally coincide with the expressions of these characteristics of the BDT model, but parameter λ here is not constant and increases exponentially with time.

14 Conclusion

As mentioned above, the process of short-term rates is the basis for building a term structure of the yield of zero-coupon bonds. This explains the interest in the analysis of the processes of short-term rates. In the literature there are many articles that made empirical attempts to find a model of short-term rates, for which a term structure closest to the actual observed structure is obtained [1–3]. On the other hand there is also empirical evidence in the literature that the famous model of short-term rates do not provide an acceptable level of goodness-of-fit [8]. Therefore there is a need for analytical studies to determine the degree of risk in the use of a particular model of short-term rates of the yield. As a necessary basis for this information is needed about the probability properties of the short-term rate processes, expressed analytically. This is the subject of this paper that shall be considered as the first stage of this work.

References

1. CKLS: Chan, K.C., Karolyi, G.A., Longstaff, F.A., Sanders, A.S.: An empirical comparison of alternative models of the short-term interest rate. J. Finance **47**, 1209–1227 (1992)
2. Ahn, D.-H., Gao, B.: A parametric nonlinear model of term structure dynamics. Rev. Finan. Stud. **12**(4), 721–762 (1999)
3. Bali, T.: An empirical comparison of continuous time models of the short term interest rate. J. Futures Markets **19**(7), 777–797 (1999)
4. Vasicek, O.A.: An equilibrium characterization of the term structure. J. Finan. Econ. **5**, 177–188 (1977)
5. CIR: Cox, J.C., Ingersoll, J.E., Ross, S.A.: A theory of the term structure of interest rate. Econometrica **53**, 385–467 (1985)
6. Duffie, D., Kan, R.: A yield-factor model of interest rates. Math. Finance. **6**, 379–406 (1996)
7. Black, F., Derman, E., Toy, W.: A one factor model of interest rates and its application to treasury bond options. Finan. Anal. J. **46**(1), 33–39 (1990)
8. Ait-Sahalia, Y.: Testing continuous-time models of the spot interest rate. Rev. Finan. Stud. **9**(2), 385–426 (1996)
9. Brennan, M.J., Schwartz, E.S.: A continuous time approach to the pricing of bond. J. Bank. Finance **3**, 135–155 (1979)
10. CIR: Cox, J.C., Ingersoll, J.E., Ross, S.A.: An analysis of variable rate loan contracts. J. Finance 35, 389–403 (1980)
11. Cox, J.C., Ross, S.A.: The valuation of options for alternative stochastic processes. J. Finan. Econ. **3**, 145–166 (1976)
12. Merton, R.C.: Theory of rational option pricing. Bell J. Econ. Manag. Sci. **4**(1), 141–183 (1973)
13. Dothan, M.: On the term structure of interest rates. J. Finan. Econ. **6**, 59–69 (1978)
14. Samuelson, P.A.: Rational theory of warrant pricing. Ind. Manag. Rev. **6**, 13–31 (1965)

Hierarchical Space Merging Algorithm for Analysis of Two Stage Queueing Network with Feedback

Agassi Melikov[1]([✉]), Leonid Ponomarenko[2], Anar Rustamov[3],
and Janos Sztrik[4]

[1] Institute of Control Systems, ANAS, F. Agayev 9, AZ1141 Baku, Azerbaijan
agassi.melikov@gmail.com
[2] International Research and Training Center for International Technologies
and Systems, National Academy of Sciences of Ukraine and Ministry of Education
and Science of Ukraine, Kyiv, Ukraine
laponomarenko@ukr.net
[3] Qafqaz University, H. Aliyev 120, Khirdalan, AZ0101 Baku, Azerbaijan
anrustemov@qu.edu.az
[4] University of Debrecen, Debrecen 4032, Hungary
sztrik.janos@inf.unideb.hu
http://www.cyber.az
http://www.ukr.net
http://www.qu.edu.az
http://www.unideb.hu

Abstract. A Markov model of two stage queuing network with feedback is proposed. Poisson flows arriving to both stages from outside and part of already serviced calls in the first node instantaneously enter to the second node (if there is free space here) while the remaining part leaves the network. At the completion of call processing in the second node there are three possibilities: (1) it leaves the network; (2) it instantaneously feeds back to the first node (if there is free space here); (3) it feeds back to the first node after some delay in orbit. All feedbacks are determined by known probabilities. Both nodes have finite capacities. The mathematical model of the investigated network is a three dimensional Markov chain (3-D MC) and hierarchical space merging algorithm to calculate its steady-state probabilities is developed. This algorithm allows asymptotic analysis of the quality of service (QoS) metrics of the investigated network as well.

Keywords: Two stage queueing network · Instantaneous and delayed feedback · Three-dimensional Markov chain · Space merging algorithm

1 Introduction

In the queueing networks, upon the completion of service in a particular node each call either is instantly sent to another node, or returns to the same node according to the routing matrix. Namely, if the network allows re-serving of

© Springer International Publishing Switzerland 2016
A. Dudin et al. (Eds.): ITMM 2016, CCIS 638, pp. 238–249, 2016.
DOI: 10.1007/978-3-319-44615-8_21

the calls in the same node, then we can assume that this network is an instantaneous feedback one. However the concept of "feedback" initially (historically) was introduced for single-station queueing systems [1], and they have been intensively studied in recent years (for further literature on work in this area, see, e.g., [2,3]).

There are several studies which investigated the two-phase model of an open queueing network with instantaneous feedback [4–8]. In these studies, the authors show that analysis of the characteristics of the integrated cellular networks and WLANs (Wireless Local Area Network, WLAN) needs to investigate similar models. However, in the queueing networks the delayed feedback is not taken into account when the already-served calls return back for repeating service after some random delay in orbit. In the available literature almost all the authors did not analyze the mentioned models in queueing networks with both types of feedback – instantaneous and delayed.

In this paper we study the model of a two-phase open queueing network with instantaneous and delayed feedback. It should be noted that taking into account the delayed feedback leads to the necessity of analyzing the three-dimensional Markov chain (3-D Markov Chain, 3-D MC). We developed an effective method for the calculation of state probabilities of the large dimension 3-D MC. There is much research on solving similar problems for the large dimension 2-D MC [9–20]. However, almost all of them have similar problems related to ill-conditioned matrices which arise during computational procedures. In this regard, we have developed a new method that uses simple explicit formulas for the calculation of the state probabilities of the constructed 3-D MC. The proposed method is based on the fundamental ideas of the theory of phase merging of stochastic systems [21]. Furthermore its counterpart for the 2-D MC was successfully used in the study of various models of single-phase queueing systems [22–25].

2   The Model

The structure of a two stage queueing network with feedback is shown in Fig. 1. For simplicity in the model it is assumed that both nodes of the network contain a single channel (server) but their service rates are not identical, i.e. the channel occupancy times of calls in node i are assumed to be independent and have identical exponential distribution with a mean $1/\mu_i, i = 1, 2$ and generally speaking $\mu_1 \neq \mu_2$. Total capacity of node i (the total number of calls in the channel and buffer) is R_i and to node i from outside arrives a Poisson flow of calls with intensity $\lambda_i, i = 1, 2$. If the node $i, i = 1, 2$, is full upon the arrival of a call then the arrived call will be lost with probability 1.

After completion of the service of the call in the first node it either leaves the network with probability θ_1 or it enters the second node with probability $\theta_2 = 1 - \theta_1$. If at the moment of completion of the service of the call in the

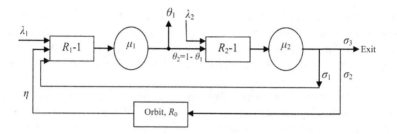

Fig. 1. The structure of the proposed model

first node the second node is full then this call will leave the network with probability 1.

After completion of the service of the call in the second node the following decisions might be made: (i) it feeds back instantaneously to the first node with probability σ_1; (ii) it enters orbit with probability σ_2; (iii) it leaves the network with probability $\sigma_3 = 1 - \sigma_2 - \sigma_1$ The orbit size is $R_0, 0 < R_0 < \infty$. It means that call arriving to orbit from the second node will be accepted if upon its arrival the number of calls in orbit is less than R_0; otherwise an arriving call will be lost. The sojourn times of calls in the orbit are independent and identically distributed random variables and they have common exponential distribution with mean $1/\eta$. It is assumed that calls from the orbit are not persistent, i.e. if upon arrival of the call the first node is full then this call is eventually lost.

3   Proposed Method for the Calculation of Steady-State Probabilities

The investigated network is described by the three-dimensional MC (3-D MC) and its states are described as 3-D vector $\boldsymbol{n} = (n_1, n_2, n_3)$ where the first (n_1) and second components (n_2), respectively, indicate the number of calls in the first and second nodes, and the third component (n_3) indicates the number of calls in orbit. The state space of the given 3-D MC is defined as follows:

$$S = \{\boldsymbol{n} : n_1 = 0, 1, ..., R_1; n_2 = 0, 1, ..., R_2; n_3 = 0, 1, ..., R_0\}. \tag{1}$$

Therefore, the geometric form of state space is depicted with integer values inside the parallelepiped whose base is a rectangle with length of sides (R_1) and (R_2); the height of the parallelepiped equal to (R_0) (see. Fig. 2).

The intensity of transition from $\boldsymbol{n}$ one state to $\boldsymbol{n}\prime$ is denoted as $q(\boldsymbol{n}, \boldsymbol{n}\prime)$, $\boldsymbol{n}, \boldsymbol{n}\prime \in S$. These parameters involve the generating matrix (Q-matrix) of the given 3-D MC. They are determined as follows:

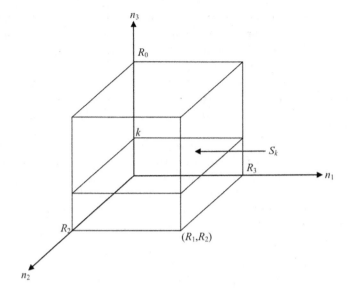

Fig. 2. The state space of the proposed model

$$
q(\boldsymbol{n},\boldsymbol{n}') = \begin{cases}
\lambda_1, & \text{if } \boldsymbol{n}' = \boldsymbol{n} + \boldsymbol{e_1} \\
\mu_1\theta_1, & \text{if } \boldsymbol{n}' = \boldsymbol{n} - \boldsymbol{e_1} \\
\mu_1\theta_2, & \text{if } \boldsymbol{n}' = \boldsymbol{n} - \boldsymbol{e_1} + \boldsymbol{e_2} \\
\lambda_2, & \text{if } \boldsymbol{n}' = \boldsymbol{n} + \boldsymbol{e_2} \\
\mu_2\sigma_3, & \text{if } \boldsymbol{n}' = \boldsymbol{n} - \boldsymbol{e_2} \\
\mu_2\sigma_1, & \text{if } \boldsymbol{n}' = \boldsymbol{n} - \boldsymbol{e_2} + \boldsymbol{e_1} \\
\mu_2\sigma_2, & \text{if } \boldsymbol{n}' = \boldsymbol{n} - \boldsymbol{e_2} + \boldsymbol{e_3} \\
\eta, & \text{if } n_1 < R_1, \boldsymbol{n}' = \boldsymbol{n} - \boldsymbol{e_3} + \boldsymbol{e_1} \text{ or } n_1 = R_1, \boldsymbol{n}' = \boldsymbol{n} - \boldsymbol{e_3} \\
0, & \text{in other cases.}
\end{cases} \tag{2}
$$

where $\boldsymbol{e_1} = (1,0), \boldsymbol{e_2} = (0,1)$.

Here e_i is the i-th unit vector of the 3-D Euclidean space, $i = 1, 2, 3$. The given 3-D MC with finite-state is irreducible since a stationary regime exists. Let $p(\boldsymbol{n})$ mean a steady-state probability of the state $\boldsymbol{n} \in S$. These probabilities are uniquely determined by solving the appropriate system of equilibrium equations (SEE) completed with a normalization condition (due to evidence of constructing the explicit form of SEE not being shown here). Unfortunately, due to the complex structure of the Q-matrix it is too complicated to find an analytical solution to the above indicated SEE. The dimension of the SEE is determined based on the dimension of the state space (1) which consists of $(R_0 + 1)(R_1 + 1)(R_2 + 1)$ states. Therefore, the above-given exact method makes it possible to calculate the steady-state probabilities only in moderate dimensions of state space (1), but in its large values it encounters great computational difficulties. Therefore the only way to solve them is to use numerical methods of linear algebra (for computational difficulties of these methods see the Introduction).

Here a hierarchical space merging algorithm (HSMA) for calculating the steady-state probabilities of the studied 3-D MC subject to the following condition is proposed: $\sigma_3 \ll \sigma_1 + \sigma_2$. In other words, it is assumed that upon completion of service in the second node, the call rarely (in comparison with the intensity of leaving the system and instantaneous return to the node 1) goes into orbit. In other words, the intensity of the call from the orbit is substantially less rather than the intensity of calls from the outside to the network nodes, i.e., $\eta \ll \min\{\lambda_1, \lambda_2\}$. Then, having this assumption we can say that the transition intensity between states inside the planes that are parallel to the base of the parallelepiped is much greater than the transition intensity between states of different planes (see Fig. 2). In that case we can consider the following splitting of the state space (1):

$$S = \bigcup_{k=0}^{R_0} S_k, \, S_k \bigcap S_{k\prime} = \varnothing, k \neq k\prime, \tag{3}$$

where $S_k = \{n \in S : n_3 = k\}, k = 0, 1, 2, ..., R_0$. In other words, it is considered that the entire state space of the network is split into different planes that are parallel to the base of the parallelepiped (see Fig. 2).

The merge function is determined based on the splitting (3) as follows:

$$U_1(n) = <k>, \text{if } n \in S_k \tag{4}$$

where $<k>$ is a merged state, which includes all states of class S_k. Let $\Omega_1 = <k> : k = 0, 1, ..., R_0$.

According to SMA [21] the steady-state probabilities of the given model are defined as follows

$$p(n) \approx \rho^k(n_1, n_2)\pi_1(<k>), \tag{5}$$

where $\rho^k(n_1, n_2)$ denote the probability of the state (n_1, n_2) within the splitting model with state space S_k, and $\pi(<k>)$ denote the probability of the merged state $(<k>) \in \Omega_1$.

From (5) we conclude that for the calculation of the steady-state probabilities given 3-D MC we need to find probability distributions of 2-D MCs (its number is $R_0 + 1$) and one 1-D MC. For large capacities of the nodes computational difficulties arise when calculating the stationary distribution of these 2-D MC with state space $S_k, k = 0, 1, ..., R_0$. Therefore in order to calculate stationary distributions within the classes $S_k, k = 0, 1, ..., R_0$, it is necessary to apply SMA to each class, in other words, we consider the hierarchy of the merged models.

All the splitting models with state spaces $S_k, k = 0, 1, ..., R_0$ involve identical 2-D MCs (see Fig. 3) since below the value of k is fixed and a splitting model with state space S_k is analyzed.

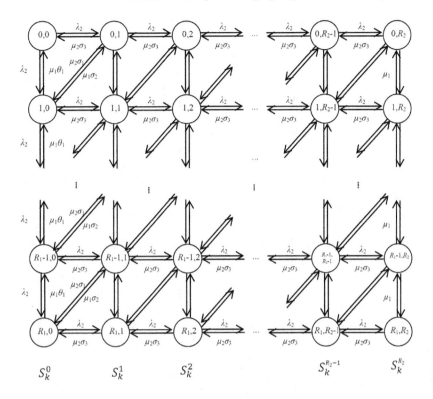

Fig. 3. The state diagram of the split model with state space S_k

The proposed method allows us to find the approximate values of state probabilities in the splitting model with state space $S_k, k = 0, 1, ..., R_0$, with asymmetric load, i.e., we distinguish two cases: (1) $\lambda_1 >> \lambda_2$, $\mu_1 >> \mu_2$; (2) $\lambda_1 << \lambda_2$, $\mu_1 << \mu_2$.

First of all let us analyze case 1. In this case, we can split the state space of S_k into the columns, i.e., in the state space S_k following splitting is considered (see Fig. 3):

$$S = \bigcup_{k=0}^{R_2} S_k^i, S_k^i \bigcap S_k^j = \varnothing, i \neq j,\qquad(6)$$

where $S_k^i = \{\boldsymbol{n} \in S_k : n_2 = i\}, i = 0, 1, 2, ..., R_2$.

Based on the splitting (6) in the state space S_k the new merged function is determined:

$$U_2(\boldsymbol{n}) = <i>, \text{if } \boldsymbol{n} \in S_k^i\qquad(7)$$

where $<i>$ is a merged state, which includes all states of class S_k^i. Let $\Omega_2 = <i> : i = 0, 1, ..., R_2$.

According to the SMA we have:

$$\rho^k(n_1, n_2) \approx \rho_{n_2}^k(n_1)\pi_2^k(<n_2>),\tag{8}$$

where $\rho_{n_2}^k(n_1)$ is the state probability of (n_1, n_2) within a splitting model with state space $S_k^{n_2}$, and $\pi_2^k(<n_2>)$ is the probability of the merged state $<n_2> \in \Omega_2$.

Let us consider the problem of calculating the state probabilities within the classes S_k^i. In the class S_k^i the second component is constant since the micro-state $(n_1, i) \in S_k^i$ can be defined only by the first component, i.e. $(n_1, i) \in S_k^i$ micro-state is just referred as $n_1, n_1 = 0, 1, ..., R_1$. From (2) we conclude that the transition intensities between the states n_1 and $n_1\prime$ of the splitting model with state space S_k^i does not depend on k. Therefore in the remaining part of the paper the subscription k is omitted in the notation of the state probabilities.

Also, from (2) we conclude that the probabilities $\rho_{n_2}^{n_1}, n_2 = 0, 1, ..., R_2 - 1$ coincide with the state probabilities of the model $M(\lambda_1)/M(\mu_1\theta_1)/1/R_1$, and when $n_2 = R_2$ then these probabilities coincide with the state probabilities of the model $M(\lambda_1)/M(\mu_1)/1/R_1$ (here and later we used a modified version of the Kendall notation where the values in brackets denote appropriate intensities).

Hence, the desired state probabilities $\rho_i(j), j = 0, 1, ..., R_1$ are calculated as follows:

$$\rho_i(j) = \begin{cases} \frac{1-\nu_1}{1-\nu_1^{R_1+1}}\nu_1^j, & \text{if } i = 0, 1, ..., R_2 - 1 \\ \frac{1-\nu_2}{1-\nu_2^{R_1+1}}\nu_2^j, & \text{if } i = R_2 \end{cases}\tag{9}$$

where $\nu_1 = \lambda_1/\mu_1\theta_1, \nu_2 = \lambda_1/\mu_1$.

Then after certain mathematical transformations we find the following relations to calculate the transition intensities $q(<i>, <j>), <i>, <j> \in \Omega_2$.

$$q(<i>, <j>) = \begin{cases} \lambda_2 + \mu_1\theta_2(1 - \rho_i(0)), & \text{if } i = 0, 1, ..., R_2 - 1, j = i + 1 \\ \mu_2(\sigma_3 + \sigma_1(1 - \rho_i(R_1))), & \text{if } i = 0, 1, ..., R_1, j = i - 1 \\ 0, & \text{in other cases.} \end{cases}\tag{10}$$

where $\Lambda_1 = \mu_2\sigma_2(1 - \pi_2(<0>))$.

Note 1. In calculating the stationary distribution of the splitting model with the state space $S_{R_0}^i$ it is necessary to set $\sigma_2 = 0$.

Thus, from (10) we get the following expression in order to calculate the probabilities of merged states $\pi_2(<n_2>), <n_2> \in \Omega_2$.

$$\pi_2(<n_2>) = \prod_{i=0}^{n_2-1} \frac{q(<i>, <i+1>)}{q(<i+1>, <i>)}\pi_2(<0>), n_2 = 1, ..., R_2,\tag{11}$$

where $\pi(<0>)$ is derived from the normalizing condition, i.e.
$\pi(<0>) = \left(\sum_{n_2=0}^{R_2} \pi_2(<n_2>)\right) = 1$.

The transition intensities between classes S_k and $S_{k'}$ are determined by the relations (2), (9) and (11) and after certain mathematical transformations we get:

$$q(S_k S_{k'}) = \begin{cases} \Lambda_1, & \text{if } k' = k+1 \\ k\eta, & \text{if } k' = k-1 \\ 0, & \text{in other cases.} \end{cases} \qquad (12)$$

where $\Lambda_1 = \mu_2 \sigma_2 (1 - \pi_2(<0>))$.

Hence, from (12) we see that the required probabilities of the merged states $\pi(<k>), <k> \in \Omega_1$ are defined as state probabilities of a classical Erlang's model $M(\Lambda_1)/M(\eta)/R_0/0$, i.e.,

$$\pi_1(<k>) = \frac{\phi^k}{k!} \left(\sum_{i=0}^{R_0} \frac{\phi^i}{i!} \right)^{-1}, k = 1, ..., R_0, \qquad (13)$$

where $\phi = \Lambda_1/\eta$.

Finally, the state probabilities of the given 3-D MC are determined as follows:

$$p(n_1, n_2, n_3) \approx \rho_{n_2}(n_1)\pi_2(<n_2>)\pi_1(<n_3>) \qquad (14)$$

Likewise, we study case 2, where we considered $\lambda_1 << \lambda_2, \mu_1 << \mu_2$ assumption. In this case it is necessary to split the state space S_k into rows, i.e., in the state space S_k following splitting is considered (see Fig. 3):

$$S = \bigcup_{i=0}^{R_1} S_k^i, S_k^i \bigcap S_k^j = \varnothing, i \neq j, \qquad (15)$$

where $S_k^i = \{ \boldsymbol{n} \in S_k : n_1 = i \}, i = 0, 1, 2, ..., R_1$.

Note 2. Here for simplicity of presentation it is better to use the same notation as used in case 1.

Next, we can implement all of the steps in the above-developed algorithm. Without repeating these steps, we describe briefly the key point of the calculation of the state probabilities. Thus, the state probabilities inside all of classes S_k^i coincide with the state probabilities of the model $M(\lambda_2)/M(\mu_2\sigma_3)/1/R_2$. Transition intensities $q(<i>, <j>), i, j \in \{0, 1, ..., R\}$ in this case are calculated as follows:

$$q(<i>, <j>) = \begin{cases} \lambda_1 + \mu_2\sigma_1(1 - \rho(0)), & \text{if } j = i+1 \\ \mu_1(\rho(R_2) + \theta_1(1 - \rho(R_2))), & \text{if } j = i-1 \\ 0, & \text{in other cases.} \end{cases} \qquad (16)$$

where $\rho(0)$ and $\rho(R_2)$ denotes the probability that system $M(\lambda_2)/M(\mu_2\sigma_3)/1/R_2$ is empty and fully respectively.

So, from (16) we conclude that the probabilities of merged states in the second stage of hierarchy $\pi_2(<n_1>), n_1 \in 0, 1, ..., R_1$ are calculated as state probabilities of the model $M(\lambda_1 + \mu_2\sigma_1(1 - \rho(0)))/M(\mu_1(\rho(R_2)) + \theta_1(1 - \rho(R_2))))/1/R_1$.

The transition intensities between classes S_k and $S_{k'}$ in this case are determined similarly to (12), but in this case the quantity Λ_1 is substituted by the quantity $\Lambda_2 = \mu_2 \sigma_2 (1 - \rho(0))$. In other words, the probabilities of the merged states in the first stage of hierarchy $\pi(<k>), k \in \{0, 1, ..., R_0\}$ are calculated by using the Erlang's formula (13).

4   QoS Metrics

After finding steady-state probabilities of the initial 3-D MC the exact values of QoS metrics of the investigated network might be determined. Thus, since the flow of calls to both nodes are Poisson ones, then the exact values of loss probabilities of calls in node 1 ($P1$) and node 2 ($P2$) are determined as follows:

$$P_i = \sum_{n \in S} p(\boldsymbol{n}) \delta(n_i, R_i), i = 1, 2 \tag{17}$$

where $\delta(i, j)$ are Kronecker's symbols.

The exact values of the average number of calls in nodes (L_1 and L_2) and retrial calls in the orbit (L_0) are defined as the expected values of the appropriate discrete random variables:

$$L_i = \sum_{j=1}^{R_i} j \Phi_i(j), \text{where } \Phi_i(j) = \sum_{n \in S} p(\boldsymbol{n}) \delta(n_i, j), i = 1, 2. \tag{18}$$

$$L_0 = \sum_{j=1}^{R_0} j \Phi_i(j), \text{where } \Phi_i(j) = \sum_{n \in S} p(\boldsymbol{n}) \delta(n_3, j). \tag{19}$$

To calculate the approximate values of the above indicated QoS metrics the following expressions are determined: For case $\lambda_1 >> \lambda_2, \mu_1 >> \mu_2$:

$$P_1 \approx \rho_0(R_1)(1 - \pi_2(<R_2>)) + \rho_{R_2}(R_1)\pi_2(<R_2>); \tag{20}$$

$$P_2 \approx \pi_2(<R_2>); \tag{21}$$

$$P_0 \approx \rho_0(R_1)(1 - \pi_2(<R_2>)) + \rho_{R_2}(R_1)\pi_2(<R_2>)(1 - \pi_1(<0>)); \tag{22}$$

$$L_1 \approx \sum_{k=1}^{R_1} k(\rho_0(k)(1 - \pi_2(<R_2>)) + \rho_{R_2}(k)\pi_2(<R_2>); \tag{23}$$

$$L_2 \approx \sum_{k=1}^{R_2} k\pi_2(<k>). \tag{24}$$

For case $\lambda_1 << \lambda_2, \mu_1 << \mu_2$:

$$P_1 \approx \pi_2(<R_1>); \tag{25}$$

$$P_1 \approx \rho(R_2); \tag{26}$$

$$P_0 \approx \pi_2(<R_1>)(1 - \pi_1(<0>)); \tag{27}$$

$$L_1 \approx \sum_{k=1}^{R_1} k\pi_2(<k>). \tag{28}$$

In this case L_2 is calculated as the average queue length in the system $M(\lambda_2)/M(\mu_2\sigma_3)/1/R_2$, i.e.

$$L_2 = \begin{cases} \frac{\omega}{1-\omega} - \frac{R_2+1}{1-\omega^{R_2+1}}\omega^{R_2+1}, & \text{if } \omega \neq 1 \\ \frac{R_2}{2}, & \text{if } \omega = 1. \end{cases} \tag{29}$$

where $\omega = \lambda_2/\mu_2\sigma_3$.

5   Conclusion

The proposed approximate method allows calculating the steady-state probabilities of the network of arbitrary dimension while the exact method might be used only for the models with a moderate size. In addition, the approximate method allows analyzing the behavior of the QoS metrics versus network parameters as well.

Note that the proposed approximate method has high accuracy for a large scale network (in order to be brief here the results which demonstrate the high accuracy of the approximate formulas are not presented). The accuracy of the proposed method is estimated by norm

$$\Delta = \max_{n \in S} |p(\boldsymbol{n}) - \tilde{p}(\boldsymbol{n})|$$

where $\tilde{p}(\boldsymbol{n})$ denotes the approximate value of probability of the state $\boldsymbol{n} \in S$. In a large interval for changing of the parameters of the network the indicated norm is acceptable in engineering practice. Moreover, numerical experiments showed that this norm asymptotically approaches zero as the dimension of the network is increased. The last fact is important since the main advantage of the proposed method is that it is developed especially for large scale networks.

References

1. Takacs, L.: A single-server queue with feedback. Bell Syst. Techn. J. **42**, 505–519 (1963)
2. Melikov, A., Ponomarenko, L., Rustamov, A.: Methods for analysis of queueing models with instantaneous and delayed feedbacks. In: Dudin, A., A Nazarov, A., Yakupov, R. (eds.) ITMM 2015. CCIS, vol. 564, pp. 185–199. Springer, Heidelberg (2015). doi:10.1007/978-3-319-25861-4_16
3. Korolyuk, V.S., Melikov, A.Z., Ponomarenko, L.A., Rustamov, A.M.: Methods for analysis of multi-channel queueing models with instantaneous and delayed feedbacks. Cybern. Syst. Anal. **52**(1), 58–70 (2016)

4. Gemikonakli, O., Ever, E., Kocyigit, A.: Approximate solution for two stage open networks with Markov-modulated queues minimizing the state space explosion problem. J. Comput. Appl. Math. **223**(1), 519–533 (2009)
5. Kirsal, Y., Gemikonakli, E., Ever, E., Mapp, G., Gemikonakli, O.: An analytical approach for performance analysis of handoffs in the next generation integral cellular networks and WLANs. In: Proceedings of 19th IEEE International Conference on Computer Communications and Networks, Zurich, Switzerland, 2–5 August 2010, pp. 1–6 (2010). doi:10.1109/ICCCN.2010.5560093
6. Ever, E., Gemikonakli, O., Kocyigit, A., Gemikonakli, E.: A hybrid approach to minimize state explosion problem for the solution of two stage tandem queues. J. Netw. Comput. Appl. **36**, 908–926 (2013)
7. Kirsal, Y., Ever, E., Kocyigit, A., Gemikonakli, O., Mapp, G.: A generic analytical modeling approach for performance evaluation of the handover schemes in heterogeneous environments. Wireless Pers. Commun. **79**, 1247–1276 (2014)
8. Kirsal, Y., Ever, E., Kocyigit, A., Gemikonakli, O., Mapp, G.: Modeling and analysis of vertical handover in highly mobile environments. J. Supercomput. **71**, 4352–4380 (2015)
9. Ciardo, G., Smirni, E.: ETAQA: an efficient technique for the analysis of QBD-processes by aggregation. Perform. Eval. **36–37**, 71–93 (1999)
10. Ciardo, G., Mao, W., Riska, A., Smirni, E.: ETAQA-MG1: an efficient technique for the analysis of a class M/G/1-type processes by aggregation. Perform. Eval. **57**, 235–260 (2004)
11. Mitrani, I., Chakka, R.: Spectral expansion solution for a class of Markov models: application and comparison with the matrix-geometric method. Perform. Eval. **23**, 241–260 (1995)
12. Mitrani, I.: Approximate solutions for heavily loaded Markov-modulated queues. Perform. Eval. **62**, 117–131 (2005)
13. Chakka, R.: Spectral expansion solution for some finite capacity queues. Ann. Oper. Res. **79**, 27–44 (1998)
14. Neuts, M.F.: Matrix-Geometric Solutions in Stochastic Models: An Algorithmic Approach, p. 332. John Hopkins University Press, Baltimore (1981)
15. Hung, T.T., Do, T.V.: Computational aspects for steady state analysis of QBD processes. Periodica Polytech. Ser. Electr. Eng. **44**(2), 179–200 (2001)
16. Baumann, H., Sandmann, W.: Numerical solution of level dependent QBD processes. Procedia Comput. Sci. **1**, 1555–1563 (2010)
17. Servi, L.D.: Algorithmic solution of two-dimensional BDP with application to capacity planning. Telecommun. Syst. **21**(2–4), 205–212 (2002)
18. Ye, J., Li, S.Q.: Folding algorithm: a computational method for finite QBD processes with level dependent transitions. IEEE Trans. Commun. **42**, 625–639 (1994)
19. De Nitto Persone, V., Grassi, V.: Solution of finite QBD processes. Appl. Probab. **33**, 1003–1010 (1996)
20. Van Do, T., Chakka, R., Sztrik, J.: Spectral expansion solution methodology for QBD-M processes and applications in future internet engineering. In: Nguyen, N.T., Van Do, T., Thi, H.A. (eds.) ICCSAMA 2013. SCI, vol. 479, pp. 131–142. Springer, Heidelberg (2013)
21. Korolyuk, V.S., Korolyuk, V.V.: Stochastic Models of Systems, p. 185. Kluwer, Boston (1999)
22. Ponomarenko, L., Kim, C.S., Melikov, A.: Performance Analysis and Optimization of Multi-traffic on Communication Networks, p. 208. Springer, New York (2010)

23. Liang, C., Luh, H.: Cost estimation queuing model for large-scale file delivery service. Int. J. Electron. Commer. Stud. **2**(1), 19–34 (2011)
24. Liang, C., Luh, H.: Optimal services for content delivery based on business priority. J. Chin. Inst. Eng. **36**(4), 422–440 (2013)
25. Liang, C., Luh, H.: Efficient method for solving a two-dimensional Markov chain model for call centers. Ind. Manag. Data Syst. **115**(5), 901–922 (2015)

Analysis of the Influence of the Subscriber Traffic Structure on the Speed of Multiplexed Connections

Pavel Mikheev$^{(\boxtimes)}$ and Sergey Suschenko

National Research Tomsk State University, Lenina str., 36, 634050 Tomsk, Russia
doka.patrick@gmail.com, ssp.inf.tsu@gmail.com

Abstract. A model for the transit node of a data communication network, which aggregates multiple incoming flows to be routed via a single outgoing direction, in the form of a single-line queuing system with discrete time, a finite buffer, non-ordinary incoming flow and batch application service is proposed. The influence of the quality of communication channels and blocks of the limited buffer memory of the transit node on the throughput of the network fragment with varying speeds of incoming and outgoing interfaces is studied.

Keywords: Traffic aggregation · Star-shaped network fragment · Throughput · Queuing system

1 Introduction

Subscriber access subnets are principal elements of computer networks. They are implemented through star-shaped routed network fragments and technologies for building wired and wireless local area networks. The most common technologies for building modern wired local area networks (LANs) are based on the method of random multiple access to the data transmission environment shared by multiple subscribers [1]. This method provides for a simple network topology; however, the network performance is deteriorated dramatically under high loads and a large number of subscribers [2]. To increase the actual LAN speed, a method of logical network structuring is applied, which is based on network segmentation using dial-up access technology [1]. In addition to the improved LAN performance, logical structuring via switches simplifies network management, increases its flexibility and security of operating with application data in different network segments. Technical implementation of switches allows for the architectures based on switching matrices, shared multiport memory, shared bus, and the arrangement of different architectures. Three modes of switching protocol data units are recognized: store-and-forward switching (intermediate and full buffering), cut-through switching with buffering frame header to the

This work is performed under the state order No. 1.511.2014/K of the Ministry of Education and Science of the Russian Federation.

© Springer International Publishing Switzerland 2016
A. Dudin et al. (Eds.): ITMM 2016, CCIS 638, pp. 250–260, 2016.
DOI: 10.1007/978-3-319-44615-8_22

destination address (on the fly) and hybrid cut-through switching with buffering of the entire header and data field of the frame of minimum standard size that provides the ability to filter conflicts [1]. Switching devices are specifically used as hubs to accumulate traffic from desktops to file servers, database servers and application servers. A balanced selection of channel speed with reference to general-purpose server systems with applications, the number of subscribers connected to such applications, as well as technical parameters of switching communication devices, are crucial to the synthesis of structure and parameters of the local data transmission networks. One of the main factors that determine the operational characteristics of the network structures is blocking of the limited buffer memory of switching nodes (at the second level of network architecture) [3] and routing nodes (at the third level) [4]. Mathematical models for a local area network, which multiplexes subscriber streams to the customer service, make possible the analysis of influence of client traffic parameters on the throughput of aggregation switch ports with limited memory, computation of the amount of buffer memory and study strategies to ensure quality of provided network services. Since the functioning of computer networks is essentially discrete in nature [5–7], the study of the influence of buffer memory blocks on the speed of network fragments using QS with finite buffer and discrete time was proposed in [8,9]. These findings were developed further in [10], where the influence of blocking buffer memory of the transit node on the proportion of the load processed by a star-shaped network fragment with traffic distribution was analyzed in terms of the discrete-time QS. This article presents a model for a star-shaped network fragment with traffic aggregation and investigates the speed of a routed star-shaped network fragments, where each subscriber has permanent access packets to the backbone network via a transit router.

2 A Mathematical Model for the Aggregating Star-Shaped Network Fragment

Let us consider a fragment of a local computer network that includes clients connected to the server via the network node. We assume that the subscribers are connected to M ports of the transit node, homogeneous in transmit-receive rate, and generate a stream of frames of equal length to the server platform connected to the node via $(M + 1)$-th port with a speed of $S \geq 1$ times greater than that of the subscriber ports. We also assume that the reliability of the subscriber connections to the network node is set by probabilities F_m, $m = \overline{1, M}$, and the reliability of server connection — by probability F. At the same time, all clients data sources are independent and operate in synch, with a period t. Duration of this period is determined by the speed of ports connecting the subscribers and the overheads associated with frame processing in the transit node. Hence, during a full cycle of frame t transmission via the subscriber ports $0 \leq i \leq S$ frames could be sent to the server port. Moreover, we assume that the transit node operates fully in the intermediate buffering mode and the frame received by the node in the current cycle t, will be transferred via the outgoing (server) port only in the next cycle. We suppose that subscriber ports always have data frames to be transmitted to the server platform. We suppose also that there is a dedicated buffer memory pool of volume $K \geq M$ for storing frames in the output queue from the network node to the server port. Then the behavior of the network fragment under consideration can be represented as a Markov Queuing System (QS) with discrete time, finite buffer size, non-ordinary incoming flow and a single batch service unit [11]. The non-ordinary incoming stream in the QS are determined by the reliability of client connections to the transit node F_m, $m = \overline{1, M}$, and the number of processed applications — by the speed of the server port S and reliability of server connection F. Dynamics of queuing to the outgoing communication channel in a given QS are described by the Markov chain. A set of probable states of the Markov chain is determined by the size of buffer memory.

The number of arrivals, or downloads, represents the most important characteristic of QS with limited storage capacity [12]. In the present case, this operating characteristic is defined as a proportion of the server connection speed reached in the conditions of aggregating traffic from M customers:

$$Z(S, M, K, F, \boldsymbol{F}) = F \left(\sum_{k=1}^{S} k P_k + S \sum_{k=S+1}^{K} P_k \right), \tag{1}$$

where $\boldsymbol{F} = \{F_1; \ldots; F_M\}$ is a vector of reliability values of subscribers connections to the transit node F_m, $m = \overline{1, M}$, and P_k — state probabilities of the Markov chain. In the case of homogeneous reliability across the client communication channels $F_m = F_*$, $m = \overline{1, M}$ the number of arrivals is denoted as $Z(S, M, K, F, F_*)$.

We define the probability U_j of arrival at the transit node $j = \overline{0, M}$ packets per cycle t as the sum of all combinations of product probabilities of successful j and unsuccessful $M - j$ transmissions via the subscriber ports:

$$U_j = \sum_{k=1}^{C_M^j} \prod_{n \in A(k,j,M)} F_n \prod_{m \in B(k,j,M)} (1 - F_m),$$

where C_M^j is the number of combinations of M elements by j, $A(k, j, M)$ — specific (k-th) combination of a subset of j elements (indices) of a set containing M elements, $B(k, j, M) = \widehat{M} - A(k, j, M)$ — a subset reverse to a subset of $A(k, j, M)$, complementing it to the set $\widehat{M} = [1, M]$.

Then, the transition probability can be written as follows:

$$
\pi_i^j = \begin{cases}
U_j, & i=0; j=\overline{0,M}; \\[2mm]
\sum\limits_{n=i-j}^{i-1} C_S^n F^n (1-F)^{S-n} U_{n+j-i} \\
\quad + F^i U_j \sum\limits_{n=0}^{S-i} C_{n+i-1}^{i-1}(1-F)^n, & i=\overline{1,S-1}; j=\overline{0,i}; \\[2mm]
\sum\limits_{n=0}^{i-1} C_S^n F^n (1-F)^{S-n} U_{n+j-i} \\
\quad + F^i U_j \sum\limits_{n=0}^{S-i} C_{n+i-1}^{i-1}(1-F)^n, & i=\overline{1,S-1}; j=\overline{i,M}; \\[2mm]
\sum\limits_{n=0}^{i-(j-M)} C_S^n F^n (1-F)^{S-n} U_{j-i+n}, & \begin{array}{l} i=\overline{1,S-1}; \\ j=\overline{M+1,i+M}; \end{array} \\[2mm]
\sum\limits_{n=0}^{j-(i-S)} C_S^n F^{S-n}(1-F)^n U_{j-(i-S)-n}, & i=\overline{S,K}; j=\overline{i-S,i-1}; \\[2mm]
\sum\limits_{n=0}^{S} C_S^n F^n (1-F)^{S-n} U_{j-i+n}, & \begin{array}{l} i=\overline{S,K-M+S-1}; \\ j=\overline{i,i+M-S}; \end{array} \\[2mm]
\sum\limits_{n=0}^{i-(j-M)} C_S^n F^n (1-F)^{S-n} U_{j-i+n}, & \begin{array}{l} i=\overline{S,K-M}; \\ j=\overline{i+M-S,i+M}; \end{array} \\[2mm]
\sum\limits_{n=0}^{i-(j-M)} C_S^n F^n (1-F)^{S-n} \\
\quad \times\left(1-\sum\limits_{k=0}^{j-i-1+n} U_k\right), & i=\overline{K-M,K-M+S}; j=K; \\[2mm]
\sum\limits_{n=0}^{S} C_S^n F^n (1-F)^{S-n} \\
\quad \times\left(1-\sum\limits_{k=0}^{j-i-1+n} U_k\right), & i=\overline{K-M+S,K}; j=K.
\end{cases}
\tag{2}
$$

3 A Router with Server Connection Speed $S=2$

When the speed of the server connection is twice as large as that of the router client connection, the throughput capacity of a router (1) can be written as follows:

$$
Z(S,M,K,F,\boldsymbol{F}) = FP_1 + 2F\sum_{k=2}^{K} P_k.
\tag{3}
$$

Let us consider a router that provides server access for two local computers ($M = 2$) using the buffer memory of capacity $K \geq M$. In this case and in view of transition probabilities (2), the system of equilibrium equations for an arbitrary finite-capacity buffer memory takes the following form:

$$P_0(F_1 + F_2 - F_1 F_2) = P_1 F(2 - F)(1 - F_1)(1 - F_2)$$
$$+ P_2 F^2(1 - F_1)(1 - F_2);$$

$$P_1\Big[F_1 + F_2 - F_1 F_2 + F(2 - F)(1 - 2F_1 - 2F_2 + 3F_1 F_2)\Big] = P_0(F_1 + F_2$$
$$- 2F_1 F_2) + P_2\Big[2F(1 - F)(1 - F_1)(1 - F_2)$$
$$+ F^2(F_1 + F_2 - 2F_1 F_2)\Big] + P_3 F^2(1 - F_1)(1 - F_2);$$

$$P_2\Big[F_1 + F_2 - F_1 F_2 + F(2 - F)(1 - 2F_1 - 2F_2 + 3F_1 F_2)$$
$$+ F^2(F_1 + F_2 - 3F_1 F_2)\Big] = P_0 F_1 F_2 + P_1\Big[(1 - F)^2(F_1 + F_2 - 2F_1 F_2)$$
$$+ F(2 - F)F_1 F_2\Big] + P_3\Big[2F(1 - F)(1 - F_1)(1 - F_2)$$
$$+ F^2(F_1 + F_2 - 2F_1 F_2)\Big] + P_4 F^2(1 - F_1)(1 - F_2);$$

$$P_i\Big[F_1 + F_2 - F_1 F_2 + F(2 - F)(1 - 2F_1 - 2F_2 + 3F_1 F_2) + F^2(F_1 + F_2$$
$$- 3F_1 F_2)\Big] = P_{i-2}(1 - F)^2 F_1 F_2 + P_{i-1}\Big[(1 - F)^2(F_1 + F_2 - 2F_1 F_2)$$
$$+ 2F(1 - F)F_1 F_2\Big] + P_{i+1}\Big[2F(1 - F)(1 - F_1)(1 - F_2)$$
$$+ F^2(F_1 + F_2 - 2F_1 F_2)\Big] + P_{i+2}F^2(1 - F_1)(1 - F_2), \quad i = \overline{3, K - 2};$$

$$P_{K-1}\Big[F_1 + F_2 - F_1 F_2 + F(2 - F)(1 - 2F_1 - 2F_2 + 3F_1 F_2)$$
$$+ F^2(F_1 + F_2 - 3F_1 F_2)\Big] = P_{K-3}(1 - F)^2 F_1 F_2$$
$$+ P_{K-2}\Big[(1 - F)^2(F_1 + F_2 - 2F_1 F_2) + 2F(1 - F)F_1 F_2\Big]$$
$$+ P_K\Big[2F(1 - F)(1 - F_1)(1 - F_2) + F^2(F_1 + F_2 - 2F_1 F_2)\Big];$$

$$P_K\Big[F(2 - F)(1 - F_1)(1 - F_2) + F^2(F_1 + F_2 - 2F_1 F_2)\Big]$$
$$= P_{K-2}(1 - F)^2 F_1 F_2$$
$$+ P_{K-1}\Big[(1 - F)^2(F_1 + F_2 - F_1 F_2) + 2F(1 - F)F_1 F_2\Big].$$

When the amount of buffer memory $K = 2$, we derive the following values for finite states of the Markov chain:

$$P_0 = \frac{F^2}{G}(1-F_1)(1-F_2)\left[(2-F)^2-(1-F)(3-F)(F_1+F_2-F_1F_2)\right];$$

$$P_1 = \frac{F}{G}\left[(2-F)(F_1+F_2-F_1F_2)-FF_1F_2-2(1-F)(F_1+F_2-F_1F_2)^2\right];$$

$$P_2 = \frac{1}{G}\left[F(2-F)F_1F_2 + (1-F)^2(F_1+F_2-F_1F_2)^2\right];$$

$$G = F^2(2-F)^2 + 2F(1-F)(1-3F+F^2)(F_1+F_2-F_1F_2)$$
$$+ 2F(1-F)F_1F_2 + (1-F)^4(F_1+F_2-F_1F_2)^2.$$

Then, in line with (3), the throughput is transformed to:

$$Z(2,2,2,F,\boldsymbol{F}) = \frac{F}{G}\left[F(2-F)(F_1+F_2-F_1F_2) + F(4-3F)F_1F_2\right.$$
$$\left. + 2(1-F)(1-2F)(F_1+F_2-F_1F_2)^2\right]. \quad (4)$$

Hence, it can be easily demonstrated that when $F = F_1 = F_2 = 1$ the throughput reaches the maximum value and becomes independent of the buffer storage size $Z(2,2,K,1,1) = 2$. For absolutely reliable server connections ($F = 1$), the throughput expression is defined by the sum of data transmission reliability via the client connection $Z(2,2,2,1,\boldsymbol{F}) = F_1 + F_2$. Under absolutely reliable client connections ($F_1 = F_2 = 1$), the throughput is determined only by the quality of the output transmission link and does not depend on the size of the buffer memory of the router $Z(2,2,K,F,1) = 2F$. In the absence of flow in one of the client ports of the router ($\boldsymbol{F} = \{F_1; 0\}$), the throughput value makes:

$$Z(2,2,2,F,\boldsymbol{F}) = \frac{FF_1\left[F(2-F)+2F_1(1-F)(1-2F)\right]}{F^2(2-F)^2+2FF_1(1-F)(1-3F+F^2)+F_1^2(1-F)^4}.$$

For statistically homogeneous client flows ($F_1 = F_2 = F_*$) the expression (4) takes the form:

$$Z(2,2,2,F,\boldsymbol{F}) = FF_*\left[F(2-F)(2-F_*)+FF_*(4-3F)+2F_*(1-F)\right.$$
$$\times (1-2F)(2-F_*)^2\left]\middle/\left[F^2(2-F)^2 + 2FF_*(1-F)\right.\right.$$
$$\times (1-3F+F^2)(2-F_*)+2FF_*^2(1-F)+F_*^2(1-F)^4(2-F_*)^2\right].$$

Under fully homogeneous network connections of a router ($F = F_1 = F_2$), the throughput expression becomes more simplified:

$$Z(2,2,2,F,F) = 2F\frac{6 - 16F + 20F^2 - 11F^3 + 2F^4}{12 - 40F + 64F^2 - 56F^3 + 28F^4 - 8F^5 + F^6}.$$

Equations for the finite state probabilities of the Markov chain and the throughput derived from (3) at $K = 3$ are similar to the case when $K = 2$, though more cumbersome. Therefore, for $K = 3$ we give only the throughput formula for router connections with homogeneous reliability transmission $(F = F_1 = F_2)$:

$$Z(2,2,3,F,F) = \Big[80F - 424F^2 + 1160F^3 - 1966F^4 + 2241F^5$$

$$- 1714F^6 + 852F^7 - 240F^8\Big] \Big/ \Big[64 - 368F + 1088F^2 - 2032F^3$$

$$+ 2618F^4 - 2382F^5 + 1535F^6 - 682F^7 + 200F^8 - 36F^9 + 3F^{10}\Big].$$

In addition to the analytical expressions for a router with two input channels $(M = 2)$ and the buffer memory $K = 2, 3$, we obtained numerical solutions for $K = 5, 7, 10, 15, 20$, as well as for three input channels $(M = 3)$ with buffer memory $K = 3, 5, 7, 10, 15, 20$.

Figures 1, 2, 3 and 4 demonstrate the findings of the router performance with the server connection speed twice exceeding that of the client ports ($S = 2$). Figure 1 shows that the throughput in the interval $F \in \big[0, \frac{\sum_{m=1}^{M} F_m}{S}\big)$ is dominated by the line $Z_*(F, \boldsymbol{F}) = SF$, and in the interval $F \in \big[\frac{\sum_{m=1}^{M} F_m}{S}, 1\big]$ below the value of $Z_*(F, \boldsymbol{F}) = \sum_{m=1}^{M} F_m$ is bounded by a parabolic curve, with a minor minimum located at about the midpoint of the segment.

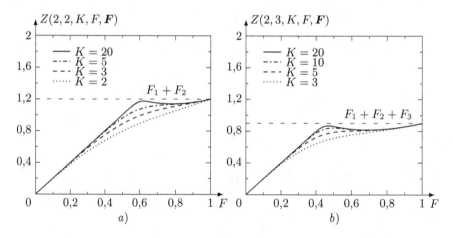

Fig. 1. Comparative curves for throughput-transmission reliability in the outgoing router channel; (a) $M = 2$, $\boldsymbol{F} = \{0,7; 0,5\}$; (b) $M = 3$, $\boldsymbol{F} = \{0,5; 0,3; 0,1\}$

The appearance of curves in Fig. 1 confirms the lowered efficiency of buffer memory usage in the transit node at $F > \frac{\sum_{m=1}^{M} F_m}{S}$, because in this interval the input channels fail to provide rapid server connection with the proper load. In the given interval of F values, the state probabilities P_i, $i = \overline{S, K}$, providing a full load on the

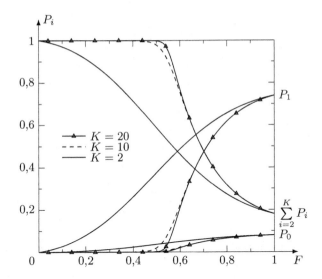

Fig. 2. Dependence of state probabilities of the Markov chain on parameter F, at $M = 2$, $\mathbf{F} = \{0{,}9; 0{,}2\}$ and varying K

output direction are dropped significantly (Fig. 2) and, on the contrary, state probabilities P_0 that leave the outgoing channels unloaded and P_i, $i = \overline{1, S-1}$, that provide only a partial load on the routers server connection increase. The greatest gain in throughput against the increase in the buffer memory capacity is observed under complying the $\sum_{m=1}^{M} F_m = SF$ equation.

Numerical studies also reveal that the more statistically homogeneous client flows in the router (the values of F_m, $m = \overline{1, M}$ approaching each other) are, the higher goes the dominating curve in the interval $F \in \left[\frac{\sum_{m=1}^{M} F_m}{S}, 1\right]$ (Fig. 3). The deepest minimum of this curve is observed at statistically significant heterogeneous incoming channels (e.g., at $F_1 = 1$, $F_2 = 0$).

Figure 3 presents a set of throughput dependencies on the quality of server connection at a given amount of a router's buffer storage and different values of F_m, $m = \overline{1, M}$, satisfying the condition $\sum_{m=1}^{M} F_m = 1$. The figures demonstrate that the maximum throughput values are observed in the areas of varying transmission reliability in the aggregating channel under significantly different sets of F_m, $m = \overline{1, M}$ with constant sum. Within the interval $F \in \left[0, \frac{\sum_{m=1}^{M} F_m}{S}\right]$ the dominating curve corresponds to the polar values of F_m, $m = \overline{1, M}$, and for the major portion of the interval $F \in \left(\frac{\sum_{m=1}^{M} F_m}{S}, 1\right]$ — to the homogeneous values: $F_m = F_*$, $m = \overline{1, M}$. This can also be explained by the fact that even a perfectly reliable communication channel ($F_n = 1$) at any one incoming direction, against absolutely inferior remaining communication channels ($F_m = 0$, $m = \overline{1, M}$, $m \neq n$), is unable to provide a full load on the outgoing direction with speed $S \geq 2$ under a low level error in the communication channel of the outgoing direction. Thus, Fig. 4 demonstrates that at polar values F_m, $m = \overline{1, 2}$

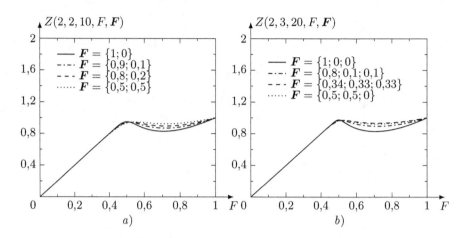

Fig. 3. Throughput-transmission reliability in the outgoing communication channel; (a) $M = 2$, $K = 10$; (b) $M = 3$, $K = 20$.

state probabilities P_i, $i = \overline{2, K}$, which ensure a full load of the outgoing direction, tend to zero when the quality of the outgoing channel is improved.

Investigation of the aggregating router with an arbitrary speed in the main channel, the number of incoming lines and the amount of buffer storage shows in certain cases that if the absolute reliability of at least $S \le M$ client connections ($F_m = 1$, $m = \overline{1, S}$) is sustained, the throughput is invariant to the number of aggregated channels and the buffer capacity, and it is determined by the transmission rate and the quality of the outgoing retransmission link:

$$Z(S, M, K, F, \boldsymbol{F}) = SF.$$

Obviously, when $F = 1$, the throughput is determined only by the physical speed of the server connection S.

For an absolutely reliable outgoing communication channel ($F = 1$) with integral speed S, coinciding with the number of sources M, the throughput, without exceeding the speed of aggregating direction, is determined by the quality of client connections, is invariant to the buffer capacity $K \ge S$ and is set by the following expression:

$$Z(S, S, K, 1, \boldsymbol{F}) = \sum_{m=1}^{M} F_m.$$

Relying on the numerical results given in Figs. 1, 2, 3 and 4, we can conclude that when constructing access networks, the similar (homogeneous) in quality subscriber lines should be selected for the border router, since in the area of high transmission reliability in the outgoing channel ($F \in \left[\frac{\sum_{m=1}^{M} F_m}{S}, 1\right]$) this choice leads to the absolutely best throughput parameter. Despite the fact that homogeneous subscriber lines provide a minimum throughput under a low-quality

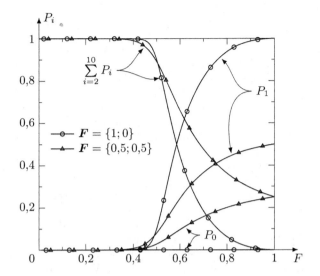

Fig. 4. Dependence of state probabilities of the Markov chain from F, parameter at varying $\boldsymbol{F}$, $M = 2$, $K = 10$

outgoing channel $(F \in \left[0, \frac{\sum_{m=1}^{M} F_m}{S}\right])$ the deviation from the potentially achievable throughput level within this range of varying F decreases rapidly with the increasing buffer capacity of the transit node.

4   Conclusion

A model for the transit node of a data communication network, which aggregates information flows from multiple sources and routes them to a single outgoing direction is proposed in the article. This study represents a continuation of researches [13]. The router model with a service rate homogeneous to the incoming and outgoing interfaces is generalized to the case when the speed of the outgoing data transmission channel exceeds manifold that of the incoming directions. A drop in the buffering efficiency of the transit node was observed at the level of transmission reliability in server connections exceeding that of the summarized reliable data transmission in incoming channels, normalized to the speed parameter of the outgoing router interface $F > \frac{\sum_{m=1}^{M} F_m}{S}$, while the drop in performance subsides against the increasing homogeneity of client flows $(F_m = F_*, m = \overline{1, M})$. A dominant for evaluating the processed flow was derived when the number of buffers exceeds manifold the number of active subscribers. For an absolutely reliable outgoing communication channel $(F = 1)$ with integral speed S, coinciding with the number of sources M, the throughput is determined by the quality of client connections, is invariant to the buffer capacity size $K \geq S$, and is set by the equation: $Z(S, S, K, 1, \boldsymbol{F}) = \sum_{m=1}^{S} F_m$.

References

1. Tanenbaum, A., Wetherall, D.: Computer Networks, 5th edn. Pearson Education, Upper Saddle River (2012). 960 p
2. Kustov, N.T., Sushchenko, S.P.: Capacity of the random multiple access method. Autom. Remote Control. **62**(1), 76–85 (2001)
3. Froom, R., Sivasubramanian, B., Frahim, E.: Implementing Cisco IP Switched Networks (SWITCH) Foundation Learning Guide. Cisco Press, Indianapolis (2010)
4. Teare, D.: Implementing Cisco IP Routing (ROUTE) Foundation Learning Guide. Cisco Press, Indianapolis (2010)
5. Boguslavskii, L.B.: Upravlenie potokami dannykh v setyakh EVM (Controlling Data Flows in Computer Networks). Energoatomizdat, Moscow (1984). 168 p
6. Ivanovskiy, V.B.: On properties of output flows in digital service systems. Autom. Remote Control. **45**(11), 1413–1419 (1984)
7. Kleinrock, L.: Queueing Systems, Theory, vol. 1. Wiley (1975). Translated under the title Vychislitel'nye sistemy s ocheredyami, 600 p. Mir, Moscow (1979)
8. Sushchenko, S.P.: On the influence of buffer memory overfilling on the operational characteristics of a data transmission link. Avtom. Vychisl. Tekh. **6**, 27–34 (1985)
9. Sushchenko, S.P.: The influence of buffer overfilling on the speed of synchronous data-transmission control procedures. Autom. Remote Control. **60**(10), 1460–1468 (1999)
10. Mikheev, P.A.: Analyzing sharing strategies for finite buffer memory in a router among outgoing channels. Autom. Remote Control. **75**(10), 1814–1825 (2014)
11. Kleinrock, L.: Queueing Systems. Wiley, New York (1976). Translated under the title Teoriya massovogo obsluzhivaniya, 432 p. Mashinostroenie, Moscow (1979)
12. Feller, W.: An Introduction to Probability Theory and Its Applications, vol. 2. Wiley, Hoboken (1970). 670 p
13. Mikheev, P.A., Sushchenko, S.P.: On aggregate data link throughput of star network. Vestnik Tomskogo Gosudarstvennogo Universiteta. Upravlenie, Vychislitel'naya Tekhnika i Informatika **4**(13), 97–105 (2010)

Regenerative Analysis of a System with a Random Volume of Customers

Evsey Morozov[1], Lyubov Potakhina[1($\boxtimes$)], and Oleg Tikhonenko[2]

[1] Institute of Applied Mathematical Research, Karelian Research Centre,
Petrozavodsk State University, Petrozavodsk, Russia
emorozov@karelia.ru, lpotahina@gmail.com
[2] Faculty of Mathematics and Natural Sciences, College of Sciences,
Cardinal Stefan Wyszyński University in Warsaw, Warsaw, Poland
tikhonenko@uksw.edu.pl

Abstract. We study a general multi-server system in which each customer has service time and a random volume. We consider two main cases: (i) the total volume of the present customers is unlimited and (ii) this volume is upper bounded by a finite constant. For this system, using the regenerative approach, we develop performance analysis. We establish a solidarity property of the basic processes: accumulated volume, waiting time (workload) and queue size. In case (i), we prove an analog of Little's formula and, provided the system is single-server and the input is Poisson, the Pollaczek-Khintchine formula. In case (ii), we suggest an approximation of the Pollaczek-Khintchine formula, which is then verified by simulation.

Keywords: Queueing system · Random volume customer · Regenerative approach · Steady-state performance · Simulation · Approximation

1 Introduction

The finite volume of computer memory is a very important aspect of modern high performance systems, which is not adequately covered by convenient queueing models with a finite buffer for waiting customers. An important and difficult feature of finite space systems is a possible dependence between service time and the volume of a given customer. On the other hand, analysis of finite space systems, where customers have both service time and a random volume, is quite important to model different problems related to the effective functioning of high performance computer and communication systems, see for instance, [7,8]. The total accumulated volume in the system, together with queue size and workload (waiting time in the queue), can be considered as the basic processes describing the dynamics of such systems. This process is still less studied, and its further analysis is a challenging problem. In this regard, we again mention previous works [7,8], and also a recent paper [5], where an asymptotic analysis of the limited space system was developed. In particular, an important role in this analysis

A. Dudin et al. (Eds.): ITMM 2016, CCIS 638, pp. 261–272, 2016.
DOI: 10.1007/978-3-319-44615-8_23

is played by an analogy between the process of the lost customers (caused by limited space) and an associated renewal process generated by the customer volumes. It has allowed a deducing of useful approximate relation between the stationary loss probability and the limiting fraction of the lost volume. In this paper, we first develop a performance analysis of a system with unlimited space. We use a unified approach based on regenerative arguments, and it is the first contribution of this paper. Moreover, we establish a solidarity property between the basic processes embedded at the arrival instants. When the stability condition holds, we obtain the tightness of all these processes. Furthermore, we present a simple regenerative proof of an analog of Little's formula, which unlike classical formula, connects the mean stationary present volume and the mean stationary waiting time in the queue. This proof covers a multiserver system in which servers may be non-identical. For the system with limited space M, a dependence between the waiting time and acceptance/rejection of a customer makes analysis much more challenging, and for this case we suggest an approximate Pollaczek-Khintchine formula. It is another important contribution of the paper. Finally, we develop regenerative simulation to estimate the unknown acceptance probability and verify the proposed approximation. In particular, we find the values of parameters (traffic intensity and the upper bound for the total volume) when this approximation turns out to be the most accurate. We stress that our approach is quite general and the only one limitation: the existence of classical regenerations of the basic processes.

The paper is organized as follows. In Sect. 2 we describe the model, then, in Sect. 3, we analyze the unlimited volume space system. For this system, in Sect. 3.1, we establish the solidarity property of the basic processes, then in Sect. 3.2, a simple proof of an analog of the famous Little's formula is given. A few examples are also presented. Then, in Sect. 4, we study the limited space system. We establish the tightness of basic processes and suggest an approximate Pollaczek-Khintchine formula. Finally, in Sect. 5, we present simulation results which demonstrate the accuracy of the proposed approximation depending on the traffic intensity ρ, total volume space M, and for various distributions of input, service time S and volume v.

2   Model Description

We consider a general multiserver $GI/G/m$-type queueing system with FIFO service discipline and random volume customers. A distinctive feature of the system is that the nth customer is described by two parameters, service time S_n and *volume* v_n, $n \geq 0$. The pairs $\{S_n, v_n\}$ are assumed to be independent identically distributed (i.i.d.), while, for given n, a dependence between S_n and v_n may exist. In particular, the volumes $\{v_i\}$ are i.i.d. with generic element v, and the service times $\{S_n\}$ are i.i.d. as well, with generic element S. We consider two main cases: (i) unlimited space for the buffered customers, and (ii) the total space of customers in the system is upper bounded by a finite constant M. The arrival instants $\{t_n\}$ form the i.i.d (renewal) sequence of the interarrival times

$\tau_n = t_{n+1} - t_n$, $n \geq 0$ ($t_0 = 0$) with the rate $\lambda := 1/\mathsf{E}\tau \in (0, \infty)$, where τ is a generic interarrival time. We introduce the accumulated volume $V(t)$, the workload (waiting time in queue) $W(t)$ and the number of customers in system (queue size) $\nu(t)$, at instant t.

3   Unlimited Space System

First we consider the system with unlimited space for the total volume of the customers in the system. We denote W_n the waiting time and ν_n the number of customers in the system at instant t_n^-, and assume (for simplicity only) zero initial state, $W_0 = \nu_0 = 0$. Also let V_n be the total accumulated volume in the system at instant t_n^-. In this section, we establish an intuitively expected *asymptotic solidarity* property between the accumulated volume V_n, queue size ν_n and workload W_n. Below $\Rightarrow$ stands for the convergence in probability.

Theorem 1. *In the system, $W_n \Rightarrow \infty$ iff $\nu_n \Rightarrow \infty$ iff $V_n \Rightarrow \infty$.*

Proof. First of all we remark that the first part of the statement, $W_n \Rightarrow \infty$ iff $\nu_n \Rightarrow \infty$, has been established in [4]. Thus it remains to establish the second part, $\nu_n \Rightarrow \infty$ iff $V_n \Rightarrow \infty$. For any $n \geq 0$, $k \geq 1$ and any fixed $x \geq 0$ we have:

$$\mathsf{P}(V_n > x) = \mathsf{P}\Big(\sum_{j=0}^{\nu_n-1} v_j > x\Big) = \mathsf{P}\Big(\sum_{j=0}^{\nu_n-1} v_j > x, \nu_n > k\Big) + \mathsf{P}\Big(\sum_{j=0}^{\nu_n-1} v_j > x, \nu_n \leq k\Big)$$

$$\geq \mathsf{P}\Big(\sum_{j=0}^{k-1} v_j > x, \nu_n > k\Big) \geq \mathsf{P}\Big(\sum_{j=0}^{k-1} v_j > x\Big) - \mathsf{P}(\nu_n \leq k). \tag{1}$$

(i) Let us assume $\nu_n \Rightarrow \infty$. By the Strong Law of Large Numbers

$$\frac{\sum_{j=0}^{k-1} v_j}{k} \to \mathsf{E}v > 0, \quad k \to \infty, \tag{2}$$

and thus $\sum_{j=0}^{k-1} v_j \to \infty$ as $k \to \infty$, with probability 1 (w.p.1). Then by (1), for arbitrary fixed x, there exists k_0 such that the probability $\mathsf{P}(\sum_{j=0}^{k_0} v_j > x)$ becomes arbitrary close to 1. On the other hand, by the assumption $\nu_n \Rightarrow \infty$, the probability $\mathsf{P}(\nu_n < k_0)$ can be made arbitrarily small for sufficiently large n. Because x is arbitrary, then $V_n \Rightarrow \infty$ by (1), and part (i) is proved.
(ii) Let us assume $V_n \Rightarrow \infty$. For each x, n, k we have the inequality,

$$\mathsf{P}(V_n > x) \leq \mathsf{P}\Big(\sum_{j=0}^{\nu_n-1} v_j > x, \nu_n > k\Big) + \mathsf{P}(\sum_{j=0}^{k-1} v_j > x), \tag{3}$$

implying

$$\mathsf{P}(\nu_n > k) \geq \mathsf{P}\Big(\sum_{j=0}^{\nu_n-1} v_j > x, \nu_n > k\Big) \geq \mathsf{P}(V_n > x) - \mathsf{P}(\sum_{j=0}^{k-1} v_j > x). \tag{4}$$

Because $\mathsf{E}v < \infty$, the last term in (4) with a fixed k, can be made arbitrarily small for some (sufficiently large) $x := x_0$, see (2). On the other hand, by the assumption $V_n \Rightarrow \infty$, the probability $\mathsf{P}(V_n > x_0)$ becomes arbitrary close to 1 for all sufficiently large n. It now follows from (4) that, for each fixed k, the probability $\mathsf{P}(\nu_n > k) \to 1$ as $n \to \infty$, and it completes the proof. ∎

Remark 1. The statement of Theorem 1 holds true also for a system with non-identical servers. (The description and stability analysis of this system can be found in [3].)

Now we define recursively the instants

$$\beta_{n+1} = \min\{k > \beta_n : V_k = \nu_k = W_k = 0\}, \quad n \geq 0, \ \beta_0 := 0, \tag{5}$$

which are the regeneration points of discrete-time processes $\{W_n\}$, $\{\nu_n\}$, $\{V_n\}$, while $T_n =: t_{\beta_n}$, $n \geq 0$, $T_0 = t_0 = 0$ are regenerations of continuous-time processes $\{V(t)\}$, $\{W(t)\}$, $\{\nu(t)\}$. We denote as T, β the generic regeneration period in continuous time and discrete time (at arrival instants), respectively. Under an extra *stability assumption* the statement of Theorem 1 becomes stronger.

Theorem 2. *If the stability condition* $\rho := \lambda \mathsf{E}S < m$ *holds and* $\mathsf{P}(\tau > S) > 0$, *then all three processes* $\{V_n\}$, $\{\nu_n\}$, $\{W_n\}$ *are tight.*

Proof. We will use the well-known fact that, under assumed conditions, the queue-size process $\{\nu_n\}$ (as well as the process $\{W_n\}$) is *tight*. Moreover, these processes are *positive recurrent regenerative*, that is $\mathsf{E}T < \infty$, $\mathsf{E}\beta < \infty$ [1,3]. Then, for any $k \geq 0$, $x \geq 0$,

$$\mathsf{P}(\nu_n \leq k) = \mathsf{P}(V_n \leq x, \nu_n \leq k) + \mathsf{P}\left(\sum_{j=0}^{\nu_n-1} v_j > x, \nu_n \leq k\right)$$

$$\leq \mathsf{P}(V_n \leq x) + \mathsf{P}\left(\sum_{j=0}^{k-1} v_j > x\right), \tag{6}$$

implying

$$\mathsf{P}(V_n \leq x) \geq \mathsf{P}(\nu_n \leq k) - \mathsf{P}\left(\sum_{j=0}^{k-1} v_j > x\right). \tag{7}$$

Now, for any $\varepsilon > 0$, we take k_0 such that $\mathsf{P}(\nu_n \leq k_0) \geq 1 - \varepsilon/2$ for all n (it is possible by the tightness of the process $\{\nu_n\}$). Then we can find x_0 such that $\mathsf{P}(\sum_{j=0}^{k_0-1} v_j > x_0) \leq \varepsilon/2$, see (2). It gives

$$\inf_n \mathsf{P}(V_n \leq x) \geq 1 - \varepsilon, \ x \geq x_0,$$

and, because ε is arbitrary, the tightness of $\{V_n\}$ follows. ∎

It is worth noting that solidarity and tightness are quite important for stability analysis.

Remark 2. If the number of servers $m = 1$, then we need not assume $P(\tau > S) > 0$ to construct classical regenerations (5).

Remark 3. In the above given statements, a dependence between S_n and v_n is allowed.

3.1   An Analog of Little's Formula

Keeping the previous notation, we now give a regenerative proof of an analog of the celebrated Little's formula, which in our case connects the mean stationary accumulated volume and the mean stationary workload. This result has been proved in [6] in another setting. We denote the input rate $\lambda = 1/E\tau$. For the zero initial state, $\beta_1 =_{st} \beta$ (stochastically), and by Wald's identity,

$$ET = Et_\beta = E\beta\, E\tau = \frac{E\beta}{\lambda}. \tag{8}$$

We introduce the weak limits $W_n \Rightarrow W_\infty$, $v_n \Rightarrow v_\infty$ and $V_n \Rightarrow V_\infty$ as $n \to \infty$, and $W(t) \Rightarrow W(\infty)$, $v(t) \Rightarrow v(\infty)$ and $V(t) \Rightarrow V(\infty)$, as $t \to \infty$, when exist. In the following statement servers may be *non-identical*.

Theorem 3. *Let us assume that the assumptions of Theorem 2 hold true. Then the following analog of Little's formula holds*

$$EV(\infty) = \lambda[EW_\infty Ev + E(Sv)]. \tag{9}$$

Proof. Under the assumptions of Theorem 2, the weak limits W_∞, v_∞, V_∞ exist, and we *assume* the existence of the limits in continuous time, $W(\infty)$, $v(\infty)$, $V(\infty)$. (It is the case, when interarrival time τ is non-lattice, see [1].) Let us recall that discrete-time regeneration period β equals the number of arrivals during the regeneration cycle and all customers arrived within a cycle depart system in that cycle. In particular, the time customer j spends in the system during the regeneration cycle is his sojourn time $W_j + S_j$. It now follows by regenerative arguments that w. p. 1, as $t \to \infty$,

$$EV(\infty) := \lim_{t\to\infty} \frac{1}{t} \int_0^t V(u)du \to \lambda \frac{E\left[\sum_{j=0}^{\beta-1} v_j(W_j + S_j)\right]}{E\beta}$$
$$= \lambda\left[Ev\, EW_\infty + E(S\,v)\right], \tag{10}$$

where the independence between the waiting time in a queue and the volume of a newly arrived customer is used. ∎

The distribution of service time is assumed to be known but for non-identical servers it depends on the assigned server.

If the service time and volume are independent, then (10) becomes

$$EV(\infty) = \lambda(EW_\infty + ES)Ev. \tag{11}$$

Finally, if $v_n \equiv 1$, then $\nu_n = V_n$ and (10) becomes the classical Little's formula:

$$E\nu(\infty) = \lambda(EW_\infty + ES).$$

Similar arguments allow us to obtain the mean stationary volume $E\hat{V}$ of customers being served. Let $\hat{V}(t)$ be the summary volume of customers present in servers at instant t. Then it follows that

$$E\hat{V} = \lim_{t\to\infty} \frac{1}{t} \int_0^t \hat{V}(u)du = \lambda \frac{E[\sum_{n=0}^{\beta-1} S_n v_n]}{E\beta} = \lambda E(Sv), \tag{12}$$

giving, for independent S_n and v_n,

$$E\hat{V} = \lambda ES\, Ev := \rho Ev. \tag{13}$$

Remark 4. It is well-known that, for the single-server system, $\rho = \lambda ES$ is the stationary busy probability.

In some cases, when S_n and v_n are dependent, it is possible to calculate the term $E(Sv)$ in an explicit form. For instance, if the service time is proportional to the volume, that is $S = cv$ where $c > 0$ is a constant, we obtain $E(Sv) = E(cv^2) = cEv^2$, implying

$$EV(\infty) = \lambda(EW_\infty Ev + cEv^2).$$

As another example, let $S = cv + \phi$, with independent v and ϕ. Then $E(Sv) = cEv^2 + E\phi Ev$, and (9) becomes

$$EV(\infty) = \lambda\Big[(EW_\infty + E\phi)Ev + cEv^2\Big].$$

3.2   Pollaczeck-Khintchine Formula

Let us consider an unlimited space single-server system with renewal input with the rate λ and condition $\rho < 1$. Then the weak limit $W_n \Rightarrow W_\infty$ exists. The Pollaczeck-Khintchine formula for this system has the same form as for the classical $M/G/1$ system because in this analysis only service times are involved. Nevertheless, we present a short and instructive (conventional) proof because it is used below for analysis of the limited space system. First, by geometrical considerations, regenerative arguments and (8), we obtain

$$EW(\infty) =: \lim_{t\to\infty} \frac{1}{t} \int_0^t W(u)du = \frac{E \int_0^T W(u)du}{ET}$$

$$= \frac{1}{ET} E\Big[\sum_{i=0}^{\beta-1} \Big(\frac{(W_n + S_n)^2}{2} - \frac{W_{n+1}^2}{2} \Big) \Big]$$

$$= \lambda \frac{E\Big[\sum_{i=0}^{\beta-1}(W_n S_n + S_n/2) \Big]}{E\beta} = \lambda(EW_\infty ES + \frac{ES^2}{2}). \tag{14}$$

Then, for the Poisson input, the weak limit $W(t) \Rightarrow W$ exists as well, and by PASTA property, $W(\infty) =_{st} W_\infty =: W$ (stochastically). Thus we obtain the classical Pollaczeck-Khintchine formula:

$$EW = \frac{\lambda E S^2}{2(1 - \rho)}. \tag{15}$$

An interesting and practically important case is proportionality, $S_n = c v_n$, where $c > 0$ is a constant, in which case we obtain from (15)

$$EW = \frac{\lambda c^2 E v^2}{2(1 - c\lambda E v)}. \tag{16}$$

4   Limited Space System

Now we consider the same m-server system in which the total accumulated volume $V(t)$ at (any) instant t is upper bounded by a finite constant M. (We call M *volume space*.) Such systems have been used to model and solve various problems occurring in computer and communication systems design, see [7,8]. Note that there is no limitation for the number of customers waiting in the queue. Thus, in this system customer n is lost if and only if

$$V(t_n^-) + v_n > M. \tag{17}$$

In this system, the accumulated volume $V(t) \leq M$, and we establish the tightness of queue size and workload.

Theorem 4. *In the limited space system, the queue-size process and the workload process are tight.*

Proof. Take any $n \geq 0$, $k \geq 1$, and write

$$0 = P(V_n > M) \geq P(\sum_{j=0}^{\nu_n - 1} v_j > M, \; \nu_n > k) \geq P(\nu_n > k) - P(\sum_{j=0}^{k-1} v_j > M),$$

implying

$$P(\nu_n > k) \leq P(\sum_{j=0}^{k-1} v_j > M) := Q_k. \tag{18}$$

However the probability Q_k can be made arbitrarily small for all sufficiently large k, because $\sum_{j=0}^{k-1} v_j \to \infty$, $k \to \infty$, w.p.1. Since the upper bound Q_k in (18) is independent of n, the required tightness of the queue-size process $\{\nu_n\}$ follows.

To prove the tightness of the workload, we denote as $S_i(t)$ the remaining service time at instant t in server $i = 1, \ldots, m$. Let us note that the number of customers waiting in the queue at instant t_n^- is defined as $Q_n = (\nu_n - m)^+$. Let us denote $\mathcal{J}(n)$ the set of occupied servers at instant t_n^-. Now we can write, for any n, k and $x \geq 0$,

$$P(W_n \leq x) = P\left(\sum_{j=0}^{Q_n} S_j + \sum_{i \in \mathcal{J}(n)} S_i(t_n^-) \leq x\right)$$

$$\geq P\left(\sum_{j=0}^{k} S_j + \sum_{i \in \mathcal{J}(n)} S_i(t_n^-) \leq x, \, Q_n \leq k\right)$$

$$\geq P\left(\sum_{j=0}^{k} S_j + \sum_{i \in \mathcal{J}(n)} S_i(t_n^-) \leq x\right) - P(Q_n > k). \qquad (19)$$

We notice that $Q_n \leq \nu_n$ and thus the process $\{Q_n\}$ is tight. It has been proved that the remaining service time process $\{S_i(t_n^-), n \geq 1\}$ is tight for each i [2]. Moreover, the cardinality of the set $|\mathcal{J}(n)| \leq m$. Hence, the sum of the remaining service times, $\sum_{i \in \mathcal{J}(n)} S_i(t_n^-), n \geq 1$, is a tight process as well. Now taking sufficiently large k_0 we can make $P(Q_n > k_0)$ arbitrarily close to 0, and then making sufficiently large $x = x_0$ we can make the probability $P\left(\sum_{j=0}^{k_0} S_j + \sum_{i \in \mathcal{J}(n)} S_i(t_n^-) \leq x_0\right)$ arbitrarily close to 1. It shows that the probability $P(W_n \leq x)$ becomes arbitrarily close to 1 for $x \geq x_0$, implying the required tightness of the workload process.   ∎

A modified Pollaczeck-Khintchine formula holds for this (single-server) system with a Poisson input, however the definition of the input rate is changed. Namely, let $N(t)$, $A(t)$ be the number of arrivals and accepted arrivals, respectively, in interval $[0, t]$. Let indicator $1_n = 1$ if customer n be accepted ($1_n = 0$, otherwise). Then in particular,

$$A(t) = \sum_{n=0}^{N(t)} 1_n, \quad N(t)/t \to \lambda \quad \text{w.p.1, as } t \to \infty.$$

One can show that the basic regenerative processes are positive recurrent, that is $ET < \infty$, $E\beta < \infty$ [3]. Then, by the regenerative arguments, the following *effective* input rate, exists w.p.1,

$$\lim_{t \to \infty} \frac{A(t)}{t} = \frac{E\beta}{ET} =: \lambda_a.$$

On the other hand,

$$\lambda_a = \lim_{t \to \infty} \frac{A(t)}{N(t)} \frac{N(t)}{t} = \lambda P_a, \qquad (20)$$

where the limiting fraction of the accepted customers, $\lim_{t \to \infty} A(t)/N(t) = P_a$, exists and there is also stationary acceptance probability $P_a = E1_a$ (because the weak limit $1_n \Rightarrow 1_a$ exists by condition $P(\tau > S) > 0$, see Theorem 2). The same geometrical arguments as in (14) together with (20) lead to the expression

$$EW(\infty) = \lambda_a(EW_\infty\, ES + \frac{ES^2}{2}).$$ (21)

The input of accepted customers is a result of a thinning of the original Poisson input (with given rate λ), and is *no longer a Poisson process* because the indicators $\{1_n\}$ are dependent. Nevertheless, we *assume* that the PASTA property holds, implying equality $W(\infty) =_{st} W_\infty := W$. Then by (21), the following (approximate) Pollaczek-Khintchine formula holds:

$$EW = \frac{\lambda_a ES^2}{2(1 - \lambda_a ES)}.$$ (22)

This result allows us to estimate the mean stationary waiting time in the original limited space system as the mean stationary waiting time in a standard $M/G/1$ system with a Poisson input with the rate λ_a. Thus we first must estimate P_a and then verify the accuracy of approximation (22). In Sect. 5 we demonstrate the accuracy for different parameters of the system.

Remark 5. One can deduce an analog of Little's formula (9) for a limited space system replacing λ by λ_a and generic volume v by the (unknown) volume of the accepted customer, however an applicability of this result is very limited.

5 Simulation Results

In this section we present simulation results to verify the accuracy of approximation (22), depending on the given parameters of the system. We notice that simulations have been carried out by means of the *system R* [10] and high performance cluster of the Karelian Research Centre [9].

We develop regenerative simulation of an $M/G/1$ limited space system combining exponential distribution, Pareto distribution,

$$F(x) = 1 - (x_m/x)^\alpha, \quad x \geq x_m > 0, \ \alpha > 0, \ F(x) = 0, \ x \leq x_m,$$

and Weibull distribution,

$$F(x) = 1 - e^{-(x/s)^i}, \quad s > 0, \ i > 0, \ x \geq 0,$$

both for service time and volume, which are assumed to be independent. Moreover, we simulate the system depending on the traffic intensity ρ in the range $[0.700, 0.999]$ and space size M (expressed in the term of mean volume size Ev), in range $[4\,Ev, 20\,Ev]$. To estimate rate λ_a, we must take into account the rate of losses. Because a loss is typically a rare event (for sufficiently large M), we apply a special construction to accumulate a sufficient number of losses. Namely, in each experiment, for fixed ρ and M, we construct 300 *extended regeneration cycles*. To obtain an extended cycle, we accumulate ordinary regeneration cycles (see (5)) until 100 losses appear. (In our setting, the ordinary cycles are typically less informative because of the rarity of losses.) We recall that (any) regeneration cycle is generated by the customers meeting an empty system. In each experiment, we calculate the sample mean $E\hat{W}$ to estimate the stationary mean workload EW in (22). Moreover, we calculate ES, ES^2 and estimate the probability P_a to obtain an alternative estimate $E\bar{W}$ of EW based on the right hand side of (22). Then we compare the obtained estimates using the (percent) *relative error* (RE)

$$\frac{|E\bar{W} - E\hat{W}|}{E\hat{W}} \times 100\%. \tag{23}$$

Figure 1 represents the results for the system with exponential service time with parameter 1 and exponential volume with parameter 0.5. Figure 2 shows the results for Pareto service time with parameters $x_m = 4$, $\alpha = 3.5$ and Weibull volume with parameters $s = 7$, $i = 2$. Finally, Fig. 3 shows the results for Weibull service time with parameters $s = 7$, $i = 2$ and Pareto volumes with parameters $x_m = 4$, $\alpha = 3.5$.

These results confirm that approximation (22) is very accurate when space size sufficiently large M is and the system is not heavily loaded, implying a rarity of losses. For instance, in the scenario described in Fig. 1, RE belongs to interval $[1.14\%, 5.35\%]$, when $M = 20\,Ev$ and $\rho \in [0.7, 0.8]$; in the scenario presented in Fig. 2, RE $\in [1.35\%, 5.47\%]$, when $M = 20\,Ev$ and $\rho \in [0.7, 0.8]$; and, for the system described by Fig. 3, RE $\in [0.38\%, 3.94\%]$, if $M = 20\,Ev$ and $\rho \in [0.7, 0.85]$. Thus under these parameters the thinning input of the accepted

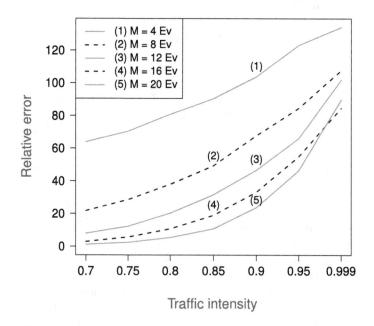

Fig. 1. Exponential S, v

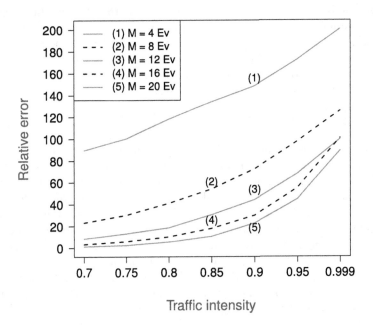

Fig. 2. Weibull S, Pareto v

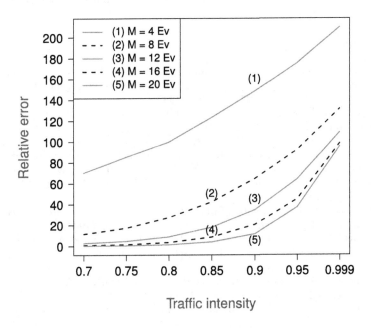

Fig. 3. Pareto S, Weibull v

customers approaches the Poisson process with the rate λ_a with high accuracy, and approximation (22) works well. For other parameters, the results become worse, and it indicates that the dependence between losses cannot be ignored. Approximation (22) is useful if the workload process is unavailable, while the process of losses is observable, however estimation of P_a may a need special speed-up simulation technique to be effective.

6   Conclusion

In this work, we apply the regenerative approach to study multiserver queueing systems in which each customer has service time and a random volume. The service time and volume of each customer may be dependent. We develop a unified analysis of this system and establish some results known for conventional queueing systems in which the volumes of customers are not considered. In particular, we establish a solidarity property of all basic processes describing the system: queue size, workload and accumulated volume. Moreover, for the system with unlimited space, we present simple proofs of an analog of the celebrated Little's formula and (for the single-server system with Poisson input) a Pollaczek-Khintchine formula. For the system with limited space, we deduce an approximate Pollaczek-Khintchine formula, and verify its accuracy by regenerative simulation.

Acknowledgements. Research is supported by Russian Foundation for Basic Research, projects 15-07-02341, 15-07-02354, 15-07-02360 and the Program of Strategy Development of Petrozavodsk State University.

References

1. Asmussen, S.: Applied Probability and Queues. Springer, New York (2003)
2. Morozov, E.: The tightness in the ergodic analysis of regenerative queueing processes. Queueing Syst. **27**, 179–203 (1997)
3. Morozov, E., Delgado, R.: Stability analysis of regenerative queueing systems. Autom. Remote Control **70**(12), 1977–1991 (2009)
4. Morozov, E., Potakhina, L., Rumyantsev, A.: Stability analysis and simulation of a state-dependent transmission rate system. ManMach. Interact. **4**, 673–683 (2015)
5. Morozov, E., Nekrasova, R., Potakhina, L., Tikhonenko, O.: Asymptotic analysis of queueing systems with finite buffer space. Commun. Comput. Inf. Sci. **431**, 223–232 (2014)
6. Tikhonenko, O.M.: An analog of littles formula for queueing systems with nonhomogeneous tasks. Autom. Remote Control **57**(1), 85–88 (1996)
7. Tikhonenko, O.M.: The problem of determination of the summarized messages volume in queueing systems and its applications. J. Inf. Process. Cybern. **32**(7), 339–352 (1987)
8. Tikhonenko, O.: Computer Systems Probability Analysis. Akademicka Oficyna Wydawnicza EXIT, Warsaw (2006). (in Polish)
9. Center for collective use of Karelian Research Centre of Russian Academy of Science. http://cluster.krc.karelia.ru/
10. R Foundation for Statistical Computing, Vienna, Austria. ISBN 3-900051-07-0. http://www.R-project.org/

Inventory Management System with Erlang Distribution of Batch Sizes

Anatoly Nazarov and Valentina Broner[✉]

Tomsk State University, Lenina avenue 36, 634050 Tomsk, Russia
nazarov.tsu@gmail.com, valsubbotina@mail.ru

Abstract. The inventory management model with On/Off control of the inventory level is considered. Input product flow is continuous with a fixed rate. Demands are a Poisson process with constant intensity and the value of purchases are independent and identically-distributed random variables having Erlang Distribution. For this model we find explicit expression for the stationary distribution of the inventory level and numerical results are given.

Keywords: Inventory management · On/off control · Erlang distribution · Mathematical modeling

1 The Problem Statement

Over the last half century, various inventory control problem are considered. There are many articles about the newsvendor problem [1–4]. It has been studied since the eighteenth century and widely used to analyse systems with perishable products in such different fields as, for example, health insurances, airlines, sports and fashion industries. Also classical single-period problem [5,6] are widely-known, and other mathematical inventory control models [7].

In this article we consider a mathematical model of inventory management (Fig. 1). Let the product flow be continuous with fixed rate $\nu = 1$.

Let $s(t)$ be the inventory level at time t. Demand occurs according to a Poisson process with piecewise constant intensity $\lambda(s)$

$$\lambda(s) = \begin{cases} \lambda_1, s < S, \\ \lambda_2, s \geq S, \end{cases} \tag{1}$$

where S is the threshold inventory level of $s(t)$.

The values of purchases are independent and identically distributed random variables having an Erlang Distribution.

$$B(x) = 1 - \sum_{k=0}^{n-1} e^{-\mu x} \frac{(\mu x)^k}{k!}, k \geq 1. \tag{2}$$

This work is performed under the state order No. 1.511.2014/K of the Ministry of Education and Science of the Russian Federation.

© Springer International Publishing Switzerland 2016
A. Dudin et al. (Eds.): ITMM 2016, CCIS 638, pp. 273–280, 2016.
DOI: 10.1007/978-3-319-44615-8_24

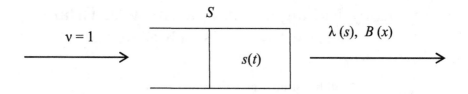

Fig. 1. Inventory management system

Note that the process $s(t)$ can be negative $s(t) < 0$, in this case the system works, and customers wait for the resource. The condition for the existence of a stationary distribution has the form

$$\lambda_1 b < 1 < \lambda_2 b, \tag{3}$$

where b is the mean of the probability distribution (2).

Accordingly, if $\lambda_1 < 1/b < \lambda_2$ and $s(t) < S$ then the stock level increases in the mean, i.e., the resources are accumulated. Otherwise $s(t)S$ and this condition means that the stock level decreases in the mean.

Taking into account (2) the mean b can be expressed as

$$b = \frac{n}{\mu}. \tag{4}$$

Based on the mathematical model analysis we conclude that $s(t)$ is a Markovian process with continuous time t and continuous state space $\infty < s < \infty$.

Let us denote the stationary density

$$P(s,t) = \frac{\partial P\{s(t) < s\}}{\partial s}.$$

The equation holds

$$P(s + \Delta t) = P(s)(1 - \lambda(s)\Delta t) + \Delta t \int_0^\infty \lambda(s + x)P(s + x)dB(x) + o(\Delta t),$$

we get

$$P'(s) + \lambda(s)P(s) = \int_0^\infty \lambda(s + x)P(s + x)dB(x), \tag{5}$$

where the boundary conditions have the form

$$P(-\infty) = P(\infty) = 0. \tag{6}$$

Let us find a solution $P(s)$ of the Eq. (5) in an explicit form, using Erlang distribution (2) as the distribution of values of purchase function. Let us denote

$$P(s) = \begin{cases} P_1(s), s < S, \\ P_2(s), s > S, \end{cases} \tag{7}$$

then we can rewrite the Eq. (6) as two equations

$$P_2'(s) + \lambda_2 P_2(s) = \lambda_2 \int_0^\infty P_2(s+x) dB(x), s > S, \tag{8}$$

$$P_1'(s) + \lambda_1 P_1(s) = \lambda_1 \int_0^{S-s} P_1(s+x) dB(x) + \lambda_2 \int_{S-s}^\infty P_2(s+x) dB(x), s < S. \tag{9}$$

We find solutions of Eqs. (8) and (9) that satisfy the boundary conditions

$$P_1(-\infty) = 0, P_2(\infty) = 0. \tag{10}$$

2   The Solution $P_2(s)$ of Equation (8)

Let solution $P_2(s), s > S$ of Eq. (8) have the form

$$P_2(s) = Ce^{-\gamma(s-S)}, s > S. \tag{11}$$

Substituting (11) into (8), we obtain the equation

$$\lambda_2 - \gamma = \lambda_2 \int_0^\infty e^{-\gamma x} dB(x), \tag{12}$$

Finding the integral in the right part (12), we have the expression

$$\lambda_2 - \gamma = \lambda_2 \left(\frac{\mu}{\mu + \gamma} \right)^n. \tag{13}$$

It is obvious that a zero root $\gamma = 0$ of Eq. (13) exists and it is an extraneous root because we have the boundary condition (10) $P_2(\infty) = 0$.

It is easy to see that under the condition (3) $\lambda_2 b > 1$ unique positive root $\gamma > 0$ of Eq. (13) for any distribution function $B(x)$ exists, then the solution of the Eq. (8) is a function (11) defined with multiplicative constant C accuracy, which value will be find later.

3   The Solution $P_1(s)$ of Equation (9)

Substituting (11) into (9), we obtain the following expression

$$P_1'(s) + \lambda_1 P_1(s) = \lambda_1 \int_0^{S-s} P_1(s+x) dB(x) + \lambda_2 Ce^{-\gamma(s-S)} \int_{S-s}^\infty e^{-\gamma x} dB(x). \tag{14}$$

Using (2), we get

$$\int\limits_{S-s}^{\infty} e^{-\gamma x} dB(x) = \int\limits_{S-s}^{\infty} e^{-\gamma x} \mu e^{-\mu x} \frac{(\mu x)^{n-1}}{(n-1)!} dx = \int\limits_{S-s}^{\infty} e^{-(\gamma+\mu)x} \frac{(\mu x)^{n-1}}{(n-1)!} d\mu x.$$

We make the change of variable $z = \mu x$

$$\int\limits_{S-s}^{\infty} e^{-\gamma x} dB(x) = \int\limits_{\mu(S-s)}^{\infty} e^{\frac{-(\gamma+\mu)}{\mu} z} \frac{(z)^{n-1}}{(n-1)!} dz = F_{n-1}.$$

Let $w = \mu(S-s)$ and $\alpha = (\mu+\gamma)/\mu$, then

$$F_n = \int\limits_{w}^{\infty} e^{-\alpha z} \frac{z^n}{n!} dz = -\frac{1}{\alpha} \frac{z^n}{n!} e^{-\alpha z}\big|_{z=w}^{\infty} + \frac{1}{\alpha} \int\limits_{w}^{\infty} e^{-\alpha z} \frac{z^{n-1}}{(n-1)!} dz$$

$$= \frac{1}{\alpha} \frac{w^n}{n!} e^{-\alpha w} + \frac{1}{\alpha} F_{n-1}.$$

Assuming $F_n = \frac{1}{\alpha^n} C(n)$ and $F_0 = \frac{1}{\alpha^n} e^{-\alpha w} = C(0)$ we obtain

$$\frac{1}{\alpha^n} C(n) = \frac{1}{\alpha} \frac{1}{\alpha^{n-1}} C(n-1) + \frac{1}{\alpha} \frac{w^n}{n!} e^{-\alpha w},$$

and

$$C(n) = C(n-1) + \alpha^{n-1} \frac{w^n}{n!} e^{-\alpha w}.$$

Then, we find

$$C(n) = C(0) + \sum_{m=1}^{n} \alpha^{m-1} \frac{w^m}{m!} e^{-\alpha w} = \frac{1}{\alpha} e^{-\alpha w} \sum_{m=0}^{n} \frac{(\alpha w)^n}{m!}$$

and

$$F_n = \frac{1}{\alpha^{n+1}} e^{-\alpha w} \sum_{m=0}^{n} \frac{(\alpha w)^n}{m!}.$$

So, the equation holds

$$\int\limits_{S-s}^{\infty} e^{-\gamma x} dB(x) = \frac{1}{\left(\frac{\mu+\gamma}{\mu}\right)^n} e^{\frac{\mu+\gamma}{\mu}\mu(s-S)} \sum_{k=0}^{n-1} \frac{[(\mu+\gamma)(S-s)]^k}{k!}.$$

From (14) we get

$$P_1'(s) + \lambda_1 P_1(s)$$
$$= \lambda_1 \int\limits_{0}^{S-s} P_1(s+x) b(x) dx + \lambda_2 e^{-\mu(S-s)} \left(\frac{\mu}{\mu+\gamma}\right)^n \sum_{k=0}^{n-1} \frac{[(\mu+\gamma)(S-s)]^k}{k!}. \tag{15}$$

Theorem 1. *Solution $P_1(s)$ of Eq. (15) has the form*

$$P_1(s) = C \sum_{\nu=1}^{n} x_\nu e^{z_\nu(s-S)}, s < S, \tag{16}$$

where $z = z_\nu$, $\nu = \overline{1,n}$ is a nonzero roots of equation

$$z + \lambda_1 = \lambda_1 \left(\frac{\mu}{\mu - z}\right)^n, \tag{17}$$

x_ν are components of the vector $\mathbf{X}$, $\mathbf{X}$ is a solution to a system of linear algebraic equations

$$\mathbf{AX} = \mathbf{E} \tag{18}$$

where $\mathbf{E}$ single column vector and $A_{k\nu}$ are elements of matrix $\mathbf{A}$

$$A_{k\nu} = \left(\frac{\mu - z_\nu}{\mu + \gamma}\right)^{k-n-1} \frac{\lambda_1}{\lambda_2}, k, \nu = \overline{1,n}, \tag{19}$$

normalizing constant C is determined by the equation

$$C = \left(\frac{1}{\gamma} + \sum_{\nu=1}^{n} \frac{x_\nu}{z_\nu}\right)^{-1} \tag{20}$$

Proof. Solution $P_1(s)$ of the Eq. (14) will be found in the form (16).
Substituting (16) into (14) we obtain the equation

$$\sum_{\nu=1}^{n} x_\nu e^{z_\nu(s-S)} \left\{ z_\nu + \lambda_1 - \lambda_1 \left(\frac{\mu}{\mu - z_\nu}\right)^n \right\}$$

$$= e^{\mu(s-S)} \left\{ \lambda_2 \left(\frac{\mu}{\mu+\gamma}\right)^n \sum_{m=0}^{n-1} \frac{[(\mu+\gamma)(S-s)]^m}{m!} \right.$$

$$\left. -\lambda_1 \sum_{\nu=1}^{n} x_\nu \left(\frac{\mu}{\mu - z_\nu}\right)^n \sum_{m=0}^{n-1} \frac{[(\mu - z_\nu)(S-s)]^m}{m!} \right\}.$$

Denote $y = (\mu + \gamma)(S - s)$, $(\mu - z_\nu)(S - s) = \frac{\mu - z_\nu}{\mu+\gamma} y$, we get

$$\sum_{\nu=1}^{n} x_\nu e^{z_\nu(s-S)} \left\{ z_\nu + \lambda_1 - \lambda_1 \left(\frac{\mu}{\mu - z_\nu}\right)^n \right\}$$

$$= \sum_{\nu=1}^{n} x_\nu e^{z_\nu(s-S)} \left\{ z_\nu + \lambda_1 - \lambda_1 \left(\frac{\mu}{\mu - z_\nu}\right)^n \right\}.$$

Equating to zero the coefficients in the linear combination of exponents $e^{z_\nu(s-S)}$ in this expression, we obtain the equality

$$z_\nu + \lambda_1 = \lambda_1 \left(\frac{\mu}{\mu - z_\nu}\right)^n, \nu = \overline{1,n}.$$

It is easy to see that this expression coincides with (17). Therefore, z_ν are the roots of the Eq. (17).

Similarly, for $e^{\mu(s-S)}y^m$ we obtain

$$\sum_{\nu=1}^{n} x_\nu \frac{\lambda_1}{\lambda_2}\left(\frac{\mu - z_\nu}{\mu + \gamma}\right)^{m-n} = 1, m = \overline{0, n}.$$

These equations are a non-homogeneous system of linear algebraic equations for x_ν, coinciding with the system (18), where the elements $A_{k\nu}$ matrix $\mathbf{A}$ defined by (19).

The value of the constant C can be found from the normalization condition

$$1 = \int_{-\infty}^{\infty} P(s)ds = \int_{-\infty}^{S} P_1(s)ds + \int_{S}^{\infty} P_2(s)ds$$

$$= C\sum_{\nu=1}^{n} x_\nu \int_{-\infty}^{S} e^{z_\nu(s-S)}ds + C\int_{S}^{\infty} e^{-\gamma(s-S)}ds$$

$$= C\sum_{\nu=1}^{n} x_\nu \int_{-\infty}^{0} e^{z_\nu x}dx + C\int_{0}^{\infty} e^{-\gamma x}dx = C\left\{\sum_{\nu=1}^{n} \frac{x_\nu}{z_\nu} + \frac{1}{\gamma}\right\}.$$

We get

$$C = \left(\sum_{\nu=1}^{n} \frac{x_\nu}{z_\nu} + \frac{1}{\gamma}\right)^{-1}.$$

It is easy to see that this expression and (20) have the same form.

The theorem is proved.

From (16) and (11), density $P(s)$ has the form

$$P(s) = \left(\sum_{\nu=1}^{n} \frac{x_\nu}{z_\nu} + \frac{1}{\gamma}\right)^{-1} \cdot \begin{cases} \sum_{\nu=1}^{n} x_\nu e^{z_\nu(s-S)}, s < S, \\ e^{-\gamma(s-S)}, s > S, \end{cases} \tag{21}$$

where z_ν is a nonzero root of Eq. (17), γ is unique positive root of Eq. (13), x_ν are components of the vector $\mathbf{X}$, $\mathbf{X}$ is a solution to a system of linear algebraic Eq. (18).

The explicit expression (21) for $P(s)$ of the Eq. (6) solves the problem of the study of mathematical models of inventory control with relay control and Erlang distribution batch sizes completely.

4　Numerical Results

Random demand has third-order Erlang distribution with parameter $\mu = 3$ and mean $b = 1$.

For the following values of the parameters $\lambda_1 = 0.8$ $\lambda_2 = 1.2$, $S = 10$ We found the roots of Eqs. (13) and (17).

Thus, the Eq. (13) has a unique positive solution γ, the Eq. (17) has one (z_1) real and two (z_2, z_3) complex conjugate roots with positive real parts. Let us find the density of the inventory level for the given parameters.

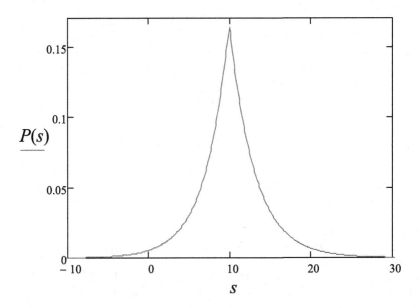

Fig. 2. Probability density function of process $P(s)$

Table 1. The numerical results

n	3	4	5	6	7
γ	0,291	0,309	0,320	0,328	0,334
z_1	0,311	0,335	0,350	0,362	0,370
x_1	1.077	1.092	1.103	1.110	1.116
z_2	$3.94 - 1.35i$	$4.28 + 2.44i$	$4.34 + 3.26i$	$4.29 - 3.82i$	$4.20 - 4.35i$
x_2	$-0.038 + 0.028i$	$-0.028 - 0.04i$	$-0.021 - 0.046i$	$-0.015 + 0.049i$	$-0.011 + 0.051i$
z_3	$3.94 + 1.35i$	$4.28 - 2.44i$	$4.34 - 3.26i$	$4.29 + 3.82i$	$4.20 + 4.35i$
x_3	$-0.038 - 0.028i$	$-0.028 + 0.04i$	$-0.021 + 0.046$	$-0.015 - 0.049i$	$-0.011 - 0.051i$
z_4		6.32	$7.59 + 1.73i$	$8.12 + 3.13i$	$8.40 + 4.65i$
x_4		-0.036	$-0.031 - 0.012i$	$-0.026 - 0.018i$	$-0.022 + 0.022i$
z_5			$7.59 - 1.73i$	$8.12 - 3.13i$	$8.40 - 4.65i$
x_5			$-0.031 + 0.012i$	$-0.026 + 0.018i$	$-0.022 + 0.022i$
z_6				9.895	$11.32 + 1.96i$
x_6				-0.028	$-0.025 - 0.006i$
z_7					$11.32 - 1.96i$
x_7					$-0.025 + 0.006i$

The parameters $x_\nu, \nu = \overline{1, n}$, of distribution (16), have the form

$$x_1 = 1.077; x_2 = -0.038 + 0.028i; x_3 = -0.038 - 0.028i,$$

and normalizing constant $C = 0.146$, the resulting distribution is shown in Fig. 2.

For different orders of Erlang distribution we compute parameters for distribution $P(s)$. Results are shown in Table 1.

Obviously, the distribution $P(s)$ of values of process $s(t)$ is continuous for all values of $s \neq S$, but also at the point $s = S$, it is not obvious.

In this work a mathematical model of the inventory management system is considered. We obtain an explicit expression for the stationary distribution of inventory levels under the following conditions: Erlang distribution of purchase values and relay control of the inventory level.

Acknowledgments. The work is supported by Tomsk State University Competitiveness Improvement Program.

References

1. Arrow, K.J., Harris, T.E., Marschak, J.: Optimal inventory policy. Econometrica **19**(3), 205–272 (1951)
2. Silver, E.A., Pyke, D.F., Peterson, R.: Inventory Management and Production Planning and Scheduling. Wiley, New York (1998)
3. Qin, Y., Wang, R., Vakharia, A., Chen, Y., Hanna-Seref, M.: The newsvendor problem: review and directions for future research. Eur. J. Oper. Res. **213**, 361–374 (2011)
4. Kitaeva, A., Subbotina, V., Zmeev, O.: The newsvendor problem with fast moving items and a compound poisson price dependent demand. In: 15th IFAC Symposium on Information Control Problems in Manufacturing INCOM 2015, Elsevier, (IFAC-PapersOnLine), vol. 48, pp. 1375–1379 (2015)
5. Khouja, M.: The single-period (news-vendor) problem: literature review and suggestions for future research. OMEGA-INT J. **27**(5), 537–553 (1999)
6. Kitaeva, A., Subbotina, V., Stepanova, N.: Estimating the compound poisson demand's parameters for single period problem for large lot size. In: 15th IFAC Symposium on Information Control Problems in Manufacturing INCOM 2015, Elsevier, (IFAC-PapersOnLine), vol. 48, pp. 1357–1361 (2015)
7. Kitaeva, A., Subbotina, V., Zmeev, O.: Diffusion appoximation in inventory management with examples of application. In: Dudin, A., Nazarov, A., Yakupov, R., Gortsev, A. (eds.) ITMM 2014. CCIS, vol. 487, pp. 189–196. Springer, Heidelberg (2014)

Markov and Non-Markov Probabilistic Models of Interacting Flows of Annihilating Particles

Anatoly Nazarov[1], Mais Farkhadov[2(✉)], and Erol Gelenbe[3]

[1] National Research Tomsk State University, TSU, 36 Lenin Prospekt,
Tomsk 634050, Russia
nazarov.tsu@gmail.com
[2] V.A. Trapeznikov Institute of Control Sciences of RAS,
Profsoyuznaya str. 65, Moscow 117997, Russia
mais.farhadov@gmail.com
[3] Intelligent Systems and Networks (ISN),
Department of Electrical and Electronic Engineering, Imperial College London,
EEE Bldg Rm 1009, South Kensington Campus, London, UK
e.gelenbe@imperial.ac.uk

Abstract. We propose Markov and non-Markov probabilistic models of how flows of annihilating particles interact, find the probability distribution of the number of positive applications in the model, and we present asymptotic results for the case of high intensity incoming flows. Then we study a system with non-exponential service where, using asymptotic analysis, we show that as the intensity of incoming flows grows, the probability distribution becomes Gaussian and find the parameters of the distribution. We also investigate flows of interacting particles as an infinitely linear queuing system with positive and negative applications of different systems and the probability distribution of the number of positive stationary applications in a system with exponential service is found. We also studied a case of arbitrary service by means of asymptotic analysis. We demonstrate that these systems are asymptotically equivalent.

Keywords: Queuing system · Method of asymptotic analysis · Flows of annihilating particles · Negative application

1 Introduction

We investigate the problem of interacting flows of elementary particles which annihilate in two ways. Flows of particles and antiparticles are counter flows; a positive particle stays in the interaction zone for a random time. The particle gets annihilated with an antiparticle that is located within the interaction zone and moves in the opposite direction. The antiparticle is not delayed in the interaction zone if positive particles are absent in it. The second case is when positive particles stay in the interaction zone for a random time. Antiparticles move into the interaction zone and enter a so-called "trap", where they are unable to continue their motion. The next positive particles that come inside the interaction zone, get annihilated with the antiparticles located in the "trap". If there are no

© Springer International Publishing Switzerland 2016
A. Dudin et al. (Eds.): ITMM 2016, CCIS 638, pp. 281–291, 2016.
DOI: 10.1007/978-3-319-44615-8_25

antiparticles, a positive particle stays in the interaction zone for a random time, and then leaves the system.

We propose Markov and non-Markovian probabilistic models of such processes of particle interactions. For the Markov model, we find the probability distribution of the number of positive applications. Asymptotic results are obtained for incoming flows of high intensity. Next, we investigate a system with non-exponential service by means of asymptotic analysis [1]. As the incoming flow grows in intensity, we see that the probability distribution tends to a Gaussian and find the parameters of the distribution.

2   Model

Let us take the following system as a model of interacting flows of annihilating particles: an infinitely linear queuing system with positive and negative applications, where the flow of particles is defined as the Poisson flow of positive applications with parameter λ^+ and an antiparticle flow as the Poisson flow of negative applications with parameter γ^-. The concept of negative applications was first introduced by Gelenbe in 1991 [2] and also in several other papers such as [10,11]. Finite linear QS with negative applications were also investigated by P.P. Bocharov, Ch.D. Apiche, R. Manzo, A.V. Pechinkin, R.V. Razumchik [3–5], Yang Woo S. [6], Quan-Lin L., Yiqiang Q.Z. [7]. According to these models, we investigate the following two classes of systems:

1. A system without waiting.
2. A system where negative applications wait for new positive ones.

In the first system, illustrated in Fig. 1, the positive and negative applications are counter flows. Any positive application that comes into the system immediately gets treated, taking any of the available instruments; a positive application gets serviced exponentially with parameter μ. If the system is empty, negative applications just leave it. On the other hand, if the system is non-empty, that is, in cases when there is at least one positive application, the positive application is destroyed by the negative one and together they leave the system.

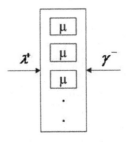

Fig. 1. System without waiting

In the second system, shown in Fig. 2, the positive and negative applications come in parallel. Any positive application that comes into the system without negative applications, immediately gets treated, taking any of the available instruments. As a negative application comes in, it waits for a new positive one to arrive, "kills" it and both leave the system. If there are no positive applications, the negative application waits for a positive application to arrive and this procedure repeats.

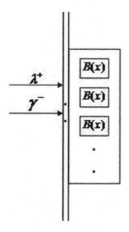

Fig. 2. System where negative applications wait for positive ones

3   System Without Waiting

Let us look at the Fig. 1, where this system is depicted. Let $i(t)$ be the number of positive applications in the system at time t. Since this process $i(t)$ is a Markov chain, the probability distribution

$$P_i(t) = P\{i(t) = i\}, \; i = \overline{0, \infty}$$

is governed by a system of Kolmogorov equations. In the stationary mode, where $P_i(t) = P_i$, the system of equations is as follows:

$$\begin{cases} -\lambda^+ P_0 + (\mu + \gamma^-)P_1 = 0, \\ \lambda^+ P_{i-1} - (\lambda^+ + i\mu + \gamma^-)P_i + [(i+1)\mu + \gamma^-]P_{i+1} = 0. \end{cases}$$

The solution of this system is

$$P_i = P_0 \prod_{n=1}^{i} \frac{\lambda^+}{n\mu + \gamma^-}. \tag{1}$$

The normalization condition gives P_0:

$$P_0 = \left(1 - \sum_{i=1}^{\infty} \prod_{n=1}^{i} \frac{\lambda^+}{n\mu + \gamma^-}\right)^{-1}. \tag{2}$$

Thus we obtained the stationary distribution of the number of positive applications in our system without waiting, with negative applications in the system (formulas (1) and (2)).

We introduce the characteristic function

$$H(u) = \sum_{i=0}^{\infty} e^{jui} P_i, \; j = \sqrt{-1}.$$

If the incoming flows are of high intensity, we examine our system by asymptotic analysis.

The condition of high intensity of the incoming flows is described by the formula

$$\lambda^+ = \lambda N, \; \gamma^- = \gamma N,$$

where λ, γ are some finite variables, while $N \to \infty$. Then asymptotic analysis of this system gives

$$h(u) = \exp\left\{juN\frac{\lambda - \gamma}{\mu} - \frac{u^2}{2}\frac{\lambda}{\mu}N\right\}, \tag{3}$$

where $h(u)$ is the asymptotic characteristic function of the number of positive applications in the system.

4   System with Waiting and Arbitrary Service

We consider a queuing system with an unlimited number of devices (Fig. 2); the system includes two incoming flows: a simple flow of positive applications with parameter λ^+ and a simple flow of negative applications with parameter γ^-. Positive applications have random service time with distribution function in $B(x)$.

Let $i(t)$ be the number of positive application in the system, while $l(t)$ is the number of negative one at time t, The process $\{i(t), l(t)\}$ is non-Markov.

4.1   Method of Sieved Flow

We investigate the following version of the sieved flow method [1]. Let $S(\tau) = 1 - B(T - \tau)$ be the probability that a positive application, which arrived at time τ, gets sieved into the flow. The applications that are still in the system by the time T are sieved.

Let us denote $n(t)$ as the number of positive applications arrived with the sieved flow up until time $t < T$. That is the number of events of the positive application flow that has not been "killed" by negative applications and has got sieved up until time t.

Also let us take a look at the sieved flow of negative applications $m(t)$, which is the number of events of the negative application flow that have not been "killed" by positive applications up until time t.

Notice also that $i(T) = n(T), l(T) = m(T)$, where the relation holds true in probability.

Our method of sieved flow is illustrated in Fig. 3.

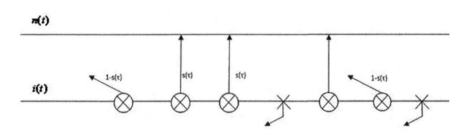

Fig. 3. Illustration of the sieved flow method, applied to our system with waiting

For our queuing system with waiting, the Δt method gives the following system of Kolmogorov equations

$$
\begin{cases}
P(n,m,t+\Delta t) = P(n,m,t)(1-\lambda^+\Delta t)(1-\gamma^-\Delta t) \\
\qquad +P(n,m-1,t)\gamma^-\Delta t + P(n,m+1,t)\lambda^+\Delta t + o(\Delta t),\ n \geq 0, \\
P(n,0,t+\Delta t) = P(n,0,t)[(1-\lambda^+\Delta t)(1-\gamma^-\Delta t) + \lambda^+\Delta t(1-S(t))] \\
\qquad +P(n-1,0,t)\lambda^+S(t)\Delta t + P(n,1,t)\lambda^+\Delta t + o(\Delta t),\ n \geq 1, \\
P(0,0,t+\Delta t) = P(0,0,t)[(1-\lambda^+\Delta t)(1-\gamma^-\Delta t) + \lambda^+\Delta t(1-S(t))] \\
\qquad +P(0,1,t)\lambda^+\Delta t + o(\Delta t).
\end{cases}
$$

Doing the necessary transformations and taking the limit $\Delta t \to 0$, the system of Kolmogorov differential equations becomes

$$
\begin{cases}
\dfrac{\partial P(n,m,t)}{\partial t} = -(\lambda^+ + \gamma^-)P(n,m,t) + \gamma^- P(n,m-1,t) \\
\qquad +\lambda^+ P(n,m+1,t),\ n \geq 0, \\
\dfrac{\partial P(n,0,t)}{\partial t} = -(\lambda^+ S(t) + \gamma^-)P(n,0,t) + \lambda^+ S(t)P(n-1,0,t) \\
\qquad +\lambda^+ P(n,1,t),\ n \geq 1, \\
\dfrac{\partial P(0,0,t)}{\partial t} = -(\lambda^+ S(t) + \gamma^-)P(0,0,t) + \lambda^+ P(0,1,t).
\end{cases}
\tag{4}
$$

4.2 First Order Asymptotics

Let us introduce the following partial characteristic functions

$$H(u, m, t) = \sum_{n=0}^{\infty} e^{jun} P(n, m, t).$$

With this taken into account, the system (4) becomes

$$\begin{cases} \dfrac{\partial H(u, m, t)}{\partial t} = -(\lambda^+ + \gamma^-) H(u, m, t) \\ \qquad\qquad + \gamma^- H(u, m - 1, t) + \lambda^+ H(u, m + 1, t), \\ \dfrac{\partial H(u, 0, t)}{\partial t} = -(\lambda^+ S(t) + \gamma^-) H(u, 0, t) \\ \qquad\qquad + \lambda^+ S(t) H(u, 0, t) e^{ju} + \lambda^+ H(u, 1, t). \end{cases} \qquad (5)$$

Let the limiting condition of high intensity flow be $\lambda^+ = \lambda N$, $\gamma^- = \gamma N$, where λ, γ are some finite variables, and $N \to \infty$. Let us denote $\dfrac{1}{N} = \varepsilon$ and substitute $u = \varepsilon w$, $H(u, m, t) = F(w, m, t, \varepsilon)$. Then, with the introduced notations taken into consideration, the system (5) becomes

$$\begin{cases} \varepsilon \dfrac{\partial F(w, m, t, \varepsilon)}{\partial t} = -(\lambda + \gamma) F(w, m, t, \varepsilon) \\ \qquad\qquad + \gamma F(w, m - l, t, \varepsilon) + \lambda F(w, m + 1, t, \varepsilon), \ m \geq 1, \\ \varepsilon \dfrac{\partial F(w, 0, t, \varepsilon)}{\partial t} = -(\lambda S(t) + \gamma) F(w, 0, t, \varepsilon) \\ \qquad\qquad + \lambda S(t) F(w, 0, t, \varepsilon) e^{jw\varepsilon} + \lambda F(w, l, t, \varepsilon). \end{cases} \qquad (6)$$

Taking the limit $\varepsilon \to 0$, we get

$$\begin{cases} -(\lambda + \gamma) F(w, m, t) + \gamma F(w, m - 1, t) + \lambda F(w, m + 1, t) = 0, \\ -\gamma F(w, 0, t) + \lambda F(w, 1, t) = 0, \end{cases} \qquad (7)$$

where

$$F(w, m, t) = \lim_{\varepsilon \to 0} F(w, m, t, \varepsilon).$$

Let us set $F(w, m, t)$ to $F(w, m, t) = \Phi(w, t) R(m)$ and substitute this product into (7). We get:

$$\begin{cases} -(\lambda + \gamma) R(m) + \gamma R(m - 1) + \lambda R(m + 1) = 0, \\ -\gamma R(0) + \lambda R(1) = 0. \end{cases} \qquad (8)$$

The system has the following solution:

$$R(m) = R(0) \left(\frac{\gamma}{\lambda} \right)^m, \quad R(0) = 1 - \frac{\gamma}{\lambda}, \qquad (9)$$

as $\gamma < \lambda$.

To find $\Phi(\omega, t)$, we use the consistency condition and get

$$\Phi(\omega, t) = \sum_{m=0}^{\infty} F(\omega, m, t) = \lim_{\varepsilon \to 0} \sum_{m=0}^{\infty} F(\omega, m, t, \varepsilon) = \lim_{\varepsilon \to 0} F(\omega, t, \varepsilon) = F(\omega, t).$$

Let us take a look at this function

$$F(\omega, t, \varepsilon) = \sum_{m=0}^{\infty} F(\omega, m, t, \varepsilon).$$

Summing up all of the Eq. (6) over $m = \overline{0, \infty}$, we get

$$\varepsilon \frac{\partial F(\omega, t, \varepsilon)}{\partial t} = \lambda S(t) F(\omega, 0, t, \varepsilon)(e^{j\omega\varepsilon} - 1).$$

Expanding the exponent in a series, dividing over ε, and taking the limit $\varepsilon \to 0$, we get the following equation for the function

$$F(\omega, t) = \lim_{\varepsilon \to 0} F(\omega, t, \varepsilon), \quad \frac{\partial F(\omega, t)}{\partial t} = \lambda S(t) F(\omega, 0, t) j\omega.$$

Taking into account the condition (9), it comes out that

$$\frac{\partial \Phi(\omega, t)}{\partial t} = \lambda S(t) \Phi(\omega, t) R(0) j\omega. \tag{10}$$

The solution of (10) is of the form $\Phi(\omega, t) = \exp\{jk_1(t)\omega\}$, where

$$k_1(t) = (\lambda - \gamma) \int_0^1 S(\tau) d\tau. \tag{11}$$

Let us return to our characteristic functions $H(u, m, t)$. For them we get

$$H(u, m, t) = F(\omega, m, t, \varepsilon) \approx F(\omega, m, t) = \Phi(\omega, t) R(m)$$
$$= \exp\{juNk_1(t)\}\left(1 - \frac{\gamma}{\lambda}\right)\left(\frac{\gamma}{\lambda}\right)^m. \tag{12}$$

We call the expression of $H(u, m, t)$, which is (12), the first order asymptotics for the QS under consideration.

4.3   Second Order Asymptotics

Let us look at the function $H(u, m, t)$ rewritten in the form

$$H(u, m, t) = H^{(2)}(u, m, t) \exp\{juNk_1(t)\}.$$

The system (5) can be rewritten as

$$
\begin{cases}
\dfrac{1}{N}\dfrac{\partial H^{(2)}(u,m,t)}{\partial t} + jk_1'(t)uH^{(2)}(u,m,t) \\
\quad = -(\lambda+\gamma)H^{(2)}(u,m,t) + \gamma H^{(2)}(u,m-1,t) + \lambda H^{(2)}(u,m+1,t),\ m \ge 1, \\
\dfrac{1}{N}\dfrac{\partial H^{(2)}(u,0,t)}{\partial t} + jk_1'(t)uH^{(2)}(u,0,t) \\
\quad = -(\lambda S(t)+\gamma)H^{(2)}(u,0,t) + \lambda S(t)H(2)(u,0,t)e^{ju} + \lambda H^{(2)}(u,1,t).
\end{cases}
$$

Let us denote $\dfrac{1}{N} = \varepsilon^2$ and substitute $u = \varepsilon\omega$, $H^{(2)}(u,m,t) = F^{(2)}(\omega,m,t,\varepsilon)$.
Then the system of equations for $F^{(2)}(u,m,t,\varepsilon)$ can be rewritten as

$$
\begin{cases}
\varepsilon^2\dfrac{\partial F^{(2)}(\omega,m,t,\varepsilon)}{\partial t} + j\omega\varepsilon(\lambda-\gamma)S(t)F^{(2)}(\omega,m,t,\varepsilon) \\
\quad = -(\lambda+\gamma)F^{(2)}(\omega,m,t,\varepsilon) + \gamma F^{(2)}(\omega,m-1,t,\varepsilon) \\
\qquad + \lambda F^{(2)}(\omega,m+1,t,\varepsilon),\ m \ge 1, \\
\varepsilon^2\dfrac{\partial F^{(2)}(\omega,0,t,\varepsilon)}{\partial t} + j\omega\varepsilon(\lambda-\gamma)S(t)F^{(2)}(\omega,0,t,\varepsilon) \\
\quad = -(\lambda S(t)+\gamma)F^{(2)}(\omega,0,t,\varepsilon) + \lambda S(t)F^{(2)}(\omega,0,t,\varepsilon)e^{j\omega\varepsilon} \\
\qquad + \lambda F^{(2)}(\omega,1,t,\varepsilon).
\end{cases}
\tag{13}
$$

$F^{(2)}(\omega,m,t,\varepsilon)$ can be expanded as

$$
F^{(2)}(\omega,m,t,\varepsilon) = \Phi^{(2)}(\omega,t)\{R(m) + j\varepsilon\omega S(t)f(m)\} + o(\varepsilon^2),
\tag{14}
$$

where $R(m)$ is determined in (9). We are yet to obtain $f(m)$.

Let us substitute (14) into the system (13), take into account (8), and take the limit $\varepsilon \to 0$. We get

$$
\begin{cases}
-(\lambda+\gamma)f(m) + \gamma f(m-1) + \lambda f(m+1) = (\lambda-\gamma)R(m),\ m \ge 1, \\
\lambda f(1) - \gamma f(0) = -\gamma R(0).
\end{cases}
\tag{15}
$$

The system (15) is a system of nonhomogeneous finite difference equations in respect to $f(m)$, where m is a discrete argument.

The general solution of the corresponding homogeneous system is of the form

$$
f_o(m) = CR(m),
$$

where C is an arbitrary constant.

The solution of the nonhomogeneous system is the sum of the general solution of the homogeneous system and the particular integral of the nonhomogeneous system (15): $f(m) = f_o(m) + f(m)$. The particular integral is of the form $f(m) = -mR(m)$.

Let us consider the functions

$$
F^{(2)}(\omega,t,\varepsilon) = \sum_{m=0}^{\infty} F^{(2)}(\omega,m,t,\varepsilon).
$$

Summing up the Eq. (13) over $m = \overline{0, \infty}$, we get:

$$\varepsilon^2 \frac{\partial F^{(2)}(\omega, t, \varepsilon)}{\partial t} + j\varepsilon\omega(\lambda - \gamma)S(t)F^{(2)}(\omega, t, \varepsilon) = \lambda S(t)F^{(2)}(\omega, 0, t, \varepsilon)(e^{j\omega\varepsilon} - 1). \quad (16)$$

Let us denote

$$f = \sum_{m=0}^{\infty} f(m).$$

It follows that

$$F^{(2)}(\omega, t, \varepsilon) = \sum_{m=0}^{\infty} F^{(2)}(\omega, m, t, \varepsilon)$$

$$= \sum_{m=0}^{\infty} \{\Phi^{(2)}(\omega, t)(R(m) + j\omega\varepsilon S(t)f(m)) + o(\varepsilon^2)\}$$

$$= \Phi^{(2)}(\omega, t)(1 + j\omega\varepsilon S(t)f) + o(\varepsilon^2).$$

Then (16) can be rewritten as

$$\varepsilon^2 \frac{\partial \Phi^{(2)}(\omega, t)}{\partial t} + j^2\varepsilon^2\omega^2(\lambda - \gamma)S^2(t)\Phi^{(2)}(\omega, t)f$$

$$= j^2\omega^2\varepsilon^2\lambda S^2(t)\Phi^{(2)}(\omega, t)f(0) + \frac{j^2\omega^2\varepsilon^2}{2}\lambda R(0)S(t)\Phi^{(2)}(\omega, t) + o(\varepsilon^3).$$

Dividing over ε^2 and taking the limit $\varepsilon \to 0$, we get

$$\frac{\partial \Phi^{(2)}(\omega, t)}{\partial t} = -(j\omega)^2(\lambda - \gamma)S^2(t)\Phi^{(2)}(\omega, t)f + \lambda(j\omega)^2 S^2(t)\Phi^{(2)}(\omega, t)f(0)$$

$$+ \frac{(j\omega)^2}{2}\lambda R(0)S(t)\Phi^{(2)}(\omega, t).$$

The solution of this equation is of the form

$$\Phi^{(2)}(\omega, t) = \exp\{-\frac{(j\omega)^2}{2}k_2(t)\},$$

where

$$k_2(t) = \left[(\lambda - \gamma)\int_0^1 S(\tau)d\tau + 2\gamma\int_0^1 S^2(\tau)d\tau\right]. \quad (17)$$

The expression for the characteristic function $H(u, t)$ becomes:

$$H(u, t) = \sum_{m=0}^{\infty} H(u, m, t) = \exp\{jk_1(t)uN\} \sum_{m=0}^{\infty} H^{(2)}(u, m, t)$$

$$= \exp\{jk_1(t)uN\} \sum_{m=0}^{\infty} F^{(2)}(\omega, m, t, \varepsilon)$$

$$= \exp\{jk_1(t)uN\}F^{(2)}(\omega, t, \varepsilon) \approx \exp\{jk_1(t)uN\}F^{(2)}(\omega, t)$$

$$= \exp\{jk_1(t)uN\}\Phi^{(2)}(\omega, t) = \exp\{jk_1(t)uN - k_2(t)u^2N\}.$$

We call this relation the second order asymptotics of our system.

4.4   Asymptotics of Our Model with Waiting and Exponential Service Time

Let us investigate our model with waiting in case of exponential service time [9] with parameter μ, where $B(x) = 1 - e^{-\mu x}$. The functions $k_1(T), k_2(T)$, obtained in (11) and (17), are as follows:

$$k_1(T) = (\lambda - \gamma) \int_0^T S(\tau)d\tau = (\lambda - \gamma) \int_0^T (1 - B(x))dx,$$

$$k_2(T) = (\lambda - \gamma) \int_0^T S(\tau)d\tau + 2\gamma \int_0^T S^2(\tau)d\tau$$

$$= (\lambda - \gamma) \int_0^T (1 - B(x))dx + 2\gamma \int_0^T (1 - B(x))^2 dx.$$

Taking the limit $T \to \infty$, we get the following relation

$$k_1 = \lim_{T \to \infty} k_1(T) = (\lambda - \gamma)b, k_2 = \lim_{T \to \infty} k_2(T) = (\lambda - \gamma)b + 2\gamma\beta,$$

where

$$\beta = \int_0^\infty (1 - B(x))^2 dx, b = \int_0^\infty (1 - B(x))dx$$

is the average service time.

For our system with waiting and exponential service time $b = \dfrac{1}{\mu}$, $\beta = \dfrac{1}{2\mu}$, the characteristic function becomes

$$H(u, t) = \exp\left\{ ju\frac{(\lambda - \gamma)}{\mu} N - u^2 \frac{\lambda}{2\mu} N \right\}.$$

The function is identical to the characteristic function (3) of the system without waiting that was investigated in part 1. This, in turn, demonstrates that these two models are asymptotically equivalent in case of exponential service.

5   Conclusion

In this work we investigated two models of interacting flows of annihilating particles. In the first model, that is without waiting, we obtained a pre-limit stationary probability distribution of applications (particles) in the model, as well as gaining asymptotic results, namely the first and second asymptotic behavior of the distribution in case of high intensity of incoming flows. The second model with expectation and arbitrary service was analyzed asymptotically. We find the first and second order asymptotics, that, if the service time is exponentially distributed, coincide with the asymptotic behavior of the first model. Thus we demonstrated that the two models are asymptotically equivalent if the service is exponential.

Similar results were obtained in [12].

Acknowledgments. The work is supported by Tomsk State University Competitiveness Improvement Program.

References

1. Nazarov, A.A.: Asymptotic Analysis in Queuing Systems. Tomsk University Publisher, Tomsk (2006). (in Russian)
2. Gelenbe, E.: Queueing networks with negative and positive customers. J. Appl. Prob. **28**, 653–656 (1991)
3. Pechinkin, A.V., Razumchik, R.V.: Discrete time queuing system with negative applications and a cache for expunged applications. Autom. Telemechanics **12**, 109–120 (1991). (in Russian)
4. Pechinkin, A.V., Razumchik, R.V.: On time characteristics of an exponential queuing system with negative applications and a cache for expunged applications. Autom. Telemechanics **12**, 75–90 (1991). (in Russian)
5. Bocharov, P.P., D'Apiche, C., Manzo, R., Pechinkin, A.V.: Analysis of a multilinear Markov queuing system with infinite accumulator and negative applications. Autom. Telemechanics **1**, 93–104 (2007). (in Russian)
6. Shin, Y.W.: Multi-server retrial queue with negative customers and disasters. Queueing Syst. **55**, 223–237 (2007)
7. Quan-Lin, L., Yiqiang, Q.Z.: A MAP/G/1 queue with negative customers. Queuing Syst. **47**, 5–43 (2004)
8. Nazarov, A.A., Ferapontov, N.M.: Investigation of infinitely linear queuing systems with negative applications that get lost in an empty system. In: Probability Theory, Mathematical Statistics and Their Applications: Proceedings of the International Conference in Minsk 2015, pp. 208–213 (2015). (in Russian)
9. Nazarov, A.A., Ferapontov, N.M.: Investigation of infinitely linear queuing systems with negative applications and their waiting. In: Information Technology and Mathematical Modeling: Proceedings of XIII A.F. Terpugov International Conference 2014, Tomsk pp. 71–76 (2014). (in Russian)
10. Gelenbe, E.: Dealing with software viruses: a biological paradigm. Inform. Secur. Tech. Rep. (Elsevier) **12**(4), 242–250 (2007)
11. Gelenbe, E.: Network of interacting synthetic molecules in steady state. Proc. R. Soc. A **464**(2096), 2219–2228 (2008)
12. Nazarov, A.A., Feropontova, N.M.: Study of the interaction of fluxes of annihilating particles. Russ. Phys. J. **58**(8), 1118–1127 (2015)

Asymptotic Analysis Retrial Queueing System M/GI/1 with Hyper Exponential Distribution of the Delay Time in the Orbit and Exclusion of Alternative Customers

Anatoly Nazarov and Yana Izmaylova$^{(\boxtimes)}$

Tomsk State University, 36 Lenina Avenue, Tomsk, Russia
nazarov.tsu@gmail.com, evgenevna.92@mail.ru
http://www.tsu.ru

Abstract. This paper deals with a retrial queueing system in which the arrival flow is described by a stationary Poisson process, the service time is random with a given distribution function, hyper exponential distribution of the delay time of customers in the orbit and exclusion of alternative customers. We examine a retrial queueing system using the method of asymptotic analysis under the condition of long delay in the orbit. For use of this method we write the system of Kolmogorov's equations for the probability distribution of the number of customers in the orbit and the server state. We have completed the transition to the system of differential equations for partial characteristic function. Using the method of asymptotic analysis we obtain two-dimensional distribution of the number of customers in the orbit in the first and second phases. This distribution can be approximated by the two-dimensional Gaussian distribution. The values of the parameters are found.

Keywords: Retrial queueing system · Hyper exponential distribution the delay time of customers in the orbit · Exclusion of alternative customers · Asymptotic analysis

1 Introduction

Retrial queueing systems are characterized by the feature that arrivals who find the server unavailable are obliged to leave the service area and to try again for their requests in random order and at random intervals [1–3]. For recent works on retrial queues, see [4–6]. Between trials a customer is called being in "orbit". This feature plays a special role in several computer, communications networks, and call centers [7]. In paper [8], the authors assume that intervals between there retrials are exponentially distributed. In real system intervals between retrials can be no exponentially distribution. S.R. Chakravarthy, A. Dudin assume that

This work is performed under the state order No. 1.511.2014/K of the Ministry of Education and Science of the Russian Federation.

© Springer International Publishing Switzerland 2016
A. Dudin et al. (Eds.): ITMM 2016, CCIS 638, pp. 292–302, 2016.
DOI: 10.1007/978-3-319-44615-8_26

the probability of a repeated attempt is independent of the number of orbiting customers (constant retrial policy) [9]. In [10] intervals retrial times are exponentially distributed with intensity depending on the number of customers in the orbit. In the paper [11] a retrial queue system M/M/1 with phase-type retrial times was studied. In this paper, we consider a retrial queueing system in which intervals between retrials are hyper exponentially distributed. We assume that a customer who finds the server to be busy then replaces the customer who is being served. We have a priority mechanism of new customers [12].

A priority mechanism is an invaluable scheduling method that allows customers to receive different qualities of service. Service priority is nowadays today a main feature of the operation of any manufacturing system. For this reason, priority queues have received considerable attention in the literature [13,14]. A review of the main results of such retrial models can be found in the survey paper of B.D. Choi and Y. Chang [15], K.B. Choi and Y.W. Lee [16].

The remainder of this paper is organized as follows. A model description is given in Sect. 2. In Sect. 3 we write a system of differential Kolmogorov's equations for the probability distribution of the number of customers in the orbit and the server state. In Sect. 4 we have obtained the system equations for the partial characteristic function. We examine a retrial queueing system using the method of asymptotic analysis [17] under the condition of long delay in the orbit. A solution of the system equations under this condition allows building an approximation for two-dimensional distribution of the system state (see Sects. 5 and 6). Some numerical examples are shown in Sect. 7. Finally, Sect. 8 concludes the paper.

2   The Mathematical Model

We consider a retrial queueing system M/GI/1 with hyper exponential distribution of the delay time in the orbit. The structure of the system is depicted in Fig. 1.

We assume that the arrival flow to the system is described by stationary Poisson process with intensity λ. A customer who finds the server to be free occupies it for service during a random time with distribution function $B(x)$. If the server is busy, then an arriving customer replaces the customer who is being served and occupies the server. The customer who was being served, moves to a so-called customer's orbit, where it performs a random delay with the duration determined by hyper exponential distribution with parameters q, σ_1, σ_2. A customer performs the exponentially random delay with intensity σ_1 (first phase) with probability q. And a customer performs the exponentially random delay with parameter σ_2 (second phase) with probability $1 - q$. From the orbit, after the random delay, the customer occupies the device again. The same goes for the customer from the orbit.

Orbit

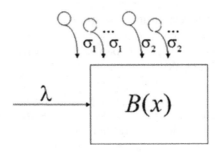

Fig. 1. Mathematical model

Let

$$B^*(\alpha) = \int\limits_0^\infty \exp(-\alpha x)\, dB(x)$$

be a Laplace-Stieltjes transform of the distribution function $B(x)$. Let $i_1(t)$ be the number of customers in the orbit in the first phase and $i_2(t)$ be the number of customers in the orbit in the second phase, $k(t)$ define the server state in the following way:

$$k(t) = \begin{cases} 0, & \text{if server is free,} \\ 1, & \text{if server is busy.} \end{cases}$$

We would like to solve a problem of computation of two-dimensional stationary probability distribution of the number of customers i_1, i_2 in the orbit for each phase and server state.

3   System of Kolmogorov Equations

We consider Markovian process $\{k(t), i_1(t), i_2(t)\}$.

Let us denote by $P\{k(t) = 0, i_1(t) = i_1, i_2(t) = i_2\} = P_0(i_1, i_2, t)$ a probability that, at the moment t, the server in the state 0 and i_1 customers are in the orbit in the first phase, i_2 customers are in the orbit in the second phase.

Let us denote by $P\{k(t) = 1, i_1(t) = i_1, i_2(t) = i_2, z(t) < z\} = P_1(i_1, i_2, z, t)$, a probability that, at the moment t, the server in the state 1, residual service time is less than z and the i_1 customers are in the orbit in the first phase, i_2 customers are in the orbit in the second phase.

We obtain the following system of Kolmogorov equations for probabilities $\{P_0(i_1, i_2, t), P_1(i_1, i_2, z, t)\}$

$$\frac{\partial P_1(i_1, i_2, 0)}{\partial z} = (\lambda + i_1\sigma_1 + i_2\sigma_2)P_0(i_1, i_2),$$

$$\frac{\partial P_1(i_1, i_2, 0)}{\partial z} - \frac{\partial P_1(i_1, i_2, z)}{\partial z} = -(\lambda + i_1\sigma_1 + i_2\sigma_2)P_1(i_1, i_2, z)$$

$$+\lambda B(z)P_0(i_1, i_2) + (i_1 + 1)\sigma_1 B(z)P_0(i_1 + 1, i_2)$$

$$+(i_2 + 1)\sigma_2 B(z)P_0(i_1, i_2 + 1) + \lambda q B(z)P_1(i_1 - 1, i_2)$$

$$+\lambda(1 - q)B(z)P_1(i_1, i_2 - 1) + i_1\sigma_1 q B(z)P_1(i_1, i_2)$$

$$+(i_1 + 1)\sigma_1 B(z)(1 - q)P_1(i_1 + 1, i_2 - 1)$$

$$+i_2\sigma_2(1 - q)B(z)P_1(i_1, i_2) + (i_2 + 1)\sigma_2 B(z)q P_1(i_1 - 1, i_2 + 1).$$

$$(1)$$

4   System of Kolmogorov Equations for Partial Characteristic Function

We introduce partial characteristic functions as follows

$$H_0(u_1, u_2) = \sum_{i_1=0}^{\infty}\sum_{i_2=0}^{\infty} e^{ju_1 i_1} e^{ju_2 i_2} P_0(i_1, i_2),$$

$$H_1(u_1, u_2, z) = \sum_{i_1=0}^{\infty}\sum_{i_2=0}^{\infty} e^{ju_1 i_1} e^{ju_2 i_2} P_0(i_1, i_2, z).$$

Here $j = \sqrt{-1}$ is an imaginary unit.

Let us denote

$$H_1(u_1, u_2, \infty) = H_1(u_1, u_2), \frac{\partial H_1(u_1, u_2, 0)}{\partial z} = \frac{\partial H_1(u_1, u_2, z)}{\partial z}\bigg|_{z=0}.$$

Using (1) write the system of equations for the partial characteristic functions in the following form

$$-\frac{\partial H_1(u_1, u_2, 0)}{\partial z} = -\lambda H_0(u_1, u_2) + j\sigma_1\frac{\partial H_0(u_1, u_2)}{\partial u_1} + j\sigma_2\frac{\partial H_0(u_1, u_2)}{\partial u_2},$$

$$-\frac{\partial H_1(u_1, u_2, z)}{\partial z} + \frac{\partial H_1(u_1, u_2, 0)}{\partial z} = -\lambda H_1(u_1, u_2, z)$$

$$+j\sigma_1\frac{\partial H_1(u_1, u_2, z)}{\partial u_1} + j\sigma_2\frac{\partial H_1(u_1, u_2, z)}{\partial u_2} + \lambda B(z)H_0(u_1, u_2)$$

$$-je^{-ju_1}\sigma_1 B(z)\frac{\partial H_0(u_1, u_2)}{\partial u_1} - je^{-ju_2}\sigma_2 B(z)\frac{\partial H_0(u_1, u_2)}{\partial u_2} \qquad (2)$$

$$+\lambda q B(z)e^{ju_1}H_1(u_1, u_2) + \lambda(1 - q)B(z)e^{ju_2}H_1(u_1, u_2)$$

$$-jq\sigma_1 B(z)\frac{\partial H_1(u_1, u_2)}{\partial u_1} - j(1 - q)\sigma_2 B(z)\frac{\partial H_1(u_1, u_2)}{\partial u_2}$$

$$-j(1 - q)\sigma_1 B(z)e^{j(u_2-u_1)}\frac{\partial H_1(u_1, u_2)}{\partial u_1} - je^{j(u_1-u_2)}q\sigma_2 B(z)\frac{\partial H_1(u_1, u_2)}{\partial u_2}.$$

In the following subsections we will solve system (2) using the method of asymptotic analysis under the condition of long delay ($\sigma \to 0$), believing that $\sigma_1 = \sigma\gamma_1$, $\sigma_2 = \sigma\gamma_2$.

5 The First-Order Asymptotic Form

Let us denote $\sigma = \varepsilon$.

Introducing following substitute

$$u_1 = \varepsilon w_1, u_2 = \varepsilon w_2, H_0(u_1, u_2) = F_0(w_1, w_2, \varepsilon),$$
$$H_1(u_1, u_2, z) = F_1(w_1, w_2, z, \varepsilon).$$

We can transform system (2) to the form

$$
\begin{aligned}
-\frac{\partial F_1(w_1, w_2, 0)}{\partial z} &= -\lambda F_0(w_1, w_2, \varepsilon) \\
+j\gamma_1 \frac{\partial F_0(w_1, w_2, \varepsilon)}{\partial w_1} &+ j\gamma_2 \frac{\partial F_0(w_1, w_2, \varepsilon)}{\partial w_2}, \\
-\frac{\partial F_1(w_1, w_2, z, \varepsilon)}{\partial z} + \frac{\partial F_1(w_1, w_2, 0, \varepsilon)}{\partial z} &= -\lambda F_1(w_1, w_2, z, \varepsilon) \\
+j\gamma_1 \frac{\partial F_1(w_1, w_2, z, \varepsilon)}{\partial w_1} + j\gamma_2 \frac{\partial F_1(w_1, w_2, z, \varepsilon)}{\partial w_2} &+ \lambda B(z) F_0(w_1, w_2, \varepsilon) \\
-je^{-j\varepsilon w_1}\gamma_1 B(z) \frac{\partial F_0(w_1, w_2, \varepsilon)}{\partial w_1} &- je^{-j\varepsilon w_2}\gamma_2 B(z) \frac{\partial F_0(w_1, w_2, \varepsilon)}{\partial w_2} \qquad (3)\\
+\lambda q B(z) e^{j\varepsilon w_1} F_1(w_1, w_2, \varepsilon) &+ \lambda(1 - q) B(z) e^{j\varepsilon w_2} F_1(w_1, w_2, \varepsilon) \\
-jq\gamma_1 B(z) \frac{\partial F_1(w_1, w_2, \varepsilon)}{\partial w_1} &- j(1 - q)\gamma_2 B(z) \frac{\partial F_1(w_1, w_2, \varepsilon)}{\partial w_2} \\
-j(1 - q)\gamma_1 B(z) e^{j\varepsilon(w_2 - w_1)} \frac{\partial F_1(w_1, w_2, \varepsilon)}{\partial w_1} \\
-je^{j\varepsilon(w_1 - w_2)} q\gamma_2 B(z) \frac{\partial F_1(w_1, w_2, \varepsilon)}{\partial w_2}.
\end{aligned}
$$

Let x_m, $m = 1, 2$ be the asymptotic mean of the number of customers in the orbit in the first and second phase, respectively. Let us denote by R_k, $k = 0, 1$ the stationary probability distribution of the server state. As will be shown below R_k, $k = 0, 1$ depend on x_m, $m = 1, 2$, we will take the keys $R_k(x_1, x_2)$, $k = 0, 1$. Let us write the following statement for an asymptotic approximation.

Theorem 1. *Limiting values ($\varepsilon \to 0$, $z \to \infty$) $\{F_0(w_1, w_2), F_1(w_1, w_2)\}$ of the solution $\{F_0(w_1, w_2, \varepsilon), F_1(w_1, w_2, z, \varepsilon)\}$ of the system (3) have the following form:*

$$F_0(w_1, w_2) = R_0(x_1, x_2) \exp\{jw_1 x_1 + jw_2 x_2\},$$

$$F_1(w_1, w_2) = R_1(x_1, x_2) \exp\{jw_1 x_1 + jw_2 x_2\},$$

where values x_1, x_2 are the solution to the system equations

$$-\gamma_1 x_1 R_0(x_1, x_2) + \lambda q R_1(x_1, x_2) - \gamma_1 x_1(1 - q)R_1(x_1, x_2)$$
$$+\gamma_2 x_2 q R_1(x_1, x_2) = 0,$$
$$-\gamma_2 x_2 R_0(x_1, x_2) + \lambda(1 - q)R_1(x_1, x_2) + \gamma_1 x_1(1 - q)R_1(x_1, x_2)$$
$$-\gamma_2 x_2 q R_1(x_1, x_2) = 0,$$

(4)

and $R_0(x_1, x_2)$, $R_1(x_1, x_2)$ are defined as follows

$$R_0(x_1, x_2) = B^*(\lambda + \gamma_1 x_1 + \gamma_2 x_2), R_1(x_1, x_2) = 1 - B^*(\lambda + \gamma_1 x_1 + \gamma_2 x_2). \quad (5)$$

6   The Second-Order Asymptotic Form

Let us denote by $H_k(u_1, u_2)$, $k = 1, 2$ the function which is determined by the following formula

$$H_k(u_1, u_2) = H_k^{(2)}(u_1, u_2) \exp\left\{j\frac{u_1}{\sigma}x_1 + j\frac{u_2}{\sigma}x_2\right\}, \ k = 1, 2.$$

Substituting this expression into Eq. (2), we obtain the following system equations

$$-\frac{\partial H_1^{(2)}(u_1, u_2, 0)}{\partial z} = -\lambda H_0^{(2)}(u_1, u_2) + j\sigma_1 \frac{\partial H_0^{(2)}(u_1, u_2)}{\partial u_1}$$
$$+j\sigma_2 \frac{\partial H_0^{(2)}(u_1, u_2)}{\partial u_2} - \gamma_1 x_1 H_0^{(2)}(u_1, u_2) - \gamma_2 x_2 H_0^{(2)}(u_1, u_2),$$
$$-\frac{\partial H_1^{(2)}(u_1, u_2, z)}{\partial z} + \frac{\partial H_1^{(2)}(u_1, u_2, 0)}{\partial z} = -\lambda H_1^{(2)}(u_1, u_2, z)$$
$$+j\sigma_1 \frac{\partial H_1^{(2)}(u_1, u_2, z)}{\partial u_1} + j\sigma_2 \frac{\partial H_1^{(2)}(u_1, u_2, z)}{\partial u_2} + \lambda B(z)H_0^{(2)}(u_1, u_2)$$
$$-je^{-ju_1}\sigma_1 B(z)\frac{\partial H_0^{(2)}(u_1, u_2)}{\partial u_1} - je^{-ju_2}\sigma_2 B(z)\frac{\partial H_0^{(2)}(u_1, u_2)}{\partial u_2}$$
$$+\lambda q B(z)e^{ju_1}H_1^{(2)}(u_1, u_2) + \lambda(1 - q)B(z)e^{ju_2}H_1^{(2)}(u_1, u_2)$$

(6)

$$-jq\sigma_1 B(z)\frac{\partial H_1^{(2)}(u_1, u_2)}{\partial u_1} - j(1 - q)\sigma_2 B(z)\frac{\partial H_1^{(2)}(u_1, u_2)}{\partial u_2}$$
$$-j(1 - q)\sigma_1 B(z)e^{j(u_2-u_1)}\frac{\partial H_1^{(2)}(u_1,u_2)}{\partial u_1}$$
$$-je^{j(u_1-u_2)}q\sigma_2 B(z)\frac{\partial H_1^{(2)}(u_1, u_2)}{\partial u_2} - \gamma_1 x_1 H_1^{(2)}(u_1, u_2, z)$$
$$-\gamma_2 x_2 H_1^{(2)}(u_1, u_2, z) + e^{-ju_1}\gamma_1 x_1 B(z)H_0^{(2)}(u_1, u_2)$$
$$+e^{-ju_2}\gamma_2 x_2 B(z)H_0^{(2)}(u_1, u_2) + q B(z)\gamma_1 x_1 H_1^{(2)}(u_1, u_2)$$
$$+(1 - q)B(z)\gamma_2 x_2 H_1^{(2)}(u_1, u_2) + (1 - q)e^{j(u_2-u_1)}B(z)\gamma_1 x_1 H_1^{(2)}(u_1, u_2)$$
$$+e^{j(u_1-u_2)}q B(z)\gamma_2 x_2 H_1^{(2)}(u_1, u_2).$$

Performing here the following changes of variables

$$\sigma = \varepsilon^2, \; u_1 = \varepsilon w_1, \; u_2 = \varepsilon w_2, \; H_0^{(2)}(u_1, u_2) = F_0(w_1, w_2, \varepsilon),$$
$$H_1^{(2)}(u_1, u_2, z) = F_1(w_1, w_2, z, \varepsilon),$$

we write the system equations

$$-\frac{\partial F_1(w_1, w_2, 0)}{\partial z} = -\lambda F_0(w_1, w_2, \varepsilon) + j\gamma_1 \frac{\partial F_0(w_1, w_2, \varepsilon)}{\partial w_1}$$
$$+j\gamma_2 \frac{\partial F_0(w_1, w_2, \varepsilon)}{\partial w_2} - \gamma_1 x_1 F_0(w_1, w_2, \varepsilon) - \gamma_2 x_2 F_0(w_1, w_2, \varepsilon),$$
$$-\frac{\partial F_1(w_1, w_2, z, \varepsilon)}{\partial z} + \frac{\partial F_1(w_1, w_2, 0, \varepsilon)}{\partial z} = -\lambda F_1(w_1, w_2, z, \varepsilon)$$
$$+j\gamma_1 \varepsilon \frac{\partial F_1(w_1, w_2, z, \varepsilon)}{\partial w_1} + j\gamma_2 \varepsilon \frac{\partial F_1(w_1, w_2, z, \varepsilon)}{\partial w_2}$$
$$-\gamma_1 x_1 F_1(w_1, w_2, z, \varepsilon) - \gamma_2 x_2 F_1(w_1, w_2, z, \varepsilon) + \lambda B(z) F_0(w_1, w_2, \varepsilon)$$
$$-je^{-j\varepsilon w_1} \gamma_1 \varepsilon B(z) \frac{\partial F_0(w_1, w_2, \varepsilon)}{\partial w_1} + e^{-j\varepsilon w_1} \gamma_1 x_1 B(z) F_0(w_1, w_2, \varepsilon)$$
$$-je^{-j\varepsilon w_2} \gamma_2 \varepsilon B(z) \frac{\partial F_0(w_1, w_2, \varepsilon)}{\partial w_2} + e^{-j\varepsilon w_2} \gamma_2 x_2 B(z) F_0(w_1, w_2, \varepsilon) \quad (7)$$
$$+\lambda q B(z) e^{j\varepsilon w_1} F_1(w_1, w_2, \varepsilon) + \lambda(1 - q) B(z) e^{j\varepsilon w_2} F_1(w_1, w_2, \varepsilon)$$
$$-jq\gamma_1 \varepsilon B(z) \frac{\partial F_1(w_1, w_2, \varepsilon)}{\partial w_1} + \gamma_1 x_1 q B(z) F_1(w_1, w_2, \varepsilon)$$
$$+\gamma_2 x_2 (1 - q) B(z) F_1(w_1, w_2, \varepsilon) - j(1 - q) \gamma_2 \varepsilon B(z) \frac{\partial F_1(w_1, w_2, \varepsilon)}{\partial w_2}$$
$$-j(1 - q) \gamma_1 \varepsilon B(z) e^{j\varepsilon(w_2 - w_1)} \frac{\partial F_1(w_1, w_2, \varepsilon)}{\partial w_1}$$
$$-je^{j\varepsilon(w_1 - w_2)} q \gamma_2 \varepsilon B(z) \frac{\partial F_1(w_1, w_2, \varepsilon)}{\partial w_2}$$
$$+\gamma_1 x_1 (1 - q) B(z) e^{j\varepsilon(w_2 - w_1)} F_1(w_1, w_2, \varepsilon)$$
$$+\gamma_2 x_2 q B(z) e^{j\varepsilon(w_1 - w_2)} F_1(w_1, w_2, \varepsilon).$$

Let us denote $a = \lambda + \gamma_1 x_1 + \gamma_2 x_2$. We write the following statement.

Theorem 2. *Limiting values* $\{F_0(w_1, w_2), F_1(w_1, w_2)\}$ *of the solution* $\{F_0(w_1, w_2, \varepsilon), F_1(w_1, w_2, z, \varepsilon)\}$ *to the system (7) have the following form:*

$$F_0(w_1, w_2) = R_0(x_1, x_2) \Phi(w_1, w_2),$$

$$F_1(w_1, w_2) = R_1(x_1, x_2) \Phi(w_1, w_2),$$

where values x_1, x_2 *are the solution of the system (4)*, $R_0(x_1, x_2)$, $R_1(x_1, x_2)$ *are given by the formulas (5). We write function* $\Phi(w_1, w_2)$ *in the following form:*

$$\Phi(w_1, w_2) = \exp \left\{ \frac{(jw_1)^2}{2} K_{11} + jw_1 jw_2 K_{12} + \frac{(jw_2)^2}{2} K_{22} \right\}, \quad (8)$$

where values K_{11}, K_{12}, K_{22} are the solution to heterogeneous system linear algebraic equations:

$$K_{11} \left(q\gamma_1 R_0 - q\gamma_1 B^*(a) + q\gamma_1 R_1^*(a) - (1-q)\gamma_1 R_1 - \gamma_1 R_0 \right)$$
$$+ K_{12} \left(q\gamma_2 R_0 - q\gamma_2 B^*(a) + q\gamma_2 R_1^*(a) + \gamma_2 q R_1 \right) = -\frac{1}{2}(1-q)\gamma_1 x_1 R_1$$
$$-\frac{1}{2}\gamma_1 x_1 R_0 - q\gamma_1 x_1 B^*(a) + q^2 a R_1 B^*(a) - \frac{1}{2}\lambda q R_1 - \frac{1}{2}q R_1 \gamma_2 x_2,$$

$$K_{11} \left((1-q)\gamma_1 R_0 - (1-q)\gamma_1 B^*(a) + (1-q)\gamma_1 R_1^*(a) + (1-q)\gamma_1 R_1 \right)$$
$$+ K_{12} \left(q\gamma_1 R_0 - q\gamma_1 B^*(a) + q\gamma_1 R_1^*(a) + (1-q)\gamma_2 R_0 - (1-q)\gamma_2 B^*(a) \right.$$
$$+ (1-q)\gamma_2 R_1^*(a) - \gamma_1 R_0 - \gamma_2 R_0 - (1-q)\gamma_1 R_1 - \gamma_2 q R_1)$$ (9)
$$+ K_{22} \left(q\gamma_2 R_0 - q\gamma_2 B^*(a) + q\gamma_2 R_1^*(a) + q\gamma_2 R_1 \right) = (1-q)\gamma_1 x_1 R_1$$
$$+ q\gamma_2 x_2 R_1 - (1-q)\gamma_1 x_1 B^*(a) - q\gamma_2 x_2 B^*(a) + 2aq(1-q)R_1 B^*(a),$$

$$K_{12} \left((1-q)\gamma_1 R_0 - (1-q)\gamma_1 B^*(a) + (1-q)\gamma_1 R_1^*(a) + (1-q)\gamma_1 R_1 \right)$$
$$+ K_{22}((1-q)\gamma_2 R_0 - (1-q)\gamma_2 B^*(a) + (1-q)\gamma_2 R_1^*(a) - q\gamma_2 R_1$$
$$-\gamma_2 R_0) = -\frac{1}{2}(1-q)\gamma_1 x_1 R_1 - \frac{1}{2}\gamma_2 x_2 R_0 - (1-q)\gamma_2 x_2 B^*(a)$$
$$+(1-q)^2 a R_1 B^*(a) - \frac{1}{2}\lambda(1-q)R_1 - \frac{1}{2}(1-q)R_1 \gamma_1 x_1.$$

We can write the following expression for the approximation of the characteristic function.

$$H(u_1, u_2) = \exp\left\{ \frac{(ju_1)^2}{2\sigma}K_{11} + \frac{(ju_2)^2}{2\sigma}K_{22} + \frac{ju_1 ju_2}{\sigma}K_{12} \right\}$$
$$\times \exp\left\{ j\frac{u_1}{\sigma}x_1 + j\frac{u_2}{\sigma}x_2 \right\}.$$

Thus

$$H(u_1, u_2) = \exp\left\{ j\frac{u_1}{\sigma}x_1 + j\frac{u_2}{\sigma}x_2 + \frac{(ju_1)^2}{2\sigma}K_{11} + \frac{(ju_2)^2}{2\sigma}K_{22} + \frac{ju_1 ju_2}{\sigma}K_{12} \right\}.$$

So, under a condition of long delay of the customers in the orbit, the two-dimensional probability distribution of the number of customers in the orbit in the first and second phase can be approximated by the two-dimensional normal distribution with parameters $x_1, x_2, K_{11}, K_{12}, K_{22}$.

7 Numerical Results

In this section, we discuss some interesting numerical examples that qualitatively describe the model under study. We find the values of the parameters of the two-dimensional Gaussian distribution of the number of the customers in the orbit in the first phase and second phase using numerical results.

In particular, we analyze the gamma distribution. Let

$$B^*(x) = \left(1 + \frac{x}{\beta} \right)^{-\alpha},$$

be a Laplace-Stieltjes transform of the distribution function of the gamma distribution, where α, β are positive parameters.

E.g., if the parameters of distribution of service time are fixed as it follows that

$$\alpha = 0.7, \ \beta = 1.$$

All other parameters are fixed as:

$$\sigma_1 = 0.02, \ \sigma_2 = 0.03, \ q = 0.6.$$

From system Eqs. (4) and (9), we find means, variances and covariance. We obtain

$$x_1 = 0.298, \ x_2 = 0.132, \ K_{11} = 0.463, \ K_{22} = 0.171, \ K_{12} = 0.079. \tag{10}$$

Graph of distribution $P(x, y)$, which is given by (10), $\lambda = 0.9$ is given in Fig. 2.

$$P(x, y) = \frac{1}{2\pi K_{11} K_{22} \sqrt{1 - r^2}}$$
$$\times \exp\left\{-\frac{1}{2(1 - r^2)}\left[\frac{(x - x_1)^2}{K_{11}^2} - 2r\frac{(x - x_1)(y - x_2)}{K_{11}K_{22}} + \frac{(y - x_2)^2}{K_{22}^2}\right]\right\}. \tag{11}$$

Here $r = \dfrac{K_{12}}{\sqrt{K_{11}K_{22}}}$ is a correlation coefficient.

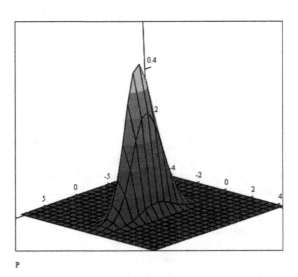

Fig. 2. Graph of distribution $P(x, y)$

Using (11) for two-dimensional distribution, we can find some characteristics of number of customers in the orbit. In particular, we could find one-dimensional distribution of the total number of customers in the orbit.

8   Conclusion

In this paper, we consider a retrial queueing system in which the arrival flow is described by the stationary Poisson process, the service time is a random with a given distribution function, hyper exponential distribution the delay time of customers in the orbit and exclusion of alternative customers. The stationary distribution of the server state has been obtained. We have found the means of the number of customers in the orbit, variances and covariance. Two-dimensional distribution of the number of the customers in the orbit is asymptotically normal with parameters x_1, x_2, K_{11}, K_{12}, K_{22}.

We consider the gamma distribution where α, β are positive parameters. We plan to consider the distribution of phase type in the future (PH-distribution).

References

1. Falin, G.I.: A survey of retrial queues. Queuing Syst. **7**, 127–167 (1990)
2. Falin, G.I., Artalejo, J.R., Martin, M.: On the single retrial queue with priority customers. Queueing Syst. **14**(3–4), 439–455 (1993)
3. Falin, G.I.: Multi-channel queueing systems with repeated calls under high intensity of repetition. J. Inf. Process. Cybern. **1**, 37–47 (1987)
4. Artalejo, J.R.: A classified bibliography of research on retrial queues: progress in 1990–1999. Top **7**(2), 187–211 (1999)
5. Artalejo, J.R.: Accessible bibliography on retrial queues. Math. Comput. Model. **30**(1–2), 1–6 (1999)
6. Artalejo, J.R.: Accessible bibliography on retrial queues: progress in 2000–2009. Math. Comput. Model. **51**, 1071–1081 (2010)
7. Pustova, S.V.: Investigation of calls centers as retrial queueing systems. Cybern. Syst. Anal. **46**, 494–499 (2010)
8. Wang, J.: On the single server retrial queue with priority subscribers and server breakdowns. J. Syst. Sci. Complex. **21**(2), 304–315 (2008)
9. Chakravarthy, S.R., Dudin, A.N.: Analysis of a retrial queuing model with MAP arrivals and two types of customers. Math. Comput. Model. **37**(3–4), 343–363 (2003)
10. Dudin, A.N., Krishnamoorthy, A., Joshua, V.C., Tsarenkov, G.V.: Analysis of the BMAP/G/1 retrial system with search of customers from the orbit. Eur. J. Oper. Res. **157**(1), 169–179 (2004)
11. Nazarov, A.A., Yakovlev, N.: Investigation of retrial queue system M/M/1 with phase-type retrial times. Vestn. Tom. State Univ. Control Comput. Eng. Comput. Sci. **27**(2), 39–46 (2014)
12. Nazarov, A., Chernikova, Y.: The accuracy of Gaussian approximations of probabilities distribution of states of the retrial queueing system with priority of new customers. In: Dudin, A., Nazarov, A., Yakupov, R., Gortsev, A. (eds.) ITMM 2014. CCIS, vol. 487, pp. 325–333. Springer, Heidelberg (2014)
13. Boutarfa, L., Djellab, N.: On the performance of the M1, M2/G1, G2/1 retrial queue with pre-emptive resume policy. Yougoslav J. Oper. Res. **25**(1), 153–164 (2015)
14. Bocharov, P.P., Pavlova, O.I., Puzikova, D.A.: M/G/1/r retrial queueing systems with priority of primary customers. Math. Comput. Model. **30**(3–4), 89–98 (1999)

15. Choi, B.D., Chang, Y.: Single server retrial queues with priority calls. Math. Comput. Model. **30**(3–4), 7–32 (1999)
16. Choi, B.D., Choi, K.B., Lee, Y.W.: M/G/1 retrial queueing systems with two types of calls and finite capacity. Queueing Syst. **19**, 215–229 (1995)
17. Nazarov, A.A., Moiseeva, S.P.: The Asymptotical Analysis Method in Queueing Theory. NTL, Tomsk (2006)

Modelling of Systems
Using a Time-Triggered Ethernet

Kirill Nikishin[✉], Nicolas Konnov, and Dmitry Pashchenko

Department of Computer Engineering, Penza State University,
40, Krasnaya street, Penza 440026, Russia
`knn@pnzgu.ru`

Abstract. This article describes the real time imitation model of Time-Triggered Ethernet. The model allows us to analyze the algorithm of functioning of a main parts system (synchronization of a global time block, the communication channel, the switch, the client). The model is created using mathematical modeling language hierarchical time colored Petri net and CPN Tools package. The authors propose verification rules for describing the fullness of the work of Time-Triggered Ethernet.

Keywords: Embedded systems · Distributed real-time systems · Time-Triggered Ethernet · Verification · Petri nets · CPN tools

1 Introduction

Embedded systems are used in avionics, home electronics, spaceships and automobile production. Complexes of embedded systems use distributed control architecture and Time-Triggered Ethernet [1,2] advanced communication protocols.

The Time-Triggered Ethernet (TTE) is a distributed safety-critical real-time system. The TTE is based on the Ethernet Protocol. It introduces two concepts: time and event. The protocol, which is developed in the TTE, is called Time-Triggered Protocol (TTP) [3–5].

Hardware real-time systems must quickly react to the error states in the system. Additionally, the correctness of results in such systems depends on the time, allocated within the system for solving the problem.

Modeling of the TTE system plays an important role in research and improvement of the TTE technology. The main emphasis was placed on safe-critical and fault-tolerant features of the system and the principle of clock synchronization between all network nodes.

Modeling and verification of the TTE as a entire real-time system using Petri nets have not previously been described and researched. Furthermore, many works research only the individual parts of the TTE [6,7]. This article describes a simulation of the entire system.

A. Dudin et al. (Eds.): ITMM 2016, CCIS 638, pp. 303–314, 2016.
DOI: 10.1007/978-3-319-44615-8_27

2   The Main Statements of a Time-Triggered Ethernet

The TTE is an extension of normal Ethernet technology (IEEE 802.3). Therefore, the TTE can be easily integrated into the architecture of existing networks. In this technology there are 3 types of traffic: time-triggered (TT), rate-constrained (RC) and best-effort (BE) traffic. TT-traffic has a higher priority of service, minimal latency and minimal jitter. RC-traffic has a lower priority than the TT-traffic, thereby getting a larger frame delay and jitter, in addition to non-availability of clock synchronization when exchanging messages. BE-traffic is compatible with the standard traffic of Ethernet technology [8,9].

The basic principle of the TTE is an allocation of a time slice for every node, while this node should accept or transfer data, i.e. create temporal parts for all nodes of the network. Message exchange is implemented with multiple access with time division (TDMA), i.e., TDMA-rounds [10]. This is achieved by synchronizing the clocks of all the network participants. There are the following main properties of TTE:

1. Fault-tolerant global time it is required for all objects that interact through a global time in the whole time interval;
2. Mixed configuration it allows the handling of the traffic as real time and standard of the Ethernet technology without the use of additional equipment;
3. Strong fault isolation if fault occurred in the network, that it is considered its processing and the system operates in normal mode;
4. Naming - each message has its name.

A clock synchronization function ensures the initialization of synchronization. All local clocks of the network must be synchronized with the global time. The mechanism of clock synchronization is based on the compression function. The compression function calculates the network delay based on the local delay of the node in order to synchronize all the nodes [1].

3   Modeling a Time-Triggered Ethernet

The TTE system can have two configurations: a standard (Fig. 1a) and a fault-tolerant (Fig. 1b). In the article a model of the TTE system only a standard configuration is built, in which each client connects to the server through a single switch to send messages to the server. In this case the client and server could be any couple of connected devices. When exchanging client-server the possibility of client error is considered, which leads to the regular sending of frames to the network.

The TTE model is developed using the mathematical apparatus of the hierarchical temporal colored Petri nets [11], which features the following dignities:

- A Petri net is universal algorithmic system, providing a description of almost any algorithm;
- colors allow the description and modeling of algorithms, which depends on content data being processed;

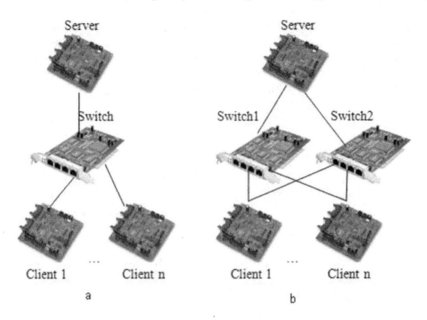

Fig. 1. The TTE configuration: a - standard, b - fault-tolerant

- the hierarchy allows the building of complex multicomponent models;
- the time property allows the modelling of dynamic properties of objects.

A CPN Tools package was selected as a freeware tool system, which allows the modelling of different aspects of the complex telecommunication systems behaviour [12–14].

The TTE model should reproduce the functioning of the system, consisting of the following parts: the block of global time synchronization, the communication channel, the client and the switch-server. This article describes only one client and one switch; however, the model can be continually expanded and complemented with n number of clients and switches.

Real time periods are not used while modeling, instead abstract time periods are used. It is convenient because this model describes the functioning of the system and is used for identifying inaccessible parts of the algorithms.

The top level of hierarchy of the TTE model is presented in Fig. 2 and consists of the parts described above – block of synchronization of global time, the communication channel, the switch, the client.

The model uses the following main colors, constants and variables:

- color **GlobTime**, which allows a variable to have a range from 0 to n, where n is equal to the sum of the number of clients and number of switches. In this case, n = 2, as it is described as colset GlobTime = int with 0..2 timed;
- variable var globtime: GlobTime, which refers to the color **GlobTime** and is a key variable in the development of the TTE model, since this variable is

responsible for the clock synchronization of the nodes of the communication network, being transferred to the clients and the switch, thereby providing synchronization of the local clock of each node;
– the constant, which is used for defining the sum of the number of clients and the number of switches in the model: val numberpart = 2;

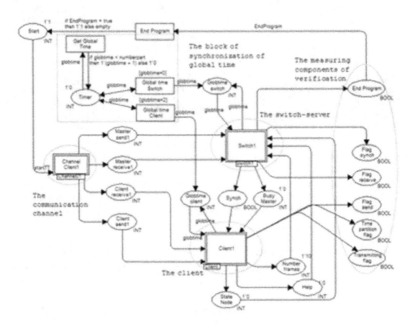

Fig. 2. Model of the TTE system

The algorithm of the system is simulated as the following sequence of events marked by tokens.

Start - to start the TTE system; **Start** is transmitted to the communication channel, so the work had begun and synchronization had been established between the client and the switch.

The communication channel produces four tokens indicating the events: the sending of the switch, the receiving of the switch, the sending of the client and the receiving of the client. The detailed algorithm of the communication channel will be described below.

In parallel with the generation of tokens from the communication channel, a global time is formed. The basis of the TTE systems is the definition of transmission real-time and time intervals. Therefore, the formation of a global time is a very important part of the TTE systems. The formation of the global time is obtained as follows: in the position of the **Timer** is the stored global time used for all nodes in the TTE system. The initial token of position **Timer** equals 0.

The calculation of the value of the global time is carried out on the transition **Get Global Time**, which provides an actually cyclic increment of variable **globtime** described above. The limit cycle is a constant **numberpart** described above. The algorithm for computing the global time is as follows: if the value of the global time is less than the number of nodes in the system, it is incrementing the value of the global time, or global time is reset, in the language CPN ML is: if globtime < numberpart then 1'(globtime + 1) else 1'0.

If the value of the global time is 0, then the processing of the switch is started, but only if two tokens of receiving/sending from a communication channel are ready. If the value of the global time is equal to 2, then the processing of the client starts, but only if two tokens of receiving/sending from a communication channel are ready. The value of the global time is equal to 1, this is an intermediate state used for research. After synchronization, all the objects will be global time.

When synchronization between the client and the switch happens, transmission of frames from the client to the switch starts. In the case of processing the last frame of switch generates the end flag of transmission and it is kept in the position **End of Program**. This flag is also one of the verification rules for the functioning of the TTE system, as will be described below. If this flag is set, a token of the start of the TTE system is produced again and the operation of the TTE system is repeated again.

We consider in more detail three subnets called **Channel01**, **Switch1**, **Client**. The subnet **Channel01** emulates a communication channel between the client and the switch; it is presented in Fig. 3.

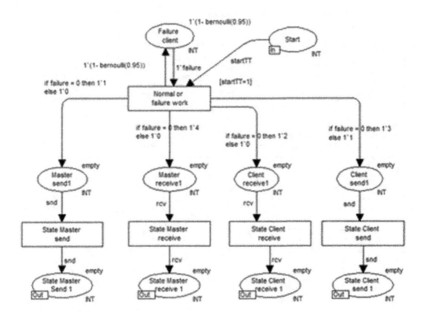

Fig. 3. Subnet of the communication channel

The model works as follows: the channel is initiated by the token at the position **Start**. A token at the position of **Failure client** sets a variable indicating the presence of an error in frame transmission. Therefore it is important that the TTE system should include the availability and handling of errors in the system. The probabilistic nature of the error function plays a special language CPN ML the function for calculating the probability according to Bernoulli's law. Thus, the initial marking in the position **Failure client** equal to 1 - bernoulli(0.95).

The error can take two integer values: 0 no error, 1 errors. The transition of **Normal work or failure** determines the normal or the erroneous work of the TTE system. The output positions of the subnet correspond to the following events:

– **State Master Send 1** the sending of the switch;
– **State Master receive 1** the receiving of the switch;
– **State Client receive 1** the receiving of the client;
– **State Client send 1** the sending of the client.

The subnet of the switch is presented in Fig. 4.

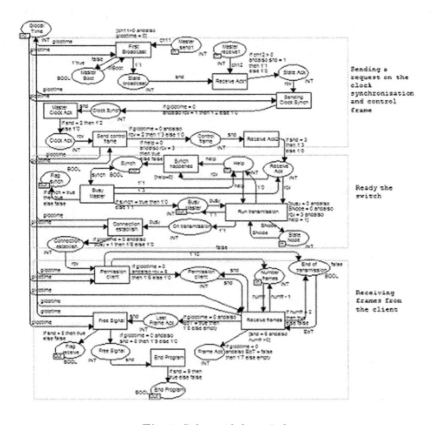

Fig. 4. Subnet of the switch

If there is no error in the channel, the colors of the tokens will have the following values in appropriate positions: 1- **State Master Send 1**, 4 - **State Master receive 1**, 2 - **State Client receive 1**, 3 - **State Client send 1**. Otherwise, all the colors of the tokens in the described positions are reset to zero, except for the color of the token of the sending client **State Client send 1**, which is set to 1. The process with an error is called "babbling idiot" [1] when the client begins to transmit frames to the network, but the network does not respond to recognized errors in the client.

The ports for subnet **Switch1** are:

- the position of **Global Time**, containing the value of the global time (the input/output port);
- the position of the **Master send1**, containing the event token of the sending of the switch (input port);
- the position of the **Master receive1**, containing the event token of the receiving of the switch (input port);
- the position **Help**, containing the token of help from the client (input port);
- the position **Synch**, containing the flag of a successful synchronization between the switch and the client (output port);
- the position **Busy Master**, containing the busy flag of the switch (output port);
- the position **State Node**, containing the value to transfer: 0 free node, 5 node free after the transfer (input port);
- the position **Number frames**, containing the number of frames that the client sends to the switch (input port);
- the position **End Program**, containing the end flag of the work of the TTE system (output).

Due to the synchronization with the communication channel a deterministic communication is provided, i.e. two messages cannot be sent at the same time, as indicated in the Time-Triggered Protocol.

Subnet **Switch1** operates as follows:

after the occurrence of the token **Master send1** and global time is equal to zero (the start of the switch), the first transmission in the network is emulated by triggering of the transition **First Broadcast**, after that an occurrence of the token **Master receive1** the acknowledgement (Ack) is accepted;

by triggering of the transition **Sending Clock Synch** the token is sent for the clock synchronization and the acknowledgement of synchronization (Ack) is received;

the token of the control frame is sent from the position **Send control frame** and the acknowledgement of control frame is accepted and the flag of the successful synchronization **synch** is set, which will also be analyzed on the client side;

the busy flag of the switch **busy** is reset, so the switch is ready to receive transmitted frames from the client. If **busy** and **SNode** equal zero (the node is free), and there is the flag **help**, the switch is in the transmission state;

the subnet **Client** is informed about the connection through the position **Run Transmission** and after the acknowledgement (Ack) the token of permission to transfer for the client is sent;

the subnet **Switch1** starts to receive frames sent from the client and their acknowledgements from the switch, but when the last frame is received, the token of the end of transmission is generated in the position **End of transmission**;

at the end of the subnet flag **End Program** is set to a true value.

The generation of two tokens of the send/receive of the switch provides a namespace of the TTE system. It is necessary to identify frames that match the format of the TTE. In this model, the actual frame format of the TTE system is ignored.

The subnet of the client **Client** is similar to the subnet of the switch. Each client has its own identification number ID, the ID is stored in the variable **SNode**. When the transmission starts, the network recognizes which client is active by using its ID. The subnet of the client is presented in Fig. 5.

The main ports on the subnet of the client **Client**: − the position **Client send1**, containing the token of sending of the client (input port); − the position **Client receive1**, containing the token of receiving of the client (input port).

This client works when global time is equal to two. Subnet **Client** operates as follows:

after receipt of the first transmission from the switch the token is sent in synchronization the frame on the clock synchronization is received with the issuance of receiving the acknowledgement;

the control frame with the issuance of the acknowledgement is received. Then the client can send the frame or can skip its turn. If the client skips its turn, then it is in a position **State send or skip**, otherwise the client sends a frame;

after setting the flags **busy** and **help** in 1 the client sends to the switch an active client's ID with receipt of the acknowledgement from the switch, i.e. in the position of the **State Node** the value is set equal to one. This position is read of the switch and is analyzed to the variable **SNode**. Thus, it provides the detection mechanism of one active client in the network. After sending the client's ID, the client receives the acknowledgement from the switch;

the token of permission to transfer is sent and sending frames begins;

when the last frame is sent, in the position **State Node** the value 5 is written (the completed transmission and the node is released).

4   Verification of the Time-Triggered Ethernet Using the Model

Verification of the TTE system could be done using verification rules, that allow us to measure the reach and the occurrence of deadlock condition in the simulated TTE system. Description of the verification rules was made using logical operations and triggering events on transitions in the model. Triggering events on transitions are described using the variable S, the index variable will have the name of the transition and the name of the subnet (client - C, the switch - M).

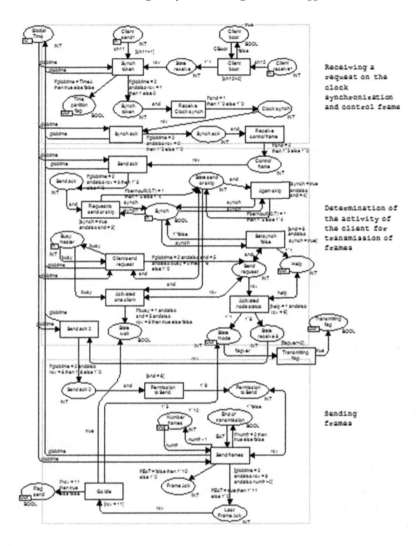

Receiving a
request on the
clock
synchronization
and control frame

Determination of
the activity of
the client for
transmission of
frames

Transmitting
flag

Sending
frames

Fig. 5. Subnet of the client

Let us consider the verification rules:

1. Transmission begins after a successful synchronization of the node.

$$S_{Synch\ happened,M}\ and\ S_{Request\ to\ send\ or\ skip,C} \rightarrow S_{Busy\ master,M}$$

If the event $S_{Synch\ happened,M}$ happens, then the flag is set to the true value
at the position **Flag synch** in the subnet of the switch. Otherwise this branch
of the algorithm is unreachable.

2. The sent frame needs to achieve its address and be accepted.

$$S_{Receive\ frames,M} \rightarrow S_{Free\ Signal,M}$$

This rule means that if the event $S_{Receive\ frames,M}$ happens and provided that the last frame is received from the client, then the flag is set to the true value at the position **Flag receive** in the subnet of the switch. Otherwise this branch of the algorithm is unreachable.

3. Transmission of frames of the client starts after the enable token for transmission.

$$S_{Pemission\ to\ send,C} \text{ and } S_{Send\ frames,C} \rightarrow S_{Go\ idle,C}$$

This rule means that if the event of token of permission to transmission and transmission of frames starts, then the flag is set to the true value at the position **Flag send** in the subnet of the client. Otherwise this branch of the algorithm is unreachable.

4. Each node has its own time partition.

In this rule are added variables denoted by P. The index at the variable will have the variable name and the name of the subnet.

$$S_{Synch\ token,C} \text{ and } P_{globtime,C} = P_{TimeA,C} \rightarrow P_{Time\ partition\ flag,C}$$

This rule means that if the event $S_{Synch\ token,C}$ happens and the variable of global time $P_{globtime,C}$ is equal to the variable $P_{TimeA,C}$ (in this case the variable of the first client $P_{TimeA,C}$ is equal to constant 2), then the variable $P_{Time\ partition\ flag,C}$ is set to the true value at the position **Time partition flag**. Otherwise this branch of the algorithm is unreachable.

5. The transmission is carried out after the detection of an active client through its identification number ID.

$$S_{Transmitting\ flag,C} \rightarrow P_{Transmitting\ flag,C}$$

This rule means that if the transition worked $S_{Transmitting\ flag,C}$, considering only the active value of the position **State Node** (0 client is not active), then the flag $P_{Transmitting\ flag,C}$ is set to the true value. Otherwise this branch of the algorithm is unreachable.

6. Each transmission must be completed.

$$S_{End\ program,M} \rightarrow P_{End\ program,M}$$

This rule means that if the transition worked $S_{End\ program,M}$, then the flag $P_{End\ program,M}$ is set to the true value. Otherwise this branch of the algorithm is unreachable.

5   Discussion of Verification Rules

Verification rules were applied to the developed model of the TTE system and were introduced either in the subnet of the client or in the subnet of the switch as measurement components of model that provide a visual verification. For convenience, the verification rules were placed on the higher level of the model hierarchy and are presented in Fig. 2.

A similar task verification of the TTE system was solved using the PRISM and built in programming language [15]. However, the verification rules accepted in this article were treated separately from the model and have a lack of visibility; there was a need for the development of additional software created with C++ language for processing these rules. Besides, it is not possible for the developer of this model to identify during the simulation if the rules are true or false.

The verification rules in this presented article are easy to percept and are included in the developed model of the TTE system, so the result of flags could be immediately seen during the simulation.

During the simulation the algorithm of functioning of the TTE system performed successfully and thereby the identification of deadlock parts is confirmed in the work of the TTE.

6   Conclusions

The model of Time-Triggered Ethernet was developed using the mathematical apparatus of hierarchical temporal colored Petri nets.

The possibility of using the model for verification of TTE protocols for inaccessibility and detection of deadlock conditions has been confirmed.

The results of research can be applied in the study and improvement of technology Time-Triggered Ethernet.

A further stage of the research is TTE system timing component accounting (delay and jitter) and traffic characteristics of switched traffic, which will perform not only as system functioning logical analysis, but also to measure its probabilistic time characteristics.

Acknowledgements. This article was written as part of the Federal Targeted Programme "Research and development in priority areas of elaboration of the science and technology complex of Russia for 2014–2020" (agreement of 06.19.2014 14.574.21.0045 UIN: RFMEFI57414X0045).

References

1. Kopetz, H.: Real-Time Systems: Design Principles for Distributed Embedded Applications. Springer, Heidelberg (2011)
2. Kopetz, H., Bauer, G.: The time-triggered architecture. In: Proceedings of the IEEE, pp. 112–126 (2003)

3. Kopetz, H., Ademaj, A., Grillinger, P., Steinhammer, K.: The time-triggered Ethernet (TTE) design. In: Proceedings of 8th IEEE International Symposium on Object-oriented Real-time Distributed Computing (ISORC), Seattle, Washington, D.C., May 2005

4. Grillinger, P., Ademaj, A., Steinhammer, K., Kopetz, H.: Software implementation of a time-triggered Ethernet controller. In: Proceedings of IEEE International Workshop on Factory Communication Systems (2006)

5. Steinhammer, K., Grillinger, P., Ademaj, A., Kopetz, H.: A time-triggered Ethernet (TTE) switch. In: Design, Automation and Test in Europe (2006)

6. Zhang, Z., Koutsoukos, X.: Modeling time-triggered Ethernet in systemC/TLM for virtual prototyping of cyber-physical systems. In: Schirner, G., Götz, M., Rettberg, A., Zanella, M.C., Rammig, F.J. (eds.) IESS 2013. IFIP AICT, vol. 403, pp. 318–330. Springer, Heidelberg (2013)

7. Brau, G., Pagetti, C.: TTEthernet-based architecture simulation with Ptolemy II. In: 6th Junior Researcher Workshop on Real-Time Computing, France, pp. 29–32 (2012)

8. Steinhammer, K., Ademaj, A.: Hardware implementation of the time-triggered Ethernet controller. In: Rettberg, A., Zanella, M.C., Dömer, R., Gerstlauer, A., Rammig, F.J. (eds.) Embedded System Design: Topics, Techniques and Trends. IFIP, vol. 231, pp. 325–338. Springer, New York (2007)

9. Zhang, Z., Eyisi, E., Koutsoukos, X., Porter, J., Karsai, G., Sztipanovits, J.: A co-simulation framework for design of time-triggered automotive cyber physical systems. In: Simulation Modelling Practice and Theory, vol. 43, pp. 16–33 (2014)

10. Nikishin, K.: The perspective of using the technology of Time-Triggered Ethernet. Actual Directions of Scientific Researches of the XXI Century: Theory and Practice. vol. 3, no. 8–1(19–1), pp. 90–92 (2015). Voronezh, 442 pages

11. Jensen, K., Kristensen, L.: Coloured Petri Nets: Modelling and Validation of Concurrent Systems. Springer, Heidelberg (2009)

12. Kizilov, E., Konnov, N., Nikishin, K., Pashchenko, D., Trokoz, D.: Scheduling queues in the Ethernet switch, considering the waiting time of frames/MATEC Web of Conferences, vol. 44, Pages 01011-p.1-01011-p (2016)

13. Kizilov, E., Konnov, N., Pashchenko, D., Trokoz, D.: Modeling of QoS in the industrial Ethernet switches. In: 2015 The 5th International Workshop on Computer Science and Engineering-Information Processing and Control Engineering (WCSE 2015-IPCE), Moscow, Russia, 15–17 April 2015, pp. 185–190 (2015)

14. Domnin, A., Konnov, N., Mekhanov, V.: Modeling EMA and MA algorithms to estimate the bitrate of data streams in packet switched networks. In: Dudin, A., Nazarov, A., Yakupov, R., Gortsev, A. (eds.) ITMM 2014. CCIS, vol. 487, pp. 91–100. Springer, Heidelberg (2014)

15. Ammar, M., Ouchani, S., Ait Mohamed, O.: Symmetry reduction of time-triggered Ethernet protocol. Procedia Computer Science 19, 273–280 (2013)

Analysis of Business Process Execution Time with Queueing Theory Models

Konstantin Samouylov, Yuliya Gaidamaka, and Elvira Zaripova$^{(\boxtimes)}$

Department of Applied Probability and Informatics,
RUDN University, 6 Miklukho-Maklaya Street,
117198 Moscow, Russian Federation
{ksam,ygaidamaka,ezarip}@sci.pfu.edu.ru

Abstract. In the paper an approach to the analysis of business process efficiency is proposed. A method for the estimation of a business process execution time as an important performance measure of business processes efficiency is developed. It represents a combination of queuing networks modelling and simplex algorithm of linear programming. The method allows the calculating of the minimum business processes execution time. A method of optimizing activity of the telecommunication company at a predetermined threshold for the business processes execution time is given. The developed technique was illustrated with an end-to-end business process flow "Request-to-Answer" with initial data close to reality.

Keywords: Optimization · Business process · Queueing theory · Execution time · Delay · Simplex algorithm

1 Introduction

In a volatile and challenging complex environment organizations have to be agile and flexible in their response to increasingly changing market conditions. To achieve and to sustain business performance the company should have an instrument to analyze and optimize their business processes in terms of performance. Optimization of business processes is a powerful feature for increasing the competitive advantage of any enterprise. After a complex analysis of an enterprise's activities, new business processes can be applied and old ones can be reorganized. Innovations are aimed to improve the performance measures of business processes, such as cost metrics and quality of service. Traditionally, for optimization of business processes the simulation approach is used including the General Purpose Simulation System (GPSS) and Petri net framework [1]. The disadvantage of this approach is the necessity for special software tools as Oracle Crystal Ball toolkits [2,3] or ARIS Business Simulator [4]. In this paper

The reported study was partially supported by the RFBR, research projects No. 14-07-00090, 15-07-03051, 15-07-03608; Vladimir Potanin Foundation.

A. Dudin et al. (Eds.): ITMM 2016, CCIS 638, pp. 315–326, 2016.
DOI: 10.1007/978-3-319-44615-8_28

we examine analytical approaches to business process modelling based on the theory of queuing networks [5], and dynamic and linear programming [6,7].

The subject of our study is enterprise activity; the problem is the estimation of efficiency of enterprise business processes; the goal is the development of a method for business process analysis and optimization in terms of process execution time. For illustration a telecommunication company is chosen as an enterprise. An end-to-end business-process flow "Request-to-Answer" is used as the example of a business process. The business process is analyzed according to Business Process Framework eTOM (enhanced Telecom Operations Map), a standard maintained by the global industry association for digital business TeleManagement Forum (TMForum) [8].

The paper is organized as follows. Section 2 is devoted to the description of the business process "Request-to-Answer" in BPMN 2.0 (Business Process Model and Notation) [9]. In Sect. 3 we built a business process model as an open queueing network with multiple classes of customers [5] and develop formulas for average process execution time calculation. In Sect. 4 we offer the method of estimating the total execution time of the above mentioned business process using a simplex algorithm (SA) [3]. Section 5 contains an algorithm for optimizing the execution time of the process flow "Request-to-Answer" at a predetermined threshold for the business processes execution time using the developed technique. The conclusion gives some recommendations on business process optimization and further study.

2   Process Flow Description Diagram

We begin with a description of the end-to-end business process flow "Request-to-Answer" using Business Process Model and Notation (BPMN) [9], a standard for business process modeling providing a graphical notation for specifying business processes in a Business Process Diagram (Fig. 1). Activity flow objects of the diagram represented with a rounded-corner rectangle depict the processes of 3rd and 4th levels of decomposition numerated according to Business Process Framework GB921, Release 15.0 [10]. The circles in the Business Process Diagram denote the event subjects such as process initiation or process termination. Diamond shapes represent gateways to determine forking and merging of paths, depending on the conditions expressed. In Fig. 1 some rectangles or combinations of rectangles are denoted as nodes; this notation will be explained in Sect. 3.

The process flow "Request-to-Answer" comprises of activities relevant to managing customer requests across all customer interfaces. The purpose of the "Request-to-Answer" process flow is to qualify and address specific information requests or product requests (sales requests) from the customer. The process could lead to the preparation of a pre-sales offer if the customer shows interest in a particular product (telecommunication service).

The subprocess "Manage contact" (1.11.1) begins an interaction between a customer and the telecommunication company. It deals with the identification of a contact, its development, enhancement and update. For queueing network

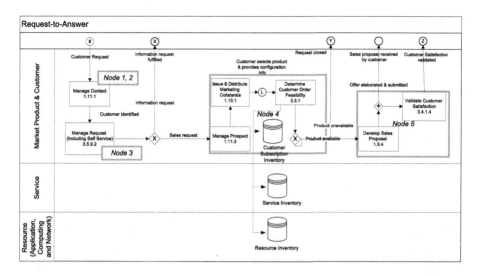

Fig. 1. Business process diagram for "Request-to-Answer" process flows

modelling in Sect. 3 two communication channels – web-interface and direct communication with a company's manager – are considered as customer interfaces.

The purpose of subprocess "Manage request" (3.5.9.2) is to manage all requests made by potential and existing customers. After receiving the request the subprocess either enables its originator to automatically fulfill it, or identifies and activates the opportune process to accomplish the request. The subprocess formally closes the request when all related activities have been terminated.

The subprocesses chain in the process flow "Request-to-Answer" includes the subprocesses aimed to increase profits at the expense of new marketing offers. They are "Issue & distribute marketing collaterals" (1.10.1) to issue and distribute marketing collateral directly to a customer, "Manage prospect" (1.11.3) to match an assigned lead with the most appropriate products and ensure that a prospect is handled appropriately, "Determine customer order feasibility" (3.3.1) to check the availability and the feasibility of providing and supporting a standard and customized product offerings where specified to a customer.

The three above-mentioned subprocesses are grouped into a single unit as a preliminary step to perform queueing network modeling in Sect. 3. The following two subprocesses can also be combined for further queueing network modeling.

The subprocess "Develop sales proposal" (1.9.4) includes selecting a service or a product in response to the customer's sales request. The Subprocess "Validate customer satisfaction" (3.4.1.4) validates that the predicted/expected value is delivered by the solution and initializes the after-sales processes (billing and assurance).

Based on the built Business Process Diagram in Sect. 3 we develop the queueing network model of end-to-end process flow "Request-to-Answer".

3   Queueing Network Model

We propose a mathematical model of end-to-end process flow "Request-to-Answer" as an open exponential queueing network (QN) with multiple class of customers $\langle \mathbf{M}, \mathbf{\Theta_0}, \lambda_0; \mu_i, i \in \mathbf{M} \rangle$. The set of queueing network nodes $\mathbf{M}$ includes nodes of two types: single-server queueing system with infinite queue $M| M| 1| \infty$ and infinite server queueing system $M| M| \infty$. A customer in this network corresponds to a request of a telecommunication company's client, each node of the network is a subprocess or a group of subprocesses (Fig. 1).

The QN model for process flow "Request-to-Answer" is shown in Fig. 2.

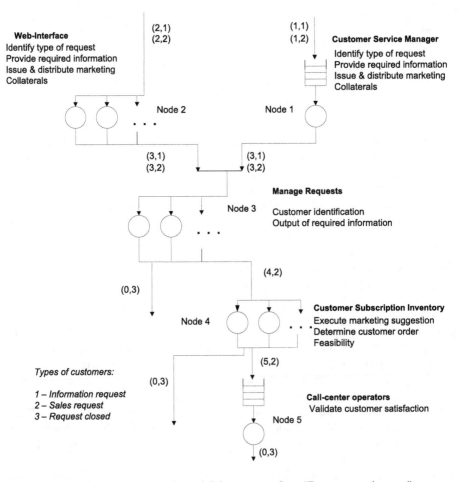

Fig. 2. Queueing network model for process flow "Request-to-Answer"

The set of queueing network nodes is determined as $\mathbf{M} = \{1, 2, ..., 5\}$, node 0 represents external source (Table 1). The set $\mathbf{R} = \{1, 2, 3\}$ of customers' classes

Table 1. Nodes of queueing network

No.	Name of node	Processes
0	External source	—
1	Customer service manager	Identify type of request
		Provide required information
		Issue & distribute marketing collaterals
		Request to CRM systems
2	Web-interface	Identify type of request
		Provide required information
		Issue & distribute marketing collaterals
		Request to CRM systems
3	CRM	Customer identification
		Output of required information
4	Customer subscription inventory	Execute marketing suggestion
		Determine customer order feasibility
5	Call-center operators	Validate customer satisfaction

Table 2. Customers' classes

No.	Name of type	Example
1	Information request	Balance enquiry
2	Sales request	Access to service request
3	Request closed	—

is represented in Table 2. The set of customers' request types is determined as follows:

$$\mathbf{L} = \{(1,1),(1,2),(2,1),(2,2),(3,1),(3,2)(4,2),(5,2),(0,3)\}.$$

The routing matrix $\mathbf{\Theta_0}$ of the queueing network is shown in Table 3.

We get the arrival flow rate $\lambda_{(i,r)}$ for each customer's request type from the following system of linear equations:

$$\lambda_{(j,s)} = \sum_{(i,r)\in\mathbf{L}} \lambda_{(i,r)}\theta_{(i,r),(j,s)} + \lambda_0\theta_{0,(j,s)}, \quad (j,s) \in \mathbf{L}. \tag{1}$$

The arrival flow rate λ_i for each node is defined as follows:

$$\lambda_i = \sum_{r\in\mathbf{R}} \lambda_{(i,r)} \tag{2}$$

with stationary balance condition [11–13]

$$\lambda_i \leq \mu_i, \quad i \in \mathbf{M}. \tag{3}$$

Table 3. The routing matrix of queueing network

Θ_0	0	(1,1)	(1,2)	(2,1)	(2,2)	(3,1)	(3,2)	(4,2)	(5,2)
0	0	a	b	d	$1-a-b-d$	0	0	0	0
(1,1)	0	0	0	0	0	1	0	0	0
(1,2)	0	0	0	0	0	0	1	0	0
(2,1)	0	0	0	0	0	1	0	0	0
(2,2)	0	0	0	0	0	0	1	0	0
(3,1)	1	0	0	0	0	0	0	0	0
(3,2)	0	0	0	0	0	0	0	1	0
(4,2)	f	0	0	0	0	0	0	0	$1-f$
(5,2)	1	0	0	0	0	0	0	0	0

The average sojourn time of a customer in the queueing system $M|\,M|\,1|\,\infty$ can be found by the formula

$$Mv_i = \frac{1}{\mu_i - \lambda_i}, \quad i \in \{1,5\}. \tag{4}$$

The average sojourn time of a customer in the queueing system $M|\,M|\,\infty$ can be defined as follows:

$$Mv_i = \mu_i^{-1}, \quad i \in \{2,3,4\}. \tag{5}$$

An average sojourn time of a customer in the queueing network – by the formula

$$Mv = \frac{1}{\lambda_0} \left[\sum_{i \in \{1,5\}} \frac{\lambda_i}{\mu_i - \lambda_i} + \sum_{i \in \{2,3,4\}} \frac{\lambda_i}{\mu_i} \right].$$

The average sojourn time of a customer corresponds to an average execution time of process flow "Request-to-Answer". This performance measure is related to QoE (Quality of Experience) parameters. A high delay in the provision of services to company's customer demonstrates the necessity for business processes reorganization. One of the ways to reduce business process execution time is to reduce the execution time of one or more subprocesses by additional resources of the company. The simplex algorithm presented in Sect. 4 allows for estimating the minimum business process execution time.

4   A Simplex Algorithm for Minimum Execution Time Estimation

Employing a simplex algorithm (SA) for the optimization of business processes is shown in [3]. Firstly, we convert the Business Process Diagram into the network graph $\langle \mathbf{V}, \mathbf{E} \rangle$, where

- the vertexes from the set **V** correspond to the states of the business process after completion of the step relevant to the subprocess or the group of sub-processes,
- the arcs (state transitions) of the set **E** determine the order of the steps,
- the weight of the arc corresponds to the transition time to the next state of the business process.

Let us consider the differences between the Business Process Diagram and the network graph in the simplex algorithm. The rectangles in BPMN sequence diagram (Fig. 1) are responsible for the activities – the subprocesses of the 3rd or 4th level (steps of the business process). The arrows indicate the transition from one step to another. For the network graph the vertexes represent states and the arcs denote activities.

The following notation shows correspondence between the vertexes of the network graph (Fig. 3), the Business Process Diagram and nodes of the QN model (Fig. 2):

1. vertex A – the state of process after execution of "Manage contact" for an information request (customer leaves Node 1 in QN model);
2. vertex B – the state of process after execution of "Manage contact" for a sales request (customer leaves Node 2 in QN model);
3. vertex C – the state of process after execution of "Manage request" for both information and sales request requests (customer leaves Node 3 in QN model);
4. vertex D – the state of process after execution of "Execute marketing suggestion" (customer leaves Node 4 in QN model);
5. vertex E – the state of process after execution "Validate customer satisfaction" (customer leaves Node 5 and the network in QN model).

In terms of the network planning, we need to find out the critical (longest) path on the network graph. Each subprocess of the process flow "Request-to-Answer" means individual activities. The activities belonging to the critical path determine the business process execution time T_Σ [3].

As a result, we get a network graph with following set of vertexes: $\mathbf{V} = \{0, A, B, C, D, E\}$.

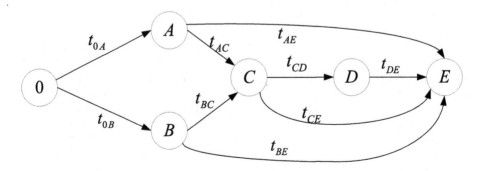

Fig. 3. Network graph for the process flow "Request-to-Answer"

The graph in Fig. 3 has 6 paths from the node 0 to the node E:

1. $0 \to A \to E$
2. $0 \to A \to C \to E$
3. $0 \to A \to C \to D \to E$
4. $0 \to B \to E$
5. $0 \to B \to C \to E$
6. $0 \to B \to C \to D \to E$

Let us introduce the following notation:
τ_{ij} – the start time of the activity ij, $\quad i, j \in \mathbf{V}$;
T_{ij} – the end time of the activity ij; $\quad i, j \in \mathbf{V}$;
t_{ij} – the duration of the activity ij, $\quad i, j \in \mathbf{V}$;
T_Σ - the total execution time of business process "Request-to-Answer".

We note that the duration of the activity ij is equal to the state i execution time, according to the network graph. The activities must be completed respectively

$$T_{ij} = \tau_{ij} + t_{ij}, \quad i, j \in \mathbf{V}. \tag{6}$$

Subprocesses are performed one after the other. The inequality (7) means that the activity ij does not start unless its predecessor activity li has finished. So we get the restrictions for optimization problem:

$$\tau_{ij} \geq T_{li}, \quad T_\Sigma \geq T_{li}, \quad i, j, l \in \mathbf{V}. \tag{7}$$

We formulate the task of optimizing the total execution time of the process flow "Request-to-Answer" as follows: find

$$\min T_\Sigma \tag{8}$$

subject to

$$\mathbf{A}x \leq \mathbf{b}, \quad x_k \geq 0, \quad k \in [1; 10]. \tag{9}$$

The components of the vector $\mathbf{x} = (x_1, x_2, ..., x_{10}) =$
$= (\tau_{0A}, \tau_{0B}, \tau_{AC}, \tau_{BC}, \tau_{CD}, \tau_{AE}, \tau_{BE}, \tau_{CE}, \tau_{DE}, T_\Sigma)$, the matrix $\mathbf{A}$ and the vector $\mathbf{b}$ are shown in Table 4.

The solution of (9) vector $\mathbf{x}$ defines the activities start time and the business process total execution time $T_\Sigma = x_{10}$.

So formulas (8)–(9) determine the method for calculating the total execution time of the process flow "Request-to-Answer" with the given duration of all component activities of the process. Note that we use the simplex algorithm on the assumption that the number of the activity executors is unlimited. Thus, the activities can start as soon as it is required. That is why as the duration of the activity in the SA model not the service time but the sojourn time of a customer in the appropriate node for QN model was used. The combination of QN and SA techniques gives us the method for business process optimization in terms of process execution time.

Table 4. Parameters of the problem

x_1	x_2	x_3	x_4	x_5	x_6	x_7	x_8	x_9	x_{10}	$\leq$	b
0	0	-1	0	0	0	0	0	0	0		$-t_{0A}$
0	0	0	0	0	-1	0	0	0	0		$-t_{0A}$
0	0	0	-1	0	0	0	0	0	0		$-t_{0B}$
0	0	0	0	0	0	-1	0	0	0		$-t_{0B}$
0	0	1	0	-1	0	0	0	0	0		$-t_{AC}$
0	0	1	0	0	0	0	-1	0	0		$-t_{AC}$
0	0	0	1	-1	0	0	0	0	0		$-t_{BC}$
0	0	0	1	0	0	0	-1	0	0		$-t_{BC}$
0	0	0	0	1	0	0	0	-1	0		$-t_{CD}$
0	0	0	0	0	1	0	0	0	-1		$-t_{AE}$
0	0	0	0	0	0	0	0	1	-1		$-t_{DE}$
0	0	0	0	0	0	0	1	0	-1		$-t_{CE}$
0	0	0	0	0	0	1	0	0	-1		$-t_{BE}$

5 Algorithm for the Estimation of Business Process Total Execution Time

The developed method for optimizing the activity of the company at a predetermined threshold for the business process execution time represents a combination of the queuing networks modelling and simplex algorithm of linear programming. The sojourn time for the appropriate node of a QN model is used in the simplex algorithm as the duration of activity. First, using a QN model for a given average throughputs μ_i of QN nodes we get the average sojourn time ν_i for the nodes of the QN model from (4) and (5). Second, we substitute the sojourn times as the durations of activities into SA to (8)–(9). Eventually we get the minimum total execution time T_Σ of the process flow Request-to-Answer for a given flow rate $\lambda_0^{\max}$ of requests to "Request-to-Answer". If the obtained value T_Σ exceeds the threshold $T^{\max}$ specified by company regulations, it is necessary to increase the throughputs of one or more sub-processes composing the process flow "Request-to-Answer".

Thus, in terms of the QN model the problem is to choose those average throughputs $(\mu_1, \mu_2, \mu_3, \mu_4, \mu_5)$ such that $T_\Sigma \leq T^{\max}$ with a given flow rate $\lambda_0^{\max}$. A restriction for the QN model is the stationary balance condition (3), which does not allow queues in the nodes with the expectation to grow indefinitely.

The Algorithm 1 below represents the approach of solving this problem. The inputs to the algorithm are the flow rate $\lambda_0^{\max}$, average throughputs $(\mu_1, \mu_2, \mu_3, \mu_4, \mu_5)$ and the predetermined threshold time $T^{\max}$ for the total execution time of "Request-to-Answer". There are two outputs of the algorithm: new average throughputs, in which the execution time will not exceed the threshold, and corresponding total execution time T_Σ. The obtained average throughputs can be used as guidelines when optimizing telecommunication company activity.

Algorithm 1. Get BP throughput values

$k = 0$
$\Lambda^{(k)} = \lambda_0^{\max}$
if $\lambda_0^{\max} < \Lambda^{(k)}$ **then**
  $i = 1$
  repeat
    QN: find $\nu_i^{(k)} = \nu_i^{(k)}\left(\lambda_0^{\max}\right)$
    $i = i + 1$
  until $i <= M$
  SA: find $T^{(k)} = T_\Sigma\left(\lambda_0^{\max}, \nu_1^{(k)}, ..., \nu_M^{(k)}\right)$
  if $T^{(k)} <= T^{\max}$ **then**
    output BP throughput values $\left(\mu_1^{(k)}, ..., \mu_M^{(k)}\right)$ and $T^{(k)}$
  end if
  $k = k + 1$
  $i = 1$
  repeat
    increase μ_i
    $i = i + 1$
  until $i <= M$
  QN: find $\Lambda^{(k)} = \Lambda\left(\mu^{(k)}\right)$
end if

The example of the performance analysis of the process flow "Request-to-Answer" is based on the initial data close to the real one:

$a = 0,1$ – the proportion of information requests received by the manager as the total number of requests;

$b = 0,05$ – the proportion of information requests received through Web-interface as the total number of requests;

$d = 0,55$ – the proportion of sales requests received by the manager as the total number of requests;

$f = 0,1$ – the probability of unavailability of a service or a product;

$\mu_1 = \frac{1}{5}$ – requests per minute, $\mu_2 = \frac{3}{4}$ requests per minute, $\mu_3 = 6$ requests per minute, $\mu_4 = 60$ requests per minute, $\mu_5 = 1$ request per minute– average throughputs of QN nodes;

$\lambda_0^{\max} = 1$ request per minute – the flow rate of requests to "Request-to-Answer".

The problem is solved for the two variants of threshold for the total execution time of "Request-to-Answer": $T_1^{\max} = 3$ min., $T_2^{\max} = 6$ min.

For these initial data the restriction on the input flow rate λ_0 from (1)–(3) is $\Lambda = 1,(3)$ requests per minute, thus the stationary balance condition is given by $\lambda_0 < 1,(3)$.

Figure 4 shows the total execution time T_Σ vs the input flow rate λ_0.

For $\lambda_0^{max} = 1$ the execution time calculated by the Algorithm 1 $T_\Sigma = 3,64$ min.

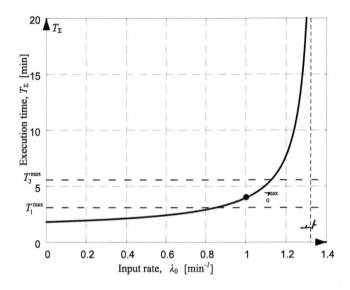

Fig. 4. Execution time for the process flow "Request-to-Answer"

With the threshold $T_2^{max} = 6\,\text{min}$ the obtained execution time T_Σ is less than the threshold value, so there is no need to change the throughputs μ_i of subprocesses. With the threshold $T_1^{max} = 3\,\text{min}$ the execution time T_Σ of "Request-to-Answer" does not match telecommunication company regulations, so it is necessary to increase the throughputs μ_i of one or more sub-processes. After changing the throughputs it is necessary to repeat the procedure for checking the efficiency of the business process. Note that in this paper we will not specify the rules by which, if necessary, the throughputs should be increased. The development of such rules is one of the tasks of further research.

6   Conclusion

The proposed technique can be applied to analyze the efficiency of business processes of a telecommunication company. In this case, the analysis of the execution time of the business process should take into account the rise in the intensities of the incoming flows in QN model nodes. The reason is the increasing of the number of subprocesses under consideration for the QN model of a company in comparison with the QN model of a stand-alone process [14].

Development of a method for grouping subprocesses for association with corresponding QN model nodes is a task for future research. The method should take into account the performance targets for process analysis and throughputs for groups of subprocesses. Another task of further research is to develop recommendations for finding the optimal solution for throughputs of subprocesses with the iterative procedure in Algorithm 1.

The reported study was partially supported by RFBR, research projects No. 14-07-00090, 15-07-03051, 15-07-03608 and Vladimir Potanin Foundation.

References

1. van der Aalst, W.M.P., Christian, S.: Modeling Business Processes: A Petri Net-oriented Approach, pp. 1–386. The MIT Press, London (2011)
2. Oracle Crystal Ball. www.oracle.com/us/products/applications/crystalball/crystal-ball-product/overview/index.html. Accessed 01 May 2016
3. Sidnev, A., Tuominen, J., Krassi, B.: Business process modeling and simulation. Helsinki University of Technology. Industrial Information Technology Laboratory Publications, pp. 1–116 (2005)
4. ARIS Business Simulator. www.ariscommunity.com/business-process-simulation. Accessed 01 Apr 2016
5. Baskett, F., Chandy, K.M., Muntz, R.R., Palacios, F.G.: Open, closed, and mixed networks of queues with different classes of customers. J. ACM **22**(2), 248–260 (1975)
6. Ficken, F.A.: The Simplex Method of Linear Programming. Dover Books, New York (2015)
7. Mantepu, T.M., Solly, M.S.: Modelling and measuring milestones in business process optimization. Probl. Persp. Manage. **12**(4), 221–224 (2014)
8. TeleManagement Forum. Enhanced Telecom Operations (eTOM) The Business Process Framework. www.tmforum.org. Accessed 01 May 2016
9. Business Process Model and Notation (BPMN). Version 2.0.2: OMG Document Number: formal.2013-12-09. Object Management Group (2013)
10. Business process framework (eTOM). End-to-end business flows. GB921 Addendum E. R15.0.0., pp. 1–110 (2016)
11. Moiseeva, S.P., Zakhorolnaya, I.A.: Mathematical model of parallel retrial queueing of multiple requests. Optoelectron. Instrum. Data Process. **47**(6), 567–572 (2011)
12. Nazarov, A.A., Moiseev, A.N.: Distributed system of processing of data of physical experiments. Russ. Phys. J. **57**(7), 984–990 (2014)
13. Bakholdina, M., Gortsev, A.: Joint probability density of the intervals length of modulated semi-synchronous integrated flow of events in conditions of a constant dead time and the flow recurrence conditions. In: Dudin, A., Nazarov, A., Yakupov, R., Gortsev, A. (eds.) ITMM 2014. CCIS, vol. 564, pp. 13–27. Springer, Heidelberg (2015). doi:10.1007/978-3-319-25861-4_2
14. Samouylov, K.E., Chukarin, A.V., Yarkina, N.V.: Business Processes and Information Technology in the Management of Telecommunications Companies, pp. 1–619. Alpina-Publishers, Randwick (2016)

On Analyzing the Blocking Probability of M2M Transmissions for a CQI-Based RRM Scheme Model in 3GPP LTE

Konstantin Samouylov[1], Irina Gudkova[2], Ekaterina Markova[1(✉)], and Iliya Dzantiev[1]

[1] RUDN University, 6 Miklukho-Maklaya Street, 117198 Moscow, Russian Federation
{samouylov_ke,markova_ev}@pfur.ru, dzonthammet@mail.ru
[2] Institute of Informatics Problems, Federal Research Center "Computer Science and Control" of Russian Academy of Sciences, RUDN University, 6 Miklukho-Maklaya Street, 117198 Moscow, Russian Federation
gudkova_ia@pfur.ru

Abstract. Machine type communications (MTC) or machine-to-machine (M2M) services become one of the drivers towards 5th generation (5G) wireless network. Various MTC devices such as smart meters and sensors form the basis of smart cites and homes. The question is how to efficiently transmit the information from MTC devices via a wireless network, which is primarily used for human-to-human (H2H) communications. Nevertheless, one of the important qualities of service (QoS) measure is still blocking probability. In the simplest case, the Erlang B formula is used to calculate the blocking probability. A more precise value can be obtained considering the MTC devices positons within a cell and applied radio resource management (RRM) mechanisms. First of all, it is expressed in the distances from the devices to eNodeB. In the paper, following the approach of including the stochastic distance in the queuing system, we propose formulas for calculating stationary probability distribution in product form considering a channel quality indicator (CQI) reported by MTC devices. Two RRM schedulers working according to the round robin policy (RRP) and full power policy (FPP) are considered. The former assumes full occupation of a time frame by all MTC devices and a variable devices' power such that to achieve a needed uplink bit rate. The latter assumes a constant power and variable time frame occupation.

Keywords: Wireless network · LTE · Machine-to-machine · M2M devices · Channel quality indicator · CQI · Round robin · Full power · Queuing system · Erlang B formula · Stochastic distance · Blocking probability

The reported study was funded by RFBR and Moscow City Government according to the research project No. 15-37-70016 mol_a_mos, and by RFBR according to the research project No. 15-07-00766 a, and the Ministry of Education and Science of the Russian Federation (President's Scholarship No. 2987.2016.5).

A. Dudin et al. (Eds.): ITMM 2016, CCIS 638, pp. 327–340, 2016.
DOI: 10.1007/978-3-319-44615-8_29

1   Introduction

According to the ITU [1], mobile technology transforms our experience of communications, entertainment, and the Internet. In May 2015, the world's population was about 7.3 billion, and there were about 7.5 billion mobile subscriptions (including machine-to-machine, M2M) and about 3.7 billion people connected. M2M devices and applications are one of the most dynamically growing segments. Many M2M devices will be connected to a smartphone or another device for their own communication; each device could have different charges/subscriptions. It is estimated that the number of mobile connected M2M devices will be around 7 billion in 2020, which is equivalent to the world's population in January 2015. By 2030, there may be 97 billion of M2M devices, and it would be more than ten times more than the estimated human population.

Consequently, the use of M2M communications will cut a good figure in everyday life, and therefore we face a need for technological development for better quality service of M2M applications, which generate a large volume of traffic. It is to be noted that different M2M applications need different technologies [2]. The first application type includes smart measurements, device position measurement, e-healthcare, etc. Such kinds of applications are defined by a small volume of data and long delay, low cost, and energy demand. The second application type is associated with the connection to a car and is characterized by a small data volume, short delay, and high reliability. The third type is related the video monitoring and is characterized by a large data volume and high data rate. These applications and services will provide a high traffic load on a wireless network.

A possible solution to the frequency coverage problem, associated with the hasty growth in the data volume generated by the M2M devices is more efficient usage of the frequency spectrum – the usage of the high frequencies range [3,4]. Also, different schedulers of radio resource management (RRM) [5], such as round robin policy (RRP) and full power policy (FPP), help to allocate the spectrum more efficiently [6–10]. These considered schedulers operate in power, frequency, and time domains and assume a constant frequency band and variable power and share of time slot. Using the RRP, the share of time slot and the power are variable resources. The time slot is divided equally between all M2M devices and the transmit power is set such as to achieve a guaranteed bit rate (GBR). Using FPP, only the share of time slot is the variable resource, but transmit power is fixed and equal to maximum transmit power of the M2M device.

An important parameter for M2M wireless networks is the distance between the device and eNodeB. Therefore, describing such a kind of networks with streaming traffic, which is characterized by GBR and fixed service duration, or with elastic traffic, which is characterized by non-guaranteed bit rate (non-GBR) and variable service duration, in case of the queuing theory we have to modify the arriving flow from active M2M devices in such a way as to take into account the distance to the eNodeB. According to the 3GPP [11], 15 different Channel Quality Indicator (CQI) levels are set for Long Term Evolution (LTE) networks. The CQI is related to the maximum Modulation and Coding Scheme (MCS) supported by the M2M device (Table 1). In this case, we may divide the cell to 15 CQI levels, and the M2M devices with the same reported CQI levels could be combined in 15 groups. All devices in the group have the same characteristics – the distance to eNodeB, the achievable bit rate, etc. We underscore that in this case the maximum distance from each device of the group to eNodeB is the same and is defined by the reported CQI level.

Table 1. CQI-MCS mapping [12]

CQI index	Modulation scheme	Code rate	Spectral efficiency [bit/s/Hz]
1	QPSK	0.076	0.1523
2	QPSK	0.120	0.2344
3	QPSK	0.190	0.3770
4	QPSK	0.300	0.6016
5	QPSK	0.440	0.8770
6	QPSK	0.590	1.1758
7	16-QAM	0.370	1.4766
8	16-QAM	0.480	1.9141
9	16-QAM	0.600	2.4063
10	64-QAM	0.450	2.7305
11	64-QAM	0.550	3.3223
12	64-QAM	0.650	3.9023
13	64-QAM	0.750	4.5234
14	64-QAM	0.850	5.1152
15	64-QAM	0.930	5.5547

The remainder of this paper is organized as follows. In Sect. 2, we propose a model of a cell of a wireless LTE network with stationary M2M devices, which generate streaming traffic. The devices may be in an active or in a passive mode, and randomly get one of the 15 CQI levels, which define their positions in the cell. We also describe a tie between the spectrum resource and the achievable bit rate and schedulers – RRP and FPP. In Sect. 3, we propose a mathematical model for two cases – the cases of infinite and finite transmit power of the active M2M devices. In Sect. 4, we conduct an analysis of a model quality of service (QoS) parameter – blocking probability – by using RRP. Finally, we conclude the paper in Sect. 5.

2 System Model

2.1 General Assumptions and Parameters

We consider a single wireless network cell of radius R and stationary M2M devices uniformly distributed within it. Let random variable $\eta = 1, \ldots, L$ be the CQI (channel quality indicator) level reported by an M2M device, then the maximum distance from it to eNodeB is $\xi_d(\eta) = RL^{-1}\eta$. The transmit power of the M2M device is a random variable $\xi_p \leq p_{\max}$. M2M devices transmit data on GBR r_0 via uplink channel. Let us suppose the maximum transmit power is the same for all the devices. We consider an ideal conditions, which are called free space. All necessary notations are given in Table 2.

Table 2. Parameters

Notation	Parameter
Cell parameters	
R	Cell radius [m]
L	Number of the CQI levels
ω	Bandwidth of uplink channel [Hz]
M2M device parameters	
η	Reported CQI level (random variable)
$p_l = \frac{2L-2l-1}{L^2}$	Probability that the reported CQI level is equal to $\eta = l$
$\xi_d(\eta)$	Maximum distance to eNodeB when the reported CQI level is equal to η (random variable) [m]
$p_{\max}$	Maximum transmit power [Hz]
ξ_p	Current transmit power (random variable) [Hz]
r_0	Guaranteed bit rate [bps]
$r(\xi_d(\eta), \xi_p)$	Achievable bit rate when an M2M device reported CQI level η and has transmit power ξ_p
λ	Arrival rate of sessions to transmit data [1/s]
μ^{-1}	Mean session duration (time interval when device is active) [s]
Space parameters	
Free space	
N_0	Noise power [Hz]
G	Propagation constant
κ	Propagation exponent

2.2 Achievable Bit Rate

According to Shannon's formula (1), the achievable bit rate $r(\xi_d(\eta), \xi_p)$ for an M2M device depends on bandwidth ω of the uplink channel, transmit power ξ_p, and distance $\xi_d(\eta)$ to eNodeB and is defined as follows:

$$r(\xi_d(\eta), \xi_p) = \omega \ln\left(1 + \frac{G\xi_p}{\xi_d^\kappa(\eta) N_0}\right) = \omega \ln\left(1 + \frac{G\xi_p}{\left(\frac{R}{L}\eta\right)^\kappa N_0}\right). \quad (1)$$

Let us suppose the M2M devices are uniformly distributed within the cell, then the cumulative distribution function (CDF) of random variable $\xi_d(\eta)$ is the following:

$$F_{\xi_d(\eta)}(d) = P\{\xi_d(\eta) \le d\} = \begin{cases} 0, & d < 0, \\ \frac{d^2}{R^2}, & d \in [0, R], \\ 1, & d > R, \end{cases} \tag{2}$$

and the corresponding probability density function (PDF) is $f_{\xi_d(\eta)}(d) = \frac{2d}{R^2}$, $0 \le d \le R$.

2.3   Schedulers: Round Robin and Full Power

So the achievable bit rate depends first of all on three parameters – the frequency band (bandwidth of uplink channel), the transmit power and the provided share of time slot, which is defined according to the distance from the device to eNodeB. These parameters changes depend on different schedulers for RRM. We consider two schedulers with a constant frequency band and variable share of time slot and transmit power – RRP and FPP. To gain a better understanding let us consider an example (Fig. 1).

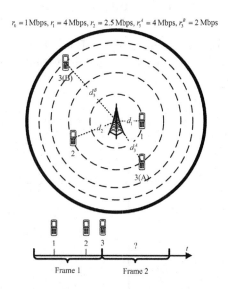

Fig. 1. The example of functioning RRP and FPP for both variants of M2M device position in the cell

Let there be placed three active M2M devices with different reported CQI levels in the cell. The first and the second are placed at the distances $\xi_d(\eta_1) = d_1$, $\xi_d(\eta_2) = d_2$ from eNodeB. The third device may be placed either near eNodeB on the distance $\xi_d(\eta_1^A) = d_3^A$, or far – $\xi_d(\eta_1^B) = d_3^B$.

Using RRP, a time slot is divided equally between all active M2M devices and the transmit power is set such as to achieve GBR r_0. In both considered cases when the third device turns to an active mode, the first two devices regulate their power to achieve the GBR. In case A, the device is placed near the eNodeB and the available power is enough for all the devices to achieve the bit rate r_0, and the device is getting served (Fig. 2).

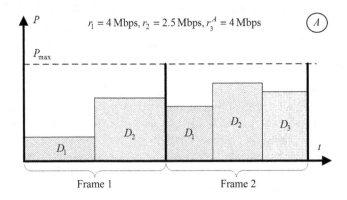

Fig. 2. RRP, case A

In case B, the device is far from the eNodeB and the power is not enough for starting the service of the third device and its data transmission will be blocked (Fig. 3).

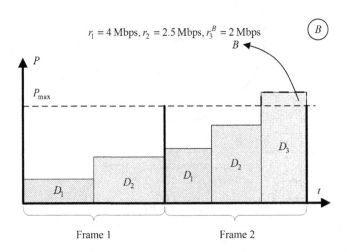

Fig. 3. RRP, case B

Unlike the RRP, using FPP all the M2M devices work with a maximum transmit power, but have a different share of the time slot, which is proportional

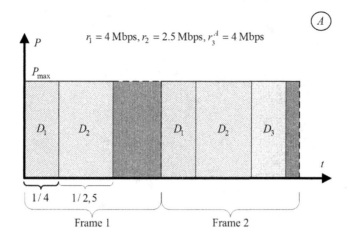

Fig. 4. FPP, case A

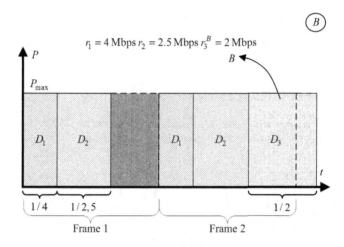

Fig. 5. FPP, case B

to the achievable bit rate r_i, $i = 1, 2, 3$. In case A the third device is getting a sufficient share of time slot to achieve the GBR r_0 and is being served (Fig. 4).

In case B the device is far away and is not able to get enough share of the time resource and it is blocked (Fig. 5).

3   Mathematical Model

We introduce the following general notation $\xi(t)$ – is the number of active M2M devices, $\eta_i(t)$, $i = 1, \ldots, \xi(t)$ – is the CQI level reported by M2M device i in moment $t \geq 0$. Then the behavior of the system is defined by the continuous

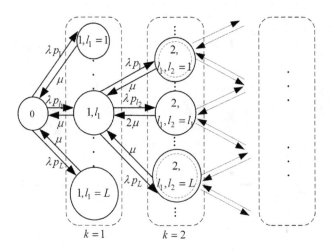

Fig. 6. The state transition diagram

Markov chain (CMC) $\{\xi(t), \eta_1(t), \ldots, \eta_{\xi(t)}(t), \ t \geq 0\}$. Accordingly, the state of the system is defined as follows $(k, l_1, \ldots, l_k)$, $\ l_i = 1, \ldots, L, \ k = 0, 1, \ldots$.

3.1   The Case of Infinite Device Transmit Power

Let us consider a case when the transmit power of active M2M devices has no restriction. Then the system state space is the following:

$$\tilde{\mathcal{L}} = \{(0), (1,1), \ldots, (1,L), (2,1,1), \ldots, (2,L,L), \ldots,$$

$$\left(k, \underbrace{1, \ldots, 1}_{k}\right), \ldots, \left(k, \underbrace{L, \ldots, L}_{k}\right), \ldots\right\} \tag{3}$$

$$= \{(0), (k, l_1, l_2, \ldots, l_k), \ l_i \in \{1, \ldots, L\}, \ i = 1, \ldots, k, \ k = 1, 2, \ldots\}.$$

Let us denote $\tilde{\pi}(k, l_1, l_2, \ldots, l_k)$ – the steady state probability that there are k active M2M devices reported CQI level l_i, $i = 1, \ldots, k$. Let us assume that the reported CQI level is equal to $\eta = l$ with probability p_l, $l \in \{1, \ldots, L\}$ which, with regard to the uniform distribution, is defined by the following: $p_l = \frac{2l-1}{L^2}$, $l = 1, \ldots, L$ (Fig. 6).

Proposition 1. The stationary probability distribution $\tilde{\pi}(k, l_1, l_2, \ldots, l_k)$, $(k, l_1, l_2, \ldots, l_k) \in \tilde{\mathcal{L}}$ that there are k active M2M devices reported CQI level l_i, $i = 1, \ldots, k$ is calculated by the formula:

$$\tilde{\pi}(k, l_1, l_2, \ldots, l_k) = \tilde{\pi}(0) \cdot \left(\frac{\lambda}{\mu}\right)^k \frac{\prod_{i=1}^{k} p_{l_i}}{\prod_{i=1}^{I(k, l_1, \ldots, l_k)} N_i(k, l_1, \ldots, l_k)!}, \tag{4}$$

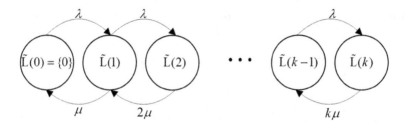

Fig. 7. The state transition diagram for agregated states of system state space $\tilde{\mathcal{L}}$

where $(k, l_1, l_2, \ldots, l_k) \in \tilde{\mathcal{L}}$,

$$\tilde{\pi}(0) = \left(\sum_{(k,l_1,l_2,\ldots,l_k) \in \tilde{\mathcal{L}}} \frac{\prod_{i=1}^{k} \lambda p_{l_i}}{\mu^k \prod_{i=1}^{I(k,l_1,l_2,\ldots,l_k)} N_i(k, l_1, l_2, \ldots, l_k)!} \right)^{-1} = G,$$

$I(k, l_1, l_2, \ldots, l_k)$, $N_i(k, l_1, l_2, \ldots, l_k)$ are defined as follows:
$(k, l_1, \ldots, l_{N_1}, l_{N_1+1}, \ldots, l_{N_1+N_2}, \ldots, l_{N_1+N_2+\ldots N_{I-1}+1}, \ldots, l_{N_1+\ldots+N_I} = l_k)$
such that $l_1 = \ldots = l_{N_1}, l_{N_1+1} = \ldots = l_{N_1+N_2}, \ldots, l_{N_1+N_2+\ldots N_{I-1}+1} = \ldots$
$= l_{N_1+\ldots+N_I} = l_k$.

The system state space $\tilde{\mathcal{L}}$ may be divided by the number of the served M2M
devices k, $\tilde{\mathcal{L}} = \bigcup_{n=0}^{\infty} \tilde{\mathcal{L}}(k)$, $\tilde{\mathcal{L}}(k) = \left\{ (k, l_1, l_2, \ldots, l_k) \in \tilde{\mathcal{L}} \right\}$, then the state transition diagram will be the following (Fig. 7):
Writing out the local balance equations for the aggregated states, we get
Note 1.

Note 1. The stationary probability distribution $\tilde{p}(k)$ of the aggregated states
is defined by the formula: $\tilde{p}(k) = \frac{\rho^k}{k!} e^{-\rho}$, where $\rho = \frac{\lambda}{\mu}$.

3.2   The Case of Finite Device Power

Practically, the transmit power of M2M devices is not infinite and
depends on the chosen scheduler. Let us consider the restriction of CMC
$\{\xi(t), \eta_1(t), \ldots, \eta_{\xi(t)}(t), \ t \geq 0\}$ to the set $\mathcal{L} \subset \tilde{\mathcal{L}}$. Certain type of $\mathcal{L}$, defined
accordingly with the RRP or FPP, is shown in Chap. 4.3. Let us divide the
system state space $\mathcal{L}$ by the number of the served M2M devices k, $\mathcal{L} = \bigcup_{k=0}^{\infty} \mathcal{L}(k)$, $\mathcal{L}(k) = \{(k, l_1, \ldots, l_k) \in \mathcal{L}\}$. Then the system probability distribution for $(k, l_1, l_2, \ldots, l_k) \in \mathcal{L}$

$$\pi(k, l_1, \ldots, l_k) = \tilde{\pi}(k, l_1, \ldots, l_k) \left(\sum_{i=0}^{\infty} \sum_{(i,l_1,\ldots,l_i) \in \mathcal{L}(i)} \tilde{\pi}(k, l_1, \ldots, l_k) \right)^{-1}.$$

Proposition 2. The system probability distribution $\pi(k, l_1, l_2, \ldots, l_k)$, $(k, l_1, l_2, \ldots, l_k) \in \mathcal{L}$ is calculated by the formula:

$$\frac{G\left(\mu^k \prod_{i=1}^{I(k,l_1,l_2,\ldots,l_k)} N_i(k, l_1, l_2, \ldots, l_k)!\right)^{-1} \prod_{i=1}^{k} \lambda p_{l_i}}{\sum_{i=0}^{\infty} \sum_{(i,l_1,\ldots,l_i) \in \mathcal{L}(i)} G\left(\mu^k \prod_{i=1}^{I(k,l_1,l_2,\ldots,l_k)} N_i(k, l_1, l_2, \ldots, l_k)!\right)^{-1} \prod_{i=1}^{k} \lambda p_{l_i}}. \tag{5}$$

With regard to the state transition diagram, shown in Fig. 6, let us compose the local balance equations.

$$\pi(0) \sum_{(1,l_1) \in \mathcal{L}(1)} \lambda p_{l_1} = \mu \sum_{(1,l_1) \in \mathcal{L}(1)} \pi(1, l_1),$$

$$\sum_{(1,l_1) \in \mathcal{L}(1)} \pi(1, l_1) \sum_{(2,l_1,l_2) \in \mathcal{L}(2)} \lambda p_{l_2} = 2\mu \sum_{(2,l_1,l_2) \in \mathcal{L}(2)} \sum_{(2,l_1,l_2) \in \mathcal{L}(2)} \pi(2, l_1, l_2),$$

$$\cdots\cdots\cdots\cdots\cdots\cdots\cdots\cdots\cdots\cdots\cdots\cdots\cdots \tag{6}$$

$$\sum_{(1,l_1) \in \mathcal{L}(1)} \cdots \sum_{(k-1,l_1,l_2,\ldots,l_{k-1}) \in \mathcal{L}(k-1)} \pi(k-1, l_1, \ldots, l_{k-1}) \sum_{(k,l_1,l_2,\ldots,l_k) \in \mathcal{L}(k)} \lambda p_{l_k}$$

$$= k\mu \sum_{(k,l_1,l_2,\ldots,l_k) \in \mathcal{L}(k)} \cdots \sum_{(k,l_1,l_2,\ldots,l_k) \in \mathcal{L}(k)} \pi(k, l_1, \ldots, l_k)$$

Let us denote p_k – probability, that there are k active M2M devices in the system, $k = 1, 2, \ldots$, then the local balance Eq. (6) are the following:

$$\lambda \sum_{(1,l_1) \in \mathcal{L}(1)} \cdots \sum_{(k,l_1,l_2,\ldots,l_k) \in \mathcal{L}(k)} \pi(k, l_1, \ldots, l_k) \lambda p_{l_k} = k\mu p_k, k = 1, 2, \ldots, \tag{7}$$

where $\sum_{(1,l_1) \in \mathcal{L}(1)} \cdots \sum_{(k,l_1,l_2,\ldots,l_k) \in \mathcal{L}(k)} \pi(k, l_1, \ldots, l_k) \lambda p_{l_k} = \text{P}\{(k-1, l_1, l_2, \ldots, l_{k-1}) \in \mathcal{L}(k-1), (k, l_1, l_2, \ldots, l_k) \in \mathcal{L}(k)\}$.

With regard to the formula of total probability:

$$\lambda \text{P}\{(k, l_1, l_2, \ldots, l_k) \in \mathcal{L}(k) \mid (k-1, l_1, l_2, \ldots, l_{k-1}) \in \mathcal{L}(k-1)\}$$
$$\cdot \text{P}\{(k-1, l_1, l_2, \ldots, l_{k-1}) \in \mathcal{L}(k-1)\} = k\mu p_k, \tag{8}$$

since $\text{P}\{(k-1, l_1, l_2, \ldots, l_{k-1}) \in \mathcal{L}(k-1)\} = 1$, then

$$\lambda \text{P}\{(k, l_1, l_2, \ldots, l_k) \in \mathcal{L}(k) \mid (k-1, l_1, l_2, \ldots, l_{k-1}) \in \mathcal{L}(k-1)\} \cdot p_{k-1} = k\mu p_k. \tag{9}$$

Let us denote $\text{P}(k-1)$ – the conditional probability $\text{P}\{(k, l_1, l_2, \ldots, l_k) \in \mathcal{L}(k) \mid (k, l_1, l_2, \ldots, l_{k-1}) \in \mathcal{L}(k-1)\}$ that the M2M device with the number k will be served under the condition that $k-1$ M2M devices are already served in the system. Then the local balance equation and state transition diagram are the following (Fig. 8):

$$\lambda \text{P}(k-1) \cdot p_{k-1} = k\mu p_k, k = 1, 2, \ldots. \tag{10}$$

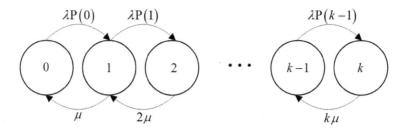

Fig. 8. The state transition diagram for the agregated states of state space $\mathcal{L}$

Proposition 3. The stationary probability distribution p_k, $k = 1, 2, \ldots$ that there are k active M2M devices is calculated by the formula:

$$p_k = p_0 \cdot \left(\frac{\lambda}{\mu}\right)^k \frac{\prod_{i=0}^{k-1} \mathrm{P}(i)}{k!}, \quad k = 1, 2, \ldots, \tag{11}$$

where $p_0 = \left(\sum_{j=0}^{k} \left(\frac{\lambda}{\mu}\right)^j \frac{\prod_{i=0}^{j-1} \mathrm{P}(i)}{j!}\right)^{-1}$.

Probability $\mathrm{P}(k-1)$ that one more M2M device could transfer data under the condition that there are already $k-1$ active M2M devices is defined by the formula of total probability:

$$\mathrm{P}(k-1) = \mathrm{P}\{(k, l_1, \ldots, l_k) \in \mathcal{L}(k) \mid (k-1, l_1, \ldots, l_{k-1}) \in \mathcal{L}(k-1)\}$$
$$= \frac{\mathrm{P}\{(k, l_1, \ldots, l_k) \in \mathcal{L}(k), (k-1, l_1, \ldots, l_{k-1}) \in \mathcal{L}(k-1)\}}{\mathrm{P}\{(k-1, l_1, \ldots, l_{k-1}) \in \mathcal{L}(k-1)\}}. \tag{12}$$

Note 2. The main probability characteristic of CQI-based RRM scheme model for M2M transmissions – blocking probability B that data transmission from an M2M device will be blocked could be computed as follows:

$$B = \sum_{k=0}^{\infty} (1 - \mathrm{P}(k)) \, p_k. \tag{13}$$

3.3 Restriction of the CMC According to the Schedulers

Let us introduce the access function for the considering schedulers. For the RRP the access function is

$$g_{\xi_d(\eta)}(k, l_1, l_2, \ldots, l_k) = \begin{cases} 1, & \frac{r_0}{r(\xi_d(\eta), p_{\max})} \leq \frac{1}{k+1}, \quad i = 1, \ldots, k+1; \\ 0, & \text{otherwise}, \end{cases} \tag{14}$$

for the FPP it is

$$g_{\xi_d(\eta)}(k, l_1, l_2, \ldots, l_k) = \begin{cases} 1, & \sum_{i=1}^{n} \frac{r_0}{r(\xi_d(\eta), p_{\max})} \leq 1; \\ 0, & \text{otherwise}. \end{cases} \tag{15}$$

Then the system state spaces $\mathcal{L}_{RR}(k) \subset \tilde{\mathcal{L}}$ or $\mathcal{L}_{FP}(k) \subset \tilde{\mathcal{L}}$, are defined with the formula (16) or (17) respectively.

$$\mathcal{L}_{RR}(k) = \left\{ 0 \le d_1 \le R, \ldots, 0 \le d_k \le R : \frac{r_0}{\omega \ln\left(1 + \frac{Gp_{max}}{d_i^\kappa N_0}\right)} \le \frac{1}{k}, \, i = 1, \ldots, k \right\}; \tag{16}$$

$$\mathcal{L}_{FP}(k) = \left\{ 0 \le d_1 \le R, \ldots, 0 \le d_k \le R : \sum_{i=1}^{k} \frac{r_0}{\omega \ln\left(1 + \frac{Gp_{max}}{d_i^\kappa N_0}\right)} \le 1 \right\}. \tag{17}$$

4   The Queue System Analysis for the RRP

4.1   The Probability Calculation of the New M2M Device to Be Served

With regard to the access function (14) and the system state space (16) the conditional probability $P(k)$ (12) of M2M device k to be served, if there are $k-1$ M2M devices in the system and the RRP is used, is defined by the formula:

$$P(k) = \left(P\left\{ \frac{r_0}{r\left(\xi_d(\eta), p_{max}\right)} \le \frac{1}{k+1} \right\} \right)^{k+1} \cdot \left(P\left\{ \frac{r_0}{r\left(\xi_d(\eta), p_{max}\right)} \le \frac{1}{k} \right\} \right)^{-k}. \tag{18}$$

The random variable $\xi_d(\eta)$ is uniformly distributed and has a distribution function defined by the formula (1), then let us consider Proposition 4.

Proposition 4. The conditional probability $P(k)$ that one more M2M device could transfer data under the condition that there are already k active M2M devices, using the RRP, is defined by the formula:

$$P(k) = \frac{1}{R^2} \left(\frac{Gp_{max}}{N_0} \right)^{\frac{2}{\kappa}} \left(e^{\frac{r_0 k}{\omega}} - 1 \right)^{\frac{2k}{\kappa}} \left(e^{\frac{r_0 k+1}{\omega}} - 1 \right)^{\frac{-2(k+1)}{\kappa}}. \tag{19}$$

4.2   The Numerical Analysis of the Blocking Probability

Let us consider the numerical analysis of the blocking probability for the queue system with the RRP.

Corollary 1. The blocking probability B_{RR} for the M2M device with the number k, using the RRP, is defined by the formula (13), with regard to the formulas (11) and (19).

Let us consider 2 cases of blocking probability dependence on the cell radius. In the first case (Fig. 9) and second case (Fig. 10) we use different values of guaranteed bit rate. We watch the system behavior with the increasing of the arrival rate of M2M requests. Also we consider different cell radiuses. All input parameters are shown in Table 3.

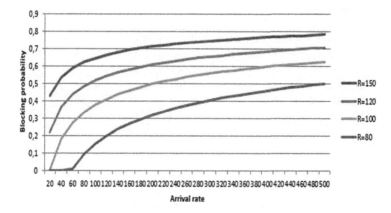

Fig. 9. The blocking probability dependence on cell radius, $r_0 = 100000\,\mathrm{kbit/s}$

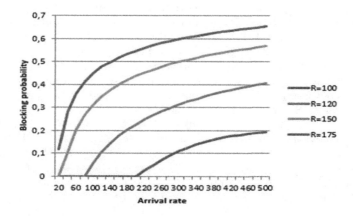

Fig. 10. The blocking probability dependence on the cell radius, $r_0 = 10000\,\mathrm{kbit/s}$

Table 3. Parameters

	Figure 9	Figure 10
R, m	80, 100, 120, 150	100, 120, 150, 175
$r_0, \mathrm{kbit/s}$	100000	10000
N_0, W	$10 * 10^{-10}$	$10 * 10^{-10}$
p_{max}, W	$3.98 * 10^{-3}$	$3.98 * 10^{-3}$
μ, s^{-1}	$1/0.14$	0.33
κ	5	5
G	197.43	197.43
ω, MHz	10	10

5   Conclusion

The paper discusses a model of a wireless LTE network with M2M devices, which are uniformly distributed in the cell and grouped by the distance to the eNodeB, defined by the reported CQI level. The devices generate streaming traffic and transmit data via an uplink channel on GBR. We considered 2 radio resource management schedulers.

References

1. ITU-R M.2370: IMT traffic estimates for the years 2020 to 2030 (2015)
2. ITU-R M.2320: Future technology trends of terrestrial IMT systems (2014)
3. ITU-R M.2083: IMT Vision - Framework and overall objectives of the future development of IMT for 2020 and beyond (2015)
4. Shorgin, S.Y., Samouylov, K.E., Gudkova, I.A., Galinina, O.S., Andreev, S.D.: On the benefits of 5G wireless technology for future mobile cloud computing. In: The 1st International Science and Technology Conference Modern Networking Technologies (MoNeTec): SDN & NFV, The Next Generation of Computational Infrastructure, pp. 151–154. MAKS Press, Moscow (2014)
5. Samouylov, K.E., Gudkova, I.A., Markova, E.V.: Formalizing set of multiservice models for analyzing pre-emption mechanisms in wireless 3GPP networks. In: Samouylov, K., Gudkova, I., Markova, E. (eds.) Distributed Computer and Communication Networks. Communications in Computer and Information Science, vol. 601, pp. 61–71. Springer, Switzerland (2016)
6. Borodakiy, V.Y., Buturlin, I.A., Gudkova, I.A., Samouylov, K.E.: Modelling and analysing a dynamic resource allocation scheme for M2M traffic in LTE networks. In: Balandin, S., Andreev, S., Koucheryavy, Y. (eds.) NEW2AN 2013 and ruSMART 2013. LNCS, vol. 8121, pp. 420–426. Springer, Heidelberg (2013)
7. Gudkova, I., Samouylov, K., Buturlin, I., Borodakiy, V., Gerasimenko, M., Galinina, O., Andreev, S.: Analyzing impacts of coexistence between M2M and H2H communication on 3GPP LTE system. In: Mellouk, A., Fowler, S., Hoceini, S., Daachi, B. (eds.) WWIC 2014. LNCS, vol. 8458, pp. 162–174. Springer, Heidelberg (2014)
8. Ahmadian, A., Galinina, O., Gudkova, I.A., Andreev, S., Shorgin, S., Samouylov, K.: On capturing spatial diversity of joint M2M/H2H dynamic uplink transmissions in 3GPP LTE cellular System. In: Balandin, S., Andreev, S., Koucheryavy, Y. (eds.) NEW2AN/ruSMART 2015. LNCS, vol. 9247, pp. 407–421. Springer, Heidelberg (2015)
9. Galinina, O., Andreev, S., Gerasimenko, M., Koucheryavy, Y., Himayat, N., Yeh, S.-P., Talwar, S.: Capturing spatial randomness of heterogeneous Cellular/WLAN deployments with dynamic traffic. Sel. Areas Commun. 32(6), 1083–1099 (2014)
10. Galinina, O., Andreev, S., Turlikov, A., Koucheryavy, Y.: Optimizing energy efficiency of a multi-radio mobile device in heterogeneous beyond-4G networks. Perform. Eval. 78, 18–41 (2014)
11. 3GPP TS 36.300: Evolved Universal Terrestrial Radio Access (E-UTRA) and Evolved Universal Terrestrial Radio Access Network (E-UTRAN) (2015)
12. 3GPP TR 38.913: Study on scenarios and requirements for next generation access technologies (2016)

Phase Transitions in Multiserver Queuing Systems

Gurami Tsitsiashvili[1]([✉]) and Marina Osipova[2]

[1] Institute for Applied Mathematics Far Eastern Branch of RAS,
Radio street 7, 690041 Vladivostok, Russia
guram@iam.dvo.ru
[2] Far Eastern Federal University, Sukhanova street 8, 690091 Vladivostok, Russia
mao1975@list.ru

Abstract. In this paper synergetic effects in queuing systems and networks are investigated. In the systems $M|M|n|0$ with failures phase transition is established in cases when competition between servers for customers is present and when it is absent. In the systems $M|M|n|\infty$ and $M|G|n|\infty$ with heavy traffic, where distribution of service times are hyperexponential, for some stationary characteristics phase transitions are established. The critical parameters of these phase transitions are defined by load coefficients. The obtained results are spread onto queuing networks with nodes which have the type $M|G|n|\infty$ and the effect of a queue disappearence is investigated. The main approach of these systems and networks analysis is their transformation into Jackson networks.

Keywords: Hyper exponential distribution · Queuing network · Product theorem

1 Introduction

The problem of calculation, analysis and application of infinite server queuing systems and networks with nodes of this type is considered in manifold articles and monographs. Now this area of research is developing intensively.

Research is devoted to information-computing networks [1], transportation systems, systems of cloud computing [2], systems of computer network design [3], systems of distributed data mining [4] and etc.

A large number of articles and monographs are devoted to mathematical methods of these systems and networks calculation and analysis: methods of accuracy calculations [5,6], and asymptotic methods [7–10] and their manifold generalizations and applications [11–14].

In this paper synergetic effects and phase transition phenomena are investigated in queuing networks with multiserver systems in their nodes. The synergetic effects and phase transitions in these networks are closer to the asymptotic analysis technique. The paper consists of four parts.

In the first part synergetic effects and phase transitions connected with these effects are analyzed for the system $M|M|n|\infty$ with heavy traffic and large input

© Springer International Publishing Switzerland 2016
A. Dudin et al. (Eds.): ITMM 2016, CCIS 638, pp. 341–353, 2016.
DOI: 10.1007/978-3-319-44615-8_30

flow intensity $n\lambda$. If the load coefficient ρ satisfies the equality $1 - \rho = n^{-\alpha}$, $\alpha > 0$, then for the stationary mean waiting time and the stationary mean queue length and the stationary probability that all servers are busy we obtain a phase transitions from zero to infinity. A critical parameter of these transitions depends on the type of efficiency index. So we obtain not only synergetic effects (a queue disappearance in the system $M|M|n|\infty$ for $n \to \infty$) but the conditions when it is true.

In the second part the obtained result is spread onto multiserver queuing systems $M|G|n|\infty$ in which the service time distribution is hyperexponential. Such distributions approximates are widely used in queuing theory with heavy tails distributions. For this aim the $M|G|n|\infty$ system is transformed into a network of the Jackson type with the same service times distribution and conditions of synergetic effects and phase transitions are established.

In the third part we consider a queuing network with multiserver nodes and hyperexponential distributions of service times. This network is transformed into a network of the Jackson type with the same service times distributions, and conditions of synergetic effects and phase transitions are also established. These results are generalized onto queuing networks with multiserver nodes and equal load coefficients.

In the fourth part the problem of maximizing the output flow intensity in an aggregated queuing system is considered. For an oneserver queuing system with service distribution function $G(t)$ and with failures a culmination of a maximal output flow intensity is accompanied by the directing of a stationary failure probability to one. But in aggregation of m oneserver systems with or without competition of servers it is possible to maximize this characteristic practically to one. When a number of aggregated systems in an aggregation without competition increases then a maximal possible output flow intensity tends to service intensity. But in an aggregation with competition it tends to an intensity of input flow.

2   Multiserver Queuing System $M|M|n|\infty$

For an n-server queuing system $M|M|n|\infty$ with the input flow intensity $n\lambda$ and the service intensity μ at a single channel the stationary probability $p_n(k)$ of k customers queue in the system is defined by the formulas [17]:

$$p_n(k) = p_n(0)\frac{n^k\rho^k}{\prod_{j=1}^{k}\min(j,n)}, \ \rho = \frac{\lambda}{\mu}, \ p_n^{-1}(0) = 1 + \sum_{k=1}^{\infty}\frac{n^k\rho^k}{\prod_{j=1}^{k}\min(j,n)}. \quad (1)$$

Our problem is to calculate the stationary mean waiting time A_n, the stationary mean queue length B_n and the stationary probability C_n that all servers are busy.

Theorem 1. *If $\rho = 1 - n^{-\alpha}$, $0 < \alpha < \infty$, then for the stationary mean waiting time A_n we have the following phase transition for $n \to \infty$:*

$$A_n \to \begin{cases} 0, & \alpha < 1, \\ 1/\mu, & \alpha = 1, \\ \infty, & \alpha > 1. \end{cases}$$

Proof. For the stationary mean waiting time A_n the limit equality

$$A_n = \sum_{k \geq n} \frac{p_n(k)(k-n+1)}{n\mu} = \frac{p_n(0)}{n\mu n!} \sum_{k \geq n} \frac{n^k \rho^k}{n^{k-n}}(k-n+1)$$

$$= \frac{p_n(0)n^n \rho^n}{n\mu n!}(1 + 2\rho + 3\rho^2 + \ldots) = \frac{p_n(0)n^n \rho^n}{n!n\mu(1-\rho)^2} \tag{2}$$

is true. From Formula (1) we obtain that

$$\sum_{k \geq n} \frac{n^k \rho^k}{n!n^{k-n}} = \frac{n^n \rho^n}{n!(1-\rho)} \leq p_n^{-1}(0) \leq \sum_{k \geq 0} \frac{n^k \rho^k}{k!} = e^{n\rho} \leq e^{n\rho} + \frac{n^n \rho^n}{n!(1-\rho)}. \tag{3}$$

For $0 < \alpha < 1$ from Formula (3) we obtain that

$$A_n \leq \frac{1}{\mu n^{1-\alpha}} \to 0, \ n \to \infty. \tag{4}$$

And for $\alpha > 1$ we have

$$A_n \geq \frac{n^n \rho^n}{n!(1-\rho)^2 n\mu}\left(e^{n\rho} + \frac{n^n \rho^n}{n!(1-\rho)}\right)^{-1} = \frac{1}{n\mu(1-\rho)} \cdot \frac{1}{1+T_n},$$

where from the Stirling formula [18, Chap. I, Sect. 2]

$$n! = n^n e^{-n}\sqrt{2\pi n} \exp\left(\frac{\psi(n)}{12n}\right), \ 0 < \psi(n) < 1, \tag{5}$$

$$T_n = P_n Q_n, \ P_n = \exp\left(\frac{\psi(n)}{12n}\right)(1-\rho)\sqrt{2\pi n}, \ Q_n = \exp(nR_n), \ R_n = \rho - \ln\rho - 1.$$

We take $\rho = 1 - n^{-\alpha}$ and using the Taylor series of R_n we obtain:

$$R_n = -n^{-\alpha} + n^{-\alpha} + \frac{n^{-2\alpha}}{2} + \frac{n^{-3\alpha}}{3} + \ldots \sim \frac{n^{-2\alpha}}{2(1-n^{-\alpha})}, \ n \to \infty, \ \alpha > 0, \tag{6}$$

consequently,

$$nR_n \sim \frac{n^{1-2\alpha}}{2(1-n^{-\alpha})} \to 0, \ Q_n = \exp(nR_n) \to 1, \ P_n \to 0, \ n \to \infty. \tag{7}$$

So $T_n = P_n Q_n \to 0$, $n \to \infty$, and

$$A_n \geq \frac{1}{\mu n^{1-\alpha}(1+T_n)} \to \infty, \ n \to \infty. \tag{8}$$

From the upper (4) and low (8) bounds of the stationary mean waiting time A_n it is easy to obtain that $A_n \to 1/\mu$, $n \to \infty$. Theorem is proved.

Theorem 2. *If $\rho = 1 - n^{-\alpha}$, $0 < \alpha < \infty$, then for the stationary mean queue length B_n and for the stationary probability C_n that all servers are busy the following limit relations are true:*

$$B_n \to \begin{cases} 0, & \alpha < 1/2, \\ \infty, & \alpha \geq 1/2, \end{cases} \quad C_n \to \begin{cases} 0, & \alpha < 1/2, \\ 1, & \alpha > 1/2. \end{cases}$$

For $\alpha = 1/2$ we have:

$$\frac{1}{1+e^{1/2}\sqrt{2\pi}} \leq \liminf C_n \leq \limsup C_n \leq \frac{1}{e^{1/2}\sqrt{2\pi}}.$$

Proof. We calculate

$$B_n = \sum_{k>n} p_n(k)(k-n) = \sum_{k>n} \frac{p_n(0)n^k \rho^k}{n!n^{k-n}}(k-n)$$

$$= \frac{p_n(0)n^n \rho^n}{n!} \sum_{k>n} \rho^{k-n}(k-n) = \frac{p_n(0)n^n \rho^n}{n!}(\rho + 2\rho^2 + 3\rho^3 + \ldots)$$

$$= \frac{p_n(0)n^n \rho^n}{n!}\left(\frac{1}{(1-\rho)^2} - 1\right) = \frac{p_n(0)n^n \rho^n}{n!} \cdot \frac{(2-\rho)\rho}{(1-\rho)^2}, \tag{9}$$

$$C_n = \sum_{k \geq n} p_n(k) = \sum_{k \geq n} \frac{p_n(0)n^k \rho^k}{n!n^{k-n}} = \frac{p_n(0)n^n \rho^n}{n!} \cdot \frac{1}{(1-\rho)}. \tag{10}$$

Let us denote $\gamma_n = \dfrac{p_n(0)n^n \rho^n}{n!}$ then Formulas (9) and (10) may be rewritten as follows

$$B_n = \gamma_n \frac{(2-\rho)\rho}{(1-\rho)^2}, \quad C_n = \gamma_n \frac{1}{(1-\rho)}. \tag{11}$$

For $0 < \alpha < 1/2$ we construct upper bounds of B_n, C_n using Formulas (3) and (5):

$$\gamma_n \leq \frac{\exp(-n\rho)n^n \rho^n}{n^n e^{-n}\sqrt{2\pi n}\exp(\psi(n)/12n)} = \frac{\exp(n(-\rho + \ln \rho + 1))}{\sqrt{2\pi n}\exp(\psi(n)/12n)}$$

$$= \frac{\exp(n(n^{-\alpha} + \ln(1 - n^{-\alpha})))}{\sqrt{2\pi n}\exp(\psi(n)/12n)} = \frac{\exp\left(n\left(n^{-\alpha} - n^{-\alpha} - n^{-2\alpha}/2 - n^{-3\alpha}/3 - \ldots\right)\right)}{\sqrt{2\pi n}\exp(\psi(n)/12n)}$$

$$\leq \frac{\exp\left(-n^{1-2\alpha}/2 - n^{1-3\alpha}/2^2 - n^{1-4\alpha}/2^3 - \ldots\right)}{\sqrt{2\pi n}\exp(\psi(n)/12n)} \leq \frac{\exp(-n^{1-2\alpha}/(2-n^{-\alpha}))}{\sqrt{2\pi n}\exp(\psi(n)/12n)}.$$

So from Formula (9) we have

$$B_n \leq \frac{\exp(-n^{1-2\alpha}/(2-n^{-\alpha}))}{\sqrt{2\pi n}\exp(\psi(n)/12n)} \cdot \frac{1-n^{-2\alpha}}{n^{-2\alpha}} \to 0, \; n \to \infty. \qquad (12)$$

Analogously from Formula (10) we obtain

$$C_n \leq \frac{\exp(-n^{1-2\alpha}/(2-n^{-\alpha}))}{\sqrt{2\pi n}\exp(\psi(n)/12n)} \cdot \frac{1}{n^{-\alpha}} \to 0, \; n \to \infty. \qquad (13)$$

Let us assume now that $\alpha \geq 1/2$ then from Formula (3) we have

$$B_n \geq \frac{n^n \rho^n \rho(1-\rho)}{n!(1-\rho)(\exp(n\rho) + n^n \rho^n/n!(1-\rho))} = \frac{n^\alpha(1-n^{-2\alpha})}{(1+T_n)}.$$

We repeat for $\alpha > 1/2$ a verification of Formula (7). Using Formula (6) we obtain that $T_n \to 0$, $n \to \infty$ and consequently

$$B_n \geq \frac{n^\alpha(1-n^{-2\alpha})}{(1+T_n)} \to \infty, \; n \to \infty. \qquad (14)$$

In turn it is not complicated to prove that for $\alpha = 1/2$ there is the convergence $T_n \to e^{1/2}\sqrt{2\pi}$, $B_n \to \infty$, $n \to \infty$. Analogously for $\alpha > 1/2$ we have

$$C_n \geq \frac{1}{(1+T_n)} \to 1, \; n \to \infty. \qquad (15)$$

If $\alpha = 1/2$ then using the limit relation $T_n \to e^{1/2}\sqrt{2\pi}$, $n \to \infty$, and Formulas (13) and (15), we end the Theorem 2 proof.

A comparison of Theorems 1 and 2 shows not only a presence of phase transitions and synergetic effects but their dependence from the system efficiency indexes.

3   Multiserver Queuing System $M|G|n|\infty$

Let us consider a multiserver queuing system in which the distribution $F(t)$ of the service times is hyperexponential, that is, the probability mixture of r exponential distributions:

$$F(t) = \sum_{k=1}^{r} p_k(1 - \exp(-\mu_k t)), \; a = \int_0^\infty \overline{F}(x)dx = \sum_{k=1}^{r} \frac{p_k}{\mu_k}, \; p_k > 0, \; \sum_{k=1}^{r} p_k = 1.$$

We assume that all positive numbers $R_k = \dfrac{p_k}{\mu_k a}$, $k = 1, \ldots, r$, are rational and are represented by the equalities $R_k = \dfrac{A_k}{B_k}$ in which the nominator A_k and the denominator B_k are mutually simple integers. We denote N the least common multiple of the integers B_k, $k = 1, \ldots, r$, and define the integers

$$n_k = nNR_k, \ k = 1, \ldots, r, \ \sum_{k=1}^{r} n_k = Nn. \tag{16}$$

Let us consider an nN-server queuing system with a Poisson input flow which has the intensity $nN\lambda$ and the distribution function of service times $F(t)$. We denote $\rho = \lambda a$ and suppose that $\rho < 1$.

Let us transform this nN-server system into opened network G' with r nodes as follows. The input flow to the network G' is Poisson with the intensity $nN\lambda$. The network G' consists of r nodes k with n_k parallel servers which have the exponential distribution functions $1 - \exp(-\mu_k t)$, $k = 1, \ldots, r$. Each customer of the input flow is directed to the node k with the probability p_k.

It is obvious that the service time distribution of each customer arriving into the network G' equals $F(t)$ and the load coefficient of each node equals ρ. The network G' is opened and has the Jackson type and so for any input flow intensity $nN\lambda$ there is a single solution to a balance linear equations system for stationary intensities of input flows to nodes of this network. So from the product theorem [15] the stationary probability $P(j_1, \ldots, j_r)$ that $j_1, \ldots, j_r$ are numbers of customers in the network nodes

$$P(j_1, \ldots, j_r) = \prod_{i=1}^{r} p_{n_i}(j_i),$$

where $p_{n_i}(j_i)$ are calculated by Formula (1).

In this network the stationary mean waiting time A_n, the stationary mean queue length B_n and the stationary probability C_n that all servers are busy are probability mixtures of appropriate characteristics $A_{n,k}$, $B_{n,k}$, $C_{n,k}$, $k = 1, \ldots, r$ of this network nodes:

$$A_n = \sum_{k=1}^{r} p_k A_{n,k}, \ B_n = \sum_{k=1}^{r} p_k B_{n,k}, \ C_n = \sum_{k=1}^{r} p_k C_{n,k}. \tag{17}$$

Theorem 3. If $\rho = 1 - n^{-\alpha}$, $0 < \alpha < \infty$, then the stationary mean waiting time A_n satisfies Theorem 1 and the stationary mean queue length B_n and the stationary probability C_n that all servers are busy satisfy Theorem 2.

Proof. As load coefficients of all network G' nodes equal ρ so the characteristics A_{n_k}, B_{n_k}, C_{n_k}, $k = 1, \ldots, r$, satisfy Theorems 1 and 2. Then from Formula (17) we have that A_n, B_n, C_n satisfy Theorems 1 and 2 also.

Remark 1. Conditions for numbers of servers in this network may be weaken assuming that the proportionality condition on R_k, $k = 1, \ldots, r$, is satisfied asymptotically.

Remark 2. The construction considered in this section may be spread to cases when service times distribution functions are completely monotone [19]. Completely monotone distributions are distributions with the density $f(t)$, which satisfy the inequalities $(-1)^k \dfrac{d^k f(t)}{dt^k} \geq 0$, $t > 0$, $k \geq 1$. Using [20, Theorem 1, Lemma 2] and their manifold generalizations it is proved that any for completely monotone distribution $F(t)$ and any $\varepsilon > 0$ there is hyperexponential distribution $G(t)$ which satisfies the inequality $\sup\limits_{t \geq 0} |F(t) - G(t)| \leq \varepsilon$. Very important completely monotone distributions are Weibull, Pareto and many others distributions with heavy tails [21,22]. An approximation of a probability that all servers are busy in the network may be realized using stability theorems for queuing networks [23].

4   A Queuing Network with Large Number of Servers in Nodes

Let us consider the opened queuing network G with a Poisson input flow which has the intensity λ_0, with m internal nodes, the indecomposable route matrix $\Theta = ||\theta_{ij}||_{i,j=0}^m$ and the service times distribution functions $F_i(t)$, $i = 1, \ldots, m$, in the network nodes. Service times in the node i are independent and identically distributed with the common distribution

$$F_i(t) = \sum_{k=1}^{r_i} p_{i,k}(1 - \exp(-\mu_{i,k}t)), \ a_i = \int_0^\infty \overline{F}_i(x)dx = \sum_{k=1}^{r_i} \frac{p_{i,k}}{\mu_{i,k}}, \ i = 1, \ldots, m,$$

which is hyperexponential and is the probability mixture of r_i exponential distributions, $p_{i,k} > 0$, $\sum\limits_{k=1}^{r_i} p_{i,k} = 1$. Here θ_{ij} is the probability to go to the node j after service at the node i, θ_{0i} is the probability that a customer arriving into the network comes to the node i, $\theta_{00} = 0$.

Let us define the vector $(\lambda_1, \ldots, \lambda_m)$ as a single solution to the system of linear algebraic equations $(\lambda_0, \ldots, \lambda_m) = (\lambda_0, \ldots, \lambda_m)\Theta$ and assume that

$$\rho_i = \lambda_i a_i < 1, \ i = 1, \ldots, m. \tag{18}$$

We suppose that all positive numbers $R_{i,k} = \dfrac{p_{i,k}}{\mu_{i,k} a_i}$, $k = 1, \ldots, r_i$, $i = 1, \ldots, m$, are rational and are represented by the equalities $R_{i,k} = \dfrac{A_{i,k}}{B_{i,k}}$, where their nominators and denominators are mutually simple integers. We denote N the least common multiple of the integers $B_{i,k}$, $k = 1, \ldots, r_i$, $i = 1, \ldots, m$, and put

$$n_{i,k} = nNR_{i,k}, \ k = 1, \ldots, r_i, \ \sum_{k=1}^{r_i} n_{i,k} = Nn, \ i = 1, \ldots, m. \tag{19}$$

Let us transform the network G with m nodes into the network G'' with $\sum\limits_{i=1}^{m} r_i$ nodes as follows. The input flow to the network G'' is Poisson with the intensity $nN\lambda_0$. The node i of the network G is replaced by the block (i) containing r_i nodes (i,k) with $n_{i,k}$ servers which have the exponential distributions $1 - \exp(-\mu_{i,k}t)$, $k = 1,\ldots,r_i$. It is obvious that the distribution function of a service time of a customer arriving the block (i) equals $F_i(t)$. From Formula (19) we obtain that the number of servers in this block equals nN.

Let us assume that on a finishing of a service in the node (i,k) a customer arrives into the node (j,l) with the probability $\theta_{i,j}p_{j,l}$ and departs from the network with the probability $\theta_{i,0}$. A customer of the input flow arrives into the node (i,k) with the probability $\theta_{0,i}p_{i,k}$. It is obvious that the route matrix of the network G' is indecomposable also.

Then in all nodes of this network from Formula (18) we have the inequalities

$$\frac{nN\lambda_i p_{i,k}}{n_{i,k}\mu_{i,k}} = \rho_i < 1, \ k = 1,\ldots,r_i, \ i = 1,\ldots,m. \tag{20}$$

Consequently for any n the network G'' has the Jackson type. And from the conditions (20) we obtain that the limit distribution of customers numbers in the nodes (i,k), $k = 1,\ldots,r_i$, $i = 1,\ldots,m$, satisfy the product theorem [15,16]:

$$P(t_{i,k}, \ k = 1,\ldots,r_i, \ i = 1,\ldots,m) = \prod_{i=1}^{m}\prod_{k=1}^{r_i} P_{i,k}(t_{i,k}), \tag{21}$$

where $P_{i,k}(t_{i,k})$ is the limit distribution of numbers $t_{i,k}$ of customers in $n_{i,k}$-server queuing system (i,k) with the type $M|M|n|\infty$ and with the input flow intensity $nN\lambda_i p_{i,k}$ and the serving intensity $\mu_{i,k}$.

Let us denote $P''_{i,k}(n)$ the stationary probability that all servers in the node (i,k) of the network G'' are busy and denote $P''(n)$ the stationary probability that all servers in some node (i,k) of the network G' are busy.

Theorem 4. *In the conditions (18) the relation $P''(n) \to 0$, $n \to \infty$, is true.*

Proof. For the n-server queuing system $M|M|n|\infty$ with the input flow intensity $n\lambda$ and with the serving intensity μ the stationary probability $p_n(k)$ that there is k in the system satisfies Formula (1). From the Stirling formula and from the inequality $f(\rho) = 1 + \ln\rho - \rho < 0$, $0 < \rho < 1$, and from Formula (3) we have obtain the stationary probability C_n that all servers are busy satisfies the relation

$$C_n = p_n(0)\sum_{k\geq n}\frac{n^k\rho^k}{n!n^{k-n}} \leq e^n\frac{p_n(0)}{\sqrt{2\pi n}}\sum_{k\geq n}\rho^k \leq \frac{\exp(nf(\rho))}{(1-\rho)\sqrt{2\pi n}} \to 0, \ n \to \infty. \tag{22}$$

From the condition (20) and the relations (21), (22) it is possible to obtain that

$$P''_{i,k}(n) \to 0, \ k = 1,\ldots,r_i, \ i = 1,\ldots,m,$$

$$P''(n) = 1 - \prod_{i=1}^{m} \prod_{k=1}^{r_i} (1 - P''_{i,k}(n)) \to 0, \ n \to \infty. \tag{23}$$

Remark 3. Consequently the network G'' satisfies the relations (23) which characterizes synergetic effect in all its blocks. it is possible to prove analogous statements for maximal stationary waiting time and for summarized number of customers in nodes queues.

Remark 4. If $\rho_1 = \ldots = \rho_m$ then is possible to spread Theorems 1 and 2 onto the network G'' with heavy traffic also.

5   Multiserver Queuing Systems with Failures

Let us consider the single server queuing system $M|M|m|\infty$ with failures and with the service time distribution $G(t)$, $b_1 = \int_0^\infty t\, dG(t)$ and with the intensity λ of Poisson input flow. We denote this system B_1. We divide the time axis $t \geq 0$ into cycles consisting of intervals when a server is busy and intervals when it is free. A length of each such interval equals the sum of two independent random variables. The first of them has exponential distribution with the parameter λ and the second summand has distribution $G(t)$. From the integral renewal theorem [24, Chap. 9] we obtain that the stationary intensity of the output flow equals

$$I(\lambda) = \left(\frac{1}{\lambda} + b_1\right)^{-1}. \tag{24}$$

and the stationary failure probability satisfies the equality

$$P(\lambda) = 1 - \frac{I(\lambda)}{\lambda}. \tag{25}$$

For the system $A_1 = M|M|1|0$ in which the service time has exponential distribution with the parameter μ the stationary failure probability and the stationary intensity of output flow equal:

$$P(\lambda) = \frac{\lambda}{\lambda + \mu} \, , \ I(\lambda) = \frac{\mu\lambda}{\lambda + \mu} \, ,$$

where the functions $P(\lambda)$, $I(\lambda)$ are monotonically increasing. Then the maximal intensity of output flow $J = \sup(I(\lambda) : \ \lambda > 0) = \lim_{\lambda \to \infty} I(\lambda) = \mu$, and $\lim_{\lambda \to \infty} P(\lambda) = 1$. So when output flow intensity in the system A_1 tends to its maximal value the stationary failure probability tends to one. But an aggregation of the systems A_1 allows us to decrease the loss of input flow customers.

Let us consider the queuing system $A_m = M|M|m|0$ with input flow intensity $m\lambda$ and with service intensity μ on all m servers. The system A_m is the aggregation of m single server systems A_1. A number of customers in the system A_m is described by the birth and death process $x_m(t)$ with birth and death intensities

$$\lambda_m(k) = m\lambda, \ 0 \leq k < m, \ \mu_m(k) = k\mu, \ 0 < k \leq m.$$

Let us denote $P^{(m)}(\lambda)$ the stationary failure probability in the system A_m and put $I^{(m)}(\lambda)$ the stationary intensity of output flow (in a recalculation for a single server), we designate

$$J^{(m)} = \sup(I^{(m)}(\lambda) : \ \lambda > 0).$$

Theorem 5. *The following relations take place for $m \to \infty$*

$$J^{(m)} \to \mu, \ P^{(m)}(\lambda) \to \begin{cases} 0, & \lambda \leq \mu, \\ 1 - \mu/\lambda, & \lambda > \mu. \end{cases}$$

Proof. Let us denote $p_m(k) = \lim\limits_{t \to \infty} P(x_m(t) = k), \ 0 \leq k \leq m$. Then

$$p_m(m-1) = p_m(m)\frac{\mu}{\lambda}\frac{m}{m}, \ p_m(m-2) = p_m(m)\left(\frac{\mu}{\lambda}\right)^2 \frac{m(m-1)}{m^2}, \ldots$$

Consequently the stationary failure probability $P^{(m)}$ from the integral renewal theorem and from the law of large numbers [24, Chap. 9, Sects. 4, 5] satisfies the equality

$$P^{(m)}(\lambda) = p_m(m) = \left(1 + \sum_{k=1}^{m}\left(\frac{\mu}{\lambda}\right)^k \prod_{j=0}^{k-1}\left(1 - \frac{j}{m}\right)\right)^{-1}. \tag{26}$$

The stationary output flow intensity in the system A_m (in a recalculation for one server) is

$$I^{(m)}(\lambda) = \lambda(1 - P^{(m)}(\lambda)), \ \lambda > 0. \tag{27}$$

From Formulas (26) and (27) we obtain that

$$I^{(m)}(\mu) \leq J^{(m)} \leq \mu. \tag{28}$$

Let us fix ε, $0,5 < \varepsilon < 1$, and denote $c = -\dfrac{\ln(1-\varepsilon)}{1-\varepsilon}$, then

$$e^{-cx} \leq 1 - x, \ 0 < x < 1 - \varepsilon,$$

$$(P^{(m)}(\mu))^{-1} \geq 1 + \sum_{1 \leq k \leq m(1-\varepsilon)} \exp\left(-c\sum_{j=0}^{k-1}\frac{j}{m}\right) > 1 + \sum_{1 \leq k \leq m(1-\varepsilon)} \exp\left(-\frac{ck^2}{2m}\right)$$

$$\geq \int_0^{m(1-\varepsilon)+1} \exp\left(-\frac{cx^2}{2m}\right) dx \sim \sqrt{\frac{\pi m}{2c}}, \ m \to \infty. \tag{29}$$

Consequently from Formulas (27), (28) and (29) we obtain the relations:

$$J^{(m)} \to \mu, \quad P^{(m)}(\mu) \to 0, \quad m \to \infty. \tag{30}$$

From Formulas (26) and (30) we obtain that $P^{(m)}(\lambda) \to 0$, $m \to \infty$, $\lambda \le \mu$.
Then as

$$P^{(m)}(\lambda) > \left(\sum_{k=0}^{m} \left(\frac{\mu}{\lambda} \right)^k \right)^{-1} = \frac{1 - \mu/\lambda}{1 - (\mu/\lambda)^{m+1}}$$

so from Formula (29) for any $0 < \gamma < 1/2$

$$P^{(m)}(\lambda) < \left(\sum_{0 \le k \le m^{1/2-\gamma}} \left(\frac{\mu}{\lambda} \right)^k \exp\left(-\frac{ck^2}{2m} \right) \right)^{-1} \le \frac{\sqrt{\exp(cm^{-\gamma})}(1 - \mu/\lambda)}{1 - (\mu/\lambda)^{m^{1/2-\gamma}}}.$$

Consequently we obtain $P^{(m)}(\lambda) \to 1 - \mu/\lambda$, $m \to \infty$, $\lambda > \mu$. Theorem 5 is proved.

Let us consider the queuing system B_m with the intensity $m\lambda$ of Poisson input flow and with a competition between m servers. The servers competition is described as follows. Let us assume that the i-th customer arrives in the moment t_i to the system B_m, if the system is empty then the customer receives the information $\eta_i^{(1)}, \dots, \eta_i^{(m)}$ of possible services at different servers and chooses a server with minimal service time

$$\zeta_i = \min(\eta_i^{(1)}, \dots, \eta_i^{(m)}), \quad i \ge 1,$$

else it receives a failure. During this customer service all the other servers do not work and do not accept other customers. Random variables $\eta_i^{(j)}, 1 \le j \le m, i \ge 1$, are independent and have the distribution $G(t)$.

Theorem 6. *Let us assume that for some $a > 0$, $b > 0$ $G(x) \sim ax^b$, $x \to 0$. Then for the system B_m and for $m \to \infty$ we have*

$$I^{(m)}(\lambda) \to \begin{cases} 0, b > 1, \\ \lambda, 0 < b \le 1, \end{cases} \quad P^{(m)}(\lambda) \to \begin{cases} 1, b > 1, \\ 0, 0 < b \le 1. \end{cases}$$

Proof. So the system B_m may be considered as single server system with Poisson input flow with the intensity $m\lambda$ and with independent and identically distributed service times $\zeta_1, \zeta_2, \dots$, which have the means

$$b_m = \int_0^\infty \overline{G}^m(t)dt > 0.$$

From Formulas (24) and (25) in the system B_m the stationary intensity of output flow (in a recalculation for single server) $I^{(m)}(\lambda)$ and the stationary failure probability $P^{(m)}(\lambda)$ satisfy the equalities

$$I^{(m)}(\lambda) = \left(\frac{1}{\lambda} + K_m \right)^{-1}, \quad P^{(m)}(\lambda) = 1 - \frac{I^{(m)}(\lambda)}{\lambda}, \quad K_m = mb_m. \tag{31}$$

It is simple to prove that if the Theorem 6 condition is true then for $m \to \infty$ the following formulas occur:

$$K_m = O(m^{1-1/b}),\ 0 < b \leq 1,\ 1/K_m = O(1/m^{1-1/b}),\ b \leq 1. \qquad (32)$$

Consequently from Formulas (31) and (32) we obtain the Theorem 6 statement.

References

1. Basharin, G.P., Tolmachev, A.L.: Queuing network theory and its applications to the analysis of information-computing networks. In: Itogi Nauki Tech. Ser. Teor. Veroiatn. Mat. Stat. Teor. Kibern, vol. 21, pp. 3–119. VINITI, Moscow (1983) (In Russian)
2. Mokrov, E.V., Samuilov, K.E.: Model of systems of claud computing as queuing systems with few queues and group arrival of customers. Telecommun. Transp. **7**(11), 139–141 (2013). (In Russian)
3. Vishnevskiy, V.M.: Theoretical Principles of Computer Network Design. Technosila, Moscow (2003)
4. Grachev, V.V., Moiseev, A.N., Nazarov, A.A., Iampolskiy, V.Z.: Multiphase model of queuing system with distributed data mining. Lect. TUSUR **2**(26), 248–251 (2012). (In Russian)
5. Ivnitskiy, V.A.: Theory of Queuing Networks. Edition of Phys. - Mat. Lit., Moscow (2004). (In Russian)
6. Ivnitskiy, V.A.: Theory of Arbitrary Input Flow. Palmarium Academic Publishing, Saarbrucken (2012). (In Russian)
7. Nazarov, A.A.: Asymptotic Analysis of Markovian Systems. Edition of Tomsk State University, Tomsk (1995). (In Russian)
8. Nazarov, A.A., Moiseeva, S.P.: Method of Asymptotic Analysis in Queuing Theory. Edition of Tomsk State University, Tomsk (2006). (In Russian)
9. Nazarov, A.A., Moiseev, A.N.: Investigation of opened NonMarkovian queuing network $GI - (GI|\infty)^K$ with high intensive recurrent input flow. Probl. Inf. Trans. **49**(2), 78–91 (2013). (In Russian)
10. Moiseev, A.N., Nazarov, A.A.: Infinite Server Queuing Systems and Networks. Edition of Tomsk State University, Tomsk (2015). (In Russian)
11. Moiseev, A.N., Nazarov, A.A.: Asymptotic analysis of high intensive half Markovian flow of events. Lect. TUSUR **3**(29), 109–115 (2013). (In Russian)
12. Moiseev, A.N., Nazarov, A.A.: Asymptotic analysis of multiphase queuing system with high intensive recurrent input flow. Autometrics **50**(2), 67–76 (2014). (In Russian)
13. Matveev, S.A., Moiseev, A.N., Nazarov, A.A.: Application of method of initial moments for investigation of multiphase queuing systems $GI/(M|\infty)^K$. Lect. TUSUR **3**(33), 129–134 (2014)
14. Lopukhova, S.V.: Investigation of MMP flow by asymptotic method of m-th order. Herald of Tomsk State Univ. Control Comput. Sci. Inf. **3**(4), 71–76 (2008). (In Russian)
15. Jackson, J.R.: Networks of waiting lines. Oper. Res. **5**(4), 518–521 (1957)
16. Serfozo, R.: Introduction to Stochastic Networks, p. 301. Springer, New York (1999)
17. Ivchenko, G.I., Kashtanov, V.A., Kovalenko, I.N.: Queuing Theory. High school, Moscow (1982). (In Russian)

18. Shiriaev, A.N.: Probability. Moscow, Science (1989). (In Russian)
19. Feldmann, A., Whitt, W.: Fitting mixtures of exponentials to long tailed distributions to analyze network perfomance models. Perfom. Eval. **31**(3–4), 245–279 (1998)
20. Vatamidou, E., et al.: On the accuracy of phase-type approximations of heavy-tailed risk models. Scand. Actuarial J. **2014**(6), 510–534 (2014)
21. Embrechts, P., Cluppelberg, C., Mikosch, T.: Modelling extremal events: for insurance and finance. Applications of Mathematics, vol. 33, p. 648. Springer, Heidelberg (1997)
22. Asmussen, S.: Ruin probabilities. In: Advanced Series on Statistical Science and Applied Probability, vol. 2. World Scientific Publishing Co. Inc., Singapore (2000)
23. Borovkov, A.A.: Limit theorems for queuing networks. Probab. Theor. Appl. **31**(3), 474–490 (1986). (In Russian)
24. Borovkov, A.A.: Probability Theory. Science, Moscow (1986)

An Open Queueing Network with a Correlated Input Arrival Process for Broadband Wireless Network Performance Evaluation

Vladimir Vishnevski[(✉)], Andrey Larionov, and Roman Ivanov

V.A. Trapeznikov Institute of Control Sciences of Russian Academy of Sciences,
Profsoyuznay. 65, 117997 Moscow, Russia
vishn@inbox.ru, larioandr@gmail.com, iromcorp@gmail.com
http://www.ipu.ru

Abstract. In the paper characteristics of an open queueing network with correlated input Markovian arrival processes and phase type distribution of service time in the nodes are studied. The developed methodology is used for the performance evaluation of broadband wireless networks with linear topology. A comparison study of numeric results of analytic modelling and simulation is carried out. To study open queueing networks with MAP input a pyQuMo library has been developed using Python 3 language. The library provides a means for model description, model properties computations by taking advantage of an analytical approach, and simulation as well as results visualisation.

Keywords: Markovian arrival process · Batch MAP · PH-distribution · Open queueing network · Wireless network

1 Introduction

The designing of a backbone network is a topical task for the systems which are as follows: intellectual transport systems (ITS), road safety systems, pico- and femtocells access systems, as well as for the development of telecommunications infrastructure along railways and pipelines. For the performance evaluation and optimal design of networks of this rank there is a need for new mathematical models describing the functionality of wireless networks. In this paper we propose a model of open homogeneous queueing networks with a correlated input arrival process (MAP), PH-distribution of service time in the network nodes and a routing matrix $\|t_{ij}\|$, where t_{ij} is the probability of packet arrival at the j-th node after its serving at the i-th node is completed.

The state space and transmission intensities of the Markovian process are described. The results of numerical computation of the queues length, packet loss probabilities and other network parameters are presented. Such results are achieved by making use of a precise analytical approach as well as simulation modelling.

© Springer International Publishing Switzerland 2016
A. Dudin et al. (Eds.): ITMM 2016, CCIS 638, pp. 354–365, 2016.
DOI: 10.1007/978-3-319-44615-8_31

2   The Model of an Open Homogeneous Network with Correlated Input Arrivals and PH-Distribution of Service Time

It is not enough to make use of traditional approaches, which are based on the models of BCMP-networks [1,2] being used extensively for the performance evaluation and optimization of computer networks characteristics, and for taking into account a number of significant features of functioning of wireless network considered. Such features are as follows:

- correlated nature of input arrival process of packets;
- limited buffer memory of base stations and respective packet loss;
- the ability to transmit packets repeatedly when the packet losses occurred as a result of, for instance, interference or strong signal attenuation;
- the service and transmission time of packets being produced by different applications can vary significantly and have various distributions.

To take account of the correlated nature of the traffic let us consider MAP (Markovian Arrival Process) and BMAP (Batch MAP) to arrive into an open homogeneous queueing network [3,4]. Let us denote MAP $A \sim MAP(D_0, D_1)$, where $D = D_0 + D_1$ is an infinitesimal generator of an appropriate Markovian chain, i.e.

$$\forall i : \sum_{j=1} d_{ij} = 0, \quad \forall i \neq j : d_{ij} \geq 0, \quad \forall i : d_{ii} \leq 0.$$

D_0 and D_1 are constrained matrices:

$$\forall i, j : \{D_1\}_{ij} \geq 0, \quad \forall i \neq j : \{D_0\}_{ij} \geq 0, \quad \forall i : \{D_0\}_{ii} \leq 0.$$

The aim of such partitioning is to divide invisible transitions leading only to a change of state and determined by matrix D_0 and visible ones determined by matrix D_1 that cause packet generation. In the case of BMAP a sequence of matrices $\{D_i : i \geq 1\}$ is set up instead of a single D_1 matrix. Each matrix from this sequence defines transition intensities resulting to packet generating. Applications of MAP and BMAP to telecommunication traffic modelling are well studied and available in the literature. It has been shown that one can simulate real traffic with sufficient precision using these arrivals [5,6]. Batch MAP allows us to describe the simultaneous arrival of an arbitrary amount of packets while MAP does not. To simplify the formulas of the suggested model, MAP is made use of, but all this reasoning is also valid for BMAP.

To take into account the errors and losses arising during transitions let us associate the error probability p_{e_i} with each station, thus a packet being served returns into the queue or leaves it irretrievably with that probability.

Service time can be modelled by PH-distribution [7]. Phase type distribution is made extensive use of and allows us to describe the service process as part of a Markovian model. Let us denote B PH-distributed random value,

$B \sim PH(S, \tau), S \in \mathbb{R}^{W \times W}, \tau \in \mathbb{R}^{W}$, that depicts the time elapsed until arrival to the accepting state. Let us define the generator of the chain as

$$\begin{bmatrix} S & -S\mathbf{1} \\ \mathbf{0} & 0 \end{bmatrix}.$$

It has $W + 1$ states, the last state is accepting. Vector τ specifies the initial distribution of probabilities of the modulating chain.

To make allowance for the limited station memory the queue of each station is supposed to have a capacity of M_i packets and each station can contain up to $K_i = M_i + 1$ packets.

Let us consider each station as associated with a user which contributes to the network with the arrival process $A_i \sim MAP(D_0^{(i)}, D_1^{(i)})$ and can be treated as the traffic sink as well: each packet accepted by the station with the probability r_i is transmitted to the user and leaves the network or it arrives to the queue with complementary probability $1 - r_i$. If the queue is full, the packet is lost irretrievably. The service time is PH-distributed: $B \sim PH(S^i, \tau^i)$.

Finally, the packet routing is carried out according to the stochastic matrix $T \in \mathbb{R}^{N \times N}$, where N is the number of stations. Element t_{ij} of the matrix equals the conditional probability of packet transmission to the j-th station after serving at the i-th station on condition of successful transmission. Since the ability to transmit the packet back to the queue having been considered, it is reasonable to suppose $t_{ii} = 0$ for all i.

Unconditional probabilities of packet transmission from the i-th to the j-th station, to its user as well as packet loss and packet retransmission probabilities are shown in Fig. 1.

The functioning of the whole network is described by a Markovian process $\mathcal{C} \sim Markov(Q)$, that is defined by a Markovian process modulating MAP and PH-distributions. Having the infinitesimal generator Q of such a process one can compute the stationary probability distribution of its states and then the distribution of queue lengths, busy coefficients of stations and other characteristics. Moreover, it is easy to build matrices $\hat{D}_0^{(i)}$, $\hat{D}_1^{(i)}$ of the MAP of packets served by the i-th station and also $\check{D}_0^{(i)}$, $\check{D}_1^{(i)}$ of overall MAP incoming at the

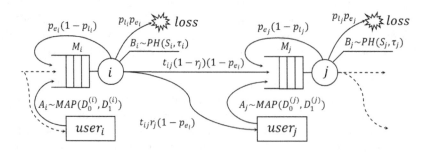

Fig. 1. The queueing network modelling wireless network with linear topology.

i-th station. Knowing these matrices allows us to compute the probability that the next packet is lost due to memory overflow.

So, the space of states and transition intensities between the states are required to be defined to describe the process $\mathcal{C}$. It is possible to completely define any state of the chain given the following values:

- the number of packets k_i at the i-th station, $k_i = 0..K_i$;
- the state a_i of a user input MAP $A_i, a_i \in 1..V_t$;
- the state b_i of a server (of its PH-distribution modulating chain), $b_i \in 1..W_i$.

It is not difficult to see that the state space of the process has size $Z = \prod_{i=1}^{N}(V_i + K_i V_i W_i)$ and grows exponentially with the growth of N. It is not feasible to express the generator matrix in explicit form due to its size. Instead of this let us take advantage of the notion of *transition classes* (e.g., see [8]) and describe them. Each class defines the set of transitions of the same type in which some subset of states changes equally while other states remain immutable.

The change of the state of the chain $\mathcal{C}$ can be caused by both the state change of a user MAP or PH-distribution of a server. If the transition does not cause either packet arrival or service completion, then the change is localized in the corresponding process. Otherwise, both the state of the corresponding process and the number of packets in one or two nodes can be changed.

Let triplet $\langle k_i, v_i, w_i \rangle$ describe the state of the i-th node, where k_i is the number of packets in the node, v_i — the state of a user input MAP, w_i — the state of a server. For the reason that the last component is meaningless when $k_i = 0$, we can omit it from time to time: $\langle 0, v_i \rangle$. The transition class is depicted by the expression of the form:

$$\langle k_i, v_i, w_i \rangle, \langle k_j, v_j, w_j \rangle \xrightarrow{\lambda} \langle \acute{k}_i, \acute{v}_i, \acute{w}_i \rangle, \langle \acute{k}_j, \acute{v}_j, \acute{w}_j \rangle$$

where the left side of the expression describes the states of the nodes before transition and the right side — the states after transition (the state with a dash sign is modified); the transition intensity is above the arrow. Other states not mentioned in the expression remain unmodified.

The transitions appearing due to the change of the state of modulating chain of input MAP-flow of the i-th node $A_i \sim MAP(D_0^{(i)}, D_1^{(i)})$:

$$\langle k_i, v_i, w_i \rangle, \xrightarrow{\{D_1^{(i)}\}_{v_i \acute{v}_i}} \langle \min(k_i + 1, K_i), \acute{v}_i, w_i \rangle \tag{1}$$

$$\langle 0, v_i \rangle, \xrightarrow{\tau_{w_i}^{(i)} \{D_1^{(i)}\}_{v_i \acute{v}_i}} \langle 1, \acute{v}_i, w_i \rangle \tag{2}$$

$$\langle k_i, v_i, w_i \rangle, \xrightarrow{\{D_0^{(i)}\}_{v_i \acute{v}_i}} \langle k_i, \acute{v}_i, w_i \rangle \tag{3}$$

$$\langle 0, v_i \rangle, \xrightarrow{\{D_0^{(i)}\}_{v_i \acute{v}_i}} \langle 0, \acute{v}_i \rangle \tag{4}$$

The transition appearing due to the change of the state of the modulating chain of PH-distribution of a server $B_i \sim PH(S^{(i)}, \tau^{(i)})$:

$$\langle k_i, v_i, w_i \rangle, \langle k_j, v_j, w_j \rangle \xrightarrow{t_{ij}(1-r_j)(1-p_{e_i})\tau_{w_i}^{(i)}\{-S^{(i)}\mathbf{1}\}_{w_i}} \langle k_i - 1, v_i, \acute{w}_i \rangle, \langle \min(k_j + 1, K_j), v_j, w_j \rangle \tag{5}$$

$$\langle k_i, v_i, w_i \rangle, \langle 0, v_j \rangle \xrightarrow{t_{ij}(1-r_j)(1-p_{e_i})\tau_{\acute{w}_i}^{(i)}\tau_{w_j}^{(j)}\{-S^{(i)}\mathbf{1}\}_{w_i}} \langle k_i - 1, v_i, \acute{w}_i \rangle, \langle 1, v_j, w_j \rangle \tag{6}$$

$$\langle 1, v_i, w_i \rangle, \langle k_j, v_j, w_j \rangle \xrightarrow{t_{ij}(1-r_j)(1-p_{e_i})\{-S^{(i)}\mathbf{1}\}_{w_i}} \langle 0, v_i \rangle, \langle \min(k_j + 1, K_j), v_j, w_j \rangle \tag{7}$$

$$\langle 1, v_i, w_i \rangle, \langle 0, v_j \rangle \xrightarrow{t_{ij}(1-r_j)(1-p_{e_i})\tau_{w_j}^{(j)}\{-S^{(i)}\mathbf{1}\}_{w_i}} \langle 0, v_i \rangle, \langle 1, v_j, w_j \rangle \tag{8}$$

$$\langle k_i, v_i, w_i \rangle, \xrightarrow{\tau_{\acute{w}_i}^{(i)}(t_{ij}r_j(1-p_{e_i})+p_{e_i}p_{l_i})\{-S^{(i)}\mathbf{1}\}_{w_i}} \langle k_i - 1, v_i, \acute{w}_i \rangle, \tag{9}$$

$$\langle 1, v_i, w_i \rangle, \xrightarrow{(t_{ij}r_j(1-p_{e_i})+p_{e_i}p_{l_i})\{-S^{(i)}\mathbf{1}\}_{w_i}} \langle 0, v_i \rangle, \tag{10}$$

$$\langle k_i, v_i, w_i \rangle, \xrightarrow{\tau_{\acute{w}_i}^{(i)}p_{e_i}(1-p_{l_i})\{-S^{(i)}\mathbf{1}\}_{w_i} + S_{w_i\acute{w}_i}^{(i)}} \langle k_i, v_i, \acute{w}_i \rangle, \tag{11}$$

The classes (1) and (2) correspond to the appearing of a packet in MAP, (3) and (4) — transition of the chain of MAP-flow without packet generating. The classes (5)–(8) correspond to transition of the chain of PH-distribution of the i-th device into an accepting state and transmission of the packet to the j-th node. The classes (9) and (10) correspond to the transition of the chain of PH-distribution into the accepting state with the error of transmission or transmission of the packet to the user of the i-th node – anyway, the packet leaves the network irretrievably. Finally, the class (11) corresponds to the transitions of the chain of PH-distribution which leads the chain into the accepting state, but the packet is retransmitted (the first summand) or the transition of the chain is carried out into a non-accepting state (the second summand) and service is continued - the amount of packets in the node is not changed.

Let $\mathbf{x} = (\mathbf{x}_1, \cdots, \mathbf{x}_i, \cdots, \mathbf{x}_N)$ be the state of the chain $\mathcal{C}$, where

$$\mathbf{x}_i = \begin{cases} \langle k_i, v_i, w_i \rangle & \text{if } k_i = 1 \cdots K_i \\ \langle 0, v_i \rangle & \text{if } k_i = 0 \end{cases}$$

— the state of the i-th node, X the set of all states of the chain, $\|X\| = \prod_{i=1}^{N}(V_i + K_i V_i W_i)$. The stationary distribution of the chain $\mathcal{C}$ can be found as the solution of the system:

$$\pi Q = 0, \pi \mathbf{1} = 1,$$

where $\pi \in [0, 1]^{|X|}$. Let $q_l^{(i)} = \mathbb{P}\{k_i = l\} = \sum_{\mathbf{x}: k_i = l} \pi_{\mathbf{x}}$ — the stationary probability that the i-th node has l packets. In particular, $q_0^{(i)}$ — the probability of the fact that the node is vacant and $q_{K_i}^{(i)}$ — the probability that the queue is full. Knowing the probabilities $\{q_K^{(i)}\}$ allows us to compute the mean number of packets $l_i = \sum_{k=0}^{K_i} k q_k^{(i)}$ in the i-th station, from which the loss probability of the arrival packet can be achieved:

$$p_{ql}^{(i)} = \mathbf{v}^{(i)} \frac{\check{D}_1^{(i)}}{\lambda^{(i)}} \mathbf{1},$$

where $\mathbf{v}^{(i)}$ is a projection of the vector $\boldsymbol{\pi}$ in which $k_i = K_i$, $\check{D}_1^{(i)}$ is the matrix of visible transitions of input MAP, and $\lambda^{(i)} = \boldsymbol{\phi}^{(i)} \check{D}_1^{(i)}$ is a mean arrival intensity of this process, $\boldsymbol{\phi}^{(i)}$ is its stationary probability distribution. The probability $\boldsymbol{\phi}^{(i)}$ and $\mathbf{v}^{(i)}$ can be achieved from $\boldsymbol{\pi}$ by summing up over the states from which the input arrival process is independent.

Knowing the mean intensity of packet arrival $\lambda^{(i)}$ and the mean number of packets in the system makes it possible to compute the mean packet delay in the station by Little's formula: $T_i = l_i / [(1 - p_{ql}^{(i)}) \lambda^{(i)}]$. Here an additional factor in the denominator appears due to the fact that the input process turns out to be filtered with the probability being equal complementary loss probability resulting from the queue overflow.

To study other characteristics of the model the expression of MAP incoming to the station (both from the user and from other stations) as well as of MAP departure may be required. Due to the existence of feedback in the general case the modulating chains will depend on the states of all network nodes. Therefore let us consider the modulating chain of the arrival process that is to be found, to have the same space of the state X as the process operating the system does. Obviously, chain $\mathcal{C}$ performs control over MAP as well, however from different points of observation the transitions being in an intensity matrix of packet generating D_i will differ. So, to determine a MAP $A^{(i,j)} \sim MAP(D_0^{(i,j)}, D_1^{(i,j)})$ describing the arriving of the packets at the j-th node after being served at the i-th node, the intensities defined by expressions (5)–(8) for the given pair (i, j) should be placed into matrix $D_1^{(i,j)}$ and the other intensities — into matrix D_0. If the MAP $\check{A}^j \sim MAP(\check{D}_0^{(j)}, \check{D}_1^{(j)})$ arriving at the station j from all other stations (excluding packet retransmission as it actually does not affect the amount of packets in the network) is in a region of interest, then all intensities (5)–(8) for all $i \neq j$ should be placed into matrix D_1. An other arrival process can be achieved similarly.

It should be noted that the matrices of the arrival process have an enormous size, which makes it difficult or even impossible to use them for analytical computations. However, it is possible to achieve more simple expressions for the arrival process matrices in some special cases. In particular, it is possible to build MAP for a tandem network that is described below as an example, with the help of the following theorems [4,9,10]:

Theorem 1. *The result of sifting of MAP $A \sim MAP(D_0, D_1)$ with probability p is MAP $A_p \sim MAP(D_0 + (1 - p)D_1, pD_1)$ (further we will denote it as pA).*

Theorem 2. *The composition of a MAP $A_1 \sim MAP(D_0^{(1)}, D_1^{(1)})$ and a MAP $A_2 \sim MAP(D_0^{(2)}, D_1^{(2)})$ is a MAP $B = A_1 \oplus A_2 \sim MAP(D_0^{(1)} \oplus D_0^{(2)}, D_1^{(1)} \oplus D_1^{(2)})$, where $\oplus$ is a Kronecker sum.*

Theorem 3. *A MAP of served packets in the system $MAP/PH/1/M$, where the interval between arrivals is distributed by $A \sim MAP(D_0, D_1)$, service time - $B \sim PH(S, \tau)$, M is the capacity of the queue, V is an order of matrix S, W is a number of input MAP states, service discipline is FIFO, is $B \sim MAP(\hat{D}_0, \hat{D}_1)$ and its matrices are defined as:*

$$
\hat{D}_0 = \begin{bmatrix}
D_0 \otimes I_V & R_0 & 0 & \cdots & 0 & 0 \\
0 & D_0 \otimes S & D_1 \otimes I_V & \cdots & 0 & 0 \\
0 & 0 & D_0 \otimes S & \cdots & 0 & 0 \\
\vdots & \vdots & \vdots & \ddots & \vdots & \vdots \\
0 & 0 & 0 & \cdots & D_0 \otimes S & D_1 \otimes I_V \\
0 & 0 & 0 & \cdots & 0 & R_A
\end{bmatrix},
$$

$$
\hat{D}_1 = \begin{bmatrix}
0 & \cdots & 0 & 0 & 0 \\
I_W \otimes C_t & \cdots & 0 & 0 & 0 \\
\vdots & \ddots & \vdots & \vdots & \vdots \\
0 & \cdots & I_W \otimes C_t & 0 & 0 \\
0 & \cdots & 0 & I_W \otimes C_t & 0
\end{bmatrix},
$$

where

$$
R_0 = D_1 \otimes (\tau \otimes 1_V),
$$
$$
R_A = (D_0 + D_1) \otimes S,
$$
$$
C_t = (-S1_V) \otimes \tau
$$

There are no state components corresponding to the number of packets in the i-th and further stations of the input MAP in construction of MAP as consistent with the given theorems. At the expense of this fact the computation is significantly simplified, in particular the computation of vectors $\phi^{(i)}$ and $\mathbf{v}^{(i)}$: the first one turns out to be a projection of the stationary distribution of generators states of the i-th station that corresponds to a full queue and the second one — the stationary distribution of the states of an input MAP generator.

Due to the exponential growth and enormously large dimension of the state space it is extremely difficult to find precise analytical expressions even for small dimensions. For practical applications of the suggested model approximation methods can be exploited. Such methods approximate both MAP and PH-distributions by the arrival process and distributions of lower dimensions and replace the large fragments of the network by much more simple chains that approximate such fragments [11]. The method of iterative search can be applied for an approximate solution by using multiplicative representation according to paper [8]. The other way to explore such systems is to apply the method of iterative modelling presented in the current work as well to simulate the networks of enormous dimensions.

3 Simulation of a Wireless Network with Linear Topology and Hot Standby Links

Wireless networks with linear topology are often used to organize connections along long-length objects (highways, railways, pipeline) when the optical fibre is not available. Up-to-date wireless communication systems allows us to build networks consisting of a large amount of retransmitters being placed from 100 m to tens of kilometres from each other and providing transmission rate from 150 Mbps (e.g. IEEE 802.11) to 1 Gbps (e.g. mmwave relay link).

The network suffers from packet losses as it is wireless. In the case of outage of one of the stations its neighbours can connect with each other if visibility conditions and the strength of the signal allow to do it. Such connections are acceptable in the case of the station working correctly as well. An example of such a network is shown in Fig. 2.

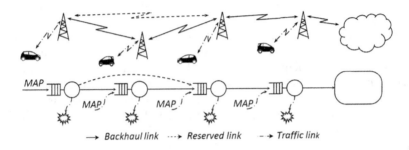

$\rightarrow$ Backhaul link $\cdots\rightarrow$ Reserved link $-\rightarrow$ Traffic link

Fig. 2. The queueing network modelling wireless network with linear topology.

A special case of a wireless network with linear topology is an uplink aggregation network where the traffic is generated by the users (e.g. video cameras) and being transmitted to a control centre. In particular, such networks are applied in road safety systems. We have studied this network in the model mentioned above as an example. The analytical computations have been performed for a simple network where the transmissions occur between neighbouring stations only, and the same network to calibrate the results as well as a network where transmissions escaping neighbouring stations are allowed is computed with the help of simulation.

As was mentioned above the construction of a chain generator faces the problem of exponential growth of the state space. In the case of an open queueing system without routing loops the scheme of chain construction can be simplified. To achieve the analytical solution of the problem the scheme of iterative construction of served packets MAP was employed [10]. This scheme is similar to the one being used to explore a more simple open network with linear topology that takes advantage of consecutive transmission without transmission losses and losses due to transmission errors. According to this scheme, starting from the first station, MAP matrices are build, that describe the intervals between the outage of served packets of each station. Output arrival processes are computed as a composition of input MAP. The iterative procedure using Theorems 1–3 mentioned above is made use of.

One of the most important characteristics of the network functioning is an end-to-end delay being equal to the time passed from the packet arrival into the network until it leaves the network in a destination node. As all packets of a data aggregation network are transmitted into the center connected with the N-th station the delay of packet transmission in the network from the i-th station can be computed as the sum of delays $\sum_{j=i}^{N} T_j$, where T_j is a delay in the j-th station, the computation of it having been described above. Let us note that in case of a computation of the mean residence time the loss probabilities must be taken into account. At the same time keeping computation of end-to-end delays we actually consider only the packets arriving successfully and take advantage of conditional probabilities of successful arrival. In the studied models the same cross-traffic arrived into all stations. Therefore a mean delay throughout the whole network is supposed to be equal to the arithmetic average of delay from each station.

The characteristics of a network with linear topology without the ability to transmit over the neighbour (i.e. all traffic is directed to the next station) have been studied with the help of an analytical model. The computation has been performed iteratively according to the scheme mentioned above. A more general case has been studied with the help of simulation. In that case the routing matrix of a data aggregation network with linear topology has an upper triangular form with zero main diagonal:

$$T = \begin{bmatrix} 0 & t_{12} & \cdots & t_{1N} \\ 0 & 0 & \cdots & t_{2N} \\ \vdots & \vdots & \ddots & \vdots \\ 0 & 0 & \cdots & 0 \end{bmatrix}$$

To simplify the computation we make a natural assumption that the station can transmit data to its neighbour or the station next to the neighbour only (that is nonzero elements of matrix T are $t_{i,i+1}$ and $t_{i,i+2}$) and these probabilities are equal for all stations excluding two border stations: $p = t_{i,i+1} \forall i < N - 1$. The input MAP $A_0 \sim MAP(D_0, D_1)$ are as follows:

$$D_0 = \begin{bmatrix} -1.724 & 0 & 0 \\ 0.172 & -1.552 & 1.293 \\ 0.086 & 1.724 & -1.811 \end{bmatrix} \quad D_1 = \begin{bmatrix} 0.862 & 0.862 & 0 \\ 0 & 0 & 0.086 \\ 0 & 0 & 0 \end{bmatrix}$$

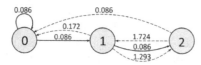

Fig. 3. A graphic representation of input MAP being used (transitions corresponding to the D_0 matrix marked by dashed lines and transitions corresponding to the D_1 matrix marked by solid lines)

The graph depicting this MAP is shown in Fig. 3. Such MAP arrives at each station as a cross-traffic. In different experiments MAP is scaled using the mean.

To simplify the computation the serving is performed according to the exponential distribution with rate $\mu = 5$ in all experiments. The number of stations in the network is supposed to be equal to $N = 5$ and the capacity of all queues is supposed to be $K = 2$.

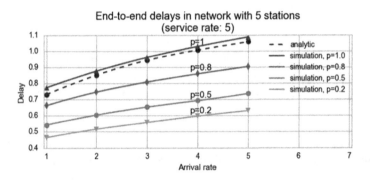

Fig. 4. Mean time of end-to-end delay depending on the mean rate of input cross-traffic. The value of the p parameter is a fraction of traffic transmitted to direct neighbour after service completion

In Fig. 4 end-to-end delays for different arrived rates of input MAP and different value of the probability p of packet transmission to the direct neighbour (the less p, the greater a fraction of the traffic transmitted escaping the direct neighbour) are shown. As we can see in the figure the delay decreases when the fraction of the traffic transmitted into direct neighbour drops. The result is expected and related to reduction of the mean path length that packet is transmitted over to the last station.

Let us note that the delay tendency to an asymptote when the intensity growths (instead of an unbound increase in the case of systems with an infinite queue) related to memory limitation — starting from some moment the stations are in a high-loaded state, all the extra packets are discarded and the delay of served packets stops increasing. The result is also confirmed by the change of delivery and loss probabilities over different stations and the change in the mean number of packets in stations as well (see Fig. 5).

As in the case of delays the losses drops when parameter p reduces, which is also related to the reduction of route lengths and, as a consequence, the reduction of input arrivals rates (see Fig. 5). At the same time starting from some moment the stations accumulate at input the maximum arrivals that they can serve whereupon the studied parameters change slightly; but if the p is small this moment comes later.

The dependence of delivery probability on the mean intensity of input outer traffic and parameter p is illustrated in the form of a heatmap in Fig. 6. Here the brighter the colour, the lower the successful delivery probability.

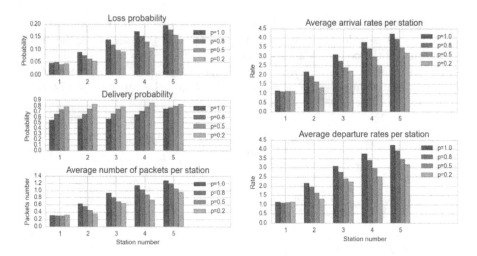

Fig. 5. The delivery and loss probabilities as well as the mean queueing lengths (left). The mean intensities of input and served arrivals over stations (right).

To carry out the calculations, analysis of results and their visualisation a pyQuMo library allowing us to work effectively with Markovian queueing models of large dimensions was developed. It is implemented in Python 3 languages and based on SciPy, NumPy and Pandas libraries. One of the advantages of pyQuMo is the storing and handling of chain generators in the form of sparse matrices which allows the handling of models having up to several million states using an ordinary laptop. As simulation requires the handling of a large amount of events the OMNeT++ system is used. The work with a simulation model is performed by pyQuMo, making it possible to compare the results of analytical modelling and simulation using a single program. A vast bulk of statistics is collected during the program work and to meet this problem SQLite is made use of.

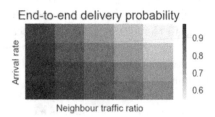

Fig. 6. The dependence of delivery probability on cross-traffic mean arrival rate (ordinate, rises from bottom to top) and the fraction of traffic transmitted to its direct neighbour (abscissa, rises from left to right).

4   Conclusion

In this paper a model of an open queueing network with correlated input Markovian arrival processes and phase type distribution of service time is presented. A network with linear topology that adequately describes wireless long-distance networks with the ability to transmit data over several stations is studied specially. A comparison study of numeric results of analytic modelling and simulation has been carried out. It is shown that to drop transmission delays as well as the fraction of packet losses it is rational to divide traffic and transmit as small a fraction as possible to the direct neighbour. We developed a pyQuMo library in Python 3 language to compute queueing systems with correlated input arrivals.

Acknowledgments. This work has been financially supported by the Russian Science Foundation and the Department of Science and Technology (India) via grant #16-49-02021 for the joint research project by the V.A. Trapeznikov Institute of control Sciences and the CMS College Kottayam.

References

1. Vishnevsky, V.M.: Fundamentals of Computer Network Designing (in Russian). Technoshera, Moscow (2003)
2. Lakatos, L., Szeide, L., Telek, M.: Introduction to Queueing Systems with Telecommunication Applications. Springer, Heidelberg (2013)
3. Neuts, M.F.: A versatile Markovian point process. J. Appl. Probab. **16**(4), 764–779 (1979)
4. Dudin, A.N., Klimenok, V.I.: Queueing Systems with Correlated Arrivals (in Russian). BSU, Minsk (2000)
5. Heyman, D., Lucantoni, D.: Modelling multiple IP traffic streams with rate limits. IEEE ACM Trans. Netw. **11**, 948–958 (2003)
6. Klemm, A., Lindermann, C., Lohmann, M.: Modelling IP traffic using the batch Marcovian arrival process. Perform. Eval. **54**, 149–173 (2008)
7. Buchholz, P., Kriege, J., Felko, I.: Input Modeling with Phase-Type Distributions and Markov Models. Springer, Heidelberg (2014)
8. Strelen, J.C.: Approximate analysis of queueing networks with Markovian arrival processes and phase type service times. In: Messung, Modellierung und Bewertung von (MMB), pp. 55–70 (1997)
9. Klimenok, V., Dudin, A., Vishnevsky, V.: Tandem queueing system with correlated input and cross-traffic. In: Kwiecień, A., Gaj, P., Stera, P. (eds.) CN 2013. CCIS, vol. 370, pp. 416–425. Springer, Heidelberg (2013)
10. Vishnevsky, V., Dudin, A., Kozyrev, D., Larionov, A.: Methods of performance evaluation of broadband wireless networks along the long transport routes. In: Vishnevsky, V., Kozyrev, D. (eds.) DCCN 2015. CCIS, vol. 601, pp. 72–85. Springer, Heidelberg (2016). doi:10.1007/978-3-319-30843-2_8
11. Horvath, G., Buchholz, P., Telek, M.: A MAP fitting approach with independent approximation of the inter-arrival time distribution and the lag correlation. In: Second International Conference on the Quantitative Evaluation of Systems (QEST 2005), pp. 124–133 (2005)
12. pyQuMo: a library for queueing models evaluation. https://github.com/larioandr/pyqumo
13. OMNeT++: Discrete Event Simulator. https://omnetpp.org

Optimization of Uniform Non-Markov Queueing Networks Using Resources and Transition Probabilities Redistribution

Vladimir N. Zadorozhnyi$^{(\boxtimes)}$

Omsk State Technical University, Omsk, Russia
zwn2015@yandex.ru

Abstract. An efficient analytic-imitational method for optimization of uniform non-Markov queueing networks is proposed. The optimization is performed through resources and transition probabilities redistribution by the minimal mean time of the request staying in the network. The convergence rate and precision of the method are evaluated experimentally. The method allows the optimization of the networks with hundreds of single- and multi-channel nodes with precision comparable to the one of the numerical methods for deterministic problems solution. Examples of the method's application in transport modeling are given. Practical application recommendations are provided.

Keywords: Queueing networks · Optimization · Gradient methods · Analytic-imitational modeling

1 Introduction

The productivity of organizational technical systems intended to process or service discrete flows of some uniform units (requests) is often measured by the time it takes for the requests to pass through the system. A unified formalized representation of such systems is a queueing network (QN) with statistically uniform requests – a uniform network [1].

Transport systems (TS) and information computation systems are traditionally represented as QNs [1–5]. In this case the requests are treated as moving cars in a TS or transmitted messages in an information computation network, i.e. user requests processed by system resources. The time it takes for the request to pass through the QN is herein referred to as response time. The mean response time E depends on:

- the distribution of the available resources among the network nodes;
- the distribution of the transition probability values at the network node output.

Recently, optimization methods for Markov QNs [1,2] – networks with exponential distributions of request servicing time and (if the network is not closed)

© Springer International Publishing Switzerland 2016
A. Dudin et al. (Eds.): ITMM 2016, CCIS 638, pp. 366–381, 2016.
DOI: 10.1007/978-3-319-44615-8_32

with Poisson input flows – have been successfully developing. The problem of optimal resource distribution for closed Markov networks with several classes of messages is solved in [6].

In [7], approximate computation methods are considered for non-Markov networks, based on approximating arbitrary distributions with those admitting the rational Laplace transform. However, this path presents many computational obstacles even for single-variant network computation with a small number of nodes (especially if one has to approximate distributions for random values restricted by a narrow range of possible values [8]). In [9], an attempt was made to solve the optimization problems for non-Markov QNs by approximating their nodes with analytic expressions that include two first moments of the servicing time and the intervals of the requests entering the nodes, considering also the node interactions. However, this work does not solve the optimization problems.

Thus, in the general case, in order to optimize non-Markov networks one has to use imitational modeling (IM). If the optimization problem contains more than 7–9 independent variables, it becomes virtually impossible to solve it without gradient methods. But computing gradients in IM is seriously impeded by the stochastic error in computed estimates $\hat{E}$ of the response time E [7,10]. Known approaches [7,11–13] for this problem's solution are either applicable only for IM of isolated queueing systems [7,12,13] and do not extend to QN, or their theoretically possible application to QN in practice leads to substantial problems. For example, using methods that imitate a large number of regeneration periods [11–14] faces the obstacle that in QN, as a rule, these periods are virtually infinite. Successful applications of such methods are restricted to classes of QNs that take into account the specifics of certain network objects in certain ranges of their parameters [3].

The application of analytic approaches towards the response E minimization is hindered by the absence of practical analytic expressions of E with varied parameters of non-Markov networks.

A similar problem is solved with analytic methods in [15]: a routing matrix allowing maximal QN throughput is found for an open non-Markov QN. The constraints for varied transition probabilities are expressed as those for the frequency of the requests visiting the network nodes. The problem is solved in terms of cooperation and competition effects that correspond to the efficient resource distribution problem.

For optimal (in terms of minimal time E) distribution of a resource on the nodes of a uniform non-Markov QN and optimal distribution of transition probabilities, we propose an analytic-imitational method that allows us to efficiently compute and use the gradients by applying special but simple separable approximation of the target function.

The problem in question is formulated in the following way.

2   Formulation of the Problem

First, let us consider an *open* network that receives a recurrent flow of requests with intensity Λ. The intervals between requests are independent random variables (r.v.) with distribution function (d.f.) $A(t)$. From the input flow, a request with probability p_{0i} proceeds to the i-th node, $i = \overline{1, n}$. In any of the K_i channels of the i-th node, the request servicing time (an independent r.v.) has d.f. $B_i(t)$. After servicing at the i-th node, the request, randomly and independently, selects another node j with the given transition probabilities p_{ij}, to continue its route or, with the probability p_{i0}, the request leaves the network. Probabilities p_{ij} $(i, j = \overline{0, n})$ are given with an indecomposable stochastic matrix $\mathbf{P} = \|p_{ij}\|$.

The mean stationary time E of a request passing through the network (the mean response time) can be represented as:

$$E = \sum_{i=1}^{n} \alpha_i u_i = \sum_{i=1}^{n} \alpha_i(w_i + b_i) = \sum_{i=1}^{n} \alpha_i \left(w_i + \frac{1}{\mu_i} \right), \tag{1}$$

where α_i is the average number of visits at the i-th node a claim makes during its path through the network,
u_i is the average time a request stays at the i-th node,
w_i is the average waiting time in the queue at the i-th node,
b_i is the average servicing time at the i-th node,
$\mu_i = b_i^{-1}$ is the intensity of the request servicing by the i-th node channel.

Coefficients α_i are uniquely determined by the system of balance equations:

$$\alpha_i = \sum_{j=0}^{n} \alpha_j p_{ji}, \quad i = \overline{0, n}, \quad \alpha_0 \equiv 1.$$

With α_i the following values are defined successively: intensities $\lambda_i = \Lambda \cdot \alpha_i$ of the nodes input flows, their load coefficients $\rho_i = \lambda_i/(\mu_i \cdot K_i)$, and stationary conditions $\rho_i \leq 1$ (or $\mu_i \geq \lambda_i/K_i$), $i = \overline{1, n}$ are checked. The values w_i for (1) are determined with IM.

Let us now consider the following generalized version of the optimization problem for a uniform non-Markov network formulated in [16].

The cost (resource) M of a uniform network as a function of the vector $\overrightarrow{\mu} = (\mu_1, \ldots, \mu_n)$ of servicing intensities in nodes $i = \overline{1, n}$ is given as $M(\overrightarrow{\mu}) = \sum_{i=1}^{n} c_i \mu_i^{\beta_i}$, where c_i are cost coefficients, $\beta_i > 0$ are nonlinearity coefficients.

The varied transition probabilities (not equal 0 or 1) being in fixed order determine some vector $\overrightarrow{p}_v$ with the dimension m. Usually m is limited by a linear function of n, since the QN nodes degree of connectedness is restricted by a constant in practice.

We need to find the vectors $\overrightarrow{\mu} = \overrightarrow{\mu}_{opt}$ and $\overrightarrow{p}_v = \overrightarrow{p}_{vopt}$ that realize the minimum of the function $E = E(\overrightarrow{\mu}, \overrightarrow{p}_v)$:

$$E(\overrightarrow{\mu}, \overrightarrow{p}_v) = \sum_{i=1}^{n} \alpha_i(\overrightarrow{p}_v) \left(w_i(\overrightarrow{\mu}, \overrightarrow{p}_v) + \frac{1}{\mu_i} \right) \to \min_{\overrightarrow{\mu}, \overrightarrow{p}_v} \tag{2}$$

and lie in the following admissible solutions region (ASR):

$$M(\overrightarrow{\mu}) = \sum_{i=1}^{n} c_i \mu_i^{\beta_i} = M^*, \quad \mu_i \geq \mu_{imin}; \quad \sum_{j=0}^{n} p_{ij} = 1, \quad (i = \overline{0,n});$$

$$0 \leq p_{vi} \leq 1, \quad (i = \overline{1,m}); \tag{3}$$

where for an open network $\mu_{imin} \geq \lambda_i / K_i$ (stationarity region boundary), and for a closed one $\mu_{imin} = 0$.

For the resource M^* in (3) it should hold that $M^* \geq M_{min}$, where for an open network $M_{min} = \sum_{i=1}^{n} c_i \mu_{imin}^{\beta_i} = \sum_{i=1}^{n} c_i (\lambda_i / K_i)^{\beta_i}$, and for a closed one $M_{min} = 0$. In the problems (2) and (3) we mean that changing any intensity μ_i leads to a change in the mean $b_i = \mu_i^{-1}$ and the corresponding scale change in the d.f. $B_i(t)$. The form of the d.f. $B_i(t)$ does not change because it is associated with a random complexity of the requests, while the varied parameter μ_i is determined by the efficiency of the i-th node channels. As a rule, the variability of transition probabilities in $\overrightarrow{p}_v$ is additionally restricted by some narrow limits that can be expressed with the intensity $\Lambda \cdot \alpha_i$, nodes load coefficient, or set as variability intervals for each probability p_{vi}.

3 General Method Structure

To find a solution $(\overrightarrow{\mu}_{opt}, \overrightarrow{p}_{opt})$ to the problems (2) and (3), is an iteration process. Each iteration starts with the sensitivity coefficient (SC) computation with the help of the advanced reduction method (ARM) [16] and the optimization of resource M distribution among QN nodes with the directing hyperbole (DH) method [17], resulting in distribution $\overrightarrow{\mu} = \overrightarrow{\mu}_{opt}$ at fixed transition probabilities p_{jk}. Then QN optimization due to probabilities at fixed $\overrightarrow{\mu}$ follows. After that either the transition to the next iteration or the completion of the optimization process (if the breakpoint condition is met) is performed. However, an analysis of the time E sensitivity to changes in transition probabilities $\partial E / \partial p_{jk}$, is necessary for optimization due to probabilities, and this analysis is a complicated task for IM itself. Such computation with the help of the small increments method faces extreme difficulties and is almost impossible in most cases. Moreover, the imitation model *precludes* from one probability p_{jk} increment with transition probabilities staying unchanged on alternative arcs, simultaneous changing of several varied parameters destroying the universal interpretation of SC as partial derivatives.

The possibility of fast and precise computation of SC $\partial E / \partial p_{jk}$ with analytic means is found due to the application of ARM [16]. Differentiation of the last expression in (1) for E with respect to any transition probability results in

$$\frac{\partial E}{\partial p_{jk}} = \sum_{i=1}^{n} \frac{\partial(\alpha_i / \mu_i)}{\partial p_{jk}} + \sum_{i=1}^{n} \frac{\partial(\alpha_i w_i)}{\partial p_{jk}}$$

$$= \sum_{i=1}^{n} \frac{1}{\mu_i} \cdot \frac{\partial \alpha_i}{\partial p_{jk}} + \sum_{i=1}^{n} \frac{\partial \alpha_i}{\partial p_{jk}} \cdot w_i + \sum_{i=1}^{n} \alpha_i \cdot \frac{\partial w_i}{\partial p_{jk}}. \tag{4}$$

It should be noted that ARM computes time E derivatives for the delay graph from the transition probabilities with precision [16], but in determining these derivatives the changes in the probabilities p_{jk} are supposed not to affect the changes in the graph nodes delays. In other words ARM computed the following "virtual" SC $\partial E_V/\partial p_{jk}$, that are determined by the omission of the third sum on the right side of the expression (4):

$$\frac{\partial E_V}{\partial p_{jk}} = \frac{\partial E}{\partial p_{jk}}\bigg|_{w_i=const} = \sum_{i=1}^{n} \frac{1}{\mu_i} \cdot \frac{\partial \alpha_i}{\partial p_{jk}} + \sum_{i=1}^{n} w_i \cdot \frac{\partial \alpha_i}{\partial p_{jk}}. \qquad (5)$$

The value of expression (5) is calculated with ARM from the given (ascribed to the graph vertices) values t_i, usually interpreted as the mean delays at the vertices. For instance, to compute SC (5) the delays $t_i = 1/\mu_i$ (at channels vertices) and $t_i = \mu_i$ (at queues vertices) are given. But such interpretation of parameters t_i is unnecessary, and the convolutions of the form (5) with partial derivatives $\partial \alpha_i/\partial p_{jk}$ are calculated with the help of ARM (with precision) at any t_i. This is to be used later.

However, applying ARM to computing SC of QN, one should take into consideration that the changes in probability p_{jk} of any arc lead to changes in the intensity of the corresponding arc flow, and consequently, affect the waiting w_i of most queues. The effect of p_{jk} on w_i, represented by the last sum in (4) is not taken into account by virtual SC (5) at ARM input.

Adjusting ARM to computation of complete SCs (4) is possible due to the relation between derivatives $\partial w_i/\partial \lambda_i$ and $\partial w_i/\partial \mu_i$, that was determined in [16] for isolated queueing systems:

$$\frac{\lambda}{w} \cdot \frac{\partial w}{\partial \lambda} + \frac{\mu}{w} \cdot \frac{\partial w}{\partial \mu} = -1. \qquad (6)$$

Using this expression it is possible to find the value of a partial derivative from the known value of the other partial derivative.

In the case of network nodes (if a partial derivative changes slightly), unlike the isolated queueing system, this expression holds only approximately, since generally the change $\partial \lambda_i$ in the intensity λ_i, caused by the change ∂p_{jk} of any partial derivative, slightly changes the static structure of the flows at the node inputs. Generally speaking, errors of expression (6), when applied to the analysis of increment ∂p_{jk} consequences are higher when d.f. $A_i(t)$ and $B_i(t)$ have a larger range of variation coefficient. If these d.f. are exponential, then formula (6) holds for the network nodes with precision. In all other cases, considering the approximate numerical values of derivatives $\partial w_i/\partial \mu_i$ are known (at each step of the DH method they are the components of the gradient being computed), formula (6) allows us to calculate (estimate) the derivatives $\partial w_i/\partial \lambda_i$ from these known $\partial w_i/\partial \mu_i$ without numerical differentiation. And as numerous tests of the DH method have shown, the solution errors are quite low: the obtained estimated minimum E differs from the precise one only by a fraction of a percent in a wide range of the method applications.

The last sum in expression (4) for complete SC is transformed as the following

$$\sum_{i=1}^{n}\alpha_i \cdot \frac{\partial w_i}{\partial p_{jk}} = \sum_{i=1}^{n}\alpha_i \cdot \frac{\partial w_i}{\partial \alpha_i} \cdot \frac{\partial \alpha_i}{\partial p_{jk}} = \sum_{i=1}^{n}\lambda_i \cdot \frac{\partial w_i}{\partial \lambda_i} \cdot \frac{\partial \alpha_i}{\partial p_{jk}},$$

with $\alpha_i = \lambda_i \cdot \Lambda^{-1}$. Having inserted this expression in (4) and taking (6) into account, we get the following formula for complete SC:

$$\frac{\partial E}{\partial p_{jk}} = \sum_{i=1}^{n}\left(\frac{1}{\mu_i}\right) \cdot \frac{\partial \alpha_i}{\partial p_{jk}} + \sum_{i=1}^{n}(w_i) \cdot \frac{\partial \alpha_i}{\partial p_{jk}} + \sum_{i=1}^{n}\left(\lambda_i \cdot \frac{\partial w_i}{\partial \lambda_i}\right) \cdot \frac{\partial \alpha_i}{\partial p_{jk}}$$

$$= \sum_{i=1}^{n}\left(\frac{1}{\mu_i}\right) \cdot \frac{\partial \alpha_i}{\partial p_{jk}} + \sum_{i=1}^{n}(w_i) \cdot \frac{\partial \alpha_i}{\partial p_{jk}} - \sum_{i=1}^{n}\left(w_i + \mu_i \frac{\partial w_i}{\partial \mu_i}\right) \cdot \frac{\partial \alpha_i}{\partial p_{jk}}. \quad (7)$$

After an imitation experiment at any step of the DH method, all components of expression (7) become known except derivatives $\partial \alpha_i/\partial p_{jk}$. However the last ones are precisely determined with ARM. Therefore, using formula (7) and ARM, it is possible to compute partial derivatives of the mean response time from the transition network probabilities effectively.

To this we can add that in general ARM automatically computes convolutions in the summation form in (7) of *any* values attributed to delay graph vertices i, with all corresponding derivatives $\partial \alpha_i/\partial p_{jk}$, $i = \overline{1,n}$, forming as a result an SC matrix of the time E to the changes in varied network probabilities. Consequently, formula (7) allows us to compute and use at least three types of SC. If one flags vertices-queues as zero and vertices-channels as mean delays $1/\mu_i$, in the delay graph, *partial* SCs corresponding to the first sum in (7) will be computed. If one substitutes the zero delays at vertices-queues by values w_i, found in the imitation experiment, ARM will determine *virtual* SCs. In adding to the delays at the vertices-queues of summand $\lambda_i \cdot \partial w_i/\partial \lambda_i$, *complete* SCs are calculated that take into account the effect of the change in the transition probabilities, induced not only by the changes in frequencies α_i of requests arriving at each node but by the changes in the mean delays w_i at each queue (depending on the intensity of the requests arriving at the nodes) as well.

All these three SC sets (partial, virtual and complete) provide plentiful information to estimate the sensitivity of the mean response time E to the changes in the transition probabilities. These SCs can be computed at any point of the factor space (FS) and used as a gradient component in solving the problem of transition probabilities optimization.

4   The Description of the QN Optimization Algorithm

4.1   General Structure of the Method

The description of the algorithm for solving problems (2) and (3), when each probability has narrow variability intervals in addition to general constraints (3), may be presented by the following sequence of steps.

The start of the algorithm.

We set iterations number $N = 1$, and the number of failed iterations $\gamma = 0$.

(1) *The procedure for optimizing the distribution of resource M among QN nodes*

At fixed $\overrightarrow{p}_v$ vector $\overrightarrow{\mu}$ is optimized with the DH method (see Subsect. 4.2), as a result we obtain optimal $\overrightarrow{\mu} = \overrightarrow{\mu}_{opt}$ and $E = E_{opt1}$.

(2) *The procedure for optimizing the probabilities $\overrightarrow{p}_v$ at fixed $\overrightarrow{\mu}$*

For each varied probability p_{jk}:

(2.1) By the sign of derivative $\partial E/\partial p_{jk}$ one can determine the direction of the changes in the current value of p_{jk}, where time E decreases. We determine the border of the variability range p_{jk}, lying in the given direction and keep its value in the variable d_{jk}. If $\partial E/\partial p_{jk} = 0$, then with a probability of 0.5 we choose and keep any of the two borders (top or bottom).

(2.2) Then we compute the steps $h_{jk} = (d_{jk} - p_{jk})/(\partial E/\partial p_{jk})$.

(2.3) We take step H equal to the least of all calculated h_{jk}, being nonzero. If all $h_{jk} = 0$, we move on to the end of the algorithm.

(The value of step H is the same at all transition probabilities p_{jk}).

(2.4) For each varied probability p_{jk} we give increment: $-H \cdot \partial E/\partial p_{jk}$, calculate new value $E = E_{opt2}$ and set $\overrightarrow{p}_{vopt} = \overrightarrow{p}_v$.

(3) If $E_{opt2} < E_{opt1}$, then we set $E_{opt} = E_{opt2}$, $N = N + 1$ and move on to step (1). Otherwise we set $\gamma = \gamma + 1$. If $\gamma \le 2$, then move on to step (1).

The end of the algorithm

To repeat all the manipulations according to the given description of the algorithm exactly, it is necessary to supplement the algorithm with a short but precise description of the optimization procedure for resource distribution carried out at step (1).

4.2   The Procedure of Optimizing Resourse Distribution Among QN Nodes

The procedure solves problems (2) and (3) at fixed transition probabilities. It includes the following two steps.

Step I: accelerated gradient search for the point $\overrightarrow{\mu}_{opt}$ by the "directing hyperbole" method (DH) that uses IM of the network and a separable approximation of the target function (the method's name reflects the role and the form of functions of one variable that are terms in this approximation).

Step II (optional): refining the obtained solution by cyclic coordinate wise (modified) descent [18] that does not use approximations.

4.3   Basic Elements of the "Directing Hyperbole" Method

The approach to solving (2) and (3) proposed in this work uses a refined version of the approximation of $E(\overrightarrow{\mu})$. On each iteration $k \ge 2$ of the optimal solution $\overrightarrow{\mu}_{opt}$ search, the approximation $E^{ap}(\overrightarrow{\mu})$ of the average response time $E(\overrightarrow{\mu})$ is computed by the network's IM results in points $\vec{\mu} = \vec{\mu}^{k-1}$ and $\vec{\mu} = \vec{\mu}^k$ and is used to determine the next point $\vec{\mu} = \vec{\mu}^{k+1}$.

Let us find the base elements of the DH method.

The center $\vec{\mu}_c$ ASR (3) for a resource $M^* > M_{\min}$ is determined by the equal load condition for the nodes: $\rho_i = \lambda_i/(\mu_i K_i) = \alpha_i \lambda_0/(\mu_i K_i) = \rho_c$, $(i = \overline{1,n})$, where λ_0 is the intensity on the terminal arc (for an open network, $\lambda_0 = \Lambda$). This implies $\mu_i = (\alpha_i \lambda_0)/(K_i \rho_c)$, $\mu_i/\mu_1 = (\alpha_i/K_i)(K_1/\alpha_1)$ and $\mu_i = (\alpha_i/K_i)(K_1/\alpha_1)\mu_1$. Substituting the latter expression μ_i into (3), we get

$$\sum_{i=1}^{n} c_i \left(\frac{\alpha_i}{K_i} \cdot \frac{K_1}{\alpha_1} \cdot \mu_1 \right)^{\beta_i} = M^*, \quad \mu_i = (\alpha_i/K_i)(K_1/\alpha_1)\mu_1, \quad i = \overline{2,n}. \quad (8)$$

From this equation, numerical methods can easily find the unique positive root μ_1 that defines all the other coordinates μ_i of the ASR center $\vec{\mu}_c$. If the network is open, we can immediately find all $\rho_i = \rho_c \leq 1$, $i = \overline{1,n}$ in the center $\vec{\mu}_c$.

If all $\beta_i = 1$, then, by (8), the ASR center coordinates can be found explicitly:

$$\mu_i = M^* \cdot (\alpha_i/K_i) \cdot \left(\sum_{j=1}^{n} c_j \cdot \alpha_j/K_j \right)^{-1}, \quad i = \overline{1,n}.$$

At the center $\vec{\mu}_c$ an open network has a maximum of the throughput $V(\vec{\mu})$, defined as $V(\vec{\mu}) = \max\{\Lambda : \rho_i = \alpha_i \Lambda/(\mu_i K_i) \leq 1, \ i = \overline{1,n}\}$.

The ASR diameter D we define as the maximal length of the ranges in which the variables μ_i lie: $D = \max\{l_i\}$, where $l_i = \mu_{i\,\max} - \mu_{i\,\min}$ and, by (3), $\mu_{i\,\max} = [(M^* - \sum_{j \neq i} c_j \mu_{j\,\min}^{\beta_i}) \cdot c_i^{-1}]^{1/\beta_i}$, $i = \overline{1,n}$.

A small step of size, say, $D \cdot 10^{-4}$ we will use for constructing and scanning trajectories on the constraint surface defined by (3). The step is chosen with regard to the optimization precision requirements.

The approximation $E^{ap}(\vec{\mu})$ of the target function $E(\vec{\mu})$, used for approximate gradient computations, is a separable function of varying variables μ_i:

$$E^{ap}(\vec{\mu}) = \sum_{i=1}^{n} \alpha_i \left(W_i(\mu_i) + \frac{1}{\mu_i} \right), \quad \text{where } W_i(\mu_i) = \begin{cases} \frac{R_i}{\mu_i - S_i}, & \text{if } \hat{w}_i^k \neq \hat{w}_i^{k-1}, \\ \hat{w}_i^k, & \text{if } \hat{w}_i^k = \hat{w}_i^{k-1}, \end{cases} \quad (9)$$

and on each optimization step k it is tuned anew (via the coefficients R_i and S_i) by the average response time estimates $\hat{w}_i^{k-1}$ and $\hat{w}_i^k$ computed for the nodes $i = \overline{1,n}$ with network IM at points $\vec{\mu} = \vec{\mu}^{k-1}$ and $\vec{\mu} = \vec{\mu}^k$. For $\hat{w}_i^k \neq \hat{w}_i^{k-1}$, the expression $R_i/(\mu_i - S_i)$ in (9) approximates the corresponding function $w_i(\vec{\mu})$ in (2) in such a way that its value at points $\vec{\mu} = \vec{\mu}^{k-1}$ and $\vec{\mu} = \vec{\mu}^k$ coincides with points $\hat{w}_i^{k-1} \approx \hat{w}_i(\vec{\mu}^{k-1})$ and $\hat{w}_i^k \approx \hat{w}_i(\vec{\mu}^k)$. Thus, we have $R_i/(\mu_i^{k-1} - S_i) = \hat{w}_i^{k-1}$, and $R_i/(\mu_i^k - S_i) = \hat{w}_i^k$, which implies

$$S_i = \frac{\hat{w}_i^k \mu_i^k - \hat{w}_i^{k-1} \mu_i^{k-1}}{\hat{w}_i^k - \hat{w}_i^{k-1}}, \quad R_i = \hat{w}_i^{k-1} \cdot (\mu_i^{k-1} - S_i), \quad (i = \overline{1,n}), \quad (10)$$

(the superscript everywhere corresponds to the optimization step).

For $\hat{w}_i^k = \hat{w}_i^{k-1}$, we do not have to compute the values R_i and S_i in (9), but to determine [L] (see below) we set $S_i = \infty$. This tuning of the function $E^{ap}(\overrightarrow{\mu})$ guarantees that it coincides with $E(\overrightarrow{\mu})$ at points $\overrightarrow{\mu}^{k-1}$ and $\overrightarrow{\mu}^k$ (up to the stochastic error of IM estimates). In other points $\overrightarrow{\mu}$ the approximation error $E^{ap}(\overrightarrow{\mu})$ grows as they move farther from the points $\overrightarrow{\mu}^{k-1}$ and $\overrightarrow{\mu}^k$, and the "less separable" $E(\overrightarrow{\mu})$ is, i.e., the more intensities μ_i change in some nodes, influences the average time w_j in other nodes ($i \neq j$).

The gradient $\nabla E^{ap}(\overrightarrow{\mu})$ in point $\overrightarrow{\mu} = \overrightarrow{\mu}^k$ is an approximation of the gradient $\nabla E(\overrightarrow{\mu})$ in this point, and it is computed with the following expression, obtained by differentiating (9):

$$\nabla E^{ap}(\overrightarrow{\mu}^k) = \left(\alpha_1 \frac{\partial W_1}{\partial \mu_1} - \frac{\alpha_1}{(\mu_1)^2}, \ldots, \alpha_n \frac{\partial W_n}{\partial \mu_n} - \frac{\alpha_n}{(\mu_n)^2} \right), \tag{11}$$

where

$$\frac{\partial W_i}{\partial \mu_i} = \begin{cases} \frac{-R_i}{(\mu_i^k - S_i)^2}, & \hat{w}_i^k \neq \hat{w}_i^{k-1}, \\ 0, & \hat{w}_i^k = \hat{w}_i^{k-1}, \end{cases} \quad (i = \overline{1, n}).$$

The valid part [L] of a trajectory L outgoing from a point $\overrightarrow{\mu}^k$ that searches for the point $\overrightarrow{\mu}^{k+1}$ lies between $\overrightarrow{\mu}^k$ and the first point $\overrightarrow{\mu}$ of L for which some coordinate μ_i reaches the ASR boundary $\mu_i = \mu_{i\,min}$ or a pole $\mu_i = S_i$ of the approximation (9).

We next give a step-by-step description of the DH method with formulas for computation.

4.4   Step I of the "Directing Hyperbole" Method

Initial optimization step. We fix the number of iterations $N > 2$ (selected with regard to available CPU time) and two points $\overrightarrow{\mu}^1 = \overrightarrow{\mu}_c$ and $\overrightarrow{\mu}^2 \neq \overrightarrow{\mu}^1$ that belong to ASR (3). With IM, we compute at the points $\overrightarrow{\mu}^1$, $\overrightarrow{\mu}^2$ estimates for the mean response time $\hat{E}^1$ and $\hat{E}^2$, and, correspondingly, mean waiting time estimates $(\hat{w}_1^1, \ldots, \hat{w}_n^1)$, $(\hat{w}_1^2, \ldots, \hat{w}_n^2)$ at the node $1, \ldots, n$. We set $k = 2$.

Base cycle. Points $\overrightarrow{\mu}^{k-1}$ and $\overrightarrow{\mu}^k$ are known, together with the estimates $(\hat{w}_1^{k-1}, \ldots, \hat{w}_n^{k-1})$, $\hat{E}^{k-1}$ and $(\hat{w}_1^k, \ldots, \hat{w}_n^k)$, $\hat{E}^k$ of the responses $(w_1^{k-1}, \ldots, w_n^{k-1})$, E^{k-1} and $(w_1^k, \ldots, w_n^k)$, E^k.

(1) Using the estimates $(\hat{w}_1^{k-1}, \ldots, \hat{w}_n^{k-1})$ and $(\hat{w}_1^k, \ldots, \hat{w}_n^k)$, we find, by formulas (10), coefficients R_i and S_i ($i = \overline{1, n}$), and approximations $E^{ap}(\overrightarrow{\mu})$.

We compute the gradient $\nabla E^{ap}(\overrightarrow{\mu}^k)$ (11) of the function E^{ap}. The direction $-\nabla E^{ap}(\overrightarrow{\mu}^k)$ of the fastest decrease of the function $E^{ap}(\overrightarrow{\mu})$ is projected onto the constraint surface (3). If $\beta_i = 1$ for all $i = \overline{1, n}$, the surface (3) is a hyperplane, and the projection L of the vector direction $-\nabla E^{ap}(\overrightarrow{\mu}^k)$ on it is the direction $-\nabla E^{ap}_{pr}(\overrightarrow{\mu}^k) = -\nabla E^{ap}(\overrightarrow{\mu}^k) + \overrightarrow{n} \cdot (\overrightarrow{n} \cdot \nabla E^{ap}(\overrightarrow{\mu}^k)) = \overrightarrow{e}$, where $\overrightarrow{n} = \overrightarrow{c}/|\overrightarrow{c}|$ is normal to the constraints hyperplane, $\overrightarrow{c} = (c_1, \ldots, c_n)$ is the vector of cost coefficients, $|\vec{x}|$ is the length of $\vec{x}$. The valid part [L] of the projection L is bounded by the points $\vec{\mu}^k$ and $\vec{\mu} = \vec{\mu}^k + h \cdot \vec{e}$, where $h = \min\{h_1, h_2\}$,

$$h_1 = \min\{h_{1i} : h_{1i} > 0; \ i = \overline{1,n}\}, \quad h_2 = \min\{h_{2i} : h_{2i} > 0; \ i = \overline{1,n}\},$$
$$h_{1i} = -(\mu_i^k - \mu_{i\,\min})/e_i, \quad\quad\quad h_{2i} = -(\mu_i^k - S_i)/e_i, \ i = \overline{1,n}.$$

If not all β_i are equal to 1, then the projection L of the antigradient direction can be computed step by step, as a polyline outgoing from the point $\vec{\mu}^k$ whose nodes $\vec{\mu}$ are projections on the surface (3) of the points of direction $-\nabla E^{ap}(\vec{\mu}^k)$ equidistant for some small step. For each node $\vec{\mu}$, we check its validity conditions $\mu_i > \mu_{i\,\min}$ and $(\mu_i - S_i) \cdot (\mu_i^k - S_i) > 0, i = \overline{1,n}$. If the differences $(\mu_i - S_i)$ and $(\mu_i^k - S_i)$ have the same signs, it means that the coordinate μ_i of the current node $\vec{\mu}$ of the trajectory L and the coordinate μ_i^k of its initial point $\vec{\mu}^k$ are on one side of the approximation pole S_i. The construction of [L] is completed either by finding and discarding the first invalid node or by finding out that the next point along the antigradient direction $-\nabla E^{ap}(\vec{\mu}^k)$ does not have a projection on the constraint surface.

(2) As the next point $\vec{\mu}^{k+1}$ we select a solution (found by scanning) of the one-dimensional optimization problem $E^{ap}(\vec{\mu}) \to \min, \ \vec{\mu} \in [L]$.

(3) We set $k = k + 1$. With IM, we compute estimates $(\hat{w}_1^k, \ldots, \hat{w}_n^k)$ and $\hat{E}^k$. If $k < N$ we proceed to step (1), otherwise–to step (4).

(4) The point $\vec{\mu}^* \in \{\vec{\mu}^1, \ldots, \vec{\mu}^N\}$ with estimate $\hat{E}(\vec{\mu}^*) = \min\{\hat{E}^1, \ldots, \hat{E}^N\}$ we take as the approximate solution of the problem. *End of the algorithm.*

4.5 Step II – Accelerated Coordinate Descent

The error of the first step (the DH method) consists of two components: stochastic and deterministic. The stochastic component is controlled by computing confidence intervals for the estimates $(\hat{w}_1^k, \ldots, \hat{w}_n^k)$ and $\hat{E}^k$, and it can be reduced by lengthening model runs. The deterministic component is caused by using the separable approximation $E^{ap}(\vec{\mu})$ for the (in general) inseparable function $E(\vec{\mu})$. This leads to the fact that the solution $\vec{\mu}^*$ found by the DH method differs from the desired $\vec{\mu}_{opt}$ even if we completely the avoid stochastic error of the IM estimates. Therefore, it is reasonable to enhance (or check) the solution $\vec{\mu}^*$ with an approach that does not employ approximation $E^{ap}(\vec{\mu})$.

To do so, at the second optimization step we define $2(n - 1)$ test points in the neighborhood of $\vec{\mu}^*$ each of which differs from $\vec{\mu}^*$ only in one of the $(n - 1)$ "free" coordinates μ_i (for example, in one of the coordinates $\mu_1, \ldots, \mu_{n-1}$) by the value $\pm \triangle\mu$. The "bound" coordinate of a test point (e.g. μ_n) is defined via its known "free" coordinates by solving Eq. (3) in order to guarantee that all the test points belong to the ASR. The difference $\triangle\mu$ is chosen with regard to admissible solution error and the possibility of reliable comparison of the responses $E(\vec{\mu})$ in corresponding points by their estimates $\hat{E}(\vec{\mu})$. Further, solution $\vec{\mu}^*$ is enhanced by performing network IM at the test points and replacing $\vec{\mu}^*$ with a test solution $\vec{\mu}$ for which $E(\vec{\mu}) < E(\vec{\mu}^*)$. If there is no such solution, the process ends, otherwise the local search step is repeated for the new solution. To avoid cycles, a return to the previous solutions is forbidden. To speed up the comparison of responses $E(\overrightarrow{\mu})$ and $E(\overrightarrow{\mu}^*)$ with given precision by their estimates $\hat{E}(\overrightarrow{\mu})$ and $\hat{E}(\overrightarrow{\mu}^*)$, we use the "generalized random numbers" approach [10].

5   Testing Optimization Method

Let us consider the application of the proposed method by the example of QN optimization (Fig. 1, left side). The total resource $M = 20$ is distributed here for $\overrightarrow{c} = (c_1, \ldots, c_n) = (K_1, \ldots, K_9) = (1, 2, 1, 1, 1, 1, 1, 3, 1)$, i.e., for each queueing network the cost coefficient equals the number of its channels. Distribution types $B_i(t)$ for the nodes $i = 1, \ldots, 9$ are defined as D, M, D, M, D, M, M, D, D respectively, where M is the exponential distribution, D is the deterministic one. The input flow of QN is assumed to be Poisson with intensity $\Lambda = 1$. Transition probabilities are shown in Fig. 1. (on the right there is the structure of a larger test network used in [17], having 100 single- and multiple-channel nodes).

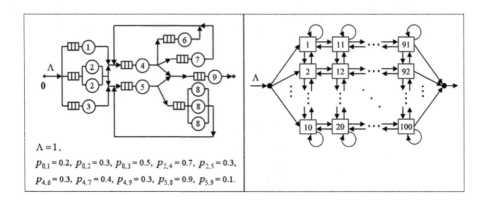

$\Lambda = 1,$

$P_{0,1} = 0.2,\ P_{0,2} = 0.3,\ P_{0,3} = 0.5,\ P_{2,4} = 0.7,\ P_{2,5} = 0.3,$

$P_{4,6} = 0.3,\ P_{4,7} = 0.4,\ P_{4,9} = 0.3,\ P_{5,8} = 0.9,\ P_{5,9} = 0.1.$

Fig. 1. QN test case

The initial values of service intensity are given in Table 1.

Table 1. Initial values of service intensity and time E

Initial values μ (at centerpoint)									E
μ_1	μ_2	μ_3	μ_4	μ_5	μ_6	μ_7	μ_8	μ_9	
0.2575	0.1931	0.6437	1.7596	7.5965	0.5278	0.7038	2.278	1.2875	25.544

Variability ranges for each transition probability are determined as given in Table 2. Optimal probabilities values found at the 4th iteration are given as well, along with the optimal resource distribution among the network nodes.

Table 2. Optimal values of transition probabilities

Probability	Initial value	Left border	Right border	Optimal value
$p_{0,1}$	0.2	0.1	0.3	0.2286
$p_{0,2}$	0.3	0.2	0.4	0.3159
$p_{0,3} = 1 - p_{0,1} - p_{0,2}$	0.5		-	0.4554
$p_{2,4}$	0.7	0.6	0.8	0.7075
$p_{2,5} = 1 - p_{2,4}$	0.3	-		0.2924
$p_{4,6}$	0.3	0.2	0.4	0.2
$p_{4,7}$	0.4	0.3	0.5	0.3
$p_{4,9} = 1 - p_{4,6} - p_{4,7}$	0.3	-		0.5
$p_{5,8}$	0.9	0.5	0.99	0.5
$p_{5,9} = 1 - p_{5,8}$	0.1	-		0.5

Approximately optimal values of the resource (service intensities) are given in the 4th row of Table 3. The values of the target function E obtained at each iteration are also given.

Table 3. The obtained optimal values of service intensities and time E

Iteration	Obtained approximations to the optimum point									E
	μ_1	μ_2	μ_3	μ_4	μ_5	μ_6	μ_7	μ_8	μ_9	
1	0.470	0.354	0.716	1.828	6.721	0.626	1.182	2.154	1.289	23.646
2	1.173	0.902	1.926	3.054	3.214	1.311	1.635	0.922	3.115	3.5535
3	1.112	0.901	2.047	2.799	3.778	1.016	1.259	0.938	3.368	3.1152
4	1.250	0.923	1.902	2.526	3.732	1.238	1.296	0.954	3.348	3.0993

Figure 2 shows a typical trajectory of the changes in the target function during the optimization process due to the method in question.

At each iteration the approximation to the optimum point undergoes the number of steps comparable to the dimension $(m+n)$ of FS. The solution determined in 4 iterations provides the decrease in the mean response time E from 25.5 to 3.1 (Table 3). For each effective iteration the number of steps in FS is less than its dimension $(m + n)$.

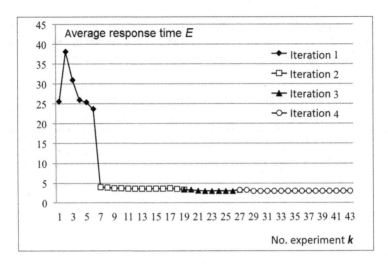

Fig. 2. The changes in time E during optimization of test QN

6   Application Example

The high efficiency of the developed method allows for it to be used for the optimization of large computation and transport networks.

An example of the modeled transport network is given in Fig. 3. The problem of optimal resource distribution for road repair (at fixed routing matrix) was being solved.

Additional experiments have shown that in transport network optimization the results of the optimization obtained with the developed method are verified (Fig. 4) by the imitation of these networks with the help of transport modeling system VISSIM, as well as by the imitation with transport cellular automation (TCA). The left part of Fig. 4 is QN with a virtual input node corresponding to the road network in Fig. 3.

In modeling road networks, the roads are represented by QN nodes. The transport delay at nodes depends on the traffic density and the features of its stochastic dynamics reproduced by the imitation model. Thanks to the fact that with an increase of the dedicated resource (improvement of the road surface) the mean delay at the nodes declines upon an approximate hyperbole, the proposed analytic-imitational method finds an approximate optimal distribution for a small number of iterations.

In road network optimization, the road (in a large range of loads) can be approximated as an infinite linear queueing system [19,20]. The optimal resource distribution for such a system is easily computed with an analytic method [21]. This distribution is convenient for use as an initial approximation for a solution to problems (2) and (3).

The efficiency of simultaneous optimization of resource and transition probabilities distribution is ensured by two key features of the developed method:

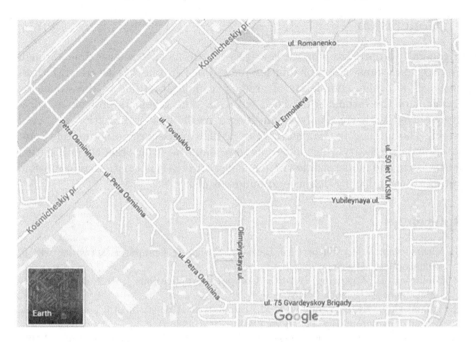

Fig. 3. Road network sector, Omsk, Russia

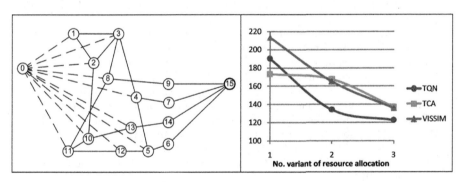

Fig. 4. Transport QN (TQN) and the testing results for its optimization with VISSIM and TCA (right side)

(*a*) effective separable approximation of the response surface taking into account the character of the resource effect on the mean delay time at nodes, and

(*b*) high precision (based on analytic transformations) of the gradient computation with the advanced graph reduction method.

7 Conclusion

In general, to optimize non-Markov QNs by the mean response time through resource and transition probabilities redistribution, one has to use IM characterized by a high complexity of gradient computation due to the stochastic character of the estimates obtained with IM. The gradients problem is effectively solved with the analytic-imitational method of optimization proposed herein. The method is characterized by a high precision and acceptable computational complexity which allows us to recommend it for practical application in the design and modernization of queueing networks (information computation, transport, etc.), having dozens and hundreds of nodes.

References

1. Vishnevskiy, V.M.: Theoretical Bases of Designing Computer Networks, 512 p. Technosphere, Moscow (2003)
2. Kleinrock, L.: Queueing Systems, Computer Applications, vol. 2, 576 p. Wiley Interscience, New York (1976)
3. Vishnevskiy, V.M., Porotskiy, C.M.: Simulation of departmental e-mail systems. Autom. Teleautomatics **12**, 48–57 (1996). (in Russian)
4. Zhozhikashvili, V.A., Vishnevskiy, V.M.: Queueing Networks. Theory and Applications to Computer Networks, 192 p. Radio i Svyaz, Moscow (1988). (in Russian)
5. Ferrari, D.: Performance Evaluation of Computer Systems, 554 p. Prentice-Hall (1978)
6. Gerasimov, A.I.: Optimization of closed queuing networks with several classes of messages. Probl. Transmitting Inform. **30**(1), 85–96 (1994). (in Russian)
7. Ryizhikov, Y.: Simulation Modeling. Theory and Technology, 384 p. KORONA print, St. Petersburg. Alteks-A, Moscow (2004). (in Russian)
8. Kleinrock, L.: Queueing Systems, Theory, vol. 1, 417 p. Wiley (1975)
9. Bitran, G.R.: Open Queueing Networks: Optimization and Performance Evaluation Models for Discrete Manufacturing Systems / Gabriel R. Bitran, Reinaldo Morabito, 47 p. URL:http://dspace.mit.edu/bitstream/handle/1721.1/2537/SWP-3743-31904719.pdf. (Accessed date, 01 Nov 2012)
10. Kleijnen, J.P.C.: Statistical Techniques in Simulation, Part 1. Marcel Dekker, New York (1974)
11. Johnson, M.E., Jackson, J.: Infinitesimal perturbation analysis: a tool for simulation. J. Oper. Res. Soc. **40**(3), 134–160 (1989)
12. Rubinstein, R.Y.: Sensitivity analysis of computer simulation models via the efficient score. Operat. Res. **37**, 72–81 (1989)
13. Suri, R., Zazanis, M.: Perturbation analysis gives strongly consistent sensitivity estimates for the M—G—1 queue. MGMT Sci. **34**, 39–64 (1988)
14. Iglehart, D.L., Shedler, G.S. (eds.): Regenerative Simulation of Response Times in Networks of Queues. LNCIS, vol. 26. Springer, Heidelberg (1980)
15. Tsitsiashvili, G.: Sh. Parametric and structural optimization of the queuing network throughput. Autom. Remote Control **68**(7), 1177–1185 (2007). doi:10.1134/S0005117907070065. PACS number: 89.75.Fb 2007
16. Zadorozhnyi, V.N.: Analytical-Simulation Research of Queuing Systems and Networks, 280 p. Omsk State Technical University Publishing House, Omsk (2010). (in Russian)

17. Zadorozhnyi, V.N.: Optimizing uniform non-markov queueing networks. Autom. Remote Control **71**(6), 1158–1169 (2010). doi:10.1134/S0005117910060172. ISSN 0005-1179
18. Bazaraa, M.S., Sherali, H.D., Shetty, C.M.: Nonlinear Programming: Theory and Algorithms, 854 p. A John Wiley and Sons, Inc. (2006)
19. Nazarov, A., Moiseev, A.N.: Investigation of the queueing network $GI(GI|\infty)^K$ by means of the first jump equation and asimptotic analysis. In: Vishnevsky, V., Kozyrev, D., Larionov, A. (eds.) DCCN 2013. CCIS, vol. 279, pp. 229–240. Springer, Heidelberg (2014)
20. Moiseev, A.A., Nazarov, A.N.: Infinitely Linear Systems and Queuing Networks, 240 p. NTL Publishing House, Tomsk (2015). (in Russian)
21. Zadorozhnyi, V.N.: Optimization highly redundant Markov chains with queues. Omskiy Nauchnyiy Vest. **3**(123), 21–25 (2014)

Methods of Simulation Queueing Systems with Heavy Tails

Vladimir N. Zadorozhnyi and Tatiana R. Zakharenkova$^{(\boxtimes)}$

Omsk State Technical University, Omsk, Russia
zwn2015@yandex.ru, ZakharenkovaTatiana@gmail.com

Abstract. Important problems of correct organization of simulation experiments for calculating fractal queueing systems are considered. Fractal systems are described asymptotically by power laws of arrival interval distribution and service time of requests and are adequate mathematical models of network devices of telecommunication systems with fractal (self-similar) traffic. We propose an effective solution to the problem for the correct realization of heavy-tailed distributions. Accuracy control techniques for calculating fractal queues by means of consecutive or repeated "parallel" runs of simulation models are developed. The application examples of the developed methods are given.

Keywords: Queueing systems · Heavy-tailed distributions · Errors · Random number generators · Estimates · Simulation

1 Introduction

Research on telecommunication systems has shown that the traffic of modern data transmission networks has a fractal (self-similar) structure [1]. In such traffic we can see a long-term dependence describing its random variables, and these variables are described by heavy-tailed distributions (HTD). Generally, the HTD tail decreases according to the power law $P[x > t] = t^{-\alpha}L(t)$ where $L(t)$ is a slow changing function [2]. The characteristic feature of HTD is that we can not neglect the probability of large observations x.

HTD properties create difficulties faced while measuring traffic and designing network devices. One of such problems is long transition processes [3], which interfere with measuring and practical application of steady-state indicators of traffic service quality by network devices. In simulation of fractal random processes, significant difficulties are referred to the HTD correct realization. Work [4] shows that HTD realized in simulation have shifted moments and you have to use random number generators (RNG) with an infinite number of digit positions, or to develop a special (for example, cascade) RNG to solve this problem.

In designing network devices at a system level, they are presented in a form of queueing systems [5,6]. Fractal systems are GI/GI/n/m systems in which intervals of request arrival and/or their service time belong to HTD and have

© Springer International Publishing Switzerland 2016
A. Dudin et al. (Eds.): ITMM 2016, CCIS 638, pp. 382–396, 2016.
DOI: 10.1007/978-3-319-44615-8_33

infinite variance. The load rate ρ of the examined systems does not exceed one: $\rho = \bar{x}/(n\bar{\tau}) \leq 1$ where $\bar{x} < \infty$ is a request service average time, $\bar{\tau} < \infty$ is an average time between request arrivals, n is a number of channels in the system. We shall call the GI/GI/n/m systems set only by light-tailed distributions (LTD) classical systems.

Typical representatives of fractal queueing systems are Pa/M/n/m, M/Pa/n/m and Pa/Pa/n/m [4] systems. Here the Pa symbol corresponds to Pareto distribution (PD):

$$F(t) = 1 - (K/t)^{\alpha}, \quad \alpha > 0, \quad K > 0, \quad t \geq K, \tag{1}$$

where α is a shape parameter, K is the smallest value of a random variable (r.v.) and simultaneously, a scale parameter. In abbreviated form we designate PD with K, α parameters as $\text{Pa}(K, \alpha)$. The range of α values, typical for fractal traffic, is defined by inequation $1 < \alpha \leq 2$. From (1) it is easy to find that with such α PD has a finite mathematical expectation (m.e.), which is equal to $\alpha K/(\alpha - 1)$, and the infinite variance. The main method of fractal system calculation is simulation [7–9].

In the article, we have elaborated a number of basic statements distinguishing the organization of simulation experiments with fractal systems from the organization of experiments with classical queues. We derive the relations allowing one to control the accuracy of fractal system calculation and to plan the duration (scope) of experiments to achieve the set accuracy. We propose the ARAND method (Accurate RAND) which is a simple and universal method of solving the problem for HTD correct realization, the method generalizing and simplifying the principles of cascade RNG. Examples of effective application of the developed statements, methods and algorithms for modeling of fractal queues are presented.

2 Asymptotic Confidence Intervals in Simulation of Queueing Systems

Generally in queues simulation to calculate the estimate $\hat{\xi}$ of some m.e. $\bar{\xi} = \text{E}(\xi)$ ("indicator") we use the sample $\xi_1, \ldots, \xi_N$ of dependent realizations for random variable ξ at the output of a simulation model. The estimate $\hat{\xi}$ is calculated as a sample average:

$$\hat{\xi} = \frac{1}{N} \sum_{i=1}^{N} \xi_i. \tag{2}$$

Variance $\text{Var}(\hat{\xi})$ of estimate $\hat{\xi}$ can be expressed in terms of $\sigma^2 = \text{Var}(\xi)$ variance and correlation coefficients between realizations $\xi_1, \ldots, \xi_N$ in the following way:

$$\text{Var}(\hat{\xi}) = \text{Var}\left(\frac{1}{N} \sum_{i=1}^{N} \xi_i\right) = \frac{1}{N^2}\left[\text{E}\left(\sum_{i=1}^{N} \xi_i\right)^2 - \text{E}^2\left(\sum_{i=1}^{N} \xi_i\right)\right] =$$

$$= \frac{1}{N}\left[\sigma^2 + \frac{2}{N} \sum_{i=1}^{N-1} \sum_{j=i+1}^{N} r_{ij}\sigma^2\right] = \frac{\sigma^2}{N}(1 + 2R_N), \tag{3}$$

where $\sigma^2 = \mathrm{Var}(\xi) = \mathrm{Var}(\xi_i)$ is a variance of sample value ξ_i does not depend on i, as we consider a steady-state sequence of $\xi_1, \ldots, \xi_N$ realizations; $r_{ij} = \mathrm{corr}(\xi_i, \xi_j) = \frac{\mathrm{E}(\xi_i \xi_j) - \mathrm{E}^2(\xi)}{\sigma^2}$ is a correlation coefficient of ξ_i, ξ_j elements, which depends only on $s = |j - i|$ distance between these elements; the R_N is an aftereffect coefficient:

$$R_N = \frac{1}{N} \sum_{i=1}^{N-1} \sum_{j=i+1}^{N} r_{ij}.$$

This coefficient characterizes a cumulative "direct" effect of $\xi_1, \ldots, \xi_N$ elements on the sample elements, which follow them. We designate r_{ij} through $r(s)$, then the expression of coefficient R_N looks like

$$R_N = \frac{1}{N} \sum_{i=1}^{N-1} \sum_{j=i+1}^{N} r_{ij} = \frac{1}{N} \sum_{s=1}^{N-1} (N - s) r(s), \tag{4}$$

where $r(s)$ is a correlation coefficient for a couple of elements, being spaced s steps apart; $(N - s)$ is a number of such couples in a sample.

The coefficient $(1 + 2R_N)$ in the Formula (3) is interpreted as the coefficient of variance increase of estimate $\mathrm{Var}(\hat{\xi})$ which in case of independent realizations $\xi_1, \ldots, \xi_N$, would merely equal σ^2/N. Considering that $|r(s)| \leq 1$ and $\mathrm{Var}(\hat{\xi}) \geq 0$, from (3) and (4) it is easy to conclude that the variance increase coefficient $(1 + 2R_N)$ ranges from 0 to N. However while modeling queues we are usually interested in indicator $\bar{\xi}$ (for example, it can be an average waiting time or a request loss probability) when all correlation coefficients $r(s)$ are positive, thus the coefficient $(1 + 2R_N)$ of variance increase takes a value greater than one. Therefore, the correlation of sampling results in the variance increase $\mathrm{Var}(\hat{\xi}) = \sigma_{\hat{\xi}}^2$ and, respectively, to an increase in the confidence interval of the estimate $\hat{\xi}$, constructed by the three-sigma rule:

$$\bar{\xi} = \hat{\xi} \pm 3\sigma_{\hat{\xi}}, \tag{5}$$

where $\sigma_{\hat{\xi}} = \sqrt{\mathrm{Var}(\hat{\xi})} = (\sigma/\sqrt{N})\sqrt{1 + 2R_N}$ is a root-mean-square deviation (r.m.s.) of the estimate $\hat{\xi}$, $\sqrt{1 + 2R_N}$ is an increase coefficient of the confidence interval.

The calculation of confidence intervals (5) for large scopes of N samples is considerably simplified, as the asymptotics of correlation coefficients $r(s)$ when calculating classical and fractal queues is not of great variety. It is defined asymptotically by the exponential decrease $r(s)$ with s growing, for classical queueing systems, for fractal systems it is defined asymptotically by the power decrease of $r(s)$. It allows us due to moderate-sized samples to find approximate analytical expressions suitable for the construction of asymptotic confidence intervals as functions of sample N volume. It also enables us to plan correctly the run length, which provides the set accuracy of results when calculating queues by the method of a long consecutive model run [10].

3   Confidence Intervals in Systems with Light Tails

Let us examine the calculation of steady-state average time $\bar{w}$ of request waiting in an M/M/1 system with arrival density $\lambda = 1$ and load rate $\rho = 0.75$ and $\rho = 0.99$ as an example of forming confidence intervals when calculating classical queues.

A statistical estimate $\hat{r}(s)$ of correlation coefficient $r(s)$ between the time w_i of waiting i-th request and the time w_{i+s} of waiting $(i + s)$-th request can be calculated on a steady-state sample $w_1, \ldots, w_N$ of a large enough size N, applying formula

$$\hat{r}(s) = \frac{\hat{E}(w_i w_{i+s}) - \hat{E}^2(w)}{\hat{Var}(w)}, \tag{6}$$

where

$$\hat{E}(w_i w_{i+s}) = \frac{1}{N-s} \sum_{i=1}^{N-s} w_i w_{i+s}, \qquad \hat{E}(w) = \frac{1}{N} \sum_{i=1}^{N} w_i,$$

$$\hat{Var}(w) = \hat{E}(w^2) - \hat{E}^2(w), \qquad \hat{E}(w^2) = \frac{1}{N} \sum_{i=1}^{N} w_i^2.$$

Estimates $\hat{r}(s)$ of correlation coefficients $r(s) = r(w_i, w_{i+s})$, calculated for two classical queues according to the Formula (6) with a model run length $N = 10$ million requests are presented as plots in Fig. 1. The relation $r(s)$ with the load rate $\rho = 0.75$ is shown on the left, on the right there is the relation with $\rho = 0.99$.

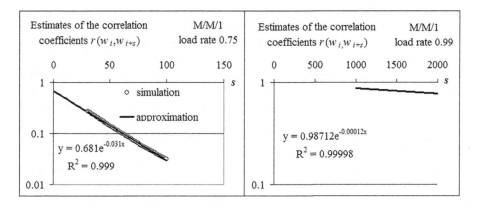

Fig. 1. Correlation of values w_i and w_{i+s} depending on s in the M/M/1 system

Generally in any classical systems the correlation coefficient $r(s)$ between two sample elements $w_1, \ldots, w_N$ shifted by s steps is described by asymptotics

$$r(s) \sim a e^{-bs}, \tag{7}$$

where a, b are constants. The relationship of the $r(s)$ function to the ae^{-bs} exponent with growing s rapidly converges to one and with large N in practice we can neglect the relative deviations $r(s)$ of ae^{-bs}. Considering (4) and (7), the aftereffect coefficient R_N can be expressed as:

$$R_N = \frac{1}{N} \sum_{s=1}^{N-1} (N-s)r(s) \sim \frac{1}{N} \sum_{s=1}^{N-1} (N-s)ae^{-bs} \sim \frac{a}{N} \int_1^{N-1} (N-s)ae^{-bs} =$$

$$= \frac{a}{Nb} \left\{ e^{-b(N-1)} \left(\frac{1}{b} - 1 \right) + e^{-b} \left(N - 1 - \frac{1}{b} \right) \right\} \sim \frac{a}{b} e^{-b}. \quad (8)$$

For length runs $N = 100$ thousand and more requests with those a, b, shown in Fig. 1 (see the trend line equations), we can use the latter asymptotic formula from (8), which represents a constant independent of s. If $\rho = 0.75$ we have $a \approx 0.681$, $b \approx -0.031$ (see Fig. 1), the aftereffect coefficient $R_N \approx \frac{a}{b} e^{-b} \approx 22$, and the confidence interval for the large N can be written as follows:

$$\bar{w} = \hat{w} \pm 3 \frac{\sigma}{\sqrt{N}} \sqrt{1 + 2R_N} \approx \hat{w} \pm \frac{3\hat{\sigma}}{\sqrt{N}} \cdot 6.7, \quad (9)$$

where $\hat{w}$ and $\hat{\sigma}$ are sample estimates of m.e. and r.m.s. of waiting time w in a steady-state condition. Similarly for $\rho = 0.99$ we define according to the Formula (8), that with the large N the aftereffect coefficient $R_N \approx 8000$ and the confidence interval looks like

$$\bar{w} = \hat{w} \pm \frac{3\hat{\sigma}}{\sqrt{N}} \cdot 130, \quad (10)$$

Expressions (9) and (10) of confidence intervals are verified by additional checks. In particular, while modeling the M/M/1 system with the load rate $\rho = 0.99$ (and the arrival density $\lambda = 1$) as a result of three runs with $N = 100$ million requests, we obtained the estimates of average service time $\hat{w} = 96.71$, $\hat{w} = 96.02$ and $\hat{w} = 99.11$, and also the r.m.s. estimate of waiting time is $\hat{\sigma} \approx 86$. The confidence interval (10) with this r.m.s. is $\bar{w} = \hat{w} \pm 3.37$. Indeed, this interval with any of the three obtained estimates $\hat{w}$ covers an accurate value of average waiting time $\bar{w} = \rho^2/(1-\rho)/\lambda = 98.01$, known for the examined M/M/1 system. The confidence intervals constructed for $\rho = 0.75$ (and also for $\rho = 0.9$) case have withstood a similar test by practical simulation.

An essentially important conclusion is that when calculating queues with light tails the correlation of sample elements at the model output does not affect the estimate convergence rate to the required accurate values of indicators. The estimate error (a half of the confidence interval length) decreases in proportion to $N^{0.5}$ and to reduce an error k-fold, the length N of a model run needs to be increased k^2 fold.

At the same time, a high correlation of sampling under the same requirements to accuracy can result in a need for a repeated length increase for the sample compared to a case of independent sample elements.

In the following section of the article the test is carried out having "the backward focus" – we examine the operation correctness of a widespread simulation system GPSS World [11] by means of correctly constructed confidence intervals. Unfortunately, in general the GPSS system has not withstood the test.

4   Accuracy Control and Model Time Defects

Let us consider GPSS World modeling results of the queue M/M/1 with the arrival density $\lambda = 1$ and the load rate $\rho = 0.9$, using asymptotic confidence intervals of estimates. The average queue length L= 8.1 and the average waiting time $\bar{w} = 8.1$ for this system are accurately known. The simulation results of the system with a various run length expressed by (average) number N of the requests arriving for modeling time T_M are presented in Table 1. The last line of the table shows a permissible error defined as a size of the asymptotic confidential half-interval, which is $\frac{3\sigma\sqrt{1+2R_N}}{\sqrt{N}} \approx 500N^{-0.5}$ (to obtain it we use an accurate value and the estimates $a = 0.95$, $b = -0.0054$ found by simulation).

Table 1. Simulation results of M/M/1 using GPSS World

Characteristic	Number of runs $N = \lambda T_M$				
	10^6	10^7	10^8	10^9	$2 \cdot 10^9$
Estimate of the average waiting time	8.332	8.061	8.087	8.043	7.799
Actual error of estimate	0.232	0.039	0.013	**0.057**	**0.301**
Permissible error	0.50	0.16	0.05	**0.016**	**0.005**

A comparison of permissible and actual errors for simulation estimates of average time $\bar{w} = 8.1$ obtained in various length runs leads to an unambiguous conclusion about the inconsistency of simulation estimates obtained in case of large N.

To clarify the reasons of estimate shift increasing with the growth of run length N we have performed a number of special experiments resulting in an unpleasant discovery: the time advance mechanism in GPSS World has latent defects.

Figure 2 shows the program whoose implementation proves the existence of time defects in GPSS World. The memory cell 1 accumulates a sum of all negative values w_i of requests waiting time in the course of running the program modeling the M/M/1 system. The number of negative w_i is counted in cell 2. An average value of negative w_i is calculated in cell W_NEG. The values of cells are derived after each of 10 pieces of a long run, executed by START. In each piece of the run the model time AC1 moves ahead per 100 mln. units of time.

The program execution results are presented in Table 2.

A small program extension allows one to find out that the first request with a negative waiting time is the request with number 21 693 107. It enters the

```
GENERATE      (Exponential(1,0,1))
ASSIGN        1,AC1
QUEUE         1
SEIZE         1
DEPART        1
SAVEVALUE     1+,((AC1-P1)#(AC1<P1))
SAVEVALUE     2+,(AC1<P1)
ADVANCE       (Exponential(1,0,0.9))
RELEASE       1
TERMINATE

GENERATE      100000000
SAVEVALUE     W_NEG,(X1/X2)
TERMINATE     1

START         1
RESET
START         1
RESET
....................
START         1
```

Fig. 2. The program revealing the existance of reverse time motion

queue at the moment of modeling time 21 693 041,901 257, then quits it at the moment 21 693 041.894 325, which is $\triangle_1 = 0.006932$ less than the entrance time (i.e. time "goes backward"). Note that the negative w_i on average grow (modulus) in proportion to the time AC1 (Table 2), and frequency of negative w_i emergence and their sum's modulus grow approximately proportional to the fourth or fifth degree of the time AC1. The existence of negative w_i proves the incorrect work of the request scheduler with lists of events. The experiments show that together with the run length the distortions of estimates and other indicators such as queue average length, request loss probability (with a limited run length), etc. also grow.

Table 2. Characteristics of negative values of waiting time

Absolute clock AC1	The number of $w_i < 0$	Sum of $w_i < 0$ values	The average value of $w_i < 0$
$1 \cdot 10^8$	344	−9.87597	−0.028709
$2 \cdot 10^8$	4961	−294.378	−0.059338
$3 \cdot 10^8$	23470	−2085.86	−0.088873
$4 \cdot 10^8$	69841	−8161.87	−0.116864
$5 \cdot 10^8$	161040	−23295.5	−0.144657
$6 \cdot 10^8$	317228	−54594.3	−0.172098
$7 \cdot 10^8$	557714	−110980	−0.198991
$8 \cdot 10^8$	901127	−202575	−0.224802
$9 \cdot 10^8$	1354436	−337396	−0.249104
10^8	1920013	−522135	−0.271943

The GPSS World developers, apparently, faced external aspects of the described defect, but did not distinguish the defect itself. We can understand it, for example, due to a vague hint in [11] that the use of RESET "can lead to some systematic mistake in the field of the smallest values". However, Table 2 clearly shows that the first 344 negative w_i appeared in our test long before the use of the first RESET thus it is not the RESET's fault here. The tests show that the reverse time motion is not also caused by rounding errors.

Thus, a rather simple technique of practical construction for asymptotic confidence intervals presented in the previous section allows us to elicit a fact of GPSS incorrect work at the long runs. It results in investigating and establishing the true reason of the unpleasant situation, which consists in the latent defects of the advance and accounting mechanism of the model time. The restrictions in the runs length used while modeling queues (no more than 100 million requests in each independent consecutive run) allow us to avoid a tangible effect of model time defects. It is expedient to increase the necessary scope of samples by enhancing the number of independent parallel runs. These restrictions are formulated, proved and brought to a wide range of the GPSS [12] system users. It is also necessary to control the total number of accesses to the RNG of the GPSS system which should not exceed 2 billion 48 million. To remove the last restriction we have developed in the GPSS language the MtRand procedure [13], which allows us to use a high-quality external RNG based on the "Mersenne twister" algorithm with a period length of random numbers sequence $2^{19\,937} - 1 \approx 10^{6000}$. Thereby the total amount of samples in queues simulation becomes almost unlimited: it should not exceed 10^{6000}.

We have also established experimentally that there are no such model time defects in simulation system AnyLogic [14], as in GPSS.

5   Confidence Intervals in Systems with HTD

Coefficients $r(s)$ of correlation elements ξ_i of processed samples in computation of fractal queueing systems are asymptotically power functions of s. This leads to the significant difference between methods for planning of experiments with fractal queues and experiments with classical queues that have the corresponding exponential asymptotics. Figure 3 shows the plot of the function $r(s)$ obtained by simulation of Pa/Pa/1 system at steady-state conditions, which is reached by running about 40 million requests. In this case, the implementations $\gamma_i \in \{0, 1\}$ of failure indicators of requests with numbers $i = 1, 2, \ldots$ are the sample elements. At steady-state conditions, the indicators mean value $\bar{\gamma}$ equals the required probability P of a request loss. The transient period was calculated by means of a large number of independent model runs.

Generally, in the calculating of any indicators as failure probability, the average waiting time, etc. of the function $r(s)$ in fractal queues computing has the following asymptotics:

$$r(s) \sim as^{-b}, \tag{11}$$

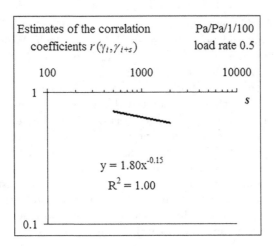

Fig. 3. Asymptotics of correlation coefficients $r(s)$ between failure indicators of i-th and $(i+s)$-th requests in Pa/Pa/1/100 system for $\alpha_1 = \alpha_2 = 1.1$, $K_1 = 1$, $K_2 = 0.5$

where a and b are some constants defined by test model runs ($a > 0$ and $0 < b < 1$).

Using the Eq. (4) and asymptotic representation (11), we define the following simple expression for confidence intervals

$$R_N = \frac{1}{N} \sum_{s=1}^{N-1} (N-s) r(s) = \frac{1}{N} \sum_{s=1}^{N-1} (N-s) a s^{-b} \sim \frac{1}{N} \int_1^{N-1} (N-s) a s^{-b} ds$$

$$= \frac{a}{1-b}(N-1)^{1-b} - \frac{a}{1-b} - \frac{a}{N}\frac{1}{2-b}(N-1)^{2-b} + \frac{a}{N}\frac{1}{2-b}$$

$$\sim \frac{a}{(1-b)(2-b)} N^{1-b}. \quad (12)$$

Therefore, in accordance with (5), we obtain the formula for confidence interval

$$\bar{\xi} = \hat{\xi} \pm 3\sigma \left(\frac{1+2R_N}{N}\right)^{1/2} \sim \hat{\xi} \pm 3\sigma \cdot CN^{-b/2} \approx \hat{\xi} \pm 3\hat{\sigma} \cdot CN^{-b/2}, \quad (13)$$

where the constant $C = \sqrt{\frac{2a}{(1-b)(2-b)}}$ which is independent of N; $\hat{\sigma}$ is the r.m.s. estimate of sample elements ξ_i.

From (13) it follows that the value of confidence half-interval declines slower than $N^{-1/2}$ as $b < 1$. For instance, if we calculate the probability P of request loss in the system with the parameters shown in Fig. 3, then the parameter b within the confidence interval (13) is equal to 0.15 (see the trend equation in Fig. 3); therefore, the value of half-interval declines in proportion to $N^{-0.075}$. For the error (half-interval) to be decreased 10-fold the run length N is to be increased

$10^{1/0.075} \approx 10^{13}$ – fold in this case. Usually, such a way of improving accuracy is almost unacceptable, and, thus, the development of accelerated methods of probability calculation is of great relevance in simulation systems with heavy tails [6–8].

The performed analysis let us conclude that in calculation of HTD systems a correlation of sample elements decreases the rate of estimates convergence; this rate becomes less than $cN^{-1/2}$ for any parameters a, b of power asymptotics of coefficients $r(s)$. This fact makes a great difference in the simulation of fractal and classical queueing systems. As a result, it is necessary to use independent parallel model runs [10], producing a large number of independent samples of requests coming through the system, for any simulation experiments of calculating fractal queues.

Besides, parallel runs allow us to solve relevant problems related to fractal queues calculation such as the determining of transient periods, detecting of convergence lack for various estimates, etc.

6   Calculation of Fractal Queues and the ARAND Method

However, before developing and testing recommendations for calculating HTD queues we should have complete awareness of the problem of HTD distortions realized in simulation [4]. To solve this problem the cascade method for generation of r.v. with HTD is proposed in [4]. We shall now give the simplest, most accurate and universal version (which was not found in [4]) of this method for the generating of standard random numbers (Fig. 4) – the ARAND method (Accurate RAND).

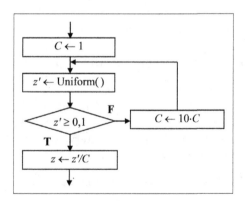

Fig. 4. The flowchart of ARAND method for GPSS World

The ARAND method transforms ordinary n-digits uniformly distributed in the interval $[0, 1]$ random numbers z' to those uniformly distributed random numbers $z \in [0, 1]$, which, no matter how small they are, have n accurate significant digits. Due to it, r.v. with LTD realized by the inverse transformation of a distribution tail have also n accurate *significant* digits, and the problem of LTD distortion is eliminated (confirming calculations performed for cascade RNG are given in [4]). For example, reversing PD tail $\bar{F}(t) = 1 - F(t) = (K/t)^\alpha$, we obtain formula $x = Kz^{-1/\alpha}$ for generation r.v. x with Pareto's distribution, and if z is generated by the ARAND method, then the r.v. x moments are not displaced. The tails $\bar{F}(t)$ of HTD are realized by the ARAND method with the same number of accurate significant digits, both for big t and small t (within t reaching 10 of several hundred or thousand in power). From this point of view, the use of ARAND is equivalent to the use of those ordinary RNG, which would give random numbers with hundreds/thousands of decimal digits after a decimal point. However, this accuracy enhance does not result in a ten/hundred-fold increase in losses concerning time or equipment, the loss is only 10 % of numbers at the outcome of the ordinary standard RNG and the rate of random numbers generation approximately decreases by 1.1-fold.

Figure 5 shows the procedure implementing the ARAND method in Plus language (here the argument Arg is the number RNG Uniform) for GPSS World.

```
PROCEDURE ARAND(Arg) BEGIN
TEMPORARY CCC,zzz;
CCC = 1;
STEP1: zzz = Uniform(Arg,0,1);
STEP2: IF (zzz < 0.1)THEN BEGIN
CCC = 10#CCC;
GOTO STEP1;
END;
STEP3: zzz = zzz/CCC;
RETURN (zzz);
END;
```

Fig. 5. Plus procedure implementing the ARAND method for GPSS World

Access to the ARAND procedure is carried out the same way as to other PLUS-procedures [11]. For instance, we can write the statement GENERATE for the generation of a request stream with time intervals between them, distributed by Pareto with parameters $K = 1$, $\alpha = 1.1$ as follows:

`GENERATE (1#ARAND(1)^(-1/1.1));` incoming stream generation.

If there is a need to generate more than 1–2 billion random numbers, the Uniform generators of the GPSS system turn out to be inefficient because of their limited period length. In this case, in the ARAND procedure we use access to the RNG MtRand ("Mersenne twister") whose procedure of connection is described in [13] instead of access to the RNG Uniform (see Fig. 5). The correct HTD implementation with a possibility of obtaining high-quality unlimited samples allows us to perform simulation experiments, controlling and providing indeed the necessary accuracy of fractal queues calculation.

The method of a single long run in the simulation of fractal queues, as a rule, is inefficient because of the low rate of estimates convergence. Therefore, it can lead to conceptually erratic conclusions, even with a very large run length. Thus, in Fig. 6, we can see the tail of empirical distribution for a waiting time of requests in Pa/Pa/1 system with Pareto distribution parameters of intervals for requests arrival $\alpha_1 = 1.1$, $K_1 = 1$ and Pareto distribution parameters of service time $\alpha_2 = 1.1$, $K_2 = 0.1$. In this system $\bar{\tau} = 11$, $\bar{x} = 1.1$, its load rate is $\rho = 0.1 \ll 1$. The empirical distribution of waiting time whose tail is given in the figure is obtained in a sequential model run according to the values of 100 million requests waiting time. The diagram rectification in Fig. 6 in a logarithmic scale on both axes of coordinates, high reliability ratio R^2 and a considerable sample length lead to the conclusion that the tail is asymptotically powerful and the parameter $\alpha = 1.188$ of the tail for the approximate equation in Fig. 6 is estimated accurately enough. As $\alpha > 1$, then the distribution has m.e., i.e. the average waiting time in the simulated system is finite. However, it is a wrong conclusion.

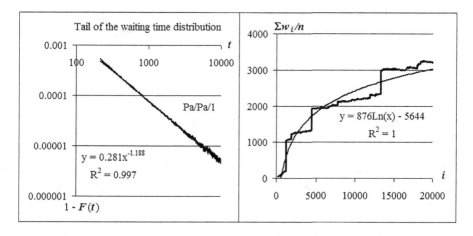

Fig. 6. Results of sequential run data processing (left) and the results of parallel runs for Pa/Pa/1 system if $\alpha_1 = \alpha_2 = 1.1, K_1 = 1, K_2 = 0.1$

Actually this is not the case. Having performed and processed several quite short runs, we can obtain graphic information (Fig. 6, to the right) that the average requests waiting time converges to infinity. This figure shows the dependence of a waiting time request on number i averaged according to $n = 10$ thousand independent process implementations (parallel runs). There is a steady-state mode in the system ($\rho < 1$), but it is not reached for the finite time. The plot of the average waiting time growth has specific jumps caused by the heavy tails of Pareto distributions, setting the system. However, the growth of average value of elements w_i for the sample with the growth of their number i has a characteristic logarithmic rate (a logarithmic trend line is a smooth curve in the right figure).

At the same time, employing parallel runs will not provide automatic obtainment of appropriate results in simulation of fractal queues unless the problem of HTD correct realization is solved. Wrong conclusions, though, could be obtained not only for $\alpha \downarrow 1$ when HTD tails are "particularly heavy", but also for $\alpha \uparrow 2$ when these tails are lighter in interval $1 < \alpha \leq 2$. The simulation for $\alpha \uparrow 2$ of M/Pa/1 queue using the ordinary standard random number generator should lead to principally mistaken results in the calculation of an average waiting time as shown in [4]. The results of such a simulation M/Pa/1 system with a large number of parallel tests is shown to the left in Fig. 7.

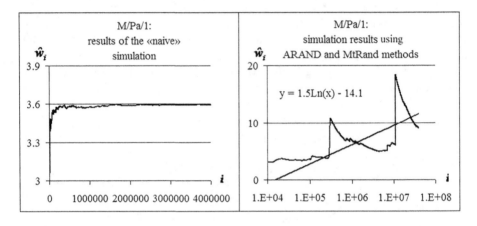

Fig. 7. Results of simulation M/Pa/1 system for $\bar{\tau} = 4$, $K = 1$, $\alpha = 2$

Each estimate $\hat{w}_i$ to the left in Fig. 7 is obtained by averaging the random variable w_i values. This averaging was performed for the specific i, when the number of independent runs of the model is big. If the ARAND method is not used, then the moments of realized HTD will displace [4] and the estimate $\hat{w}_i$ will converge to 3.6 with the growth of i.

The curve on the right hand side in Fig. 7 shows the estimates obtained in the same experiment by using the ARAND method instead of the ordinary standard RNG. Now, we see that the dependence of $\hat{w}_i$ on i does not converge to any finite value. The curve on the right has typical jumps for non-convergent estimates. These jumps result in the growth of estimate approximately with a logarithmic speed (axis of abscissas is in the logarithmic scale in Fig. 7). The ARAND method allows us to see the true state of affairs, when there is a steady-state average waiting time and when this time is equal to infinity (because the estimate $\hat{w}_i$ grows as $Ln(i)$ and is not proportionate to i). The example demonstrates the complementarity of ARAND and parallel run methods. These methods complement each other and their combined usage provides a real opportunity for calculating fractal queues with control of errors and achieving appropriate accuracy of calculations.

7   Conclusion

The analysis of fractal queues simulation errors and recommendations based on that analysis provide a real opportunity for fractal queues calculation with an appropriate accuracy. However, computational efforts increase noticeably in comparison with the simulation of classical queues, but they are within acceptable limits.

The main feature of fractal queues simulation is the need to use systematically the multiple parallel independent runs of the model (including test runs) along with the methods of correct HTD realization (the ARAND method described in the article is one of the most effective).

The technique proposed in the article of asymptotic confidence intervals calculation allows us not only to plan experiments providing given calculation accuracy but also to test simulation systems used. In particular, by such means the hidden defects of model time in the GPSS system were revealed.

In general, the research performed in the article, the recommendations developed, and the examples of their usage make it possible to reduce the calculation of fractal queues by means of their simulation to the sequence of operations though relatively difficult, but typical for applying numerical analysis methods.

References

1. Leland, W.E., Taqqu, M.S., Willinger, W., Wilson, D.V.: On the self-similar nature of Ethernet traffic. ACM/SIGCOMM Comput. Commun. Rev., pp. 146–155 (1993)
2. Czachorski, T., Domanska, J., Pagano, M.: On stochastic models of internet traffic. In: Dudin, A., Nazarov, A., Yakupov, R. (eds.) Information Technologies and Mathematical Modeling, vol. 564, pp. 289–303. Springer, Switzerland (2015)
3. Crovella, M.E., Lipsky, L.: Simulation with heavy-tailed workloads. In: Park, K., Willinger, W. (eds.) Self Similar Network Traffic and Performance Evaluation, pp. 89–100. Willey, New Jersey (2000)
4. Zadorozhnyi, V.N.: Fractal queues simulation peculiarities. In: Dudin, A., Nazarov, A., Yakupov, R. (eds.) Communications in Computer and Information Science, vol. 564, pp. 413–432. Springer, Switzerland (2015)
5. Kleinrock, L.: Queueing systems. In: Computer Applications, vol. 2, p. 576. Wiley Interscience, New York (1976)
6. Zwart, A.P.: Queueing systems with heavy tails, p. 227. Eindhoven University of Technology, Eindhoven (2001)
7. Asmussen, S., Binswanger, K., Hojgaard, B.: Rare events simulation for heavy-tailed distributions. Bernoulli **6**(2), 303–322 (2000)
8. Boots, N.K., Shahabuddin, P.: Simulating GI/GI/1 queues and insurance risk processes with subexponential distributions. In: Proceedings of the 2000 Winter Simulation Conference, pp. 656–665 (2000). Unpublished manuscript, Free University, Amsterdam. Shortened Version

9. Zadorozhnyi, V.N.: Simulation modeling of fractal queues. In: Dynamics of Systems, Mechanisms and Machines (Dynamics), pp. 1–4, December 2014. doi:10.1109/Dynamics.7005703

10. Kleijnen, J.P.C.: Statistical Techniques in Simulation, Part 1. Marcel Dekker, New York (1974)

11. GPSS World reference manual: Minuteman Software, 5th Edn., Holly Springs, NC, U.S.A (2009)

12. Zadorozhnyi, V.N.: Realization of big samples in simulation of queueing systems in GRSS World. In: IMMOD 2015, Moskow, pp. 225–230 (2015). (in Russian)

13. Zadorozhnyi, V.N.: The improving of GPSS models accuracy by employing the random variables generator - "Mersene twister". Omskiy nauchnyiy vestnik 1(145), 90–94 (2016). (in Russian)

14. Karpov, J.: Simulation Modeling of Systems. Introduction to Simulation in AnyLogic, vol. 5. BHV-Peterburg, St. Petersburg (2005)

Author Index

Printed in the United States
By Bookmasters